Aphids.

$157.25

DATE			

APHIDS
THEIR BIOLOGY, NATURAL ENEMIES AND CONTROL

World Crop Pests

Editor-in-Chief
W. Helle
University of Amsterdam
Laboratory of Experimental Entomology
Kruislaan 302
1098 SM Amsterdam
The Netherlands

Volumes in the Series

1. Spider Mites. Their Biology, Natural Enemies and Control
   Edited by W. Helle and M.W. Sabelis
      A. 1985   xviii + 405 pp.   ISBN 0-444-42372-9
      B. 1985   xviii + 458 pp.   ISBN 0-444-42374-5
2. Aphids. Their Biology, Natural Enemies and Control
   Edited by A.K. Minks and P. Harrewijn
      A. 1987 xx + 450 pp.   ISBN 0-444-42630-2
      B. 1988 xix + 364 pp.   ISBN 0-444-42798-8
      C. 1989 xvi + 314 pp.   ISBN 0-444-42799-6
3. Fruit Flies. Their Biology, Natural Enemies and Control
   Edited by A.S. Robinson and G. Hooper
      A. ISBN 0-444-42763-2
      B. ISBN 0-444-42750-3
4. Armoured Scale Insects. Their Biology, Natural Enemies and Control
   Edited D. Rosen
      A. ISBN 0-444-42854-2
      B. ISBN 0-444-42902-6
5. Tortricoid Pests
   Edited by L.P.S. van der Geest and H.H. Evenhuis

World Crop Pests, 2C

APHIDS THEIR BIOLOGY, NATURAL ENEMIES AND CONTROL

Volume C

Edited by

A.K. MINKS and P. HARREWIJN

*Research Institute for Plant Protection,
Wageningen, The Netherlands*

ELSEVIER

Amsterdam – Oxford – New York – Tokyo   1989

ELSEVIER SCIENCE PUBLISHERS B.V.
Sara Burgerhartstraat 25
P.O. Box 211, 1000 AE Amsterdam, The Netherlands

Distributors for the United States and Canada:

ELSEVIER SCIENCE PUBLISHING COMPANY INC.
52, Vanderbilt Avenue
New York, NY 10017

ISBN 0-444-42799-6 (Vol. 2C)
ISBN 0-444-42373-7 (Series)

Printed in The Netherlands

Preface

Aphids undoubtedly are the most important pest insects in the agriculture of the temperate climatic zones. Because over the years a huge amount of data on these insects has been collected, only few people have been able to get a general picture of the aphids and most research workers had to specialize and become experts in one narrow area of aphidology. This explains the lack of a general book on aphids with detailed coverage of the whole area from anatomy and morphology to control. Therefore, when the Editor-in-Chief of this series, W. Helle, approached us with the question how many comprehensive books had been published so far, we soon came to realize that there is hardly any treatise that can be considered as such.

Of course, there are some smaller books available, such as *Biology of Aphids* by A.F.G. Dixon (1973), treating the way of life of aphids, aphid–plant relations, including virus transmission and population development, and *Aphids* by R.L. Blackman (1974), covering such topics as life cycles and polymorphism, and natural enemies. These books, however, treated these subjects in a small compass. Also some books with more specific information appeared: *Aphid Technology*, edited by H.F. van Emden (1972), and *Survey of the World's Aphids*, by V.F. Eastop and D. Hille Ris Lambers (1976). Recently, *Aphids on the World's Crops: An Identification Guide*, by R.L. Blackman and V.F. Eastop (1984) and *Aphid Ecology*, by A.F.G. Dixon (1985) were published. In these, the reader can find more detailed information on pest aphid species, species variability, population dynamics and community structure.

We hope that the present book can fill the gap in the available information on aphids and make it more accessible. We fully realize that such a book, containing the updated knowledge on more than 4000 species of these sap-sucking insects, could comprise many volumes and still suffer from incompleteness and generalization. Nevertheless we thought it worthwhile to ask the best specialists in the area of aphidology to contribute to a book that could be useful to those who want to understand the interrelationships between aphid anatomy, evolution, physiology, behaviour and host plant effects, as this is the only basis on which to develop the right means of controlling aphids as pests. Scientific solutions frequently come from relatively unexpected directions and this is why much attention must be given to fundamental aspects of aphid biology. Thus the separate sections written by specialists are brought together in this book and intended to be a synthesis of pure and applied research.

Inevitably a multi-author book will show some overlap in the various sections. We have tried to reduce this to an acceptable level, but often a particular subject is discussed from different points of view, and we think this form of overlap not to be disturbing, but even stimulating. Biological control, for instance, can be treated for the sake of the biology of natural enemies, but just as well for better control of aphids. For similar reasons the reader will come across galling aphids in more than one place, and a specific adaptation such as soldier larvae is discussed by at least four authors.

Aphid systematics is still a controversial area, as is reflected by section 1.3, where F.A. Ilharco and A. van Harten discuss the most important aphid classifications, some of which are no longer used. It became apparent to us that various authors were using different classifications and nomenclatures. As we feel that standardization of terminology is essential to avoid any possibility of confusing our readers, we have arbitrarily chosen to use, throughout the book, the classification proposed by O.E. Heie in 1980. As to the Latin nomenclature of aphid species, we followed *Survey of the World's Aphids* by V.F. Eastop and D. Hille Ris Lambers. We also decided not to include a key for the identification of aphid species, as several good keys can be found elsewhere in the literature, and we considered this as too much of a detail in the context of this book.

We realise that some areas, e.g. host plant relations, polymorphism, behaviour and aphid control, go through such a rapid development that the present data will soon become outdated; therefore we attempt to pay extra attention to future developments. For this reason much emphasis is laid on alternative control measures, such as plant resistance breeding (section 11.4, Volume C) and biological control (section 11.2, Volume C), with an extensive treatment of natural enemies in Chapter 9 (Volume B).

We are very much indebted to the late D. Hille Ris Lambers, who gave us important advice during the early preparations for this book. Some of the authors did a great deal more than "just" writing their own contribution: we owe special gratitude to R.L. Blackman, A.F.G. Dixon and O.E. Heie for their editorial assistance. A great number of colleagues also helped us, which we gratefully acknowledge: A.B.R. Beemster, C.A.D. de Kort, M. Llewellyn, W.P. Mantel, P.A. Oomen, D. Peters, J.D. Prinsen, D.J. Sullivan, W.F. Tjallingii and N. Wilding. We would also like to mention with special thanks C.A. Koedam, the photographer of our Institute.

Finally we gratefully acknowledge the generous financial support by the Foundation "Fonds Landbouw Export Bureau 1916/1918", Wageningen, and the "Uyttenboogaart/Eliasen" Foundation, Amsterdam, for publication of the colour plates.

P. Harrewijn   A.K. Minks

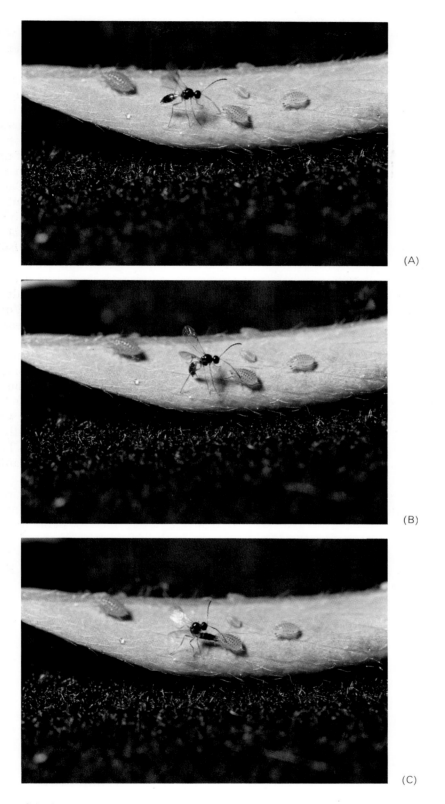

Stinging sequence of the ovipositioning parasite, *Trioxys complanatus* Quilis, in the aphid *Therioaphis trifolii* f. *maculata* (Buckton). (A) Parasite approaching its prey. (B) Parasite touching its prey with the antenna and bending its abdomen. (C) Parasite completely bending its abdomen and stinging its prey with the ovipositor. (Photographs by L.T. Woolcock, CSIRO, Canberra, A.C.T., Australia.)

Contents

Contents of Volume A
Chapter 1. Morphology and Systematics
Chapter 2. Anatomy and Physiology
Chapter 3. Reproduction, Cytogenetics and Development
Chapter 4. Biology
Chapter 5. Aphids and their Environment
Chapter 6. Evolution
Chapter 7. Organization (Structure) of Population and Species, and Speciation

Contents of Volume B
Chapter 8. Techniques
Chapter 9. Natural Enemies

Contents of Volume C

Chapter 10. Damage 1

**10.1 The Responses of Plants to the Feeding of Aphidoidea: Principles, by   1
   P.W. Miles**. . . . .
   Conservation of the food resource 1
   Factors influencing the specificities of responses 2
   Interaction of mechanical and chemical stimuli. 2
   Immediate physiological responses. 3
   Some uncertainties concerning salivary influence. 3
   Phytotoxic reactions: to salivary toxins or depletion of assimilate?. 4
   Systemic and localized reactions 5
   Cecidogenic redifferentiation 5
   The insect as a localized nutrient "sink". 6
   Respiratory consequences of a nutrient sink 7
   Defensive reactions in galled tissues. 7
   Phytotoxic reactions characterized 7
   Experimental simulation of natural responses 8
   Hormones as cecidogens 8
   Amino acids as cecidogens 10
   Enzyme systems as cecidogens 10
   Enzyme inhibitors as cecidogens 11
   Respiratory influences in cecidogenesis 12
   Behavioural influences in cecidogenesis 13
   Rejection of involvement of neoplasia and microbial intervention 13
   Interactions of the responsiveness of the plant with behavioural and chemical specifi-
   cities of the insect 14
   Cecidogenesis seen as a progressive influence on biochemical equilibria 14
   References. 16

10.1.1 Specific Responses and Damage caused by Aphidoidea, by P.W. Miles . . . 23
   Introduction. 23
   Damage caused by nutrient drain 23
   Damage related to the sensitivity of the plant 24
   Chlorosis caused by aphidids 25
   Systemic effects ascribed to aphidid toxins. 25
   Localized effects ascribed to aphidid toxins. 26
   Growth distortions caused by aphidids. 27
   Damage by aphidids summarized 27
   Galls caused by the woolly aphid of apple 28
   Pemphigid galls on *Populus*. 29
   Damage by pemphigids to secondary hosts 29
   Adelgid galls on spruce. 32
   Adelgid attacks on needles and shoots of secondary hosts 34

Attacks by the balsam woolly aphid on its secondary hosts   35
Defensive reactions in woody tissues to the feeding of adelgids   36
Galls of the grape phylloxera   36
Galling by other phylloxerids   40
Galls due to other Aphidoidea   40
Cecidogenesis summarized   40
References   42

10.2 Crop Loss Assessment, by P.W. Wellings, S.A. Ward, A.F.G. Dixon and   49
R. Rabbinge   49
Introduction   49
Defining yield and loss   52
Production levels and production systems   52
Production levels   52
Production systems   52
Injury and damage   53
Assessing crop loss   54
Laboratory, glasshouse and cage experiments   54
Field experiments   55
Explaining and forecasting crop loss   57
Conclusions   60
References   61
Appendix – Production loss attributable to aphids   63

10.3 Viruses Transmitted by Aphids, by E.S. Sylvester   65
Introduction   65
Non-persistent transmission   65
Potyviruses   67
Potexviruses   67
Carlaviruses   68
Multicomponent RNA viruses   69
Semi-persistent transmission   71
Bimodal transmission   71
Closteroviruses   72
Caulimoviruses   73
Unclassified viruses   73
Persistent transmission   74
Luteoviruses   74
Unclassified viruses   75
Propagative transmission   76
Plant rhabdoviruses   76
Epidemiology and management of aphid-borne viruses   77
Elimination of virus sources   79
Isolation from virus sources   80
Crop manipulation   81
Miscellaneous methods   82
References   83

Chapter 11. Control of Aphids   89

11.1 Chemical Control, by A. Schepers   89
Introduction   89
Aphids and aphicides   89
Selectivity   90
Systemic activity   90
Residual activity and persistence   90
Rapid action   90
Low phytotoxicity   91
Application of aphicides   91
Chemical control to reduce direct damage   91
Solanaceae   91
Gramineae   93
Chenopodiaceae   96
Leguminosae   97
Cruciferae   99
Compositae 100

   Rosaceae 101
   Malvaceae. 103
   Miscellaneous 103
  Chemical control to reduce virus diseases 104
   Solanaceae 104
   Gramineae. 106
   Chenopodiaceae 107
   Leguminosae 109
   Cruciferae. 110
   Compositae 110
   Rosaceae 110
   Malvaceae. 111
   Miscellaneous 111
  References. 112
   Solanaceae 112
   Gramineae. 114
   Chenopodiaceae 115
   Leguminosae 116
   Cruciferae. 118
   Compositae 119
   Rosaceae 119
   Malvaceae. 120
   Miscellaneous 121

11.1.1 Resistance to Aphids to Insecticides, by A.L. Devonshire. 123
 Occurrence of insecticide-resistant aphids 123
 The biochemical nature of resistance in aphids. 125
  Myzus persicae. 125
  Phorodon humuli. 128
  Aphis gossypii 128
 Karyotype and instability of resistance in *M. persicae*. 129
 Detailed studies of resistant populations of *M. persicae* 131
 The future. 134
 References. 135

11.2 Biological Control of Aphids, by M. Carver. 141
 Definition and scope 141
 Some procedures and principles of biological control 142
 History of biological control, with special reference to aphids 143
 Increased need for biological control 149
 Aphids as candidate pests for biological control 151
  Population dynamics. 151
  Limitations of adventives. 151
  Heteroecy. 151
  Virus vectors 152
  Dispersive vehicles. 153
 The control agents 154
 Multilateral control 154
 Multiple introductions 156
 Ants 156
 The exotic environment. 156
  Existent entomophaga 156
  Invading species 158
  An example: Australia 158
 Aphids as controllers of weeds 160
 Assessment 160
 References. 161

11.2.1 Biological Control in the Open Field, by R.D. Hughes. 167
 Introduction. 167
 The practice of biological control of aphids 168
  Step 1. Characterisation of target aphid and its environment 168
  Step 2. Appropriateness of biological control, types of enemy and specific approach 173
  Step 3. Selection of the natural enemy. 175
  Step 4. Import, quarantine procedures, identification and examination 175

Step 5. Release of natural enemies. 177
Step 6. Establishment. 181
Step 7. Assessment of impact 183
The ecological attributes of a successful natural enemy 186
An analysis of attempts to biologically control target aphid species 187
Conclusions 192
References. 194

11.2.2 Biological Control in Greenhouses, by P.M.J. Ramakers 199
Introduction. 199
Parasitoids 200
Predators 202
Fungi 204
Summary 205
Acknowledgements. 206
References. 206

11.3 Modifying Aphid Behaviour, by R.W. Gibson and A.D. Rice 209
Introduction. 209
Preventing flying aphids from colonizing plants 209
The effects of light on aphid behaviour 209
Plant odours and aphid behaviour. 211
Aphid alarm pheromone 211
Repelling aphids that have alighted 212
Aphid alarm pheromone 213
Other repellent chemicals. 213
Polygodial. 214
Carboxylic acids 215
Host plant resistance. 215
Effect on behaviour 215
Host plant resistance and virus control 216
Effects of aphicides on aphid behaviour and virus spread 217
Effects on aphid probing 217
Effects on aphid movement 218
Conclusions 221
References. 221

11.4 Host Plant Resistance, by J.L. Auclair 225
Introduction. 225
Breeding plants resistant to aphids 226
Screening approaches for detecting resistance to aphids. 227
Aphid resistance in particular host plants 230
Leguminosae 230
Gramineae. 236
Compositae 243
Cruciferae. 244
Cucurbitaceae 245
Rosaceae 246
Solanaceae 248
Miscellaneous host plants 251
Conclusion 253
Acknowledgements. 254
References. 254

11.5 Integrated Control 267

11.5.1 Integrated Aphid Management: General Aspects, by P. Harrewijn and   267
    A.K. Minks. . . . .
Introduction. 267
How to reduce aphid numbers. 268
Control measures in the overwintering stage. 269
Control measures during migration and host-plant finding. 269
Control during probing and feeding 270
Control during larviposition and reproduction 271
References. 271

11.5.2 Integrated Control of Cereal Aphids, by G.W. Ankersmit 273
  Introduction 273
  Agricultural practice and control 274
    Change of sowing date 274
    Increase of sowing density 274
    Simulation of parasite development by undersowing 274
    Manipulation of nitrogen fertilization 275
    Breeding for host-plant resistance 275
    Use of fungicides 275
    Use of insecticides 275
    Use of insect pathogens for control 276
  EPIPRE: a special case of supervised control of cereal aphids 276
  References 277

11.5.3. Integrated Control of Potato Aphids, by P. Harrewijn 279
  Introduction 279
  Ware potatoes 280
  Seed potatoes 281
  Concluding remarks 283
  References 284

11.5.4 Integrated Control of Aphids in Field-Grown Vegetables, by J. Theunissen 285
  Introduction 285
  Brevicoryne brassicae 285
    Tolerance levels 285
    Other elements in IPM 287
  References 288

11.5.5 From Integrated Control to Integrated Farming, by P. Vereijken 291
  Motivation 291
  Integrated farm management 292
  Perspectives and recommendations 293
  References 295

Epilogue 297

General Index 299

Index to the Aphids 309

Index to the Aphid Parasites/Pathogens 311

Index to the Aphid Predators 314

Contributors to this Volume

G.W. ANKERSMIT
Department of Entomology, Agricultural University, P.O. Box 8031, 6700 EH
Wageningen, The Netherlands

J.L. AUCLAIR
Department of Biological Sciences, University of Montréal, Case Postale 6128
– Succursale A, Montréal, Qué. H3C 3J7, Canada

M. CARVER
CSIRO Division of Entomology, GPO Box 1700, Canberra, A.C.T. 2601, Australia

A.L. DEVONSHIRE
Insecticides and Fungicides Department, Rothamsted Experimental Station,
Harpenden, Herts. AL5 2JQ, Great Britain

A.F.G. DIXON
School of Biological Sciences, University of East Anglia, Norwich NR4 7TJ,
Great Britain

K.G. GIBSON
Plant Pathology Department, Rothamsted Experimental Station, Harpenden,
Herts. AL5 2JG, Great Britain

P. HARREWIJN
Research Institute for Plant Protection, P.O. Box 9060, 6700 GW Wageningen,
The Netherlands

R.D. HUGHES
CSIRO Division of Entomology, GPO Box 1700, Canberra, A.C.T. 2601, Australia

P.W. MILES
Department of Entomology, Waite Agricultural Research Institute, University
of Adelaide, Glen Osmond, S.A. 5064, Australia

A.K. MINKS
Research Institute for Plant Protection, P.O. Box 9060, 6700 GW Wageningen,
The Netherlands

R. RABBINGE
Department of Theoretical Production Ecology, Agricultural University, P.O.
Box 430, 6700 EH Wageningen, The Netherlands

P.M.J. RAMAKERS
Research Institute for Plant Protection, P.O. Box 9060, 6700 GW Wageningen,
The Netherlands

A.D. RICE
Insecticides and Fungicides Department, Rothamsted Experimental Station,
Harpenden, Herts. AL5 2JQ, Great Britain

A. SCHEPERS
Mecklenburglaan 30, 3843 BP Harderwijk, The Netherlands

E.S. SYLVESTER
Entomology and Parasitology Department, University of California, Berkeley,
CA 94720, U.S.A.

J. THEUNISSEN
Research Institute for Plant Protection, P.O. Box 9060, 6700 GW Wageningen,
The Netherlands

P. VEREIJKEN
Centre for Agrobiological Research, P.O. Box 14, 6700 AA Wageningen, The
Netherlands

S.A. WARD
Department of Zoology, La Trobe University, Bundoora, Vic. 3083, Australia

P.W. WELLINGS
CSIRO Division of Entomology, GPO Box 1700, Canberra, A.C.T. 2601, Australia

Chapter 10 Damage

10.1 The Responses of Plants to the Feeding of Aphidoidea: Principles

P.W. MILES

CONSERVATION OF THE FOOD RESOURCE

Aphidoidea, aside from any role they may play in the transmission of virus diseases, mostly cause little damage to their food plants (Van Emden et al., 1969; Dixon, 1971b; Gutierrez et al., 1971). They have been considered to cause less damage than other sucking insects (Horne and Lefroy, 1915; Davidson, 1923; West, 1946), and certainly less than chewing insects (Pollard, 1973); in contrast to the latter, Aphidoidea do not generally destroy their food resource (Llewellyn, 1972). Even mass feeding by aphids may cause insidious rather than obvious damage, expressed only as an overall reduction of growth of the plant (Ortman and Painter, 1960; Van Emden et al., 1969; Dixon, 1971a, b).

Such changes as occur often appear to improve the plant as a nutrient source for the insect (Kennedy, 1951; Dixon and Wratten, 1971; Forrest, 1971). As will be discussed below, the drain on the plant's assimilate (and possibly inter-ference with its translocation system) tends to cause a breakdown of insoluble reserves in tissues near the insects and a mobilization of free amino and amide nitrogen (Kennedy and Stroyan, 1959; Kloft, 1960a, b; Van Emden et al., 1969).

These effects are particularly noticeable in galled tissues; the leaf galls of the grape phylloxera, *Viteus vitifoliae* (Fitch), contain more soluble nitrogen and carbohydrate than normal leaf tissue despite their much lower rate of photosynthesis (Schaefer, 1972; Rilling et al., 1975; Rilling and Steffan, 1978; Steffan and Rilling, 1981; see also sections 5.4 and 10.1.1). This increase in the supply of compounds readily assimilable by the insects has been likened to changes in senescing leaves, for which some aphidids show a particular preference (Kennedy, 1951; Way, 1968; Van Emden et al., 1969; Pollard, 1973).

At the same time, some authors suggest that the feeding of aphidids may, in appropriate circumstances, increase the well-being of their food plants. Harrewijn (1976) has shown that the feeding of *Myzus persicae* (Sulzer) at low densities increases the growth of potato plants growing in media containing relatively low concentrations of phosphorus, by stimulating the plants' nu-trient uptake. Way and Cammell (1970) showed that infested leaves assimilated more CO_2 than uninfested, and postulated that aphidids remove accumulations of carbohydrates that would otherwise inhibit further photosynthesis. It has also been claimed that the honeydew of aphids, on falling into the soil, stimu-lates the growth of nitrifying bacteria (Owen and Weigert, 1976).

When feeding on long-lived plants, aphids may cause an increase in the nutrient content of leaves at leaf-fall (Dixon, 1971b). By several means, therefore, including their own participation in food chains, aphids may hasten recycling in the ecosystem of nutrient elements that are limiting for plant growth (Mattson and Addy, 1975; Springett, 1978).

Section 10.1 references, p. 16

FACTORS INFLUENCING THE SPECIFICITIES OF RESPONSES

The complexity of chemical and physical activities that accompanies the feeding of the Aphidoidea (see section 5.3) has provided a broad scope for speculation and debate concerning the biochemical and physiological causes of the effects these insects have on plants. At the same time, the underlying similarity in their feeding processes can make specificities in such effects difficult to explain convincingly. In the analysis of sometimes conflicting results and views that constitutes this section, the following guidelines have been applied:
– Correlations, however statistically significant, need a testable physiological rationale before they can be accepted as definitive indicators of direct causality (Miles et al., 1982).
– The timing and location of some chemical exchanges may be at least as significant as the exact nature of the substances exchanged (Boysen-Jensen, 1948; Newcomb, 1951; Miles, 1968b; Parry, 1971; Chan and Forbes, 1975).
– Different, even adjacent, plant tissues may have very different thresholds of response to stimuli (Prat, 1955; Dunn, 1960; Meyer, 1962) and such thresholds may provide the key to specificities of interaction when a purely qualitative analysis of the stimuli fails to provide an adequate basis.

Some degree of specificity in the plant's response will reside in the choice of feeding site by the insect. Visual stimuli play an important role in eliciting landing by alatae (Moericke, 1969; Van Emden, 1973; see also section 4.2), and any obstacle that prevents walking by alatae or apterae may elicit probing as an antagonistic reflex (Ibbotson and Kennedy, 1959), but continued penetration of the plant is most probably influenced by arrestant and/or phagostimulant substances either on the surface (Klingauf, 1971, see also section 4.2; Jördens and Klingauf, 1977; Jördens-Röttger, 1979a, b) or within the epidermis (Pollard, 1973).

Even so-called polyphagous species, which are less likely to be affected by host-specific chemical stimuli, may be no less discerning of physiological markers and would seem to select parts and/or stages of development of their food plants where particular conditions prevail (Van Emden, 1978). Insects that select feeding sites in close proximity to meristematic tissue are more likely *a priori* to affect subsequent growth of the food plant than those that feed on mature or senescent parts.

A useful generalization to emerge is that readily recognizable physiological effects on plants, especially localized "toxic" reactions such as the formation of galls, are more likely to be the result of parenchyma-feeding than of phloem-feeding.

Some phloem-feeders have been shown to cause systemic toxic effects, presumably through interference with the plant's translocation system but, with very few exceptions, the effects of phloem-feeding are either too diffuse to be immediately obvious or can more simply be ascribed to loss of photosynthate. Conversely, parenchyma-feeders more often cause noticeable reactions in plants, perhaps because the effects of their feeding will be concentrated in tissues immediately surrounding the feeding puncture; it is also possible that feeding on parenchyma requires continual release of saliva into the plant (Diehl and Chatters, 1956; Kloft, 1960a, b; McLean and Kinsey, 1968).

INTERACTION OF MECHANICAL AND CHEMICAL STIMULI

Many authors have considered the mechanical aspects of the feeding of individual Aphidoidea to have few if any consequences (Pollard, 1973; Gibson,

1974), possibly no more than the histologically detectable death of a few cells that have suffered direct and irreparable physical damage. This view has led authors to relate effects of the feeding of Aphidoidea to introduction of exogenous substances in the saliva. Nevertheless, it must be expected that the mechanical aspects of feeding may themselves give rise to abnormal diffusion of endogenous compounds into tissues surrounding the feeding puncture (Spiller et al., 1985; see also section 5.3). Indeed, some plant tissues are sufficiently sensitive to wounding that multiple pricking with pins alone causes development of localized swellings (Rose, 1939, 1941a, b); less dramatic effects, such as transient increases in cytoplasmic streaming and cell permeability (Marek, 1961), could well be general.

Possibly mechanical effects on their own depend on the phenology of the tissues affected and on the frequency of occurrence of the damage, meristematic tissues being the most likely to respond. Even in tissues where a mechanical puncture in itself does not cause obvious or rapid effects, some degree of wounding by the insect may be a necessary preliminary to chemical influences of the insect on the plant (Maresquelle and Meyer, 1965).

In this context, it should be noted that although salivary secretions have generally been thought of as instigators of plant responses it has also been suggested that the stylet sheath may serve to moderate the effects of mechanical wounding, some of which appear to be harmful to the insects (see sections 5.3 and 10.1.1).

IMMEDIATE PHYSIOLOGICAL RESPONSES

Careful studies of the immediate physiological effects of feeding by Aphidoidea have shown that cells close to the feeding puncture are affected in a variety of ways. Some authors have described plasmolysis, increased cytoplasmic streaming, and accumulation of cytoplasm nearest the origin of the disturbance. Enlargement of the nucleus and nucleoli and/or nuclear disorganization, increases in mitochondria ("chondriosomes"), degenerative changes in the size and/or number of plastids, and changes in the thickness of cell walls have all been documented (Pollard, 1973); however, the extent to which such effects are responses to mechanical stimulation, including rupture of plasmodesmata or the cell wall (Spiller et al., 1985), or to the diffusion or direct entry of salivary components, is by no means certain.

Much of the detailed evidence for the immediate physiological responses of plants to their penetration by Aphidoidea is due to pioneering studies by Kloft (1954, 1960a, b, c), who found that aphids can affect water uptake and transpiration of their food plants. Kloft was among the first to show that the saliva of aphids contained amino acids (see section 5.3) and plant cells, when treated experimentally with solutions of amino acids, exhibit increased permeability and cytoplasmic streaming (Kloft, 1960a, b, c; Marek, 1961), increased respiration and decreased photosynthesis (Kloft, 1960c).

SOME UNCERTAINTIES CONCERNING SALIVARY INFLUENCE

Kloft (1960a, b) found that the immediate physiological effect of feeding by an aphid ceased when the stylets reached the phloem but returned briefly when they withdrew; he concluded that the effects coincided with discharge of saliva into the plant tissue, which ceased while the insects ingested phloem sap. Yet there is a convincing report by Lamb et al. (1967) that *Aphis fabae* Scopoli, labelled with ^{86}Rb, secreted radioactivity into a bean plant continuously over

Section 10.1 references, p. 16

a 6-h period, and they assumed that the insect must have been feeding during, stylet insertion of such duration. There is also difficulty in accounting for an effect of salivary amino acids on plant tissues during the withdrawal of the stylets, for it would appear that salivation at that time merely fills the central channel of the stylet sheath, the walls of which, according to McLean and Kinsey (1967), become impervious once they have formed.

There are a number of ways in which these apparently divergent findings can be reconciled, albeit on a speculative basis. It appears that Aphididae, although predominantly phloem-feeding, may sometimes feed on parenchyma (e.g. some stages of development or some biotypes, or perhaps even some species when feeding on a particular food plant: see discussion of such possibilities in section 5.3); hence it is just possible that the individuals on which Lamb et al. (1967) made their observations had not tapped the phloem. Indeed, in order to avoid contamination of samples with radioactive excreta, the experimenters deliberately excluded from their determinations leaves bearing evidence of massive deposits of honeydew, although such excretion has since been interpreted as an indication that phloem has been tapped (Hoad and Bowen, 1968; Hoad et al., 1971).

There remains the difficulty of accounting for the influence of salivary components secreted during stylet withdrawal. Perhaps the effects that Kloft observed at this time were all the result of a brief discharge of saliva into the phloem at the moment withdrawal commenced; alternatively, perhaps plants do indeed respond to mechanical trauma, including the vibration of the stylet sheath that presumably occurs during withdrawal of the stylets. In the absence of such explanation it would seem necessary to conclude that the sheath is either less impervious or less continuous than has so far been claimed (e.g. by McLean and Kinsey, 1967; Miles, 1972; Pollard, 1973).

PHYTOTOXIC REACTIONS: TO SALIVARY TOXINS OR DEPLETION OF ASSIMILATE?

The muted response of plants to the feeding of most Aphidoidea (the Aphididae in particular) has made exceptions all the more noteworthy. Such obvious responses as occur have been described as "toxic" reactions and, not unnaturally, the insects concerned have been considered to secrete toxins into their food plants.

It is important, however, to distinguish the effects of mass attack from those ascribable to toxins *per se*. Horsfall (1923) described the collapse of cells of *Rumex crispus* attacked by *A. fabae* but, as pointed out by Pollard (1973), the sections illustrated in Horsfall's paper show an extreme concentration of stylet tracks. Lawson et al. (1954) were uncertain how much of the damage to tobacco by *M. persicae* was due to mass removal of nutrients and how much "to injected salivary secretion absorbed and translocated by the plant". Kantack and Dahms (1957) showed a clear relation between the capacity of species of aphidids to cause damage to various cereals and the respective rates of multiplication of the insects on the plants. Similarly, some at least of the "toxicity" of the spotted alfalfa aphid, *Therioaphis trifolii* f. *maculata* (Buckton), to *Medicago sativa* was eventually ascribed to the insect's prodigious rate of multiplication and the heavy fluid and nutrient drain imposed by it (Mittler and Sylvester, 1961).

Nevertheless, detailed analysis has sometimes revealed an overall reduction of plant performance due to the feeding of aphids that cannot be explained by nutrient drain alone. Dixon (1971a) observed that the effect of the sycamore aphid, *Drepanosiphum platanoidis* (Schrank), on the transport system, root growth, and overall distribution of resources in the sycamore, *Acer pseudo-platanus*, appeared excessive in relation to the amount of photosynthate consumed by the aphids. He believed it must be caused by the infiltration into the plant of components of the insects' saliva. Similar conclusions were reached by Mittler and Sylvester (1961) with respect to the effects of the spotted alfalfa aphid and the pea aphid, *Acyrthosiphon pisum* (Harris) on alfalfa.

Of the effects that have been ascribed to salivary toxins of aphids, many are degenerative but the effects of cecidogenesis (gall formation) also include stimulation or alteration of growth, and Norris (1979) has proposed the term "phytoallactin", meaning "causing change in the plant", as more appropriate than "toxin" to cover all the phenomena involved.

SYSTEMIC AND LOCALIZED REACTIONS

Whatever their precise cause, the effects on plants ascribable to aphids can be considered as falling into two general categories: (i) those that act systemically, thereby affecting the plant at some distance from the insect; and (ii) those that cause growth anomalies in the tissues immediately surrounding it.

Systemic toxins have been assumed as causal factors when aphidids feeding above ground have been shown to bring about reduced growth of xylem vessels, and/or of roots, thereby reducing overall growth in height, leaf size, etc. (Ortman and Painter, 1960; Van Emden et al., 1969; Dixon, 1971a; Puritch, 1977). Systemic toxins have similarly been said to cause or increase susceptibility to wilting (Van Emden et al., 1969; Jones et al., 1973). The feasibility of systemic influences of aphids on plants has been demonstrated experimentally by the transport in all directions of radioactivity from labelled insects (Lawson et al., 1954; Peel and Ho, 1970; Forrest and Noordink, 1971).

Localized growth anomalies caused by insects, whether due to inhibition and/or stimulation, have all been called "galls" (Maresquelle and Meyer, 1965); although examples of curling, rolling or pouching of leaves without much thickening have also been termed "pseudogalls" (e.g. Forrest, 1971). As already noted, the more complex galls caused by Aphidoidea seem to be produced by the parenchyma-feeders (see also section 10.1.1).

Kloft (1955, 1960b) considered that the local accumulation of protein (and peptides and amino acids) that surrounds the feeding punctures of *Adelges* spp. on *Abies alba*, even when not accompanied by any increase in growth in the plant tissues, represented an incipient state of cecidogenesis that he terms a "physiological gall". Otherwise, Maresquelle and Meyer (1965) have described various kinds and degrees of growth abnormality caused by Aphidoidea and by other cecidogenic insects (see also section 5.4).

CECIDOGENIC REDIFFERENTIATION

Maresquelle and Meyer (1965) point out that in galls of animal origin there is always inhibition of growth of some cells, whatever the degree of simulation of others. They relate the form of the gall to diffusion through the affected tissues of exogenous substances, presumed to be of salivary origin in the Aphidoidea. They conclude that such substances form gradients of concentration modified by histological discontinuities in the plant and perhaps also by the placing of multiple injections by the insect. Where the insect's influence is strongest, gross disorganization and necrosis may occur; successively thereafter there will be decreasing degrees of reduced growth and development, grading into hypertrophy and hyperplasia.

Section 10.1 references, p. 16

The precise nature of the response of individual plant cells will depend both on the effective concentration of the stimulus and on the degree of morphological development – and associated threshold of sensitivity – of the cell. Thus even a uniform gradient of cecidogenic influence will tend to have stepwise effects within the tissues (Prat, 1955; Dunn, 1960; Meyer, 1962). The younger the tissues are phenologically, the more likely they are to respond, to have their morphological development inhibited or redirected and to become involved in hyperplasia. Conversely, older tissues, when they respond to a growth stimulus, are more likely to become hypertrophic.

Some degree of dedifferentiation of plant cells under the influence of a variety of cecidogenic insects, including Aphidoidea, is also possible (Maresquelle and Meyer, 1965). Such cells are likely to contribute to a special tissue subserving the nutrition of the insect; typically they have enlarged nuclei and nucleoli, abundant mitochondria, and a cytoplasm containing much RNA and showing evidence of the activity of hydrolyzing enzymes; the vacuole may be enlarged or fragmented and the water content is higher than in normal cells, so that the affected tissue has a lower nutrient (and "ash") content than normal on a fresh weight basis, although the proportions of soluble protein, amino and amide nitrogen, and oligosaccharide are usually enhanced on a dry weight basis (Maresquelle and Meyer, 1965). When such tissue is composed of morphologically undifferentiated cells with a "precocious" physiological development as described above, Meyer (1951) has designated it a *"plastème nourricier"*.

THE INSECT AS A LOCALIZED NUTRIENT "SINK"

The insect represents a "sink" for assimilate (Kennedy and Stroyan, 1959; Baron and Guthrie, 1960; Chang and Thrower, 1981), and if the drain is sufficiently strong and localized, the plant reacts to it in some respects as if it were a bud (Hill, 1962; Way, 1968; Way and Cammell, 1970). Individuals of *A. fabae* grow faster in communities than when isolated (Dixon and Wratten, 1971) and this is probably true of aphids generally, arising from the cooperative establishment of a sink of sufficient magnitude to attract significant amounts of assimilate from other parts of the plant.

Thus a single individual of *A. fabae* benefits from the concurrent feeding of others of its species grouped on the other side of the leaf (Dixon and Wratten, 1971) and, when feeding on apple leaves, from the proximity of a pseudogall in which another species, *Dysaphis devecta* (Walker), is feeding (Forrest, 1971). A similar phenomenon occurs with leaf galls of *V. vitifoliae*, which, when crowded, accumulate more assimilate per gall than when isolated (Rilling and Steffan, 1978).

Although most investigators of cecidogenesis attribute the initial accumulation and eventual breakdown of reserves – especially breakdown of cell structure – to direct intervention by the insect's secretions, some of the activities of the plant cells could well be a direct response to the strong and persistent flow of assimilate created by the continual removal of metabolites by a sedentary insect. Cells at an early stage of formation of the gall, especially those in a peripheral region surrounding the insect, may build up starch and protein reserves but, under the increasing influence of the insect, reserves become progressively broken down and removed (Kloft, 1960a, b; Maresquelle and Meyer, 1965; Eichhorn, 1968; Schaefer, 1972; Zotov and Gadiev, 1975; Jones et al., 1976; Rilling and Steffan, 1978; Rohfritsch, 1981; see also section 10.1.1).

RESPIRATORY CONSEQUENCES OF A NUTRIENT SINK

Wherever a nutrient sink has been established, whether or not a gall is formed, the increasing dedication of cells to mobilization and transport functions is associated with a reduction of photosynthesis in the tissues so affected; chloroplasts degenerate and varying degrees of chlorosis result (Smith and Brierley, 1948; Severin and Tompkins, 1950; Diehl and Chatters, 1956; Kantack and Dahms, 1957).

At the same time, there is a rise in metabolic activity and oxygen-binding compounds accumulate, sometimes producing pigments seemingly specific to the insect–plant relationship (Forrest, 1971; Forrest and Dixon, 1975). As discussed below, some of these compounds would appear to be defensive in nature.

Decreased photosynthesis, increased metabolism and accumulation of reducing compounds increase respiratory demand in the affected tissues, and the respiratory quotient in galls may rise to levels that indicate partial asphyxia of cells away from the external surface (Newcomb, 1951).

DEFENSIVE REACTIONS IN GALLED TISSUES

Superimposed on the breakdown of reserves and the flow of assimilate, other changes occur, in galled tissues in particular, that appear to be defensive reactions against wounding and/or the presence of a foreign organism. These include the accumulation of secondary defensive substances such as triterpenes (Monaco et al., 1974; Caputo et al., 1979), tannins and their phenolic precursors (Mandl, 1957; Denisova, 1965; Maresquelle and Meyer, 1965; Sobestkii and Derzhavina, 1973; Puritch, 1977; Rohfritsch, 1981), and phytoalexins (Nielson and Don, 1974; Puritch and Nijholt, 1974; Argandoña et al., 1983). Defensive necrosis and/or the formation of wound periderm have also been reported (Bramstedt, 1938; Kloft, 1955; Oechssler, 1962; Denisova, 1965; Van Emden et al., 1969; Thalenhorst, 1972; Mullick and Jensen, 1976; Zotov, 1976; Rohfritsch, 1981).

PHYTOTOXIC REACTIONS CHARACTERIZED

In the foregoing analysis of phytotoxicoses, four separate influences are discernable, even if their effects tend to overlap:
– Firstly, common to all the relationships is the occurrence of a drain of assimilate towards the insect and away from the rest of the plant. In so far as this drain exceeds the capacities of the plant for compensatory growth, it will result in an apparently systemic reduction of overall growth. In addition, there may be more localized degenerative changes, such as chlorosis and regional inhibition of growth, due to the stronger application of the same influence closer to the insect.
– Secondly, in some relationships the systemic effect on overall growth of the plant seems greater than can be ascribed to nutrient drain alone. The exact nature of the assumed, causal "toxins" is unknown; affects on the respiration and water translocation of leaves obtained by infiltration with amino acids (see section 5.3) may implicate these compounds. On the other hand, individual aphids would seem likely to reinject significantly smaller amounts of amino acids than they have already removed by ingestion! Amino acids are only some of the compounds that have been implicated in cecidogenesis (see sections 5.3, 5.4 and 10.1.1) and it is possible that systemic phytotoxicoses are reactions of the whole plant to general release throughout its transport system (e.g. by

Section 10.1 references, p. 16

phloem-feeders) of those same compounds that cause galls when more locally concentrated (e.g. by parenchyma-feeders).

– Thirdly, there may occur localized redifferentiation (with or without prior dedifferentiation) of tissues surrounding a sedentary insect, as a result of which plant cells undergo cytological changes and appear to be subserving the nutritive and protective requirements of the insects (Maresquelle and Meyer, 1965). Such effects are most readily ascribable to a local influence of secretions released during parenchyma-feeding.

– Fourthly, also in tissues surrounding sedentary parenchyma-feeders, particularly gall-forming species, there may occur defensive responses by the plant, triggered by chemical changes induced by the insect. Where such changes lead to visible necrosis and seal-off of the affected tissues they may lead to death of the insect and confer on the plant some degree of resistance to attack. Investigation of the physiology of galls of *V. vitifoliae* has indicated that, even when visible histological signs of defense are absent or not yet apparent, the insect continually controls and/or modifies the content of the surrounding tissues, presumably suppressing the accumulation of toxic products, before ingesting exudates therefrom (Sobetskii and Derzhavina, 1973; see also section 10.1.1).

EXPERIMENTAL SIMULATION OF NATURAL RESPONSES

Several authors have shown that breis or extracts or collected secretions of cecidogenic insects can raise intumescences on plants when injected or applied externally in pastes. Plumb (1953) injected extracts of the fundatrices of *Chermes abietis* (Linnaeus) into buds of its host plant, *Picea alba*, initiating simple swellings thereby.

Also noteworthy was the experiment of Kloft (1950) who grew seedlings of cress (*Lepidium sativa*) in aqueous extracts of the grape phylloxera and of the woolly aphid of apple, *Eriosoma lanigerum* (Hausmann), and obtained either stimulation or inhibition of longitudinal root growth, depending on the dilution of the extracts. Cress is not a food plant of these insects and the results obtained were not in any sense galls, but the experiment is significant in that it clearly showed the hormone-like activity of the extracts while avoiding any possible interference from physical trauma.

Positive results with extracts of cecidogenic insects led to experiments with other materials, including specific compounds. An unexpected side result of such tests was the discovery that intumescences could be raised on cabbage and pelargonium stems by injections of such unlikely "cecidogenic" agents as urine or bee venom (Rose, 1939), and that subapical swellings on vine rootlets, similar to the nodosities caused by *V. vitifoliae*, were produced when vine cuttings were grown in slightly alkaline solutions of potassium phosphate (Miles, 1968a). Demonstrations of this kind presumably do no more than indicate that some plant tissues will produce anomalous growth in response to a variety of influences, whether natural or unnatural; it is not surprising, therefore, that other substances having somewhat greater a priori claims of relevance to cecidogenesis have been employed with similar success.

HORMONES AS CECIDOGENS

Early candidates as cecidogenic influences were the plants' own growth substances, auxins in particular. Nystérakis (1948) drew analogies between growth deformities produced experimentally with indole-3-acetic acid (IAA) and the natural deformities caused by a variety of Aphidoidea, including *V. vitifoliae*, pemphigids, and aphidids. Kloft (1950), as mentioned above, demonstrated the hormone-like activity of extracts of such insects. IAA has been extracted from various cecidogenic Aphidoidea (Link et al., 1940; Nystérakis, 1948; Maxwell and Painter, 1962) and from their saliva (Schäller, 1968a; see also section 5.3, Table 5.3.1). Duspiva (1954) showed that the salivary glands of aphids were capable of converting tryptophan to IAA.

Mandl (1957) simulated relatively simple galls, such as those formed by some pemphigids, using pastes containing IAA or 2,4-dichlorophenoxyacetic acid (2,4-D) but pointed out that very high concentrations of the natural hormone had to be applied compared with the synthetic compound (2,4-D); other authors have since attributed this to the ability of plants to degrade excess amounts of natural growth substances especially when applied externally (Kefeli and Turetskaya, 1965; Van Overbeek, 1966).

IAA very probably is involved in one way or another in the growth of galls, as it has been shown to occur in higher concentrations in galled than in normal tissues (Brain, 1957; Schäller, 1968a), but the relation between the IAA found in the plant and any that may be found in the insect is problematical (see section 5.4). Schäller found more salivary IAA in cecidogenic Aphidoidea than in less phytotoxic species, and he was able to simulate galls of the grape phylloxera with solutions containing inter alia $10^{-6} M$ IAA (Schäller, 1968b), which is a concentration that falls within the range found in normal tissues of plants (summarized by Miles and Hori, 1977 as being about 10^{-3} to $10^{-4} M$ in flowers, 10^{-5} to $10^{-6} M$ in shoots, 10^{-7} to $10^{-8} M$ in buds, and 10^{-10} to $10^{-11} M$ in roots).

It follows that the amount of IAA injected by the insects could well be such as could be dealt with by the plant's own regulatory system (a point made by Brain (1957) in relation to cecidogenic microorganisms); indeed, it has been suggested that the IAA in the saliva of sucking insects in general is most probably recycled from the plant and cannot therefore participate in cecidogenesis unless the insect were to change its food plant or feeding site (Nuorteva, 1962; Schäller, 1965; Miles and Hori, 1977; Hori et al., 1979). Even then, salivary IAA could at most be involved as an adjunct to the initiation of galls, but not to keep them growing (Hori, in press).

Schäller (1968b) initiated swellings in leaves and petioles of the grape vine by inserting into them caniculi containing solutions of IAA. He also developed nodosities on the rootlets of vine seedlings by germinating them in solutions containing IAA.

Schäller obtained his most realistic simulations of leaf galls when the amino acids found in the insect's saliva were also present in the solutions, but cautioned against an oversimplistic interpretation of these results: although his most effective solutions contained components of the saliva of *V. vitifoliae* at approximately correct concentrations, the total quantities of exogenous chemicals continuously introduced from the caniculi in the course of the experiments were unlikely to have bourne a simple relation to the limited quantities of metabolites that must be exchanged during the natural recycling processes between insect and plant.

Alternatively, compounds other than IAA that have auxin-like effects could be involved in cecidogenesis. Puritch (1977) considered that the changes induced by adelgids in the sapwood of firs were caused by ethylene generated within the plant by auxin-like compounds in the insects' saliva. Other authors have suggested that components of the saliva may release or spare the plant's own IAA, and these possibilities will be considered further below.

Section 10.1 references, p. 16

AMINO ACIDS AS CECIDOGENS

V. vitifoliae has no anus and the salivary glands are, inter alia, modified to perform an excretory role (Anders, 1957). When removed from galls, the fundatrices sometimes continue to secrete drops of fluid from their stylets, and Anders (1957, 1958, 1961) found that this fluid contained a high concentration of free amino acids. He showed that when grape seeds were germinated in solutions of the specific amino acids that he believed to be present in the insect's secretion, the rootlets developed subapical nodosities (Anders, 1958, 1960b, 1961).

Other work confirmed the presence of amino acids in the saliva of *V. vitifoliae* and demonstrated them in the saliva of other Aphidoidea (Kloft and Ehrhardt, 1959; Kloft, 1960a; Schäller, 1961, 1963b, 1968a). Amino acids were also found in the saliva of phytophagous Heteroptera and probably occur in the saliva of all sucking insects that produce a stylet sheath (Miles, 1968b). This widespread occurrence of the compounds should, therefore, be taken into account when assessing their significance in specific reactions such as the production of galls.

Anders (1957, 1958, 1961) had apparently misidentified the compounds present in the saliva of *V. vitifoliae*, and the amino acids he had "shown" to be cecidogenic, namely tryptophan, lysine, and histidine, are, in fact, in very low concentrations in the saliva of this and other Aphidoidea (Schäller, 1960); tryptophan, the most active compound of all according to Anders, was not found by Schäller (1963a) in the saliva of any of the races of *V. vitifoliae* he tested (see also section 5.3, Table 5.3.1).

Anders (1960a, b, 1961) believed that the amino acids he had specified acted as polyploidizing agents in meristematic tissues, with an effect similar to that of colchicine. However, results similar to those of Anders are obtainable with other compounds, including potassium phosphate, and Miles (1972) suggested that the phenomenon might be due to disturbance of the permeability of cell membranes in the subapical meristem, regardless of the relevance of the cause to natural cecidogenesis.

It would now seem that the development of nodosities on vine roots can be induced by too many compounds to be of much diagnostic value in elucidating cecidogenesis. In so far as amino acids are involved in the production of galls, they are more likely to be modifying factors that initiating stimuli (Schäller, 1968b; Hori, 1986).

ENZYME SYSTEMS AS CECIDOGENS

Weidner (1957) suggested that cecidogenesis could be caused by the release of bound IAA in plant tissues brought about by the insect's feeding activity. Anders (1961) stated that the saliva of *V. vitifoliae* contained a protease and thought it might be capable of splitting IAA from its inactive conjugate with protein.

Sobetskii and Derzhavina (1965) and Sobetskii (1967) did not find protease in the saliva of *V. vitifoliae* (or of any other cecidogenic Aphidoidea, whether parenchyma- or phloem-feeding); they thought that the insect's gut contents (which had both amylolytic and proteolytic activity) must be regurgitated from time to time to bring about the differences they noted between the tissues of the gall and the exudates actually ingested by the insect.

Schäller (1965, 1968a, b) found proteases in the saliva of *V. vitifoliae* and *E. lanigerum*; he pointed out that up to 95% of the IAA in plants is bound to protein; he also considered that these enzymes could well contribute to cecidogenesis by releasing IAA from conjugates.

Zotov (1976), quoting a number of sources, considered the saliva of the grape phylloxera to contain tryptophan, protease, RNA-ase, DNA-ase, and plant growth substances, which together produced the growth deformities and other physiological changes observable in the host plant.

Another enzyme system that has been considered as potentially cecidogenic is the combination of polyphenol oxidase and diphenol that is found in the saliva of apparently all Homoptera and phytophagous (as well as some other) Heteroptera (Miles, 1964, 1972). The product of such a system is a quinone, which can produce IAA from tryptophan (see section 5.3). The salivary glands of aphids had already been shown to carry out this conversion (Duspiva, 1954) and, in an apparent demonstration that this process could be involved in cecidogenesis, Miles (1968a) caused a non-cecidogenic plant bug to produce simple leaf galls around its feeding punctures by artificially boosting the amounts of tryptophan and phenylalanine in its haemolymph and thereby of tryptophan and DOPA in its saliva.

Despite the apparent success of this experiment, its implications remain obscure. Phenolic compounds and the quinones formed from them enter into many biochemical reactions. In excess, quinones cause necrosis. In subnecrotic quantities (and if not consumed in conversions such as that indicated above), quinones combine with and inactivate IAA (Tomaszewski, 1959; Pilet, 1960; Leopold and Plummer, 1961); for this reason Schäller, when seeking an explanation for the relative phytotoxicities of species of aphids, suggested that the phenolic compounds he had found in their saliva served to reduce the potential cecidogenic effects of other components in some species. It may be noted, however, that the non-cecidogenic aphids in Schäller's experiments were also phloem-feeders and for this reason alone less likely to cause local symptoms.

A further complication is introduced by the activity of phenols towards IAA in the absense of phenolases. Monophenols reduce the activity of IAA in naturally occurring systems, whereas polyphenols enhance it (Pilet, 1960; Schantz, 1966; Tomaszewski and Thimann, 1966). Plant tissues contain a range of such substances in varying amounts, although normally the metabolites are concentrated in the vacuole and the enzymes in the cytoplasm; presumably the outcome of mechanical and chemical intervention by a sucking insect in the cytochemistry of plant cells will to an extent depend on the nature and the relative concentrations of vacuolar and cytoplasmic reactants at the time they are brought together.

As with the other demonstrations of artificial cecidogenesis described so far (Anders, 1960a; Schäller, 1968b), that of Miles (1968a) would have failed to simulate natural phenomena if the concentrations and/or quantities of reactants used to achieve the effects were much in excess of those normally interchanged in the insect/plant interaction. Possibly such experiments should be regarded only as indications that the growth-regulating system of plants is sensitive to several kinds of interference, especially if they are applied with sufficient vigour.

ENZYME INHIBITORS AS CECIDOGENS

A number of authors have come to the conclusion that the cecidogenic influence most likely to be effective would be one that inhibited the enzymes that regulate the concentration of growth substances in the plant – a probable target enzyme, cited in a number of studies, is IAA-oxidase (Sterling, 1952; Pilet, 1960; Ignoffo and Granovsky, 1961). An inhibitor of this enzyme would function as an "IAA-synergist": Hori (1974, 1975, 1976; Hori and Miles, 1977) has demonstrated just such a system in the salivary glands of a number of

Section 10.1 references, p. 16

Heteroptera, and there are sufficient similarities between the Homoptera and phytophagous Heteroptera to make the occurrence of a similar system in the Aphidoidea a possibility.

The synergists investigated by Hori, although separated on chromatograms, have not yet been identified chemically. In the Heteroptera, ingested IAA and other indoles seem to be mostly metabolized and/or conjugated in the gut and do not appear to be transferred in any great quantities to the salivary glands (Hori and Endo, 1977; Hori et al., 1979; Hori, 1979a, b, 1980). Nevertheless, the analyses of Schäller (1965) would seem to indicate that indoles do occur in the salivary glands of Aphidoidea, and it is possible that non-auxinic indoles or related metabolites could function as inhibitors of IAA-oxidase, as could polyphenols (see above).

The experimental evidence presented above for the existence of such inhibitors in sucking insects has been derived, not from aphids, but from non-cecidogenic Heteroptera! Nevertheless, a lygaeid bug has been made "cecidogenic" experimentally (Miles, 1968a), and nearly all the cecidogenic influences suggested in the literature have turned out to be features of cecidogenic and non-cecidogenic species alike. As discussed below ("Behavioural influences in cecidogenesis"), it may not be association with a potential stimulus so much as where, for how long and with what intensity it is applied that makes one species cecidogenic and another not or accounts for the specific features of individual galls. Certainly inhibition of IAA-oxidase has that potential, as the paragraphs immediately following illustrate.

RESPIRATORY INFLUENCES IN CECIDOGENESIS

Madden and Stone (1984) have shown that, on the leaves of eucalypts, blockage of stomata and/or general reduction of permeability to air of the epidermis, whether due to the physical presence of sedentary insects or their excreta, or to artificial means, can give rise to pseudogalls. On young leaves, when reduction of permeability was unaccompanied by a nutrient drain, i.e. where the cause was nonliving, or immediately following premature removal of a sedentary insect, eruptive hypertrophy of the underlying pallisade tissue could also occur.

The authors produced evidence that reduced respiration, by inhibiting oxidation reactions, spared IAA which in turn promoted production of ethylene. The morphogenetic and histological consequences included increased mitotic activity, the eruptive hypertrophy in young tissues mentioned above, and formation of lignin and accumulation of phenolics in surrounding regions of the leaf. In older tissues there was less hypertrophy, or the only response was accumulation of phenolics.

These observations were made in relation to the formation of pouch galls by psyllids, but clearly the work is potentially of wider relevance. According to Puritch (1977), auxin-generated ethylene is responsible for the formation of adelgid galls in firs (see section 10.1.1). Also, as discussed above, studies of cecidogenesis by *V. vitifoliae* have indicated an apparent lack of oxygen in galled tissue, although it has so far been interpreted not as a cause of cecidogenesis but as the result of an increased respiratory requirement of tissues responding to some more fundamental cecidogenic stimulus.

It would seem that a relative shortage of oxygen is to be expected in galled tissue, but whether this shortage is the initiator or a consequence of specific instances of cecidogenesis has still to be settled.

BEHAVIOURAL INFLUENCES IN CECIDOGENESIS

A major generalization to emerge from studies of the interactions between cecidogenic Aphidoidea and their host plants is that the observed effects are always under the direct influence of the insect. As many investigators have noted, a gall ceases to develop following removal of the insect (Maresquelle and Meyer, 1965). Nowhere is this better illustrated than by the phenology of adelgid galls, the formation of which on the Norway spruce, *Picea abies*, is described in section 10.1.1. Removal of the fundatrix or of the larvae at appropriate times shows that each in turn takes over control of development of the gall. The larvae determine the final specificity of appearance of the gall (Rohfritsch and Meyer, 1966; Rohfritsch, 1967). They are able to invade chambers initiated by another species, however, and thus it is possible to develop a hybrid gall, started by the fundatrix of one species and completed by larvae of another (Rohfritsch, 1966), or a composite gall in which some chambers are inhabited by and are morphologically characteristic of one species and some are inhabited by and are characteristic of another (Eichhorn, 1975).

Direct evidence of a strong behavioural component in the determination of the morphology of galls is provided by the account by Dunn (1960) of the formation of the various galls produced by *Pemphigus* spp. on the leaf petioles of the Lombardy poplar, *Populus nigra* var. *italica*. For some days, the fundatrix performs a specific pattern of stylet insertions over a more or less restricted area, which causes reduced growth immediately below the insect and swelling of tissues further away, until a gall with the shape typical for the species has been formed.

According to Dunn, the insect does not begin to feed until the swollen tissues close over her. He assumed the inhibitory injections made by the fundatrix to be more concentrated than the stimulatory injections, and that the instinctive patterns he observed in the placing, timing and frequency of the injections, together with varying concentrations of the substances injected and the changing responsiveness of the plant tissues as they matured, all contributed to the morphological specificity of the galls.

REJECTION OF INVOLVEMENT OF NEOPLASIA AND MICROBIAL INTERVENTION

Because the development of galls does not continue if the insects' influence is removed, reviewers have dismissed suggestions that cecidogenic effects are due to neoplastic transformation or to the transfer of microorganisms (Maresquelle and Mayer, 1965; Carter, 1973). If polyploidization occurs in gall tissue as described by Anders (1960a, 1961), clearly it does not lead to independent growth, such as is observed in crown gall (caused by *Agrobacter tumefasciens* (Smith and Townsend) Conn). For much the same reason, claims that the galls caused by Aphidoidea are due to the transmission of microorganisms have also been discounted. The only serious evidence put forward for involvement of microbial pathogens is the claim that a trans-ovarially transmitted, virus-like "cecidogen" that multiplies only in the insect is responsible for the galls produced on whitch hazel (*Hamamelis virginiana*) by *Hormaphis hamamelidis* Fitch (Lewis and Walton, 1958) and on red spruce (*Picea rubens*) by *Pineus floccus* (Patch) (Walton, 1980). Although the evidence for this "cecidogen" is based on meticulous histological studies, they are basically interpretations of visual observations that could be given alternative explanations – certainly there has been no experimental verification of the existence of a microbial pathogen such as would satisfy Koch's postulates.

Section 10.1 references, p. 16

INTERACTIONS OF THE RESPONSIVENESS OF THE PLANT WITH BEHAVIOURAL AND CHEMICAL SPECIFICITIES OF THE INSECT

From the foregoing it appears that cecidogenic effects are strongly dependent on the nature of the substrate. Some tissues are clearly sensitive to morphogenetic disturbance, while others are refractory. Madden and Stone (1984) found that "eruptive" galling due to reduction of gaseous permeability could be induced only on the young leaves of certain, susceptible eucalypts. Sterling (1952) pointed out that although *V. vitifoliae* is able to initiate a gall on either surface of a leaf, only galls on the adaxial surface proceed to completion. Some aphidids cause deformation of leaves or shoots of trees early in the season, but not later (Müller, 1957; Schäller, 1965); and *Macrosiphum euphorbiae* (Thomas) causes small protuberances on the lamina of leaves of the potato plant but not on stems or veins (Gibson, 1974). As will be illustrated in section 10.1.1, the resistance of plant varieties to at least some of the cecidogenic parenchyma-feeders depends on the sensitivity of the defensive responses of the plant.

At the same time, there is also evidence of essential differences between the phytotoxicity of different species. Once account has been taken of the greater cecidogenic capabilities associated with parenchyma-feeding compared with phloem-feeding and of the part played by specific behaviour patterns, as in the effects of *Pemphigus* spp. on poplar, there remain differences in phytotoxic effects that seem most likely due to specificities in the chemical nature of the insects' feeding activities. Thus, Horsfall (1923) pointed out that *M. persicae* does not produce galls on a variety of plants on which other insects do cause galls. *Myzus cerasi* Passerini produces deformations of *Prunus* shoots on which *Myzus ascalonicus* Doncaster and *Megoura viciae* Buckton do not (Schäller, 1965); *Dysaphis devecta* (Walker) and *D. plantaginea* (Passerini) produce differently coloured, as well as differently oriented, "pseudogalls" on apple (Forrest and Dixon, 1975); resins in the galls on *Pistacia terebinthus* produced by *Pemphigus cornicularius* Passerini and *P. semilunarius* Passerini contain different triterpenes, ascribable to the action of stereospecific enzymes specific to the activities of each insect (Caputo et al., 1975).

In the simplest of galls, the degree of cytochemical change in surrounding tissues that makes them more suitable as food resources for the insect may conceivably be a simple response of sensitive cells to the drain of assimilate imposed by the insect. But in the more complex, "organoid" galls, where specialized nutrient tissues can be distinguished, at least part of the form and function of the component cells has been assumed to be under some direct chemical influence on the part of the insect, although again dependent on the sensitivity of the plant tissues, as discussed above ("Cecidogenic redifferentiation"; see also sections 5.4 and 10.1.1).

CECIDOGENESIS SEEN AS A PROGRESSIVE INFLUENCE ON BIOCHEMICAL EQUILIBRIA

Most attempts to simulate cecidogenic activities, by concentrating on particular candidate processes, have probably been too simplistic whatever their apparent success; for, as Forrest and Dixon (1975) point out, the morphological development of the plant is under the control of a number of growth factors, and it may be naive to suppose that phytotoxic insects affect only one of them.

Some authors have tried to avoid the confusion created by independent demonstration of a variety of cecidogenic influences by considering the cecidogenic stimulus not as addition or withdrawal of specific metabolites, but as a resetting of the cytochemical environment in some more general way, thereby shifting the metabolic equilibria that progressively determine the speed and direction of development.

– Beck (1946), in relation to the *Solidago* gall (caused by a caterpillar, *Gnorimoschema gallaesolidaginis* Riley), believed that raising the pH of the plant tissues would cause or maintain meristematic activity by favouring enzymes that release IAA.

– Turian (1958), in relation to fungal galls, believed that any influence that increased the activity of catalase in cells would set in motion a series of effects. IAA-oxidase would be inhibited, causing an increase of IAA; the IAA would promote synthesis of RNA and thereby of proteins (including phosphatase); nuclear hypertrophy and hyperplasia would follow, and also increases in phosphate and other osmotically active chemicals that in turn would cause cytoplasmic hyperhydration and hypertrophy.

– Maresquelle and Meyer (1965) pointed out that all cecidogenesis appeared to pass through three stages: (i) a stimulation of growth; (ii) an inhibition of differentiation; (iii) a redifferentiation of the tissues into the organization specific to the gall. They considered that if these stages – or the galls of specific insects – were due to different substances, they had yet to be identified; if due to varying manifestations of the same basic cause, it had yet to be demonstrated; the explanation of the complex etiology of galls still remained "*absolument mystérieuse.*" Nevertheless, their analysis of the progressive nature of cecidogenesis was in itself significant.

– Miles (1968b) drew attention to the phenol–phenolase system in the saliva and the increased oxidation state of mature galls (Newcomb, 1951). He suggested that any stimulus that altered the redox potential of plant cells could have morphogenetic effects by influencing the progressive composition and oxidation states that the phenolic contents of the cells undergo (after Tomaszewski and Thimann, 1966). At the initial, reducing state characteristic of meristematic cells, the preponderance of monophenols tends to decarboxylate IAA and to suppress enlargement of individual cells in favour of cell division; at an intermediate stage, when diphenols preponderate, IAA-oxidase tends to be inactivated and growth and differentiation is maximized, although any tipping of the metabolic balance in favour of oxidation of the phenols to quinones would initiate defensive reactions, such as hypersensitive necrosis; finally, with polymerization of phenols to inactive (often coloured) compounds, the cells mature and generally become unresponsive to further external influence. He supposed that the insect intervened in this cytochemical progression of oxidation states, thereby halting or reversing cell phenology. Madden and Stone (1984) have since demonstrated cecidogenic and other physiological effects of reduced oxygenation of responsive tissues.

– Hori (in press) suggests that true cecidogenesis should consist of successive, complementary influences: a "conditioner" (possibly consisting of amino acids) that would make the cells more receptive; a true "inducer" (possibly IAA, another plant hormone, or a hormone-sparing synergist) that would induce hypertrophy and/or hyperplasia; and a "maturator'" (possibly consisting of the continued action of the inducer under the control of the insect's behaviour) that would maintain and control the abnormal growth and differentiation specific to the gall. He saw these three external influences as integrating with phenolic and other conditioners of cell chemistry, such as those described above, in a step-wise or progressive fashion during development of the specific tissues (cf. Maresquelle and Meyer, 1965, cited above).

When viewed in the context of the differences between plant tissues and the complexity of their phenology, it is perhaps small wonder that, despite seeming

---

Section 10.1 references, p. 16

commonalities in the feeding processes of aphids, variation in detail in the chemical and physical interactions of insect and plant should produce a divergence of results. These will be considered on a species by species basis in the next section, which also contains summaries of current views on the causation of specific phytotoxicoses due to phloem- and parenchyma-feeders respectively.

REFERENCES

Anders, F., 1957. Über die gallenerregenden Agenzien der Reblaus (*Viteus (Phylloxera) vitifolii* Shimer). Vitis, 1: 121–124.

Anders, F., 1958. Aminosäuren als gallenerregende Stoffe der Reblaus (*Viteus [Phylloxera] vitifolii* Shimer). Experientia, 14: 62–63.

Anders, F., 1960a. Untersuchungen über das cecidogene Prinzip der Reblaus (*Viteus vitifolii*) Shimer. I. Untersuchungen an der Reblausgalle. Biologisches Zentralblatt, 79: 47–58.

Anders, F., 1960b. Untersuchungen über das cecidogene Prinzip der Reblaus (*Viteus vitifolii* Shimer). II. Biologische Untersuchungen über das galleninduzierende Sekret der Reblaus. Biologisches Zentralblatt, 79: 679–700.

Anders, F., 1961. Untersuchungen über das cecidogene Prinzip der Reblaus (*Viteus vitifolii* Shimer). III. Biochemische Untersuchungen über das galleninduzierende Agens. Biologisches Zentralblatt, 80: 199–233.

Argandoña, V.H., Corcuera, L.J., Niemeyer, M.H. and Campbell, B.C., 1983. Toxicity and feeding deterrency of hydroxamic acids from Gramineae in synthetic diets against the greenbug, *Schizaphis graminum*. Entomologia Experimentalis et Applicata, 34: 134–138.

Baron, R.L. and Guthrie, F.E., 1960. A quantitative and qualitative study of sugars found in tobacco as affected by the green peach aphid, *Myzus persicae*, and its honeydew. Annals of the Entomological Society of America, 53: 220–228.

Beck, E.G., 1946. A study of the *Solidago* gall caused by *Eurosta solidaginis*. American Journal of Botany, 33: 228.

Boysen-Jensen, P., 1948. Formation of galls by *Mikiola fagi*. Physiologia Plantarum, 1: 95–108.

Brain, P.W., 1957. The effect of some microbial metabolic products on plant growth. Symposia of the Society for Experimental Biology, 11: 166–182.

Bramstedt, F., 1938. Der Nachweis der Blutlausunanfälligkeit der Apfelsorten auf histologischer Grundlage. Zeitschrift für Pflanzenkrankheiten, Pflanzenpathologie und Pflanzenschutz, 48: 480–488.

Caputo, R., Mangoni, L., Monaco, P. and Palumbo, G., 1975. Triterpenes of galls of *Pistacia terebinthus*: Galls produced by *Pemphigus utricularius*. Phytochemistry, 14: 809–811.

Caputo, R., Mangoni, L., Monaco, P. and Palumbo, G., 1979. Triterpenes from the galls of *Pistacia palestina*. Phytochemistry, 18: 896–898.

Carter, W., 1973. Insects in Relation to Disease. Wiley, New York, 2nd edn., 759 pp.

Chan, C-K. and Forbes, A.R., 1975. Life cycle of a spiral gall aphid *Pemphigus spirothecae* (Homoptera: Aphididae), on poplar in British Columbia. Journal of the Entomological Society of British Columbia, 72: 26–30.

Chang, L-H. and Thrower, L.B., 1981. The effect of *Uromyces appendiculatus* and *Aphis craccivora* on the yield of *Vigna sesquipedalis*. Phytopathology, 101: 143–152.

Davidson, J., 1923. Biological studies of *Aphis rumicus* Linn. The penetration of plant tissues and the source of the food supply of aphids. Annals of Applied Biology, 10: 35–54.

Denisova, T.V., 1965. The phenolic complex of vine roots infested by *Phylloxera* as a factor in resistance. Vestnik Sel'sko-khozyaistvennoi Nauki, Moscow, 5: 114–118 (in Russian, with English summary).

Diehl, S.G. and Chatters, R.M., 1956. Studies on the mechanics of feeding of the spotted alfalfa aphid. Journal of Economic Entomology, 49: 589–591.

Dixon, A.F.G., 1971a. The role of aphids in wood formation. I. The effect of the sycamore aphid, *Drepanosiphum platanoides* (Schr.) (Aphididae), on the growth of sycamore, *Acer pseudoplatanus* (L.). Journal of Applied Ecology, 8: 165–179.

Dixon, A.F.G., 1971b. The role of aphids in wood formation. II. The effect of the lime aphid, *Eucallipterus tiliae* L. (Aphididae), on the growth of lime *Tilia x vulgaris* Hayne. Journal of Applied Ecology, 8: 393–399.

Dixon, A.F.G. and Wratten, S.D., 1971. Laboratory studies on aggregation, size and fecundity in the black bean aphid, *Aphis fabae* Scop. Bulletin of Entomolgoical Research, 61: 97–111.

Dunn, J.A., 1960. The formation of galls by some species of *Pemphigus*. Marcellia (Supplement) 30: 155–167.

Duspiva, F., 1954. Weitere Untersuchungen über stoffwechsel-physiologische Beziehungen zwischen Rhynchoten und ihren Wirtspflanzen. Mitteilungen der Biologischen Zentralanstalt für Land- und Forstwirtschaft, 80: 155–162.

Eichhorn, O., 1968. Problems of the population dynamics of silver fir woolly aphids, Genus *Adelges* (= *Dreyfusia*), Adelgidae. Zeitschrift für Angewandte Entomologie, 61: 157–214.

Eichhorn, O., 1975. Über die Gallen der Arten der Gattung *Dreyfusia* (Adelgidae), ihre Erzeuger und Bewohner. Zeitschrift für Angewandte Entomologie, 79: 56–76.

Forrest, J.M.S., 1971. The growth of *Aphis fabae* as an indicator of the nutritional advantage of galling to the apple aphid *Dysaphis devecta*. Entomologia Experimentalis et Applicata, 14: 477–483.

Forrest, J.M.S. and Dixon, A.F.G., 1975. The induction of leaf-roll galls by the apple aphids *Dysaphis devecta* and *D. plantaginea*. Annals of Applied Biology, 81: 281–288.

Forrest, J.M.S. and Noordink, J.P.W., 1971. Translocation and subsequent uptake by aphids of ^{32}P introduced into plants by radioactive aphids. Entomologia Experimentalis et Applicata, 14: 133–134.

Gibson, R.W., 1974. The induction of top-roll symptoms on potato plants by the aphid *Macrosiphum euphorbiae*. Annals of Applied Biology, 76: 19–26.

Gutierrez, A.P., Morgan, D.J. and Haverstein, D.E., 1971. The ecology of *Aphis craccivora* Koch and subterranean clover stunt virus. I. The phenology of aphid populations and the epidemiology of virus in pastures in southeast Australia. Journal of Applied Ecology, 8: 699–721.

Harrewijn, P., 1976. Balance studies on the role of phosphate in host plant–aphid relationships. Proceedings of the International Congress on Soilless Culture IV, Las Palmas, pp. 339–346.

Hill, G.P., 1962. Exudation from aphid stylets during the period of dormancy to bud break in *Tilia americana* (L.). Journal of Experimental Botany, 13: 144–151.

Hoad, G.V. and Bowen, M.R., 1968. Evidence for gibberellin-like substances in the phloem exudate of higher plants. Planta, 82: 22–32.

Hoad, G.V., Hillman, S.K. and Wareing, P.F., 1971. Studies on the movement of indole auxins in willow (*Salix viminalis* L.). Planta, 99: 73–88.

Hori, K., 1974. Plant growth-promoting factor in the salivary gland of the bug *Lygus disponsi*. Journal of Insect Physiology, 20: 1623–1627.

Hori, K., 1975. Plant growth-regulating factor, substances reacting with Salkovski reagent and phenoloxidase activities in vein tissue injured by *Lygus disponsi* Linnavouri (Hemiptera: Miridae) and surrounding mesophyll tissues of sugar beet leaf. Applied Entomology and Zoology, 10: 130–135.

Hori, K., 1976. Plant growth-regulating factor in the salivary gland of several heteropterous insects. Comparative Biochemistry and Physiology, 53B: 435–438.

Hori, K., 1979a. Metabolism of ingested auxins in the bug *Lygus disponsi*: conversion of several indole compounds. Applied Entomology and Zoology, 14: 56–63.

Hori, K., 1979b. Metabolism of ingested indole-3-acetic acid in the gut of various heteropterous insects. Applied Entomology and Zoology, 14: 149–158.

Hori, K., 1980. Metabolism of ingested auxins in the bug *Lygus disponsi*: indole compounds appearing in excreta of bugs fed with host plants and the effect of indole-3-acetic acid on the feeding. Applied Entomology and Zoology, 15: 123–128.

Hori, K., in press. Insect secretions and their effect on plant growth, with special reference to hemipterous insects. In: J.D. Shorthouse and O. Rohfritsch (Editors), Biology of Insect and Acarina Induced Galls.

Hori, K. and Endo, M., 1977. Metabolism of ingested auxins in the bug *Lygus disponsi*: conversion of indole-3-acetic acid and gibberellin. Journal of Insect Physiology, 23: 1075–1080.

Hori, K. and Miles, P.W., 1977. Multiple plant growth-promoting factors in the salivary glands of plant bugs. Marcellia, 39: 399–400.

Hori, K., Singh, D.R. and Sugitani, A., 1979. Metabolism of ingested indole compounds in the gut of three species of Heteroptera. Comparative Biochemistry and Physiology, 64C: 217–222.

Horne, A.S. and Lefroy, H.M., 1915. Effects produced by sucking insects and red spider upon potato foliage. Annals of Applied Biology, 1: 370–386.

Horsfall, J.L., 1923. The effects of feeding punctures of aphids on certain plant tissues. Bulletin of the Pennsylvania Agricultural Experimental Station, 182: 23 pp.

Ibbotson, A. and Kennedy, J.S., 1959. Interaction between walking and probing in *Aphis fabae* Scop. Journal of Experimental Biology, 36: 377–390.

Ignoffo, C.M. and Granovsky, A.A., 1961. Life history and gall development of *Mordwilkoja vagabunda* (Homoptera: Aphididae) on *Populus deltoides*. Part II. Gall development. Annals of the Entomological Society of America, 54: 635–641.

Jones, M.E., Jr., Green, E.A. and Williams, R.L., 1973. Morphological and histological changes in the leaf of *Juglans nigra* following infection by *Phylloxera* sp. Bulletin of the Georgia Academy of Science, 31: 67.

Jones, M.E., Jr., Green, E.A. and Chester, W.J., Jr., 1976. A comparative morphological and histological study of changes in the leaf of *Juglans nigra* and *Carya illinoises* following infection by *Phylloxera* sp. Bulletin of the Georgia Academy of Science, 34: 50–51.

Jördens, D. and Klingauf, F., 1977. Der Einfluss von L-Dopa auf Ansiedlung und Entwicklung von *Aphis fabae* Scop. an synthetischer Diät. Mededelingen van de Faculteit Landbouwwetenschappen, Rijksuniversiteit te Gent, 42: 1411–1419.

Jördens-Röttger, D., 1979a. Das Verhalten der Schwarzen Bohnenblattlaus, *Aphis fabae* Scop. gegenüber chemischen Reizen von Pflanzenoberflächen. Zeitschrift für Angewandte Entomologie, 88: 158–166.

Jördens-Röttger, D., 1979b. The role of phenolic substances for host selection behaviour of the black bean aphid, *Aphis fabae*. Entomologia Experimentalis et Applicata, 26: 49–54.

Kantack, E.J. and Dahms, R.G., 1957. A comparison of injury caused by the apple grain aphid and greenbug to small grains. Journal of Economic Entomology, 50: 156–158.

Kefeli, V.I. and Turetskaya, R.K., 1965. Participation of phenolic compounds in the inhibition of auxin activity and in the growth of willow shoots. Plant Physiology, Washington, 12: 554–566.

Kennedy, J.S., 1951. Benefits to aphids from feeding on galled and virus-infected leaves. Nature, 168: 825–826.

Kennedy, J.S. and Stroyan, H.L.G., 1959. Biology of aphids. Annual Review of Entomology, 4: 139–160.

Klingauf, F., 1971. Die Wirkung des Glucosids Phlorizin auf das Wirtswahlverhalten von *Rhopalosiphum insertum* (Walk.) und *Aphis pomi* De Geer (Homoptera: Aphididae). Zeitschrift für Angewandte Entomologie, 68: 41–55.

Kloft, W., 1950. Vergleichende Untersuchungen an eigenen Cocciden und Aphiden. Verhandlungen der Deutschen Zoologischen Gesellschaft, 50: 290–296.

Kloft, W., 1954. Über Einwirkungen des Saugaktes von *Myzus padellus* HRL. and Rogers (Aphidinae, Mycini CB.) auf den Wasserhaushalt von *Prunus padus* L. Phytopathologische Zeitschrift, 22: 454–458.

Kloft, W., 1955. Untersuchungen an der Rinde von Weisstannen (*Abies pectinata*) bei Befall durch *Dreyfusia (Adelges) piceae* Ratz. Zeitschrift für Angewandte Entomologie, 37: 340–348.

Kloft, W., 1960a. Wechselwirkungen zwischen pflanzensaugenden Insekten und den von ihnen besogenen Pflanzengeweben. Teil I. Zeitschrift für Angewandte Entomologie, 45: 337–381.

Kloft, W., 1960b. Wechselwirkungen zwischen pflanzensaugenden Insekten und den von ihnen besogenen Pflanzengeweben. Teil II. Zeitschrift für Angewandte Entomologie, 46: 42–70.

Kloft, W., 1960c. Nachweis freier Aminosäuren als phytopathologisch wirksame Stoffe im Speichel pflanzensaugender Insekten. Symposium of the International Congress of Entomology XI, Vienna, 3: 141–143.

Kloft, W. and Ehrhardt, P., 1959. Untersuchungen über Saugtätigkeit und Schadwirkung der Sitkafichtenlaus *Liosomaphis abietina* (Walk.) (*Neomyzaphis abietina* Walk.). Phytopathologische Zeitschrift, 35: 401–410.

Lamb, K.P., Ehrhardt, P. and Moericke, V., 1967. Labelling of aphid saliva with rubidium-86. Nature, 214: 602–603.

Lawson, F.R., Lucas, G.B. and Hall, N.S., 1954. Translocation of radioactive phosphorus injected by the green peach aphid into tobacco plants. Journal of Economic Entomology, 47: 749–752.

Leopold, A.C. and Plummer, J.H., 1961. Auxin–phenol complexes. Plant Physiology, Lancaster, Pennsylvania, 36: 589–592.

Lewis, I.F. and Walton, L., 1958. Gall-formation on *Hamamelis virginiana* resulting from material injected by the aphid *Hormaphis hamamelidis*. Transactions of the American Microscopical Society, 77: 146–200.

Link, G.K.K., Eggers, V. and Moulton, J.E., 1940. *Avena* coleoptile assay of ether extracts of aphids and their hosts. Botanical Gazette, 101: 928–939.

Llewellyn, M., 1972. The effects of the lime aphid, *Eucalipterus tiliae* L. (Aphididae) on the growth of the lime *Tilia X vulgaris* Hayne. I. Energy requirements of the aphid population. Journal of Applied Ecology, 9: 261–282.

McLean, D.L. and Kinsey, M.G., 1967. Probing behavior of the pea aphid, *Acyrthosiphon pisum*. I. Definitive correlation of electronically recorded waveforms with aphid probing activities. Annals of the Entomological Society of America, 61: 400–406.

McLean, D.L. and Kinsey, M.G., 1968. Probing behavior of the pea aphid, *Acyrthosiphon pisum*. II. Comparison of salivation and ingestion in host and non-host plant leaves. Annals of the Entomological Society of America, 61: 730–739.

Madden, J.L. and Stone, C., 1984. Induction and formation of pouch and emergence galls in *Eucalyptus pulchella* leaves. Australian Journal of Botany, 32: 33–42.

Mandl, L., 1957. Wachstum von Pflanzengallen durch synthetische Wuchsstoffe. Österreichische Botanische Zeitschrift, 104: 185–208.

Marek, J., 1961. Die Wirkung von Aphidenstichen auf pflanzliche Zellen. Entomologia Experimentalis et Applicata, 4: 20–34.

Maresquelle, H.J. and Meyer, J., 1965. Physiologie et morphogenèse des galles d'origine animale (zoocéidies). In: E. Ashby, J. Bonner, M. Geiger-Huber, W.O. James, A. Lang, D. Müller and M.G. Stålfelt (Editors), Handbuch der Pflanzenphysiologie – Encyclopaedia of Plant Physiology, 15/2. Springer, Berlin, Heidelberg, New York, pp. 279–329.

Mattson, M.J. and Addy, N.D., 1975. Phytophagous insects as regulators of forest primary production. Science, 190:515–522.

Maxwell, F.G. and Painter, R.H., 1962. Plant growth hormones in ether extracts of the greenbug,

Toxoptera graminum, and the pea aphid *Macrosiphum pisi*, fed on selected tolerant and susceptible host plants. Journal of Economic Entomology, 55: 57–62.

Meyer, J., 1951. Origine maternelle des principaux tissus et du plastème nourricier des galles laraires d'*Adelges abietis* sur *Picea excelsa* Lk. Comptes Rendus Hebdomedaires des Séances de l'Académie des Sciences, Paris, 233: 886–888.

Meyer, J., 1962. Croissance labiale et limites cécidogènes de la galle d'*Adelges abietis* Kalt. sur *Picea excelsa* Lam.: Notion de seuil d'action cécidogène. Marcellia, 30: 225–235.

Miles, P.W., 1964. Studies on the salivary physiology of plant bugs: oxidase activity in the salivary apparatus and saliva. Journal of Insect Physiology, 10: 121–129.

Miles, P.W., 1968a. Studies on the salivary physiology of plant bugs: experimental induction of galls. Journal of Insect Physiology, 14: 97–106.

Miles, P.W., 1968b. Insect secretions in plants. Annual Review of Phytopathology, 6: 137–164.

Miles, P.W., 1972. The saliva of Hemiptera. Advances in Insect Physiology, 9: 183–255.

Miles, P.W. and Hori, K., 1977. Fate of ingested β-indolyl acetic acid in *Creontiades dilutus*. Journal of Insect Physiology, 23: 221–226.

Miles, P.W., Aspinall, D. and Rosenberg, L., 1982. Performance of the cabbage aphid, *Brevicoryne brassicae* (L.), on water-stressed rape plants in relation to changes in their chemical composition. Australian Journal of Zoology, 30: 337–345.

Mittler, T.E. and Sylvester, E.S., 1961. A comparison of the injury to alfalfa by the aphids *Therioaphis maculata* and *Macrosiphum pisi*. Journal of Economic Entomology, 54: 615–622.

Moericke, V., 1969. Host plant specific colour behaviour by *Hyalopterus pruni* (Aphididae). Entomologia Experimentalis et Applicata, 12: 524–534.

Monaco, P., Caputo, R., Palumbo, G. and Mangoni, L., 1974. Triterpene components of galls on the leaves of *Pistacia terebinthus*, produced by *Pemphigus semilunarius*. Phytochemistry, 13: 1992–1993.

Müller, F.P., 1957. Die Hauptwirte von *Myzus persicae* (Sulz.) und von *Aphis fabae* Scop. Nachrichtenblatt für Deutschen Pflanzenschutzdienst, Berlin, 11: 21–27.

Mullick, D.B. and Jensen, G.D., 1976. Rates of non-suberized impervious tissue development after wounding at different times of the year in three conifer species. Canadian Journal of Botany, 54: 881–892.

Newcomb, E.H., 1951. Comparative studies of metabolism in insect galls and normal tissues. In: F. Skoog (Editor), Plant Growth Substances. University of Wisconsin Press, Madison, WI, pp. 417–427.

Nielson, M.W. and Don, H., 1974. Probing behavior of biotypes of the spotted alfalfa aphid on resistant and susceptible alfalfa clones. Entomologia Experimentalis et Applicata, 17: 477–486.

Norris, D.M., 1979. How insects induce disease. In: Plant Disease, 4. Academic Press, New York, San Fransisco, London, pp. 239–255.

Nuorteva, P., 1962. Studies on the causes of phytopathogenicity of *Calligypona pellucida* (F.) (Hom. Araeopidae). Annals of the Zoological Society Vanamo, 23: 1–58.

Nystérakis, F., 1948. Phytohormones et inhibition de la croissance des organes végétaux attaqués par des aphides. Comptes Rendus Hebdomedaires des Séances de l'Académie des Sciences, Paris, 226: 746–747.

Oechssler, G., 1962. Studien über die Saugschäden mitteleuropäischer Tannenläuse im Gewebe einheimischer und ausländischer Tannen. Zeitschrift für Angewandte Entomologie, 50: 408–454.

Ortman, E.E. and Painter, R.H., 1960. Quantitative measurements of damage by the greenbug, *Toxoptera graminum*, to four wheat varieties. Journal of Economic Entomology, 53: 798–802.

Owen, D.F. and Weigert, R.G., 1976. Do consumers maximize plant fitness? Oikos, 27: 488–492.

Parry, W.H., 1971. Differences in the probing behaviour of *Elatobium abietum* feeding on Sitka and Norway spruces. Annals of Applied Biology, 69: 177–185.

Peel, A.J. and Ho, L.C., 1970. Colony size of *Tuberolachnus salignus* (Gmelin) in relation to mass transport of ^{14}C labelled assimilates from the leaves of willow. Physiologia Plantarum, 23: 1033–1038.

Pilet, P.E., 1960. Auxin content and auxin catabolism of the stems of *Euphorbia cyparissias* L. infested by *Uromyces pisi* (Pers.). Phytopathologische Zeitschrift, 40: 75–90.

Plumb, G.H., 1953. The formation and development of the Norway spruce gall caused by *Adelges abietis*. Bulletin of the Connecticut Agricultural Experimental Station, 566, 77 pp.

Pollard, D.G., 1973. Plant penetration by feeding aphids (Hemiptera, Aphidoidea): A review. Bulletin of Entomological Research, 62: 631–714.

Prat, H., 1955. Relations existant entre les notions de gradient chimique et de rayon d'activité cécidogénétique. Marcellia, 30 (Supplement): 51–53.

Puritch, G.S., 1977. Distribution and phenolic composition of sapwood and heartwood in *Abies grandis* and effects of the balsam woolly aphid. Canadian Journal of Forestry Research, 7: 54–62.

Puritch, G.S. and Nijholt, W.W., 1974. Occurrence of juvabione-related compounds in grand fir and pacific silver fir infested by balsam woolly aphid. Canadian Journal of Botany, 52: 585–587.

Rilling, G. and Steffan, H., 1978. Versuche über die CO_2-Fixierung und den Assimilatimport durch Blattgallen der Reblaus (*Dactylosphaera vitifolii* Shimer) an *Vitis rupestris* 187G. Angewandte Botanik, 52: 343–354.

Rilling, G., Rapp, A. and Reuther, K.-H., 1975. Veränderungen des Aminosäurengehaltes von Rebensorten bei Befall durch die Reblaus (*Dactylosphaera vitifolii* Shimer). Vitis, 14: 198–219.

Rohfritsch, O., 1966. Action spécifique des gallicoles de Chemesidae sur la maturation et l'ouvertures des galles. Mise en évidence par la réalisation de galles mixtes. Comptes Rendus Hebdomadaires des Séances de l'Académie des Sciences, Paris, 262: 370–372.

Rohfritsch, O., 1967. Conditions histo-cytologiques nécessaires à la base des aguilles d'Épicéa pour permettre le développement des larves de Chermesidae. Comptes Rendus Hebdomadaires des Séances de l'Académie des Sciences, Paris, 265: 1905–1908.

Rohfritsch, O., 1981. A "defense" mechanism of *Picea excelsa* L. against the gall former *Chermes abietis* L. (Homoptera, Adelgidae). Zeitschrift für Angewandte Entomologie, 92: 18–26.

Rohfritsch, O. and Meyer, J., 1966. Action déterminante des gallicoles sur la déhiscence des galles d'*Adelges abietis* Kalt. Comptes Rendus Hebdomadaires des Séances de l'Academie des Sciences, Paris, 262: 248–250.

Rose, M., 1939. Recherches expérimentales sur la cécidogenèse et les neoplasies chez les végétaux. Bulletin Biologique de la France et de la Belgique, 73: 336–366.

Rose, M., 1941a. Production expérimentale de tumeurs par piqures multiples sur la feuille de Chou cultivé. Compte rendu des Séances de la Société de Biologie, 135: 1485–1487.

Rose, M., 1941b. Production expérimentale de tumeurs sur la feuille de Chou par la technique du "Puceron artificiel". Compte rendu des Séances de la Société de Biologie, 135: 1491–1492.

Schaefer, H., 1972. Über Unterschiede im Stoffwechsel von reblausvergallten und gesunden Rebenblättern. Phytopathologische Zeitschrift, 75: 285–314.

Schäller, G., 1960. Untersuchungen über den Aminosäuregehalt des Speicheldrüsensekretes der Reblaus (*Viteus [Phylloxera] vitifolii* Shimer), Homoptera. Entomologia Experimentalis et Applicata, 3: 128–136.

Schäller, G., 1961. Aminosäuren im Speichel und Honigtau der grünen Apfelblattlaus, *Aphis pomi* Deg. – Homoptera. Entomologia Experimentalis et Applicata, 4: 73–85.

Schäller, G., 1963a. Biochemische Rassentrennung bei der Reblaus (*Viteus vitifolii* Shimer) durch Speichelanalysen. Zoologische Jahrbücher, Abteilung Allgemeine Zoologie und Physiologie der Tiere, 70: 278–283.

Schäller, G., 1963b. Papierchromatographische Analyse der Aminosäuren und Amide des Speichels und Honigtaues von 10 Aphidenarten mit unterschiedlicher Phytopathogenität. Zoologische Jahrbücher, Abteilung Allgemeine Zoologie und Physiologie der Tiere, 70: 399–406.

Schäller, G., 1965. Untersuchungen über den β-Indolylessigsäuregehalt des Speichels von Aphidenarten mit unterschiedlicher Phytopathogenität. Zoologische Jahrbücher, Abteilung Allgemeine Zoologie und Physiologie der Tiere, 71: 385–392.

Schäller, G., 1968a. Biochemische Analyse des Aphidenspeichels und seine Bedeutung für die Gallenbildung. Zoologische Jahrbücher, Abteilung Allgemeine Zoologie und Physiologie der Tiere, 74 54–87.

Schäller, G., 1968b. Untersuchungen zur Erzeugung künstlicher Pflanzengallen. Marcellia, 35: 131–153.

Schantz, E.M., 1966. Chemistry of naturally occurring growth-regulating substances. Annual Review of Plant Physiology, 17: 409–438.

Severin, H.H.P. and Tompkins, C.M., 1950. Symptoms induced by some species of aphids feeding on ferns. Hilgardia, 20: 81–92.

Smith, F.F. and Brierley, P., 1948. Simulation of lily rosette symptoms by feeding injury of the foxglove aphid. Phytopathology, 38: 849–851.

Sobetskii, L.A., 1967. Some features of the physiology of the feeding of aphids. Dissertation for the degree of candidate of biological sciences, Academy of Sciences of the Moldavian SSR, 22 pp (in Russian).

Sobetskii, L.A. and Derzhavina, M.A., 1965. Determination of protease in the salivary glands and intestines of some aphids. Izvestiya Akademii Nauk Moldavskoi SSR, 5: 89–97 (in Russian).

Sobetskii, L.A. and Derzhavina, M.A., 1973. Contribution to the study of the physiology of the feeding of the vine phylloxera, *Viteus vitifolii* Fitch (Homoptera, Phylloxeridae). Entomological Review, Washington, 52: 357–361.

Spiller, N.J., Kimmins, F.M. and Llewellyn, M., 1985. Fine structure of aphid stylet pathways and its use in host plant resistance studies. Entomologia Experimentalis et Applicata, 38: 293–295.

Springett, B.P., 1978. On the ecological role of insects in Australian *Eucalyptus* forests. Australian Journal of Ecology, 3: 129–139.

Steffan, H. and Rilling, G., 1981. Der Einfluss von Blatt- und Würzelgallen der Reblaus (*Dactylosphaera vitifolii* Shimer) auf das Verteilungsmuster der Assimilate in Reben (*Vitis rupestris* 187G). Vitis, 20: 146–155.

Sterling, C., 1952. Ontogeny of the *Phylloxera* gall of grape leaf. American Journal of Botany, 39: 6–15.

Thalenhorst, W., 1972. Zur Frage der Resistenz der Fichte gegen die Gallenlaus *Sacchiphantes abietis* (L.). Zeitschrift für Angewandte Entomologie, 71: 225–249.

Tomaszewski, M., 1959. Chlorogenic acid–phenolase as a system inactivating auxin isolated from leaves of some *Prunus* L. species. Bulletin de l'Académie Polonaise des Sciences, Classe II, Série des Sciences Biologiques, 7: 127–130.

Tomaszewski, M. and Thimann, K.V., 1966. Interaction of phenolic acids, metallic ions and chelating agents on auxin-induced growth. Plant Physiology, 41: 1443–1454.

Turian, G., 1958. Recherches anatomo-physiologiques sur les cécidies. Revue Générale de Botanique, 65: 279–293.

Van Emden, H.F., 1973. Aphid host plant relationships: some recent studies. Bulletin of the Entomological Society of New Zealand, 2: 54–64.

Van Emden, H.F., 1978. Insects and secondary plant substances – An alternative viewpoint with species reference to aphids. In: J.B. Harbourne (Editor), Biochemical Aspects of Plant and Animal Coevolution. Academic Press, New York, San Francisco, London, pp. 309–323.

Van Emden, H.F., Eastop, V.F., Hughes, R.S. and Way, M.J., 1969. The ecology of *Myzus persicae*. Annual Review of Entomology, 14: 197–270.

Van Overbeek, J., 1966. Plant hormones and regulators. Science, 152: 721–731.

Walton, L., 1980. Gall formation and life history of *Pineus floccus* (Patch) (Homoptera: Adelgidae) in Virginia. Virginia Journal of Science, 31: 55–60.

Way, M.J., 1966. Intraspecific mechanisms with special reference to aphid populations. In T.R.E. Southwood (Editor), Royal Entomological Society of London, Symposium 4: Insect Abundance. Blackwell, Oxford, Edinburgh, pp. 18–36.

Way, M.J. and Cammell, M., 1970. Aggregation behaviour in relation to food utilization by aphids. In: A. Watson (Editor), British Ecological Society, Symposium 10: Animal Populations in Relation to their Food Resources. Blackwell, Oxford, London, pp. 229–247.

Weidner, H., 1957. Neuere Anschauungen über die Entstehung der Gallen durch die Einwirkung von Insekten. Zeitschrift für Pflanzenkrankheiten, Pflanzenpathologie und Pflanzenschutz, 64: 287–309.

West, F.T., 1946. Ecological effects of an aphid population upon weight gains of tomato plant. Journal of Economic Entomology, 39: 338–343.

Zotov, V.V., 1976. On the nature of phylloxera-resistance in grape plants. Sel'sko-khozyaistvennaya Biologiya, 11: 277–285 (in Russian, with English summary).

Zotov, V.V. and Gadiev, R.Sh., 1975. Stimulators of growth of gall tissues in phylloxera-infected grape plants. Sel'sko-khozyaistvennaya Biologiya, 10: 241–244 (in Russian, with English summary).

10.1.1 Specific Responses and Damage Caused by Aphidoidea

P.W. MILES

INTRODUCTION

In this section, the effects on plants of those aphids most commonly recognized as pests are categorized. Coverage is not intended to be exhaustive, but it is hoped that the effects of species not mentioned by name will, in most instances, be recognizable as belonging to one or another category.

DAMAGE CAUSED BY NUTRIENT DRAIN

Some of the most economically serious aphidid pests have little effect as individuals on their food plants and are important agriculturally only because they occur in such numbers that they impose a serious nutrient and fluid drain on the plants they attack, or because they are vectors of virus disease.

The black bean aphid, *Aphis fabae* Scopoli, does little noticeable damage in small numbers; nevertheless it is one of the most serious pests of sugar beet in England and Wales (Dunning, 1974). Much of its effect is due to a preference for the growing tip of the plant, where its mass feeding depresses overall growth. In southern Sweden it has been recorded as reducing both root growth and sugar yield of beet by about 16% (Möllerström, 1963).

The green peach aphid, *Myzus persicae* (Sulzer), is an agricultural and horticultural pest mainly because it is a major vector of diseases, associated with which it has a remarkably wide host range (Van Emden et al., 1969). It seldom occurs in very large numbers. On cruciferous crops and potato it tends to feed on senescent leaves, where its visible effects are negligible; on other crops, depending to some extent on plant phenology, aphid biotype and the season, it can feed on young tissues – but here too, the immediate impact of feeding is slight (Van Emden et al., 1969).

According to Müller (1957), one subspecies of *M. persicae* produces local distortions of growth (pseudogalls) in spring on young peach leaves, another (named by him *Myzus (Nectarosiphon) persicae* subsp. *dyslycialis* F.P. Müller) produces them instead on *Lycium halimifolii*, but neither causes growth disturbances on any but the one host plant, whatever the season. Where other specific effects of the feeding of *M. persicae* have been noted, they are on highly sensitive tissues such as the buds of *Cymbidium* orchids, where almost any aphid that feeds causes chlorosis and/or distortions (Jensen, 1954). More serious symptoms, such as widespread chlorosis, rosetting, and leaf curling, e.g. on *Antirrhinum* (Baker and Tompkins, 1942), which have sometimes been claimed as evidence of the toxicity of the insects, are probably due to undiagnosed virus diseases (Van Emden et al., 1969).

Section 10.1.1 references, p. 42

Aphis craccivora Koch is another species that seems to affect its host more by removal of nutrients than by causing any specific symptoms of disease. When 25 adults were caged on red beaked long bean, *Vigna sesquipedalis*, for 28 days, the insects and their progeny removed up to 61% of the plant's production of photosynthate. Of what remained, nearly all (97.7%) stayed in the leaves, whereas in uninfested plants 18% of leaf photosynthate was transported to the pods and 11% to other parts of the plant. The pods produced their own photosynthate, but small pods were in danger of losing even this to a heavy infestation of the aphids. Overall, they reduced the mean fresh weight of pods to less than half that in the uninfested controls (Chang and Thrower, 1981).

Cinaran aphids (Lachnidae) feed on conifers. They do little if any immediately recognizable damage, but *Cinara atlantica* (Wilson) and *C. watsoni* Tissot, attacking seedlings of the loblolly pine, *Pinus taeda*, in the United States, have been found to reduce growth. The insects selected the most vigorous of the 2-year-old seedlings and on 3-year-olds halved the incremental growth of diameter and decreased terminal growth (Fox and Griffith, 1977).

The lime aphid, *Eucallipterus tiliae* (Linnaeus), although not a commercial pest, provides a carefully researched example of the effects of nutrient drain per se (Dixon, 1971b; Llewellyn, 1972). While concurrent, above-ground growth of infested lime trees was not seriously affected, reduction of root growth was consistent with the calculated loss of assimilate. Compared with uninfested trees, they shed leaves earlier and with a heavier dry weight per unit area, and the next year's leaves were smaller.

The rose aphid, *Macrosiphum rosae* (Linnaeus), usually causes relatively little effect on rose buds; it feeds on them only when they are young and, in warm weather, walks off as the sepals open (Maelzer, 1976); nevertheless, they can cause the buds to become distorted or to abort if they feed in overwhelming numbers (Maelzer, 1977). There are no immediately distinguishable, superficial effects of feeding.

It is tempting to generalize that near monophagous species such as *E. tiliae* and *M. rosae* have become so well adapted to their hosts as to cause them minimal damage; and that *A. fabae* and *M. persicae* are so well adapted to polyphagy that they manage to have little effect on any plant. In view of the specific effects of *Myzus* spp. on peach and *Lycium* perhaps the relationship is not that simple; nevertheless, there can be no doubt that the species described in this section so far do little damage to their food plants in their own right, apart from removal of assimilate. Their pest status, if any, is due to their numbers or (as in *M. persicae*) their propensity to transmit viruses and is more a function of their quantitative ecology than physiology of feeding.

DAMAGE RELATED TO THE SENSITIVITY OF THE PLANT

According to Jensen (1954), leaves of *Cymbidium* orchids are avoided by aphids, but some species will feed on the ventral sepal of the developing bud. Here, even *M. persicae* and *M. ornatus* Laing, which show minimal effects on other plants, cause persistent chlorotic spots around their feeding punctures. Similarly, the lily aphid, *Neomyzus (Aulacorthum) circumflexus* (Buckton) – single individuals of which are said to cause little or no externally visible, local effects on potato (Smith, 1926) or on ornamental ferns (Severin and Tompkins, 1950) – is described as causing pale circular spots around its feeding puncture on *Cymbidium*, and twisting and rolling of the sepal, which later turns yellow and dies prematurely. The foxglove aphid, *Aulacorthum solani* (Kaltenbach), which causes noticeable symptoms on other food plants (see below), is not surprisingly reported to have a particularly severe effect on *Cymbidium* (Jensen, 1954).

CHLOROSIS CAUSED BY APHIDIDS

One of the responses of plants to the feeding of sucking insects is the degeneration and disappearance of chloroplasts in the vicinity of the feeding puncture, leading to formation of a yellowish spot. Sometimes chlorotic spots are restricted to a region only a few cells deep surrounding the stylet track, such as those produced by *M. persicae* and *M. ornatus* on the *Cymbidium* bud. When reported as a specific symptom of the feeding of aphids, a somewhat more widespread effect is usually meant, however.

A. solani, as mentioned above, causes a yellow area with a rose-coloured border to develop around its feeding puncture on *Cymbidium*, and the whole bud eventually becomes discoloured (Jensen, 1954). According to Severin and Tompkins (1950), the continued feeding of this aphid on the bird's-nest fern, *Asplenium nidis*, caused chlorotic areas that spread towards the midrib and fused, and new fronds were also chlorotic; on the holly fern, *Cyrtomium falcatum*, the bead-like chlorotic spots attributed to the feeding of the insect joined up in streaks and the whole plant became affected, so that when it was cut back, the new fronds that emerged were also chlorotic.

Despite the precautions taken by the experimenters to exclude the possibility of transmission of viruses, some of the symptoms on ferns they ascribed to *A. solani* are suspiciously like those of virus disease, but *A. solani* has also been reported to cause yellow spots, with or without leaf curling, on African violet, geranium, potato, tomato, cucumber, turnip and lily (Smith and Brierley, 1948). D. Hille Ris Lambers (personal communication, 1970), noting such effects up to 50 cells from one feeding puncture, believed saliva-induced histolysis enabled *A. solani* to imbibe from parenchyma.

Further examples of chlorosis are discussed inter alia below.

SYSTEMIC EFFECTS ASCRIBED TO APHIDID TOXINS

Some authors describe a depression of growth caused by the feeding of aphidids that appears disproportionate to their numbers. The sycamore aphid, *Drepanosiphum platanoidis* (Schrank), feeding on *Acer pseudoplatanus* causes a decrease in the size of leaves and in the width of the annual ring of xylem, so that the total growth of the plant is reduced more than can be accounted for simply by the nutrient drain, according to Dixon (1971a). The leaves are heavier per unit area at leaf fall and have a higher nitrogen content than leaves from uninfested trees; the next year's leaves are noticeably paler. The author ascribes the effect to a systemic action of the insect's saliva acting inter alia on the growth of the xylem.

The pea aphid, *Acyrthosiphon pisum* (Harris), can have a devastating effect on lucerne (alfalfa), *Medicago sativa*. As a result of a heavy attack, top growth is stunted, buds are killed in the crown and the plant is unable to recover within one year. The affected crop, when cut for hay, has a lower carotene content, but this is because of the lower proportion of leaves to stems and the lower carotene content of the latter (Harvey et al., 1972). *A. pisum* is inherently more damaging to susceptible cultivars of lucerne than even the spotted alfalfa aphid, *Therioaphis trifolii* f. *maculata* (Buckton), a species especially recognized for its toxic effects (see below); *A. pisum* drains off twice as much sap per insect; it also gains weight four times as fast because it retains more assimilate but, for the same reason, it excretes less honeydew and so causes less fouling with sooty mould. Clearly the prodigious feeding of this species is a major factor in its effects on food plants, but Mittler and Sylvester (1961) did not rule out the possibility of additional, generalized toxic effects.

Section 10.1.1 references, p. 42

T. trifolii f. *maculata* causes chlorosis and a distinctive vein banding at the growing tips of *M. sativa* while feeding lower down the stem, as well as reducing growth of the whole plant. The insects begin feeding on the lower leaves, kill them and move up the plant. If growth of the plant is slowed by other causes, such as drought or cold, the aphid can kill the plant. Seedlings are especially vulnerable and therefore the regeneration of the crop as a whole is endangered (Dickson et al., 1955).

T. trifolii f. *maculata* is generally considered to be a phloem-feeder; however, it is also said to feed on the parenchyma, and to this has been ascribed the small chlorotic spots that form immediately around the feeding puncture (Diehl and Chatters, 1956). From the studies of Paschke and Sylvester (1957) and Nickel and Sylvester (1959), it is clear that the symptoms induced by this aphid are not due to a virus. The systemic effects that are apparent at the tip increase with prolonged feeding but are arrested when the insects are removed, and regrowth from plants that are cut back are symptomless; the characteristic vein banding of the young leaves is presumed to be due to salivary toxins infiltrating the mesophyll.

LOCALIZED EFFECTS ASCRIBED TO APHIDID TOXINS

The greenbug, *Schizaphis graminum* (Rondani), feeds on a variety of cereals, on which it causes localized chlorotic spots that begin to merge as the attack continues. The insect can overwhelm the plants and has been considered the worst of the aphid pests of cereals (Kantack and Dahms, 1957). In a heavy attack, it can cause a 55% reduction of root growth with a consequent decrease of above-ground growth (Ortman and Painter, 1960). Although the insect has been considered toxic, alternative explanations for its effects are possible. The species does not feed solely on the phloem; especially on resistant varieties of cereals, some biotypes are known to feed almost exclusively on mesophyll (Saxena and Chada, 1971). With the removal of the contents of epidermal and mesophyll cells, as described by Saxena and Chada (1971), localized chlorosis would seem to be inevitable, and the effects of mass attack could be either that of nutrient drain or the destruction of photosynthetic regions or a combination of the two.

The cabbage aphid, *Brevicoryne brassicae* (Linnaeus), has also been described as having a toxic effect on its food plants: it causes tissue damage and the appearance of pale spots surrounding its feeding punctures on cruciferous crops (Smith, 1951). In England and Wales, the insect is a pest particularly on Brussels sprouts. Strickland (1957), who reported that most growers seemed to regard it as merely a cosmetic pest, nevertheless estimated that over a 10-year period the insect had on average caused a 16% annual loss of production in untreated crops. In southern California, high populations have similarly been reported as causing serious damage to cabbage crops in winter and spring (Oatman and Platner, 1969). Ibbotson (1953) described attacked leaf tissue as showing thickened mesophyll and the presence of air bubbles; thus, although the damage has been ascribed to toxic saliva, it appears that, like *S. graminum*, *B. brassicae* may feed on more than phloem sap, and some of its effects on plants could be due to removal of mesophyll, and consequent impairment of photosynthesis.

Elatobium abietinum (Walker) feeds on the needles of both the sitka spruce, *Picea sitchensis* and the Norway spruce, *P. abies* causing chlorotic banding and death of the individual needles. Kloft and Ehrhardt (1959) investigated the symptoms on the sitka spruce and ascribed them to salivary toxins. The effect is less on the Norway spruce, however, and Parry (1971) explained the difference as due to the feeding of the insects through the stomata. On the Norway spruce these are on both sides, whereas on the sitka spruce all the stomata are on the one side. Parry showed that, in the sitka spruce, the insects produced a heavier concentration of stylet tracks, and the tracks were more branched. He believed that the insect's salivary "toxin" is thus injected in greater local concentration in the sitka than in the Norway spruce. Nevertheless, the effect of the insects on the respiration and translocation of the needles, as a result of blockage of stomata and phloem, needs to be taken into account in assessing causes for the reported damage.

GROWTH DISTORTIONS CAUSED BY APHIDIDS

Associated with chlorosis, hypersensitive reactions, and the appearance of unusual pigmentation of plant tissues, all of which could be responses to physical damage and removal of cell contents, there may also be distortions of growth, some of which clearly involve growth stimulation as well as inhibition.

Macrosiphum euphorbiae (Thomas) causes black necrotic spots when it feeds on the stems and leaf veins of potato, presumably by inducing a hypersensitive reaction in the plant: it also produces small intumescences in the leaf laminae and causes leaf rolling and lower production of tubers. Gibson (1974) pointed out that these effects tend to be reversed if the insect is removed and are thus not likely to be due to virus; he suggested that both the necrotic and growth effects might be due to interference with the plant's phenol–phenolase system. Overall, the insect is not very damaging. It is thought to have no toxic effect on tobacco (Roberts, 1940). According to West (1946), it required more than 40 individuals per gram of tomato plant to cause any reduction of weight gain.

The apple aphids *Dysaphis devecta* (Walker) and *D. plantaginea* (Passerini) cause leaf rolling on apple, and the "pseudogalls" thus formed are of species-specific orientation and coloration. *D. plantaginea* can cause the effect on a nearby leaf when feeding on the stem some centimetres away (see section 5.4); the plant's response thus seems to be to a localized systemic disturbance. According to Savary (1953), the growth of a branch of a pear tree heavily infested with this species may be reduced by as much as 75%.

M. persicae causes deformation on young peach or *Lycium halimifolii* leaves, depending on the subspecies of the insect, according to Müller (1957). *Myzus cerasi* Passerini similarly causes deformations on young shoots of *Prunus* spp. (Schäller, 1965), as does *Brachycaudus helichrysi* (Kaltenbach) on young plum leaves (Dawish, 1984). *Cryptomyzus ribis* (Linnaeus) causes yellow-red pouch galls (Kennedy, 1951, 1958) and also disappearance of the guard cells of stomata (Sienicka, 1959). In all such growth deformations, other than those caused by overwhelming mass attack, specific interference by the insect with the plant's growth regulating system seems to be indicated, inducing greater activity on the outwardly curved (usually adaxial) surface, as well as reduced growth on the inner surface (Maresquelle and Meyer, 1965). The means by which this may be achieved have been explored in section 10.1.

DAMAGE BY APHIDIDS SUMMARIZED

Phloem-feeding has been considered characteristic of the Aphididae (see section 5.3), and the following generalizations have been made with this premise in mind:
− Some plants are specially sensitive to disturbances (e.g. *Cymbidium* buds) and may well produce symptoms in responses to even relatively mild physical and chemical stimuli associated with phloem-feeding.

Section 10.1.1 references, p. 42

– Phloem-feeders do not entirely avoid damage to parenchymal cells (Spiller et al., 1985). Also some may occasionally or when at high densities feed on mesophyll, and perhaps this is part of the reason for the damage caused by *B. brassicae*. Some aphidids, such as *M. ornatus*, may feed mainly on mesophyll, at least on particular food plants (Lowe, 1967a, b); others, such as biotypes of *S. graminum*, may overcome resistance of their food plants (presumably when this centres on the nature or contents of the vascular bundles) by feeding apparently exclusively on mesophyll (Saxena and Chada, 1971). Whenever this occurs, loss of cell contents, including chloroplasts, and resultant local chlorosis is inevitable.

– The heavy nutrient drain of mass attack must be expected to have effects on distribution of assimilate and hence growth of roots and thereafter of the rest of the plant; it is difficult to be certain in such circumstances where nutrient drain ends and "toxic" responses begin.

– When mass attack affects a significant proportion of the photosynthetic surface, the effects of nutrient drain will be exacerbated.

– After all such considerations have been taken into account, there remain specific growth disturbances and production of specific pigments that are caused by some species of Aphididae and not by others feeding on the same plant. Strictly localized effects are possibly due to direct influences on parenchyma, as discussed above. Other instances appear to be systemic and, in some, widespread interference with the growth of conductive tissue may be inferred. All these effects may be presumed to result from specific interactions between the salivary secretions and feeding behaviour of the insect, on the one hand, and the physiology of the plant on the other, and to provide at least prima facie evidence for the involvement of a toxin or "phytoallactin" (in sensu Norris, 1979).

GALLS CAUSED BY THE WOOLLY APHID OF APPLE

The woolly aphid of apple, *Eriosoma lanigerum* (Hausmann), forms relatively unstructured intumescences on the woody, aerial parts and on the roots of apple trees. Susceptible trees may be attacked to such an extent that they are seriously debilitated. Staniland (1924) described the formation of the galls as an influence of the insect's saliva on the cambium, which is stimulated to abnormal meristematic activity, producing unusually large parenchymatous cells; at the same time, differentiation of xylem and sclerenchyma is depressed, and these woody elements become scattered in a disorganized fashion throughout the relatively undifferentiated parenchyma; continued production of the latter causes pressure to build up on the poorly lignified tissues of the gall, resulting in their eventual collapse into a pulpy mass.

The insects' stylets do not reach the phloem, and they feed on a *"plastème nourricier"* (see section 10.1), formed by the hypertrophied parenchyma. Young insects feed selectively on previously galled tissues. Staniland, by allowing the insects to probe into agar gells containing starch, was able to show that the insects' saliva contained an amylase; he considered it must also dissolve the midlammellae between cells, assisting in the breakdown of organization in the tissues. Wartenberg (1954a, b) ascribed the abnormal growth of the trees' tissues to interference by the saliva with the plant's growth regulating system.

Resistant cultivars of apple have been bred, notably at the East Malling Research Station in the United Kingdom (Roach and Massee, 1931; Greenslade et al., 1933). Bramstedt (1938) investigated the basis of the resistance and concluded that it was due to a hypersensitive reaction on the part of the plant, which resulted in a necrotic region surrounding the insects' stylets, thus halting further penetration into the plant of the biochemical effects of the insects' feeding (see sections 5.3 and 11.4).

According to Sengupta and Miles (1975), although *E. lanigerum* appeared to show a greater preference for cultivars with higher amounts of soluble nitrogenous compounds in comparable tissues, the insects did not necessarily feed at sites within the trees that had the highest nitrogen content. Instead, preferences for feeding sites both between and within trees could be interpreted as selection of tissues with low ratios of phenolic to nitrogenous water-soluble contents. The authors suggested that the phenolic content of the tissues of the apple tree were in some measure toxic to *E. lanigerum*; that the insect was adapted to deal with such compounds – they presumed by reaction with the salivary phenol-phenolase system – so that the apple phenolics did not prohibit feeding but limited it to a rate required for detoxification of the ingesta. Thus the higher the phenolic content, the greater the concentration of nitrogenous nutrients needed to compensate for the lower feeding rate.

PEMPHIGID GALLS ON *POPULUS*

Pemphigus spp. form petiole galls on poplars and cottonwoods (Dunn, 1960; Alleyne and Morrisson, 1974, 1977a, b; Chan and Forbes, 1975; Faith, 1979). The fundatrix hatches in spring from an overwintering egg on the bark of the tree and inserts her stylets into the petiole at the base of a leaf. There is no evidence that she feeds at this time; rather the insect performs a more or less complex pattern of insertion of the stylets, as described previously in sections 5.4 and 10.1, which builds up a gall of taxonomically definitive shape (Dunn, 1960; Chan and Forbes, 1975; Alleyne and Morrison, 1977b), although, according to Faith (1979), *Pemphigus populitransversus* Riley produces somewhat differently shaped galls on *Populus deltoides* depending on the stage of development of the leaf when the gall is initiated.

Once the gall has closed over the fundatrix, she begins to feed, develop, and eventually lay eggs, up to 100 or more per fundatrix, according to species (Chan and Forbes, 1975; Alleyne and Morrison, 1977a). Her offspring, the fundatrigeniae, also develop within the gall and in turn give rise to alate migrants that escape from the gall in summer, when the galls mature, dry somewhat, and tiny pores open up along the original line of closure. The galled leaves are not necessarily smaller or lighter than normal, but are among the first to fall (Alleyne and Morrisson, 1974; Chan and Forbes, 1975).

The migrants of most species fly to the soil and infest roots of various secondary hosts, but an exception is *Pemphigus spirothecae* Passerini, which is found only on primary hosts, usually the Lombardy poplar. This pemphigid produces remarkable spiral galls composed of three turns of the petiole (Fig. 10.1.1.1). The fundatrigeniae give rise to sexuparae within the gall; the latter emerge at the end of the final larval stadium in late summer, moult into alate adults and either remain on the same tree or fly to another poplar to complete the cycle.

DAMAGE BY PEMPHIGIDS TO SECONDARY HOSTS

The secondary hosts of the pemphigids that fly from *Populus* spp. are mostly Compositae, but a number of crop plants in other families are also attacked: *Pemphigus bursarius* (Linnaeus) occurs on lettuce (Dunn, 1959; Alleyne and

Section 10.1.1 references, p. 42

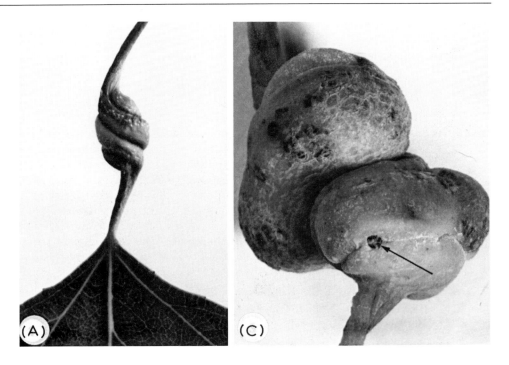

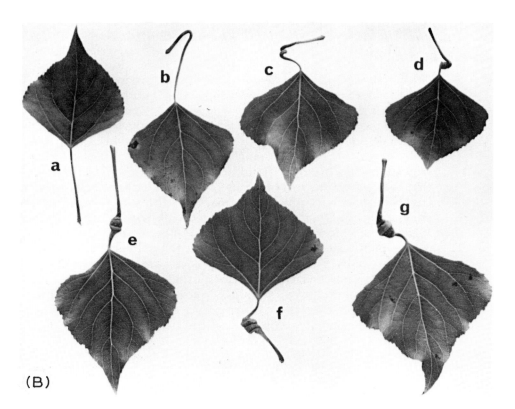

Fig. 10.1.1.1. Spiral gall on the petiole of Lombardy poplar, *Populus nigra italica*, formed by *Pemphigus spirothecae* Passerini. (A) Fully formed spiral gall. (B) Stages in the formation of the gall: (a) a non-galled leaf, (b) bending of the petiole, (c, d) spiralling of the petiole, (e, f, g) swelling of the gall. (C) Mature gall with an ostiole (arrow).

Fig. 10.1.1.2. "Pineapple" galls of *Adelges* spp. on *Picea orientalis*, due to (A) *A. merkeri* Eichhorn, (B) *A. prelli* Grossmass, (C) *A. nordmannianae* Eckstein (from Eichhorn, 1975).

Morrison, 1977a) and endives (Herfs, 1973); *P. populitransversus* occurs on turnip and cabbage (Chalfant, 1969). Although the insects seem to do little damage to their primary hosts, they can be serious pests in vegetable crops, on which they do not form galls but disfigure the roots and cause stunting, general chlorosis and wilting; attacked lettuce may die during a drought (Dunn, 1959).

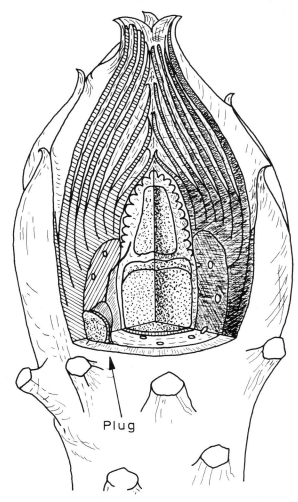

Plug

Fig. 10.1.1.3. Diagrammatic representation of the woody plug that forms around the initial penetration by an adelgid into a spruce bud (from Thalenhorst, 1972).

Section 10.1.1 references, p. 42

In early autumn, winged sexuparae appear on the secondary host and fly to the primary host, where they give birth to wingless, non-feeding sexuales. After mating, each female lays a single overwintering egg.

ADELGID GALLS ON SPRUCE

Most *Adelges* and *Pineus* spp. have complex, polymorphic life cycles and an alternation of hosts. The primary host, on which the sexual cycle is completed, is always a spruce (*Picea* sp.); on it the fundatrices attack the young buds, forming characteristic "pineapple galls" (Fig. 10.1.1.2; see also Fig. 5.4.3). The secondary hosts belong to other genera of conifers. Exceptionally, some species of adelgid are found to occur only on primary or only on secondary hosts – perhaps because they have spread into areas with the one kind of host beyond the distribution of the other (Mordvilko, 1935; Balch, 1952).

On a *Picea* sp., the fundatrix hatches from an egg laid near a bud in autumn and begins feeding at the base of the bud, at first forming a local "physiological gall" (*in sensu* Kloft, 1955): an accumulation of nutritive substances without any immediate swelling which Rohfritsch (1981) has termed the "nutritive tissue I". The insect apparently feeds here for a short time, and the plant usually shows signs of defensive reaction; cells at the periphery of the tissue begin to take on characteristics similar to those of cells surrounding resin canals, and eventually wall off the region with wound periderm, forming a woody "plug" (Fig. 10.1.1.3).

Sometimes the defense appears to be successful in that the insect fails in the next phase of its activity, either being unable to penetrate the bud further, or losing its aim (Fig. 10.1.1.4), but often the insect manages to negotiate the necrotic areas and continues an intercellular penetration up into the bud, finally bringing the stylet tips to rest in the pad of collenchyma that occurs just below the embryonic first two leaves. Here the insect injects more saliva and

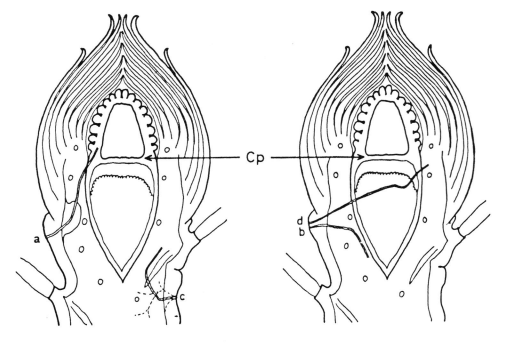

Fig. 10.1.1.4. Diagrammatic representation of successful (a) and unsuccessful (b–d) final stages of penetration of an adelgid to the base of the bud prior to cecidogenesis proper; Cp, collenchyma plate (from Thalenhorst, 1972).

induces development of a "nutritive tissue II" before going into diapause over winter (Thalenhorst, 1972; Eichhorn, 1975; Rohfritsch, 1981).

In spring, the insect stirs the bud into activity somewhat earlier than uninfested buds (Campbell and Thielges, 1973; Eichhorn, 1975) but in doing so inhibits the elongation and stimulates lateral growth of the bud. The needles also are prevented from elongating; they thicken, mostly proximally, but their bases remain constricted, forming "false petioles", between which small cavities are left, and these become the "larval chambers" (Fig. 10.1.1.5) (Meyer, 1951, 1962; Maresquelle and Schnell, 1963; Eichhorn, 1975; Rohfritsch, 1976, 1981; Walton, 1980; see also sections 5.4 and 10.1).

Under the influence of the fundatrix, the cells surrounding the larval chambers do not undergo normal morphological differentiation. Instead the nuclei enlarge, the cytoplasm accumulates starch and lipids, the vacuoles become fragmented and the plastids and mitochondria remain undifferentiated (Rohfritsch, 1976, 1981). Again, the plant may show evidence of a hypersensitive reaction, in which some of the nutritive cells are killed; if the necrosis is extensive the insect also perishes (see also section 5.4). The defensive response is a specific reaction between insect and plant, since some trees resistant to *Chermes abietis* (Linnaeus) remain susceptible to *C. strobilobius* Kaltenbach (Rohfritsch, 1981).

Once the development of the gall has begun, the fundatrix lays eggs, and the larvae, on hatching in late spring, migrate into the larval chambers that have formed. The influence of the fundatrix on the growth of the gall extends to initiation of hyperplasia in the needles and development of hairs (trichomes) at the lips of the larval chambers (Meyer, 1962), and both the fundatrix and her

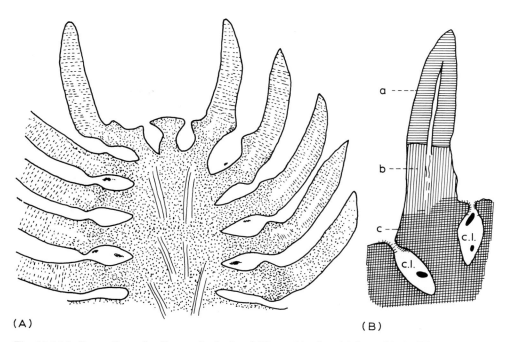

(A)   (B)

Fig. 10.1.1.5. Formation of galls on the buds of *Picea abies* by *Adelges abietis* (Linnaeus). (A) Longitudinal section of bud at the stage just after migration of the larvae into the larval chambers. (B) Diagrammatic representation of a longitudinal section through a needle of a galled bud showing (a) unaffected tissue, (b) outer region under cecidogenic influence (increased tannin and starch, reduced elongation and increased lateral expansion), (c) inner region under cecidogenic influence showing hyperplasia of the lips surrounding the openings; c.l., larval chambers (after Meyer, 1962.)

Section 10.1.1 references, p. 42

offspring (gallicolae) influence the accumulation of nutrients and their autolysis in the gall (Rohfritsch, 1976).

Further development of the gall is controlled by the feeding of the gallicolae: if they are killed or removed, growth ceases and the gall remains small and green (Rohfritsch and Meyer, 1966; Eichhorn, 1975). The tissues surrounding the larval chambers stay responsive to the initial influence of the migrating gallicolae for a period of some three weeks (Rohfritsch, 1967). The larvae cause the truncated needles to expand laterally into flattened scales, giving the gall its characteristic "pineapple" appearance and imparting to it specificities of coloration and final form (Meyer, 1962; Eichhorn, 1975). By late summer, the galls ripen: still under the influence of the gallicolae, the scales turn brown and part slightly, allowing the larvae to escape (Eichhorn, 1975; Rohfritsch, 1981).

After emergence from the gall, the life histories of different species diverge. The migrants of most species fly to other genera of conifers on which parthenogenetic generations feed either on the needles or in the cortex of the trunk, branches or twigs. *A. abietis* has only primary hosts, however: larvae emerge from the galls (which can be formed on a variety of *Picea* spp.), moult to alatae and either lay eggs in situ or after flying to another *Picea* tree. Most adelgids damage their primary hosts only marginally, but *C. abietis* is a pest in plantations of the European "christmas tree" (Norway spruce, *Picea abies*), where it can prevent development of a large proportion of the growing tips (Campbell and Thielges, 1973; Nielsen and Balderston, 1977).

ADELGID ATTACKS ON NEEDLES AND SHOOTS OF SECONDARY HOSTS

Adelgids do not form galls on the needles or shoots of conifers other than spruce, but they can cause severe defoliation of some secondary hosts, particularly in northern America.

Pineus pinifoliae (Fitch) usually has little effect on its primary host, the red spruce *Picea rubens*. Although there are records that up to 80% of shoots have been killed in some years in parts of Indiana, little permanent damage seems to have resulted. The migrants leave the galls in midsummer in northern America and fly to the white pine, *Pinus strobus*, feeding mostly on the previous year's needles; here they lay eggs that hatch into larval apterae which crawl to the axes of new shoots, insert their stylets and soon after go into diapause, but not before initiating considerable damage, apparently as a result of injection of toxic saliva. Because of this, and their feeding the following season, affected shoots of the white pine droop and show reduced growth or die. Young trees may succumb if many shoots are affected; on trees that survive the attack, overall growth is reduced (Balch and Underwood, 1950).

The Cooley spruce aphid, *Adelges cooleyi* Gillette, has several species of primary host in North America but, again, one main secondary host: the Douglas fir, *Pseudotsuga menzeii*. Besides being a shade and timber tree, the Douglas fir is grown in plantations as a North American "christmas tree". Several generations of *A. cooleyi* complete their development on the fir during the warmer months, e.g. between April and October in Ohio (Nielsen and Balderston, 1977). Different morphs show an alternation of preference for needles of the current and previous year and thus all parts of the tree can be affected (Spires, 1981). The insect causes distortion of shoots and chlorosis and premature abscission of the needles, making ornamental trees unsightly and those grown as christmas trees unsaleable.

ATTACKS BY THE BALSAM WOOLLY APHID ON ITS SECONDARY HOSTS

Most adelgids that feed on the bark of their secondary hosts do not usually cause much damage. An exception, however, is *Adelges piceae* (Ratzeburg), known in America as the balsam woolly aphid. This species is thought to have spread further than its original primary hosts (in Asia); in Europe it occurs only on secondary hosts, as a wholly parthenogenetic species, attacking the woody tissues of various *Abies* spp.; from there it was introduced into North America, probably at the turn of the century (Mordvilko, 1935; Balch, 1952).

European firs are relatively tolerant of *A. piceae*, but serious damage can be caused to the balsam fir, *Abies balsamea*, and to other *Abies* spp. indigenous to northern, maritime America. Balch (1952) and Balch et al. (1964) described how the insect attacks the parenchyma of various parts of the trunk and branches, causing thickening of cell walls, enlargement of nuclei, hypertrophy and hyperplasia, formation of secondary periderm and a local enlargement of the xylem. The tracheids in the affected region are histologically similar to so-called "compression wood", which is hard, brittle, and unsuited for lumber. Attacks on twigs cause development of foreshortened, swollen regions, symptoms designated as "gout disease" (Fig. 10.1.1.6). In heavy attacks, bud growth is inhibited, large parts of the outer bark die, translocation is impeded, and branches die back; if the infestation is persistent, young trees may succumb.

Fig. 10.1.1.6. Swellings and inhibition of buds in *Abies balsamea*, known as "gout disease", produced by *Adelges piceae* (Ratzeburg) (from Balch et al., 1964).

Section 10.1.1 references, p. 42

Puritch (1977) described the progress of similar damage caused by *A. piceae* to *Abies grandis*. Auxin-induced production of ethylene resulted in an accumulation of phenolics and the abnormal production of a relatively superficial layer of tissue similar to heartwood. The xylem involved was less permeable to water and this effect, if sufficiently extensive, could kill the tree.

Attack by *A. piceae* on *Abies fraseri* does not cause much immediately apparent damage (Amman, 1970) but is reported to make the seed smaller, less viable, and more vulnerable to the seed chalcid, *Megastigmus specularis* Walley so that the proportion of infested seed rose from 3% to 31% (Fedde, 1973).

DEFENSIVE REACTIONS IN WOODY TISSUES TO THE FEEDING OF ADELGIDS

Investigators of the relative degrees of resistance of various European conifers to attacks on their woody tissues by adelgids have drawn attention to the importance of the formation of wound periderm (Balch et al., 1964; Oechssler, 1962; Eichhorn, 1968). Oechssler (1962) showed that when *Adelges nordmannianae* (Eckstein) and *Adelges merkeri* (Eichhorn) attacked *Abies alba*, both species caused about as much damage per insect, although *A. merkeri* induced a diagnostically greater degree of hypertrophy of cells and swelling of twigs. Of greatest significance to the health of the tree was the speed of formation of the wound periderm; once it had coalesced around the feeding site, the parenchyma accessible to the insect died, and the insect could no longer feed. The dead scales thus formed on the tree remained for up to ten years and were impervious to further attack.

Kloft (1955) showed that the scale formed when *A. piceae* attacked *Abies alba* persisted similarly. Older trees have a considerable resistance based on a hypersensitive reaction that walls off growth disturbances before they reach the cambium and xylem (Balch et al., 1964). Young trees are more at risk, however, since there may not be enough healthy parenchyma left to support normal growth. Also conducting tissues are closer to the surface than in older trees and more likely to be damaged in the wound responses (Oechssler, 1962).

Mullick and Jensen (1976) noted that resistance to attack varied between individual trees but also with environmental conditions and the season; they concluded that the rate of formation of "non-suberized impervious tissue" (NIT), which forms in response to any wounding and is prerequisite to necrophytic periderm formation, determined the resistance of *Abies amabilis, Tsuga heterophylla* and *Thuja plicata* to *A. piceae*. The rate of NIT formation was reduced by water stress and low temperatures, which also reduced the resistance of the trees. However, the degree of resistance to the insects was not correlated in a simple fashion with the rate at which NIT formed in response to mechanical wounding alone, and the authors presumed that biochemical interactions between insect and plant were also involved.

GALLS OF THE GRAPE PHYLLOXERA

Viteus vitifoliae (Fitch) is a native of North America, where its natural hosts are indigenous *Vitis* spp., on which parthenogenetic apterae may be found in pouch galls on the leaves and on the surface of intumescences on the roots. Many descriptions have been made of the insect's complex life cycle and its effects on host plants; that given below is based primarily on the work of Nystérakis (1948), Coombe (1963), and Zotov and Gadiev (1975).

In the complete life cycle, an overwintering egg is laid on the bark of a vine stem and, in spring, a fundatrix hatches, crawls to a developing leaf and, if it is of an appropriate species or variety of vine (see below), initiates a gall. The fundatrix gives rise to further generations of apterous offspring. These may be of two kinds, gallicolae, which remain on the leaf, either feeding in the maternal gall or founding new leaf galls, and radicolae, which migrate down to the roots and form root galls. In the autumn, the gallicolae die when the leaves fall, but the radicolae remain on the roots and hibernate.

In those parts of the world with an appropriate (cold temperate) climate, some of the radicolae of holocyclic biotypes give rise in autumn to alate sexuparae, which emerge from the soil and fly to vine trunks, giving rise to apterous, non-feeding sexuales; these mate and the female lays a single overwintering egg. It should be noted, however, that in the absence of the sexual or any other of the above ground generations the insect is able to maintain an indefinite succession of parthenogenetic generations on the roots.

On the natural American hosts, the galls of the gallicolae can be initiated on either side of the leaf (Sterling, 1952) or on stems or even tendrils (Meyer, 1958), but only those on the upper (adaxial) surface of the leaf mature successfully (Sterling, 1952). The radicolae settle in preference on the tips of young roots, forming subapical "nodosities" (i.e. swellings just behind the root cap), but they also form "tuberosities" on the sides of older roots.

Whether found on the leaf, stem or root, galls are characterized by hyperplasia of younger cells and hypertrophy of somewhat older cells surrounding the feeding puncture, the tissues so formed constituting a *plastème nourricier*. Tissues a little away from the insect are stimulated to grow more vigorously than those immediately beneath, which are the first to show signs of eventual degeneration and necrosis. On the leaf, the galls sink through the leaf surface, due to elongation of the cells near the lip, forming a pouch in the leaf with lips slightly raised above the surface and enclosed with trichomes (Fig. 10.1.1.7).

The original hosts of *V. vitifoliae* do not appear to suffer greatly from the presence of the insect, but when it was accidentally introduced into Europe on American vines in the mid-nineteenth century, the effects of the insect on plantings of *Vitis vinifera* were devastating; for a time, the collapse of a considerable part of the European grape growing industries seemed likely. On *V. vinifera*, the insect appears unable to form leaf galls (although it can do so on some hybrids between *vinifera* and American species), and the climate of the major grape growing areas of Europe seems seldom to induce completion of the sexual cycle; but the radicolae have a particularly severe effect on

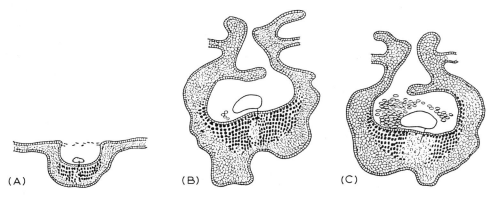

(A)   (B)   (C)

Fig. 10.1.1.7. Formation of the leaf gall of *Viteus vitifoliae* (Fitch). (A) Beginnings of the gall; (B) normal gall after 12–13 days, beginning of egg production; (C) gall at the end of 40 days, expansion of the zone of hydrolysis. Dark areas indicate accumulation of nutrient reserves; the light area indicates breakdown of cell structure and contents. (After Zotov and Gadiev, 1975).

Section 10.1.1 references, p. 42

vinifera-type roots, perhaps not so much because of the insects' sucking, but because the roots rot as the root galls break down, and secondary pathogens progressively destroy the infested root system. A previously healthy *vinifera*-type vine can be killed in as little as three years after infestation, although well established, vigorously growing vines can last longer and an occasional plant may show no visible effects.

The European wine industry was eventually saved by adopting the practice of grafting commercial cultivars onto phylloxera-tolerant rootstocks of American species or their hybrids with *V. vinifera*. Some tolerant hybrids that themselves produce wine grapes have also been bred, but so far have not played a major role in wine production. Rilling et al. (1974, 1975) provide useful lists of some tolerant and susceptible varieties in relation to the physiology of their interaction with the grape phylloxera (see also section 11.4).

Much has been done to determine the exact nature of the different responses of species and cultivars of *Vitis* to the insect. *V. cinerea* is genuinely resistant, in that phylloxera do not infest it at all. In other species, gall formation on the roots begins with a pattern similar to that already described in relation to pemphigids and adelgids; there is hypertrophy of some cells and hyperplasia of meristem; accumulation of starch and protein occurs in the young galls and on the periphery of the insect's influence; hydrolysis occurs later and closer to the insect; the gall becomes a major sink, drawing assimilate from the rest of the plant (Sterling, 1952; Schaefer, 1972; Sobetskii and Derzhavina, 1973; Zotov and Gadiev, 1975; Zotov, 1976; Rilling and Steffan, 1978; Steffan and Rilling, 1981).

Defensive reactions become evident in the tissues surrounding root galls: phenolic materials accumulate, oxidizing enzymes become active, and wound periderm begins to be laid down. In susceptible varieties, however, this reaction does not proceed fast enough to protect the meristematic and vascular central cylinder of the root, and, as the gall tissues eventually break down, microorganisms invade, rotting the root and killing it from that point on. In tolerant plants, on the other hand, the wound periderm forms sufficiently quickly to protect the root cylinder.

According to Denisova (1965), the naturally occurring phenolic compounds in vine roots include chlorogenic acid and free quercetin, which tend to stimulate formation of the galls; seasonal change brings about conversion of the one to caffeic and quinic acids and the other to glycosides, which tend to inhibit gall formation. In tolerant plants, production of the inhibitory compounds is accelerated in response to the presence of the insect and aids formation of wound periderm in its vicinity, whereas, in susceptible vines, phenolic compounds tend to remain dispersed uniformly within the tissues and do not provide the same degree of localized protection. Thus, in tolerant vines, the root galls and the insects that initiate them are eventually isolated; some insects die, but others and their offspring move on to other parts of the root system (Zotov, 1976).

A similar process occurs during formation of leaf galls, although, superimposed on the histological development of the nutritive tissues, a greater degree of morphological organization of surrounding tissues gives rise to a well-structured "organoid gall". It is interesting to note that the greater structural complexity of the leaf gall is a result of the greater degree of natural morphogenetic organization of the leaf and not to any greater organizing influence of the gallicolae as opposed to the radicolae. At one time it was thought that the leaf and root inhabiting forms of *V. vitifoliae* represented distinctly different stages in the alternation of generations, and that gallicolae could arise from the root-inhabiting generations only through intervention of the sexual generation; it is now known from the work of Rilling (1964) that this is untrue; just as gallicolae can give rise to offspring that descend to become radicolae,

so, under long day length and high temperatures especially, some of the offspring of the radicolae can ascend and become gallicolae. There would appear
to be no reason to regard the two forms as having any essentially different
cecidogenic propensities.

An apparent problem is the inability of the insect to produce galls on the
leaves of the "susceptible" *vinifera* vines, whereas it readily causes both leaf
and root galls on tolerant vines (often termed "resistant" in the literature).
According to Zotov (1976), however, the insect does initiate galls on the leaves
of *V. vinifera* cultivars, but the swellings quickly degenerate into a pulpy mass,
i.e. the tissues are too susceptible to the insect's influence and break down
before they can provide a stable *plastème nourricier*.

On leaves, galling on tolerant foliage causes a severe drop in photosynthesis
combined with an increased hydration of the galled tissues, so that their rate
of photosynthesis falls to 5% of normal by dry weight and only 2.5% of normal
by fresh weight. At the same time, there is a breakdown of local reserves and
an inflow of assimilate from other parts of the plant, so that a single gall
accumulates assimilate at the rate of an apical leaf. As discussed in section
10.1, this sink effect, like that of individual aphids, is cooperative in that galls,
when crowded, can accumulate up to three times as much per gall as when
isolated.

Not all the changes that occur in the vicinity of the insect are to its
advantage. Sobetskii and Derzhavina (1973) observed that tissue sap from leaf
galls differed from the exudate that welled up when one of the insects was
removed from its feeding puncture. They found that the gall as a whole was
characterized by its high content of phenolic compounds compared with normal leaf tissue, whereas the exudate was low in phenols and differed in other
ways from sap expressed from the whole gall, such as in the identity of the
oligosaccharides present. It has also been shown that phylloxera galls are
unpalatable to chewing insects (Lawrence, 1977). Sobetskii and Derzhavina
concluded that *V. vitifoliae* of necessity exerted control over the composition
of the tissue fluids on which it fed.

The possible role of IAA and amino acids in the initiation of galls has
already been mentioned (section 10.1). Zotov and Gadiev (1975) isolated three
unidentified water-soluble compounds from galled but not healthy tissues of
vine roots and leaves; two of them also occurred in phylloxera saliva, but they
were not amino acids, phenols or indoles; when bioassayed their activity was
closest to that of cytokinins. Within galls, there is an increase in enzymes,
including those involved in nucleic acid synthesis (Zotov, 1976), in protein and
starch synthesis and later hydrolysis (Schaefer, 1972), and in oxidative reactions (Henke, 1963; Schaefer, 1972; Rilling et al., 1975); esterases also increase
and there is de novo synthesis of isoperoxidases (Schaefer, 1972).

Oxidation reactions occur more vigorously in tolerant than in susceptible
tissues, possible because the latter possess more quinone reductase relative to
polyphenol oxidase activity, and the balance of oxidation–reduction reactions
could well be decisive both in the plant's defense (Henke, 1963) and in gall
production, as discussed in section 10.1. Amino acids in galled tissues may be
subject to oxidative deamination (Henke, 1963; Rilling et al., 1975), possibly
giving rise to ammonia toxicity (Rilling et al., 1975); on the other hand, there
may be a compensatory production of reducing substrates in tolerant tissues
(Mirzaev et al., 1972). Zotov (1976) believed that oxidative phosphorylation is
uncoupled in the susceptible varieties.

There are thus a number of pointers to the physiological disturbances that
give rise to phylloxera galls, particularly with respect to hydrolyzing and
oxidation–reduction systems, and there are a number of contributory explanations of how the eventual histological degeneration of tissues is brought

Section 10.1.1 references, p. 42

about, but how exactly all such processes are initiated or are interrelated remains, not perhaps quite so *"absolument mystérieuse"* as Maresquelle and Meyer found in 1965, but still in need of considerable clarification.

GALLING BY OTHER PHYLLOXERIDS

Phylloxerids cause galls on the leaves of various Caryaceae in America, including the black walnut, *Juglans nigra*, and the pecan, *Carya illinoensis* (Jones et al., 1973, 1976; Stoetzel, 1981, Stoetzel and Tedders, 1981). Most serious commercially is the pecan phylloxera, *Phylloxera devastatrix* Pergande, which attacks pecan in many parts of the U.S.A., causing deformation of shoots, petioles, leaves, and young nuts (Calcote and Hyder, 1980). *P. notabilis* Pergande and, increasingly, *P. russellae* Stoetzel also cause damage in Georgia (Stoetzel and Tedders, 1981).

Phylloxera spp. on walnut and pecan appear to live and cause damage entirely on the aerial parts of the plant. The pouch galls open on the ventral surface of the leaves. The insect is holocyclic and each year's alternation of generations begins with the hatching of a fundatrix from the overwintering egg. By midsummer, the galls open and dry off and necrosis extends some millimetres around them. If infestations are heavy, premature leaf-fall occurs, nuts are destroyed, and the effect on the vigour of the trees extends to a reduction of the following year's crop.

On both walnut and pecan, *Phylloxera* spp. cause pallisade cells to become oval shaped and lose chloroplasts; the resultant tissue becomes spongy, with distorted cytoplasm. Fragments of vascular tissues become scattered through the gall, the cells of which resemble those found in the sheath of the vascular bundles of normal leaves (Jones et al., 1973). On walnut, the cells become crowded and filled with starch, but little organic matter was found in the cells of galls on pecan (Jones et al., 1976). Even so, Payne and Schwartz (1971) found that pecan galls were more nutritive than unaffected leaves to the Japanese beetle – unlike the galls of *V. vitifoliae* on vine leaves (Lawrence, 1977).

GALLS DUE TO OTHER APHIDOIDEA

Mention should also be made of the gall-forming Thelaxidae. Although these insects do not constitute an acknowledged commercial problem and there has been little if any physiological investigation of them, they are worthy of note because of the histological investigations, mentioned in section 10.1, by Lewis and Walton (1958) on cecidogenesis by *Hormaphis hamamelidis* Fitch on witch hazel, *Hamamelis virginiana*, on the basis of which they first put forward their concept of a virus-like "cecidogen" transmitted by this and other cecidogenic insects (Lewis and Walton, 1958, 1964; Walton, 1980).

CECIDOGENESIS SUMMARIZED

Not all parenchyma-feeders produce galls, but of the galls formed by Aphidoidea, most that involve marked thickening of tissues seem to be produced by parenchyma-feeding members of families other than Aphididae. Possibly even the "pseudogalls" produced by some of the mainly phloem-feeding aphidids are also due to direct influences on parenchyma, as discussed earlier in this section.

Certain features seem to be common to the physiology of several if not all examples of cecidogenesis by Aphidoidea:

– The fact that relatively sedentary species are more likely to cause true galls is in itself indicative that the cecidogenic influence needs to be prolonged and concentrated in the tissues surrounding the insect.

– The influence does not appear necessarily related to feeding as such, since the initial activities, leading to gall formation, of the fundatrices of pemphigid species on poplars involve injection of saliva with little if any withdrawal of substances from the plant (Dunn, 1960). These observations seem to dispose of the possibility that the nutrient sink effect is alone responsible for cecidogenesis.

– Whatever the insect injects into the plant causes a gradient of reaction from inhibition of growth at the highest concentration (most noticeably in thin tissues such as leaves) to hyperplasia of young, especially of meristematic, tissues, and hypertrophy of older tissues (Maresquelle and Meyer, 1965).

– The cecidogenic influence hinders normal morphological differentiation of tissues and stimulates activities typical of a *plastème nourricier*. "Reserve substances" accumulate some distance from the insect, but are progressively broken down and mobilized under the influence of the insect (Maresquelle and Meyer, 1965). These effects, if not caused, are at least maintained by the flow of metabolites towards the nutrient sink that the insect represents.

– At the same time, the plant produces a defensive reaction, possibly as a direct response to substances in the insect's saliva, possibly more indirectly, in response to the breakdown of cells or of the constituents of the cell walls, as discussed in section 5.3. Phenolic materials, including tannins, begin to collect around and in the tissues influenced by the insect. These substances are involved in formation of wound periderm and, in excess, cause necrosis; some of them appear to be directly toxic to the insect, as described in section 10.1. Even if the insects are adapted to deal with the toxicity of defensive compounds specific to the host plant as indicated by Sobetskii and Derzhavina (1973), development of necrotic and lignified layers, if sufficiently rapid, protects underlying tissues, and may isolate the insect so that it eventually perishes.

– The location of attack by sucking insects within the plant as well as the degree of resistance (or tolerance) of cultivars may thus be related to the capacity of the insects' counter defenses (Sengupta and Miles, 1975; Miles, 1985). Perhaps, as suggested in section 5.3, the sulphydryl groups of the sheath material can immobilize defensive phenolics, a reaction that the salivary phenolase would no doubt assist. Yet some parts of even susceptible plants deter attack despite a seemingly favourable nutritive content; possibly because these are tissues where the demands of an effective counter to local defenses would outweigh the insect's potential rate of nutritional gain (Sengupta and Miles, 1975).

– It also follows that the responsiveness of tissues may be of as much significance as their initial composition in determining where or whether a sucking insect will settle and feed. A quantitative defensive response to components of the insect's saliva, such as can be inferred from the work of Puritch (1977) on adelgid galls, would mean that the strength and rate of the plant's reaction to the minimum feeding activity required of the insect for its adequate nutrition would determine the balance between the nutritive and defensive potential of that particular tissue.

– Various physiological changes that occur in galls can be ascribed either to increased metabolic activity or increased mobilization of oxidizable substrates or both (Henke, 1963; Schaefer, 1972). Tissue respiration and respiratory quotients rise; oxidized compounds accumulate, sometimes producing a characteristic coloration. In photosynthetic tissues, chloroplasts degenerate and photosynthesis is reduced (Rilling and Steffan, 1978).

– The total oxygen requirements of galled tissues thus tend to outstrip supply

Section 10.1.1 references, p. 42

and their partial asphixia may result (Newcomb, 1951). The relation may not be a simple one, however, since oxygen-lack has itself been demonstrated as a cause of eruptive growth in some tissues (Madden and Stone, 1984).

– Specificities of form seem to be closely linked with the instinctive behaviour of the insect (Maresquelle and Meyer, 1965), and this is especially well demonstrated in the formation of petiole galls of *Pemphigus* spp. on poplar (Dunn, 1960; Chan and Forbes, 1975).

– Much of the effect of the insects involves stimulation of nuclear activity. Often there is hypertrophy of the nuclei and nucleoli; polyploidization may occur (Anders, 1960, 1961; Maresquelle and Meyer, 1965) and RNA and protein synthesis increase. In this respect, it is conceivable that amino acids in the saliva of the insects could constitute not a direct cecidogenic stimulus, but a localized initial increase in raw materials for protein synthesis, providing a "conditioner" for some other primary "inducer" of cecidogenesis, as proposed by Hori (in press). But such an effect must be at most very temporary, as amino acids of plant origin soon appear to swamp any injected by the insect (Rilling et al., 1975).

– Auxins are almost inevitably associated with the hypertrophy observed in galls, but whether as cause or effect of the initiation of cecidogenesis is not clearly established. Although some insects inject IAA into the plant (Schäller, 1965), salivary proteases that would release IAA from conjugates (Anders, 1961; Schäller, 1965, 1968), or inhibitors of IAA-oxidase (Hori, 1974, 1975, 1976; Hori and Miles, 1977; Madden and Stone, 1984), would seem to have a greater potential to stimulate growth. Alternatively, it is possible that the insects may inject substances with cytokinin-like activity and that these stimulate production of IAA in the gall (Forrest and Dixon, 1975; Zotov and Gadiev, 1975). The most invasive effects in cecidogenesis may be caused by ethylene, produced as a result of the action of auxins (Puritch, 1977; Madden and Stone, 1984).

– Finally, it must be emphasized that although various attempts to demonstrate specific initiators of cecidogenesis have seemingly been successful, the agents thus demonstrated – mechanical wounding, specific amino acids, auxins, and the sealing of the stomata – are so diverse as to leave the impression that some more fundamental cause has been overlooked. On the other hand, perhaps the control of growth and differentiation in plants is sufficiently complex and progressive that it can be interfered with at different stages and by several means, whether natural or artificial.

REFERENCES

Alleyne, E.H. and Morrisson, F.O., 1974. *Pemphigus spirothecae* (Homoptera: Aphidoidea), an aphid which causes spiral galls on poplar in Quebec. Canadian Entomologist, 106: 1229–1231.

Alleyne, E.H. and Morrisson, F.O., 1977a. The natural enemies of the lettuce root aphid, *Pemphigus bursarius* (L.) in Québec, Canada. Annals of the Entomological Society of Québec, 22: 181–187.

Alleyne, E.H. and Morrisson, F.O., 1977b. Some Canadian poplar aphid galls. Canadian Entomologist, 109: 321–328.

Amman, G.D., 1970. Phenomena of *Adelges piceae* populations (Homoptera: Phylloxeridae) in North Carolina. Annals of the Entomological Society of America, 63: 1727–1734.

Anders, F., 1960. Untersuchungen über das cecidogene Prinzip der Reblaus (*Viteus vitifolii* Shimer). I. Untersuchungen an der Reblausgalle. Biologisches Zentralblatt, 79: 47–58.

Anders, F., 1961. Untersuchungen über das cecidogene Prinzip der Reblaus (*Viteus vitifolii* Shimer). III. Biochemische Untersuchungen über das galleninduzierende Agens. Biologisches Zentralblatt, 80: 199–233.

Baker, K.F. and Tompkins, C.M., 1942. A virus-like injury of snapdragon caused by feeding of the peach aphid. Phytopathology, 32: 93–95.

Balch, R.E., 1952. Studies on the balsam woolly aphid, *Adelges piceae* (Ratz.) (Homoptera: Phylloxeridae) and its effects on balsam fir, *Abies balsamea* (L.) Mill. Publications of the Canadian Department of Agriculture, Dominion Entomological Laboratory, Fredericton, N.B., No. 867, 76 pp.

Balch, R.E. and Underwood, G.R., 1950. The life-history of *Pineus pinifoliae* (Fitch) (Homoptera: Phylloxeridae) and its effects on white pine. Canadian Entomologist, 82: 117–123.

Balch, R.E., Clark, J. and Bonga, J.M., 1964. Hormonal action in production of tumours and compression wood of an aphid. Nature, 202: 721–722.

Bramstedt, F., 1938. Der Nachweis der Blutlausunanfälligkeit der Apfelsorten auf histologischer Grundlage. Zeitschrift für Pflanzenkrankheiten, Pflanzenpathologie und Pflanzenschutz, 48: 480–488.

Calcote, V.R. and Hyder, D.E., 1980. Pecan cultivars tested for resistance to pecan phylloxera. Journal of the Georgia Entomological Society, 15: 428–431.

Campbell, R.L. and Thielges, B.A., 1973. Spruces resistant to gall aphids – background and prospect. Arborist's News, 38: 1–4.

Chalfant, R.B., 1969. Control of the poplar petiole gall aphid on turnip roots. Journal of Economic Entomology, 62: 1519.

Chan, C-K. and Forbes, A.R., 1975. Life cycle of a spiral gall aphid *Pemphigus spirothecae* (Homoptera: Aphididae), on poplar in British Columbia. Journal of the Entomological Society of British Columbia, 72: 26–30.

Chang, L-H. and Thrower, L.B., 1981. The effect of *Uromyces appendiculatus* and *Aphis craccivora* on the yield of *Vigna sesquipedalis*. Phytopathology, 101: 143–152.

Coombe, B.G., 1963. Phylloxera and its relation to South Australian viticulture. Technical Bulletin of the Department of Agriculture of South Australia, No. 31, 95 pp.

Dawish, E.T.E., 1984. Biology and seasonal activity of the plum leaf curling aphid, *Brachycaudus helichrysi* (Kalt.) on plum trees and sunflower in Hungary. Proceedings of the International Conference on Integrated Plant Protection, Budapest, 2: 68–71.

Denisova, T.V., 1965. The phenolic complex of vine roots infested by Phylloxera as a factor in resistance. Vestnik Sel'sko-khozyaistvennoi Nauki, Moscow, 5: 114–118 (in Russian, with English summary).

Dickson, R.C., Laird, E.F., Jr. and Pesho, G.R., 1955. The spotted alfalfa aphid (yellow clover aphid on alfalfa). Hilgardia, 24: 93–118.

Diehl, S.G. and Chatters, R.M., 1956. Studies on the mechanics of feeding of the spotted alfalfa aphid. Journal of Economic Entomology, 49: 589–591.

Dixon, A.F.G., 1971a. The role of aphids in wood formation. I. The effect of the sycamore aphid, *Drepanosiphum platanoides* (Schr.) (Aphididae), on the growth of sycamore, *Acer pseudoplatanus* (L.) Journal of Applied Ecology, 8: 165–179.

Dixon, A.F.G., 1971b. The role of aphids in wood formation. II. The effect of the lime aphid, *Eucallipterus tiliae* L. (Aphididae), on the growth of lime *Tilia x vulgaris* Hayne. Journal of Applied Ecology, 8: 393–399.

Dunn, J.A., 1959. The biology of the lettuce root aphid. Annals of Applied Biology, 47: 475–491.

Dunn, J.A., 1960. The formation of galls by some species of *Pemphigus*. Marcellia, 30 (Supplement): 155–167.

Dunning, R.A., 1974. Arthropod pest damage to sugar beet in England and Wales 1947–74. Report of the Rothamsted Experimental Station, 2: 171–185.

Eichhorn, O., 1968. Problems of the population dynamics of silver fir woolly aphids, Genus *Adelges* (= *Dreyfusia*), Adelgidae. Zeitschrift für Angewandte Entomologie, 61: 157–214.

Eichhorn, O., 1975. Über die Gallen der Arten der Gattung *Dreyfusia* (Adelgidae), ihre Erzeuger und Bewohner. Zeitschrift für Angewandte Entomologie, 79: 56–76.

Faith, D.P., 1979. Strategies of gall formation in *Pemphigus* aphids. Journal of the New York Entomological Society, 87: 21–37.

Fedde, G.F., 1973. Impact of the balsam woolly aphid (Homoptera: Phylloxeridae) on cones and seed produced by infested Fraser fir. Canadian Entomologist, 105: 673–700.

Forrest, J.M.S. and Dixon, A.F.G., 1975. The induction of leaf-roll galls by the apple aphids *Dysaphis devecta* and *D. plantaginea*. Annals of Applied Biology, 81: 281–288.

Fox, R.C. and Griffith, K.H., 1977. Pine seedling growth loss caused by cinaran aphids in South Carolina. Journal of the Georgia Entomological Society, 12: 29–34.

Gibson, R.W., 1974. The induction of top-roll symptoms on potato plants by the aphid *Macrosiphum euphorbiae*. Annals of Applied Biology, 76: 19–26.

Greenslade, R.M., Massee, A.M. and Roach, W.A., 1933. A progress report of the causes of immunity to the apple woolly aphis (*Eriosoma lanigerum* Hausm.). Report of the East Malling Research Station, A17: 220–224.

Harvey, T.L., Hackerott, H.L. and Sorensen, E.L., 1972. Pea aphid resistant alfalfa selected in the field. Journal of Economic Entomology, 65: 1661–1663.

Henke, O., 1963. Über den Stoffwechsel reblausanfälliger und -unanfälliger Reben. Phytopathologische Zeitschrift, 47: 314–326.

Herfs, W., 1973. Untersuchungen zur Biologie der Salatwurzellaus *Pemphigus bursarius* (L.). Zeitschrift für Angewandte Entomologie, 74: 223–245.

Hori, K., 1974. Plant growth-promoting factor in the salivary gland of the bug *Lygus disponsi*. Journal of Insect Physiology, 20: 1623–1627.

Hori, K., 1975. Plant growth-regulating factor, substances reacting with Salkovski reagent and phenoloxidase activities in vein tissue injured by *Lygus disponsi* Linnavouri (Hemiptera: Miridae) and surrounding mesophyll tissues of sugar beet leaf. Applied Entomology and Zoology, 10: 130–135.

Hori, K., 1976. Plant growth-regulating factor in the salivary gland of several heteropterous insects. Comparative Biochemistry and Physiology, 53B: 435–438.

Hori, K., in press. Insect secretions and their effect on plant growth, with special reference to hemipterous insects. In: J.D. Shorthouse and O. Rohfritsch (Editors), Biology of Insect and Acarina Induced Galls.

Hori, K. and Miles, P.W., 1977. Multiple plant growth-promoting factors in the salivary glands of plant bugs. Marcellia, 39: 399–400.

Ibbotson, A., 1953. Studies on cabbage aphid infestations on brussels sprouts. Plant Pathology, 2: 25–30.

Jensen, D.D., 1954. The effect of aphid toxins on *Cymbidium* orchid flowers. Phytopathology, 44: 493–494.

Jones, M.E., Jr., Green, E.A. and Williams, R.L., 1973. Morphological and histological changes in the leaf of *Juglans nigra* following infection by *Phylloxera* sp. Bulletin of the Georgia Academy of Science, 31: 67.

Jones, M.E., Jr., Green, E.A. and Chester, W.J., Jr., 1976. A comparative morphological and histological study of changes in the leaf of *Juglans nigra* and *Carya illinoises* following infection by *Phylloxera* sp. Bulletin of the Georgia Academy of Science, 34: 50-51.

Kantack, E.J. and Dahms, R.G., 1957. A comparison of injury caused by the apple grain aphid and greenbug to small grains. Journal of Economic Entomology, 50: 156–158.

Kennedy, J.S., 1951. Benefits to aphids from feeding on galled and virus-infected leaves. Nature, 168: 825–826.

Kennedy, J.S., 1958. Physiological conditions of the host-plant and susceptibility to aphid attack. Entomologia Experimentalis et Applicata, 1: 50–65.

Kloft, W., 1955. Untersuchungen an der Rinde von Weisstannen (*Abies pectinata*) bei Befall durch *Dreyfusia (Adelges) piceae* Ratz. Zeitschrift für Angewandte Entomologie, 37: 340–348.

Kloft, W. and Ehrhardt, P., 1959. Untersuchungen über Saugtätigkeit und Schadwirkung der Sitkafichtenlaus *Liosomaphis abietina* (Walk.) (*Neomyzaphis abietina* Walk.). Phytopathologische Zeitschrift, 35: 401–410.

Lawrence, K.O., 1977. Japanese beetles: feeding habits on phylloxera infested grape foliage. Environmental Entomology, 6: 507–508.

Lewis, I.F. and Walton, L., 1958. Gall-formation on *Hamamelis virginiana* resulting from material injected by the aphid *Hormaphis hamamelidis*. Transactions of the American Microscopical Society, 77: 146–200.

Lewis, I.F. and Walton, L., 1964. Gall-formation on leaves of *Celtis occidentalis* L. resulting from material injected by *Pachypsylla* sp. Transactions of the American Microscopical Society, 33: 62–78.

Llewellyn, M., 1972. The effects of the lime aphid, *Eucallipterus tiliae* L. (Aphididae) on the growth of the lime *Tilia x vulgaris* Hayne. I. Energy requirements of the aphid population. Journal of Applied Ecology, 9: 261–282.

Lowe, H.J.B., 1967a. Interspecific differences in the biology of aphids (Homoptera: Aphididae) on leaves of *Vicia faba*. I. Feeding behaviour. Entomologia Experimentalis et Applicata, 10: 347–357.

Lowe, H.J.B., 1967b. Interspecific differences in the biology of aphids (Homoptera: Aphididae) on leaves of *Vicia faba*. II. Growth and excretion. Entomologia Experimentalis et Applicata, 10: 413–420.

Madden, J.L. and Stone, C., 1984. Induction and formation of pouch and emergence galls in *Eucalyptus pulchella* leaves. Australian Journal of Botany, 32: 33–42.

Maelzer, D.A., 1976. A photographic method and a ranking procedure for estimating numbers of the rose aphid, *Macrosiphum rosae* (L.), on rose buds. Australian Journal of Ecology, 1: 89–96.

Maelzer, D.A., 1977. The biology and main causes of changes in numbers of the rose aphid, *Macrosiphum rosae* (L.), on cultivated roses in South Australia. Australian Journal of Zoology, 25: 269–284.

Maresquelle, H.J. and Meyer, J., 1965. Physiologie et morphogenèse des galles d'origine animale (zoocécidies). In: E. Ashby, J. Bonner, M. Geiger-Huber, W.O. James, A. Lang, D. Müller and M.G. Stålfelt (Editors), Handbuch der Pflanzenphysiologie – Encyclopaedia of Plant Physiology, 15/2. Springer-Verlag, Berlin, Heidelberg, New York, pp. 279–329.

Maresquelle, H.J. and Schnell, R., 1963. Étude expérimentale des phases de l'action cécidogène dans une galle. Comptes Rendus Hebdomedaires des Séances de l'Académie des Sciences, Paris, 203: 270–272.

Meyer, J., 1951. Origine maternelle des principaux tissus et du plastème nourricier des galles larvaires d'*Adelges abietis* sur *Picea excelsa* Lk. Comptes Rendus Hebdomedaires des Séances de l'Académie des Sciences, Paris, 233: 886–888.

Meyer, J., 1958. Études de cécidogenèse comparée. Marcellia, 30 (Supplement): 69–91.

Meyer, J., 1962. Croissance labiale et limites cécidogènes de la galle d'*Adelges abietis* Kalt. sur *Picea excelsa* Lam.: Notion de seuil d'action cécidogène. Marcellia, 30: 225–235.

Miles, P.W., 1985. Dynamic aspects of the chemical relation between the rose aphid and rose buds. Entomologia Experimentalis et Applicata, 37: 129–135.

Mirzaev, M.N., Kitlaev, V.N., Perov, N.N. and Mammaev, A.T., 1972. Autoregulatory role of antioxidants in the growth of grapevines infested by phylloxera. Sel'sko-khozyaistvennaya Biologiya, 7: 628–630 (in Russian).

Mittler, T.E. and Sylvester, E.S., 1961. A comparison of the injury to alfalfa by the aphids *Therioaphis maculata* and *Macrosiphum pisi*. Journal of Economic Entomology, 54: 615–622.

Möllerström, G., 1963. Different kinds of injury to leaves of the sugar beets and their effect on yield. Meddelanden från Stadens Växtskyddsanstalt, 12: 299–309.

Mordvilko, A., 1935. Die Blattläuse mit unvollständigem Generationszyklus und ihre Entstehung. Ergebnisse und Fortschritte der Zoologie, 8: 36–328.

Müller, F.P., 1957. Die Hauptwirte von *Myzus persicae* (Sulz.) und von *Aphis fabae* Scop. Nachrichtenblatt für Deutschen Pflanzenschutzdienst, Berlin, 11: 21–27.

Mullick, D.B. and Jensen, G.D., 1976. Rates of non-suberized impervious tissue development after wounding at different times of the year in three conifer species. Canadian Journal of Botany, 54: 881–892.

Newcomb, E.H., 1951. Comparative studies of metabolism in insect galls and normal tissues. In: F. Skoog (Editor), Plant Growth Substances. University of Wisconsin Press, Madison, WI, pp. 417–427.

Nickel, J.L. and Sylvester, E.S., 1959. Influence of feeding time, stylet penetration, and developmental instar on the toxic effect of the spotted alfalfa aphid. Journal of Economic Entomology, 52: 249–254.

Nielsen, D.G. and Balderston, C.P., 1977. Control of eastern spruce and Cooley spruce gall aphids with soil-applied systemic insecticides. Journal of Economic Entomology, 70: 205–208.

Norris, D.M., 1979. How insects induce disease. In: Plant Disease, 4. Academic Press, New York, San Fransisco, London, pp. 239–255.

Nystérakis, F., 1948. Phytohormones et inhibition de la croissance des organes végétaux attaqués par des aphides. Comptes Rendus Hebdomedaires des Séances de l'Académie des Sciences, Paris, 226: 746–747.

Oatman, E.R. and Platner, G.R., 1969. An ecological study of insect populations on cabbage in southern California. Hilgardia, 40: 1–40.

Oechssler, G., 1962. Studien über die Saugschäden mitteleuropäischer Tannenläuse im Gewebe einheimischer und ausländischer Tannen. Zeitschrift für Angewandte Entomologie, 50: 408–454.

Ortman, E.E. and Painter, R.H., 1960. Quantitative measurements of damage by the greenbug, *Toxoptera graminum*, to four wheat varieties. Journal of Economic Entomology, 53: 798–802.

Parry, W.H., 1971. Differences in the probing behaviour of *Elatobium abietum* feeding on Sitka and Norway spruces. Annals of Applied Biology, 69: 177–185.

Paschke, J.D. and Sylvester, E.S., 1957. Laboratory studies on the toxic effects of *Therioaphis maculata* (Buckton). Journal of Economic Entomology, 50: 742–748.

Payne, J.A. and Schwartz, P.H., Jr., 1971. Feeding of Japanese beetles on *Phylloxera* galls. Annals of the Entomological Society of America, 64: 1466–1467.

Puritch, G.S., 1977. Distribution and phenolic composition of sapwood and heartwood in *Abies grandis* and effects of the balsam woolly aphid. Canadian Journal of Forestry Research, 7: 54–62.

Rilling, G., 1964. Die Entwicklungspotenzen von Radicolen- und Gallicoleneiern der Reblaus (*Dactylosphaera vitifolii* Shimer) in Beziehung zu Umweltfaktoren. Vitis, 4: 144–151.

Rilling, G. and Steffan, H., 1978. Versuche über die CO_2-Fixierung und den Assimilatimport durch Blattgallen der Reblaus (*Dactylosphaera vitifolii* Shimer) an *Vitis rupestris* 197G. Angewandte Botanik, 52: 343–354.

Rilling, G., Rapp, A., Steffan, H. and Reuther, K.H., 1974. Freie und gebundene Aminosäuren der Reblaus (*Dactylosphaera vitifolii* Shimer) und Möglichkeiten ihrer Biosynthese aus Saccharose-$^{14}C(U)$. Zeitschrift für Angewandte Entomologie, 77: 195–210.

Rilling, G., Rapp, A. and Reuther, K.-H., 1975. Veränderungen des Aminosäurengehaltes von Rebensorten bei Befall durch die Reblaus (*Dactylosphaera vitifolii* Shimer). Vitis, 14: 198–219.

Roach, W.A. and Massee, A.M., 1931. Preliminary experiments on the physiology of the resistance of certain rootstocks to attack by woolly aphis. Report of the East Malling Research Station, 1928–1930, Supplement II: 111–120.

Roberts, F.M., 1940. Studies on the feeding methods and penetration rates of *Myzus persicae* Sulz., *Myzus circumflexus* Buckt., and *Macrosiphum gei* Koch. Annals of Applied Biology, 27: 348–358.

Rohfritsch, O., 1967. Conditions histo-cytologiques nécessaires à la base des aguilles d'Épicéa pour permettre le développement des larves de Chermesidae. Comptes Rendus Hebdomadaires des Séances de l'Académie des Sciences, Paris, 265: 1905–1908.

Rohfritsch, O., 1976. Étude ultrastructurale du tissu nourricier de la fondatrice du *Chermes strobilobus* Kalt. (Aphidoidea) dans les bourgeons de *Picea excelsa* L. Journal de Microscopie et de Biologie Cellulaire, 26: 24a.

Rohfritsch, O., 1981. A "defense" mechanism of *Picea excelsa* L. against the gall former *Chermes abietis* L. (Homoptera, Adelgidae). Zeitschrift für Angewandte Entomologie, 92: 18–26.

Rohfritsch, O. and Meyer, J., 1966. Action déterminante des gallicoles sur la déhiscence des galles d'*Adelges abietis* Kalt. Comptes Rendus Hebdomadaires des Séances de l'Académie des Sciences, Paris, 262: 248–250.

Savary, A., 1953. Le puceron cendré du poirier (*Sappaphis pyri* Fonsc.) en Suisse romande. Landwirtschaftliches Jahrbuch der Schweiz, 67: 247–314.

Saxena, P.N. and Chada, H.L., 1971. The greenbug, *Schizaphis graminum*, 1. Mouth parts and feeding habits. Annals of the Entomological Society of America, 64: 897–904.

Schäller, G., 1965. Untersuchungen über den β-Indolylessigsäuregehalt des Speichels von Aphidenarten mit unterschiedlicher Phytopathogenität. Zoologische Jahrbücher, Abteilung Allgemeine Zoologie und Physiologie der Tiere, 71: 385–392.

Schäller, G., 1968. Biochemische Analyse des Aphidenspeichels und seine Bedeutung für die Gallenbildung. Zoologische Jahrbücher, Abteilung Allgemeine Zoologie und Physiologie der Tiere, 74: 54–87.

Schaefer, H., 1972. Über Unterschiede im Stoffwechsel von reblausvergallten und gesunden Rebenblättern. Phytopathologische Zeitschrift, 75: 285–314.

Sengupta, G.C. and Miles, P.W., 1975. Studies on the susceptibility of varieties of apple to the feeding of two strains of woolly aphis (Homoptera) in relation to the chemical content of the tissues of the host. Australian Journal of Agricultural Research, 26: 157–168.

Severin, H.H.P. and Tompkins, C.M., 1950. Symptoms induced by some species of aphids feeding on ferns. Hilgardia, 20: 81–92.

Sienicka, A., 1959. Anatomical and cytological changes brought about by *M. ribis* on the leaves of currants (*Ribes*) and their effect on fruit yields. Zeszyty Naukowe Wyższej Szkoły Rolniczej w Szczecinie, 2: 91–117 (in Polish, with English summary).

Smith, F.F. and Brierley, P., 1948. Simulation of lily rosette symptoms by feeding injury of the foxglove aphid. Phytopathology, 38: 849–851.

Smith, K.M., 1926. A comparative study of the feeding methods of certain Hemiptera and of the resulting effects on plant tissue, with special reference to the potato plant. Annals of Applied Biology, 13: 109–136.

Smith, K.M., 1951. A Textbook of Agricultural Entomology. Cambridge University Press, Cambridge, 289 pp.

Sobetskii, L.A. and Derzhavina, M.A., 1973. Contribution to the study of the physiology of the feeding of the vine phylloxera, *Viteus vitifolii* Fitch (Homoptera, Phylloxeridae). Entomological Review, Washington, 52: 357–361.

Spiller, N.J., Kimmins, F.M. and Llewellyn, M., 1985. Fine structure of aphid stylet pathways and its use in host plant resistance studies. Entomologia Experimentalis et Applicata, 38: 293–295.

Spires, S., 1981. Effects of host needle age on the settling behaviour of two morphs of *Adelges cooleyi* (Gill.). Ecological Entomology, 6: 205–208.

Staniland, L.N., 1924. The immunity of apple stocks from attacks by woolly aphis (*Eriosoma lanigerum*, Hausmann). Part II. The causes of the relative resistance of the stocks. Bulletin of Entomological Research, 15: 157–170.

Steffan, H. and Rilling, G., 1981. Der Einfluss von Blatt- und Wurzelgallen der Reblaus (*Dactylosphaera vitifolii* Shimer) auf das Verteilungsmuster der Assimilate in Reben (*Vitis rupestris* 187G). Vitis, 20: 146–155.

Sterling, C., 1952. Ontogeny of the *Phylloxera* gall of grape leaf. American Journal of Botany, 39: 6–15.

Stoetzel, M.B., 1981. Two new species of *Phylloxera* (Phylloxeridae: Homoptera) on pecan. Journal of the Georgia Entomological Society, 16: 127–144.

Stoetzel, M.B. and Tedders, W.L., 1981. Investigation of two species of *Phylloxera* on pecan in Georgia. Journal of the Georgia Entomological Society, 16: 144–150.

Strickland, A.H., 1957. Cabbage aphid assessment and damage in England and Wales, 1946–55. Plant Pathology, 6: 1–9.

Thalenhorst, W., 1972. Zur Frage der Resistenz der Fichte gegen die Gallenlaus *Sacchiphantes abietis* (L.). Zeitschrift für Angewandte Entomologie, 71: 225–249.

Van Emden, H.F., Eastop, V.F., Hughes, R.S. and Way, M.J., 1969. The ecology of *Myzus persicae*. Annual Review of Entomology, 14: 197–270.

Walton, L., 1980. Gall formation and life history of *Pineus floccus* (Patch) (Homoptera: Adelgidae) in Virginia. Virginia Journal of Science, 31: 55–60.

Wartenberg, H., 1954a. Histologische Studien über Blutlausgallen, Blutlausabwehrnekrosen, Parenchymholzbinden und Markflecke bei *Malus*-sorten. Wissenschaftliche Zeitschrift der Friedrich Schiller-Universität, Jena, 3: 409–430.

Wartenberg, H., 1954b. Über pflanzenphysiologische Ursachen des Massenwechsels der Apfelblattlaus (*Eriosoma lanigerum*) auf *Malus pumila*. Mitteilungen der Biologischen Zentralanstalt für Land- und Forstwirtschaft, 75: 53–56.

West, F.T., 1946. Ecological effects of an aphid population upon weight gains of tomato plant. Journal of Economic Entomology, 39: 338–343.

Zotov, V.V., 1976. On the nature of phylloxera-resistance in grape plants. Sel'sko-khozyaistvennaya Biologiya, 11: 277–285 (in Russian, with English summary).

Zotov, V.V. and Gadiev, R.Sh., 1975. Stimulators of growth of gall tissues in phylloxera-infected grape plants. Sel'sko-khozyaistvennaya Biologiya, 10: 241–244 (in Russian, with English summary).

10.2 Crop Loss Assessment

P.W. WELLINGS, S.A. WARD, A.F.G. DIXON and R. RABBINGE

INTRODUCTION

Aphids are one of the most widespread groups of pests in agricultural systems and may cause crop losses in forest trees, field crops and horticultural crops (see Appendix to this section). Plants in these production systems may be affected directly or indirectly as a result of the presence of aphid populations. Direct effects come about through suction of feeding aphids, injection of active substances in saliva, interference with the physiological performance of crops and the removal of amino-nitrogen from the plants. Indirect effects are caused by virus transmission, honeydew excretion and changes in the microflora communities on plant surfaces which may influence crop physiology. The extent of these effects depends on the production level at which the crop is being grown and the nature of the crop itself. In this section, we provide an overview of the types of damage which can be classified as crop loss, but do not consider those losses which occur as a result of virus transmission by aphids (see section 10.3).

DEFINING YIELD AND LOSS

Methods of assessing crop loss require a fundamental understanding of definitions of yield (James, 1980). Indeed this is perhaps the central issue in interpreting and synthesizing the information on losses caused by aphids. Zadoks (1980) described five types of yield:

(a) Theoretical yield (THY) – the yield under the best possible conditions, as determined by crop physiologists.

(b) Attainable yield (ATY) – the FAO reference value which refers to crops grown under optimal conditions using modern technology (Chiarappa, 1971).

(c) Primitive yield (PRY) – the yield of crops grown under subsistence conditions without any modern inputs.

(d) Actual yield (ACY) – the yield returned to the farmer under current crop husbandry practices.

(e) Economic yield (ECY) – the yield that gives the highest economic return on expenditure.

These ideas on yield result in three concepts of loss which are defined as:

Theoretical loss = THY − ACY

Crop loss = ATY − ACY

Economic loss = ECY − ACY

Section 10.2 references, p. 61

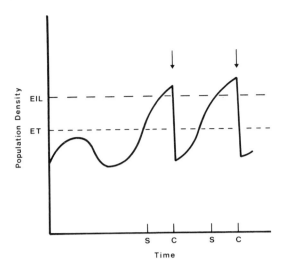

Fig. 10.2.1. Simplified representation of the economic threshold (ET) and economic injury level (EIL). If, at the time of sampling (S), the population density is greater than the ET, then by the time control can be implemented (C) the EIL will have been exceeded. S, sampling dates on which the population is greater than the ET; C and vertical arrows, control measures.

The purpose of defining and quantifying damage is to determine whether control measures are economically justified, i.e. whether the benefits (from the increase in yield) exceed the cost of control. In this context it is necessary to estimate the "economic damage", the "economic injury level" (EIL) and the "economic threshold" (ET). The economic damage is the amount of preventable damage which causes a financial loss equal to the cost of the control measures. Conventionally, the EIL is the lowest population density resulting in economic damage, i.e. the density of the smallest population worth controlling, and the ET is the density above which the decision should be taken to implement control (Fig. 10.2.1) (Stern, 1967, 1973; Stone and Pedigo, 1972; Ogunlana and Pedigo, 1974). However, as emphasized by Plant (1986) and Onstad (1987),

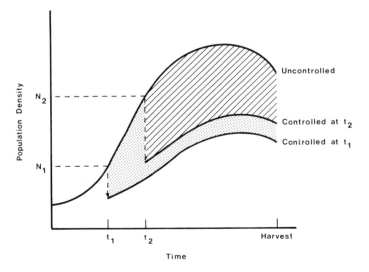

Fig. 10.2.2. The development of a hypothetical aphid population, and the benefits of control at times t_1 and t_2, to illustrate changes in the EIL (see text for further details).

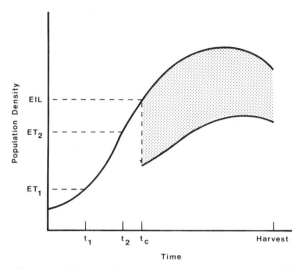

Fig. 10.2.3. The relation between the action threshold (ET) and the time at which the pest population is monitored.

neither EIL nor ET should be regarded as constant; both must be considered in the context of the system's dynamics, the control scheme and the sampling programme.

Factors affecting the EIL are illustrated in Fig. 10.2.2. Here, an insecticide causing 50% mortality can be applied at either time t_1 or time t_2. Assuming, for simplicity, that the crop is equally sensitive to aphids throughout the season and the damage is linearly related to accumulated density, the benefits of control are proportional to the sum of the shaded and diagonally hatched area (for application at t_1) or the hatched area (for t_2). If the benefit of a spray at time t_1 equals the cost of the application, then the EIL at t_1 equals N_1. Clearly, the benefits of control at t_2 are smaller than those of the earlier spray, so control at t_2 is not economically justified: To obtain the same effect of control at t_2 as at t_1, in terms of money, the density N_2 should be much higher, because then the diagonally hatched area below the population density curve would equal the value attained at t_1. Thus the EIL at t_2 is greater than N_2, and much greater than at t_1.

Turning now to the ET (sometimes called "action threshold"), we can see from Fig. 10.2.3 that this depends not only on the (time-dependent) EIL but also on the time at which the population is sampled. In fact, it should be regarded not as an actual density (which the EIL is) but as part of a sampling protocol (Onstad, 1987). Since control need not be implemented immediately, but requires some time to become effective (t_c), the use of ET's rather than EIL's means that the monitoring scheme can be optimized independently of the control tactics. ET's are thus much more easily incorporated into crop protection strategies than EIL's (Onstad, 1987).

Most research on damage caused by aphids measures loss in relation to estimates of actual yield in the absence of the pest herbivore. This makes much of the literature difficult to interpret within the framework of the above definitions of loss.

In addition, many experimental approaches do not place these estimates of actual yield in the context of the environment in which crop plants are commercially produced. Thus, they do not provide any insight into the way that damage effects change with crop production level.

Section 10.2 references, p. 61

PRODUCTION LEVELS AND PRODUCTION SYSTEMS

Production levels

Classifications of crop growth and production emphasize the importance of dry matter production, and Penning de Vries and Van Laar (1982) distinguish four broad production levels:

Production level 1: Growth occurs in conditions with ample plant nutrients and soil water all the time. Growth and yield are determined by weather conditions and crop characteristics. Under these conditions the growth rates of C_3 plants amount to approximately 200 kg dry matter per ha per day and that of C_4 plants* may vary between 200–400 kg dry matter per ha per day, depending on latitude. Actual yields can approach attainable yields under these circumstances.

Production level 2: Growth may be limited by water shortage for part of the growth period. However, if water is available the growth rates rise to the maximum set by weather conditions. Thus, at this production level, growth rates are scaled by the ratio between the actual and potential transpiration. Under these circumstances growth rates are in general much lower than the maximum value and vary between 100–200 kg dry matter per ha per day.

Production level 3: Growth is limited by a shortage of nitrogen for at least part of the season and by a water shortage or unsuitable weather conditions for the remainder of the growth period. These conditions result in growth rates which are usually lower than 1000 kg dry matter per ha per day.

Production level 4: Growth is limited by the low availability of phosphorus or other minerals for part of the growing period and by a lack of nitrogen, shortage of water or unsuitable weather for the remainder of the season. Growth rates usually vary between 12–50 kg dry matter per ha per day and yield levels are about 100 kg per ha dry matter in each growing season (i.e. actual yields approach primitive yields).

This framework of production levels outlines the factors which limit the growth and yield of crops. The detailed way in which these factors interact with each other is the domain of crop physiologists and is reviewed and discussed in a number of recent books (e.g. Carlson, 1980; Johnson, 1981; Pearson, 1984). In any discussion of crop loss assessment it must be borne in mind that aphids may reduce plant growth and yield, but that the level of growth and yield in the absence of aphids is constrained by prevailing environmental conditions and their influence on crop productivity.

More than 60% of the world's agricultural production takes place at the lowest production level, and only about 1% of agricultural crops are classified in production level 1. Many horticultural and some cash crops fall into the highest level. There are very few examples of field crops belonging to this group, perhaps only the extremely intensive arable cropping situation in Western Europe approaches level 1. Most field crops and pastures are at levels 3 and 4 and a minority can be classified in production level 2. Aphids occur and may cause crop losses at all production levels. However, the relative loss in yield may differ markedly between production levels.

Production systems

The occurrence of either crop loss or economic loss as a result of the presence of aphids is highly dependent on the type of crop and production level.

---

*The expression "C_3 and C_4 plants" refers to the CO_2 assimilation system in photosynthesis: the first molecule formed in the biochemical pathway contains 3 or 4 C atoms respectively.

For example, consumers have a very low tolerance for the presence of aphids in ornamental crops (e.g. cut flowers) sold for aesthetic purposes. In addition the occurrence of any aphids on such crops produced for export may cause them to be rejected by countries with rigorous quarantine regulations. Horticultural crops grown at the highest production levels also have very low tolerance for aphid attack. These crops may be so valuable that even the losses caused by small infestations may be greater than the cost of control methods.

The majority of production systems are less sensitive to crop losses caused by aphids. Annual plants grown as food and fiber crops may be able to tolerate moderate aphid populations before crop losses become important, and these damage relationships may change with the production level at which the crop is grown. Because aphid populations have high rates of increase, these populations should be closely monitored and their effects minimized. The influence of aphids on perennial crop plants is more difficult to quantify because of problems in assessing annual yield increment. Large aphid populations may be needed to cause significant damage levels, and their occurrence may be infrequent because aphid abundance is markedly affected by changes in weather (Wellings and Dixon, 1987). The influences of aphids on perennial crops may depend on the age structure of the host population at the time of attack, and aphids may cause long-term losses due to growth distortion of economically desirable characteristics in these plants (e.g. *Dysaphis plantaginea* (Passerini) on apple).

Many of the aphids which feed on perennial hosts have strict host plant relationships; they tend to be monophagous and holocyclic (e.g. *Elatobium abietinum* (Walker) on *Picea* spp., *Drepanosiphum platanoidis* (Schrank) on *Acer pseudoplatanus*). In contrast, some aphids which attack annual and horticultural plants are polyphagous species, exhibiting heteroecious or anholocyclic life-cycles (e.g. *Myzus persicae* (Sulzer), *Acyrthosiphon pisum* (Harris) and *Aphis fabae* Scopoli. The above species can cause considerable crop losses.

INJURY AND DAMAGE

Aphids are able to induce a large number of effects on their host plants, as the patterns of translocation and growth of the plant may be altered as a consequence of infestation (see sections 10.1 and 10.1.1). Not all effects on plants are solely attributable to the aphids' removing nutrients, amino-acids and carbohydrates from the phloem; there is clear evidence that aphids have a direct impact on the physiological processes of plants. Infestation may cause severe disruption of plant tissues or even the death of plants (Forrest et al., 1973). However, the sub-lethal effects occur most frequently.

Many physiological changes have been recorded in plants following aphid infestation. These include a reduction in water permeability and the level of carbohydrate reserves in *Abies grandis* after infestation by *Adelges piceae* (Ratzeburg) (Putritch, 1971; Putritch and Talmon-de l'Armee, 1971) and higher and lower levels of growth-inhibiting and growth-promoting substances, respectively, in the radish, *Raphanus sativus*, after infestation by *M. persicae* (Hussain et al., 1974).

Mallott and Davy (1978) described some effects of infestation by *Rhopalosiphum padi* (Linnaeus) on growth and photosynthesis in barley plants (*Hordeum vulgare*). This study demonstrated that infestation caused a significant reduction in dry weight yield, leaf area, the number of tillers and the number of leaves of barley plants. Detailed growth analysis indicated that the major effect was to reduce unit leaf rate. The rates of net photosynthesis were

Section 10.2 references, p. 61

similar in uninfested and infested leaves, indicating that the reduction in unit leaf rate did not result from an effect on photosynthesis. Mallott and Davy hypothesized that the decreased growth of infested plants was attributable to the cumulative assimilate loss due to aphid feeding. The effects of aphid feeding on growth are reflected in grain yield; infested cereal plants may have smaller ears and smaller grains at harvest, and grain quality may be reduced (Wratten, 1975, 1978; Bhatia and Singh, 1977; Ba-Angood and Stewart, 1980; Lee et al., 1981).

Similar effects of infestation have been recorded in many aphid species. However, different aphid species may have different effects on the same host plant. For example, *A. fabae* causes a reduction in the weight of seeds and the number of seeds per pod in *Vicia faba*, while *A. pisum* may only reduce seed weight (Bouchery, 1977). Other studies have shown that different species of host plant respond to infestations by the same species of aphid in different ways (e.g. *E. abietinum* on Norway spruce, *Picea abies*, and Sitka spruce, *Picea sitchensis*; Parry, 1974).

The structure of the host plant can also be affected by infestation; for example, the galls of *Cryptomyzus ribis* (Linnaeus) on currants and *Pemphigus bursarius* (Linnaeus) on poplar, and the pseudogalls of *Dysaphis devecta* (Walker) on apple and *Myzus cerasi* (Fabricius) on cherry (Dixon, 1975). These changes may be brought about by the action of aphid saliva, which contains physiologically active substances known to influence plant growth (Nuorteva, 1956). Galling by aphids on fruit trees may cause the distortion of shoots and the production of small fruit, often of inferior quality.

The effects of infestations by aphids may also be indirect. Harper and Freyman (1979) investigated the survival of infested and uninfested lucerne, *Medicago sativa*, after exposure to freezing temperatures. Infestation by *A. pisum* significantly reduced cold-hardiness, and these authors suggested that physiological changes as a result of heavy aphid infestation could result in a higher incidence of winterkill. Rabbinge et al. (1981), in a series of laboratory and field experiments, demonstrated that, as well as damaging wheat plants directly, the cereal aphids *Sitobion avenae* (Fabricius) and *Metopolo-phium dirhodum* (Walker) cause indirect damage by their excreta. This study showed that honeydew on the leaf surfaces causes a significant reduction in light use efficiency and the maximum rate of assimilation of the leaves; in addition, honeydew promotes the rate of ageing of leaves. These effects are also increased by the presence of saprophytic fungi feeding on the honeydew. More complex types of indirect effects have been suggested in which aphid infestation may predispose their host plants to disease. For example, Bergstrom et al. (1982) showed that the presence of *Aphis gossypii* Glover on cucumber, *Cucumis sativus*, may induce susceptibility to the fungus *Mycosphaerella melonis* (Passerini).

ASSESSING CROP LOSS

Laboratory, glasshouse and cage experiments

There are many experimental studies in this category which demonstrate that aphids may influence the yield of their host plants. However, they are difficult to interpret, as it is almost impossible to extrapolate measures of yield from individual or small groups of plants to give estimates of losses likely to be due to aphids in field conditions. So these approaches provide a starting point for evaluating crop loss due to aphids; however, they are not an end in themselves.

TABLE 10.2.1

The effect of an *Aphis craccivora* Koch infestation on the yield of Los Banos bush sitao (*Vigna unguiculata* spp. *sesquipedalis*) (after Bernabe, 1972)

	Duration of aphid infestation (days)	Trial 1: May–July		Trial 2: July–September	
		Yield (g)	Yield loss due to aphids (%)	Yield (g)	Yield loss due to aphids (%)
No aphids		301.1	–	359.0	–
Aphids present	–				
Sprayed after	15	246.6	18.1	284.6	20.7
	22	241.0	20.0	260.5	27.4
	29	156.1	48.2	229.8	36.0
Not sprayed	–	145.5	51.7	153.0	57.4

Three experimental approaches are presented in the literature: (a) spot yield estimates, which compare the performance of infested and uninfested plants; (b) fixed infestation level/variable infestation duration yield estimates, in which plants are infested with a constant number of aphids at a particular growth stage and colonies are allowed to develop for a range of pre-determined times before spraying with an insecticide; (c) variable infestation levels/fixed infestation duration, in which plants are infested with variable numbers of aphids at particular growth stages and these colonies are allowed to develop for a fixed time period.

Approach (a) has been used to demonstrate that aphids have direct effects on their host plants. For example, *A. pisum* infestation reduced the weight and height of crowns of *M. sativa* (Harper and Freyman, 1979), and defoliation of Scots pine trees, *Pinus sylvestris*, by *Schizolachnus pineti* (Fabricius) resulted in a significant reduction in shoot and needle elongation in the following season (Thompson, 1977). Approach (b) has demonstrated that yield loss may be a function of the duration of aphid infestation (e.g. Bernabe, 1972; Lee et al., 1981; Rohitha and Penman, 1983) (Table 10.2.1). However, this approach is difficult to analyze, as the plants are of different ages at the end of aphid infestation. This problem is overcome in approach (c), which may be used to study the effects of different levels of infestation at different plant growth stages. Ba-Angood and Stewart (1980) demonstrated that the quality and quantity of barley grain depends on the level of infestation and the form of the relationship between yield and infestation changes with the growth stage of the host plant at the time of initial infestation (Fig. 10.2.4). Yield losses in *V. faba* due to *A. fabae* have also been shown to depend on the timing and intensity of colonization (Hinz and Daebeler, 1981). In a detailed study of the effects of *A. pisum* on field peas, **Pisum sativum**, Maiteki and Lamb (1985a) also demonstrated that the host **plant stage is** important.

Field experiments

Experiments performed in the field are likely to give more reliable indications of the nature of crop losses caused by aphids, especially when conducted in crops managed under prevailing husbandry practices. However, some caution is required, as plot size and shape may influence yield estimates and infestation levels are difficult to repeat. Two general groups of experiments can be recognized. Firstly, those which investigate the interactions between individual factors and aphid infestation on yield loss. Secondly, those which attempt to find a relationship between yield and some measure of aphid infestation under specified crop conditions.

Section 10.2 references, p. 61

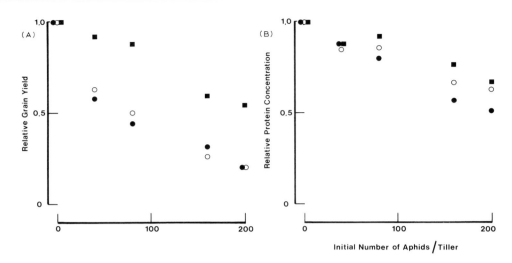

Fig. 10.2.4. The effect of different population densities of cereal aphids on (A) the relative grain yield of barley (*Hordeum vulgare*) and (B) the relative protein concentration of barley grains. Aphids were initially introduced at one of three crop growth stages: flowering (●), milky ripe (○) and mealy ripe (■) (after Ba-Angood and Stewart, 1980.).

The first category is well documented, and here we present a brief outline of the major types of factors which interact with aphid infestation and influence yield. These factors include:

(a) Plant growth stage at the time of infestation (e.g. Daebeler and Hinz, 1976; Wilde and Ohiagu, 1976; Wratten, 1978; Gräpel, 1982).

(b) Sowing date of the host plant (e.g. Lammerink and Banfield, 1980; see Table 10.2.2).

(c) Water availability during aphid infestation (e.g. Foott and Timmins, 1973; Daebeler and Hinz, 1977).

(d) Host plant cultivar (e.g. Harvey et al., 1971, see Table 10.2.3; Teetes and Johnson, 1974; Bhatia and Singh, 1977; Lloyd et al., 1985).

(e) Nitrogen fertilization (e.g. Archer et al., 1982).

(f) Indirect effects due to honeydew production (e.g. Rabbinge et al., 1981).

These studies indicate that the crop losses due to aphids may be a complex function of a large number of variables. Experimental studies are unlikely to resolve the way in which these multifactorial systems function, especially if the variables covary with each other. However, simulation studies may provide considerable insight into the mechanisms causing yield loss. In general, the patterns seen in field experiments appear to correspond with parallel experiments conducted under constrained laboratory/glasshouse situations. Thus,

TABLE 10.2.2

The effect of sowing date and infestations of *Brevicoryne brassicae* (Linnaeus) on rapeseed (*Brassica napus*) yield and quality (after Lammerink and Banfield, 1980)

Sowing date	Aphid infestation	Plant height (cm)	Seed yield (kg per ha)	Seeds per pod	1000-Seed weight (g)	No. pods per m²	Moisture in seed (%)	Oil in seed (%)	Total N (oil-free basis) (%)
11 October	Infested	89	440	7.1	3.52	1720	6.4	40.0	8.1
	Uninfested	101	1480	14.1	3.42	3120	6.0	41.9	8.5
4 November	Infested	81	190	4.7	3.13	1200	6.8	33.9	7.2
	Uninfested	97	1850	18.8	3.96	2540	6.2	41.0	8.3

TABLE 10.2.3

The influence of *Acyrthosiphon pisum* (Harris) infestation on pea-aphid resistant and pea-aphid susceptible cultivars of *Medicago sativa* (after Harvey et al., 1971).

Cultivar	Aphid infestation	Plant height (cm)	Dry matter (%)	Dry matter (% yield in g per m²)
Resistant	Uninfested	64	32	367
	Infested	58	32	278
Susceptible	Uninfested	66	30	368
	Infested	30	51	162

shifts in host plant phenology, ontological changes in host plants and differences in production level may all influence crop loss.

Other field studies have described the relationship between plant production and the size of aphid populations. For example, leaf size in *Acer pseudoplatanus* is inversely related to the spring abundance of *D. platanoidis*, and this has an impact on timber production (Dixon, 1971). Some investigators have used this approach, or minor modifications of it, in an attempt to estimate economic damage due to aphids and economic thresholds (e.g. Capinera, 1981; Bishop et al., 1982; Cuperus et al., 1982; Maiteki and Lamb, 1985b).

Explaining and forecasting crop loss

Clearly, the two groups of experiments outlined on p. 55 yield very different results. Those aimed at description of the relation between aphid density (or, for example, accumulated aphid-days) and yield loss may generate simple expressions which can be used directly as a basis for decisions about the need for control.

Thus, estimates of economic injury levels (EIL) are frequently derived from the regression $y = a - bx$, where y is the estimated yield, x is a measure of infestation (e.g. aphid numbers, aphid-days) and a and b are constants. The constant a represents the expected yield in the absence of aphid infestation (which should lie somewhere between attainable yield and actual yield). The economic damage resulting from an infestation of aphids is equal to $(a - y)$. The EIL (measured in terms of x) is then calculated as the point at which the value of economic damage is exactly equal to the cost of control. Finally, economic thresholds (ET) may be derived. This requires estimates of the rate of increase of the aphid population (r) and the time need to implement control operations (t); thus $ET = EIL/e^{rt}$.

There are a number of criticisms which can be levelled at this approach. For example, (a) there are no a priori reasons to support the use of the linear regression model, although such models do yield estimates of the extent to which crop losses are affected by aphid infestations; (b) the model assumes a fixed damage relationship, and yet crop losses due to aphids are dependent on many variables; (c) the estimates of EIL and ET do not always take into account variable external economic forces (although in principle there is no reason why they should not). This final point has been examined by Way and Cammell (1980). They note that a fixed "working" economic threshold for forecasting outbreaks of *A. fabae* on *V. faba* has been in place in the United Kingdom since 1970. This working economic threshold was based on the cost of chemical control, the sales price of beans and the average bean yield per ha in the year 1968. Of course, these costs change with time, and some may be unknown at the time of decision making. In a retrospective analysis, the

Section 10.2 references, p. 61

working economic threshold was compared with the "calculated" economic threshold using exact estimates of these unpredictable variables over the period 1970–1976. Way and Cammell demonstrated that the working economic threshold overestimated the need for chemical control in each year and that individual factors (e.g. control costs, sale price of crop, crop yield) have variable effects on calculated economic thresholds. Nevertheless, this study demonstrated the value of adopting forecasting systems; while the economic returns were never maximized, they were significantly greater than would have been achieved by alternative strategies such as no-control of aphids and prophylactic spraying. Similar insights into the value of forecasting cereal aphid damage are provided by George and Gair (1979), Watt (1983) and Watt et al. (1984).

While in principle economic factors can be incorporated into simple management models, the complex interactions revealed by experiments on the *mechanisms* of yield reduction are often more problematical. Most assessments of crop loss have been conducted on crops grown under specific conditions, and when economic loss assessment has been attempted most studies have used average values for crop value and control costs. However, particular crops may be grown under a wide range of conditions, and there may be temporal variations in yield as a result of varying climatic conditions. This means that yield loss and aphid abundance will inevitably be poorly correlated (Fig. 10.2.5). What is really needed is a dynamic approach which integrates information from experimental studies on crop loss with the development of aphid populations. This is a major function of simulation modelling.

Comprehensive simulation of these aphid–plant interactions requires a detailed model of crop growth, a model simulating the aphids' population dynamics, and a coupling sub-model of the effects of the aphids on the crop's physiology. Models of aphid populations (e.g. Carter et al., 1982; Rabbinge et al., 1979) are discussed in section 8.2. In this section we consider the use of simulation to elucidate the effects of aphids on yield; we concentrate here on a model developed to simulate the effects of *S. avenae* on the growth of winter wheat (Van Roermund et al., 1986).

The crop growth model simulates the accumulation of dry matter in roots, shoots and grain and the plant's nitrogen budget (Rabbinge and Vereijken, 1980). The accumulation of carbohydrate depends on the gross rate of

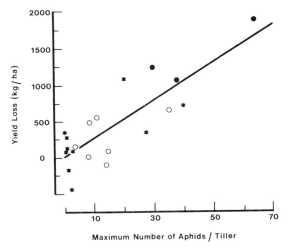

Fig. 10.2.5. Yield loss of winter wheat (kg per ha) as a function of the peak number of aphids per ear. (●) *Metopolophium dirhodum* (Walker); (○) *Sitobion avenae* (Fabricius) (★), both species (after Rabbinge and Mantel, 1981).

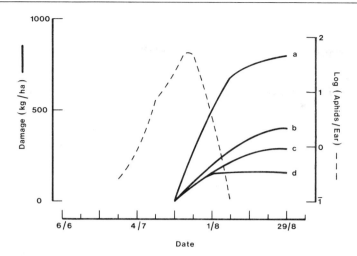

Fig. 10.2.6. The population density (– – –) of *Sitobion avenae* (Fabricius) on winter wheat, and the simulated damage (——) caused by this population. Four damage components are illustrated in cumulative curves: (a) total aphid damage; (b) damage in the absence of honeydew effects on light use efficiency; (c) direct aphid damage through nitrogen and carbohydrate withdrawaal; (d) direct damage with carbohydrate withdrawl only (see text for further details).

photosynthesis (a function of light intensity, day length and the nitrogen content of the leaves) and the temperature-dependent rate of maintenance respiration.

To study the nature of aphid damage, the aphid's population dynamics is input as a driving variable describing the development of a measured field population. Four (potential) components of damage are simulated: the removal of (1) carbohydrates and (2) nitrogen by feeding aphids; and the effects of honeydew on (3) light use efficiency (in synthesizing carbohydrates at low light intensity) and (4) the maximum rate of photosynthesis. The relations between aphid weight and nutrient uptake, and honeydew and photosynthesis were estimated from laboratory experiments.

Damage due to aphids can be partitioned into several component parts (Fig. 10.2.6), whose relative importance may depend on other factors determining crop growth (Rabbinge et al., 1983; Van Roermund et al., 1986). To generate the results presented here, the development of the aphid population was forced to pass through field sample data. The infestation started at the end of June and peaked on 25 July, causing a total loss of 810 kg per ha (Fig. 10.2.6a). The simulation results suggest that most of the damage occurs after, but is caused before, the peak in aphid abundance. Stem resources appear to be depleted and thus grain growth is source limited. Direct and indirect components of aphid damage can be identified. It appears that direct damage (300 kg per ha) (Fig. 10.2.6c) is mainly attributable to two factors: the withdrawal of carbohydrates and nitrogen. These factors are equally important, but the time at which they have an effect on crop loss differs. When aphids are present on the crop, direct damage is attributable to carbohydrate withdrawal. It is only at the beginning of August, when the aphid population has collapsed, that the effect of nitrogen withdrawal (which took place during the infestation) becomes apparent. Simulations which eliminate this effect result in low estimates of direct damage (Fig. 10.2.6d).

The simulation suggests that most of the damage to the crop is a result of indirect effects. This comes about because honeydew falling on the leaves reduces the maximum rate of photosynthesis and light use efficiency, both of which influence crop photosynthesis. Fig. 10.2.6 illustrates the effect of

Section 10.2 references, p. 61

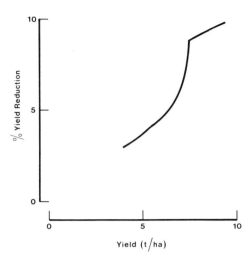

Fig. 10.2.7. Simulated damage by *Sitobion avenae* (Fabricius) on winter wheat. Damage is presented as a percentage of expected yield at harvest.

eliminating the effect of honeydew on light use efficiency (i.e. the difference between a and b in Fig. 10.2.6 is the indirect effect). It is clear that this is an important damage component. Apparently the effect of light-capturing by the honeydew blanket is more important than the effect of honeydew on CO_2-assimilation at light satiation. How these effects exactly work and what the exact quantitative relation is, remains to be studied in more detail.

Simulation studies of the type described here are an important vehicle with which to integrate and interpret ideas on the contributions of various components of damage and to quantify the distribution of damage in time. In addition, it is possible to extend these models by using parameters corresponding to a wide range of production levels. Studies of this sort suggest that crop loss, due to the same intensity of aphid infestation, is unlikely to be a constant proportion of yield (Fig. 10.2.7). In this example, relative yield loss increases in cereal crops grown under high nitrogen levels (which is the main factor limiting yield). Increase in the nitrogen level reduces the rate of leaf senescence and raises the photosynthesis rate. However, because of the persistence of green leaves, the indirect effects of honeydew are more damaging. The relative damage curve eventually begins to level off because green plant material does not increase linearly with nitrogen availability. Dynamic simulation models can be used to calculate economic thresholds for artificial control. These thresholds, however, differ from those generated by regression models in that they vary during the growing season, and are dependent on the time and place of presence of aphids. Such thresholds are also affected by sowing date, nitrogen fertilization and water availability. The use of dynamic simulation models may thus prove invaluable in overcoming the problems associated with the simpler regression methods of estimating economic thresholds.

CONCLUSIONS

Aphids are able to cause a large number of physiological changes in their host plants and in many cases reduce plant productivity. Experiments that attempt to assess crop loss have demonstrated that aphids may have considerable effects on yield and that both direct and indirect effects of aphid infestation are significant factors. Aphid feeding is responsible for most of the direct damage, which arises because of nitrogen and carbohydrate removal and injection of physiologically active substances in saliva. Indirect damage is mainly attributable to aphid excretion. Honeydew falling on leaf surfaces may cause decreases in light use efficiency and the maximum rate of photosynthesis and increases in the rate of leaf senescence. A number of laboratory and field experiments indicate that crop loss caused by aphids varies under different crop and environmental conditions. This suggests that damage relations are unlikely to be well correlated with aphid abundance. Because crop loss is multifactorial, future investigations should rigorously define the production level operating in the cropping system. This means that investigators should attempt to determine attainable and actual yields over a wide range of conditions. Simulation studies provide insights into these problems and have the additional advantage of being able to resolve the relative importance of damage components. At present the assessment of crop loss is constrained by a lack of detailed, accurate models of aphid populations. These simulations should be integrated with the development of models of crop growth and productivity. If this approach is to be relevant to forecasting and controlling aphids in crop systems, future studies need to be interdisciplinary, with inputs from agronomists, applied entomologists and plant physiologists.

REFERENCES

Anonymus, 1985, FAO Production Year Book, 38: 1–326.

Archer, T.L., Onken, A.B., Matheson, R.L. and Bynum, E.D. Jr., 1982. Nitrogen fertilizer influences on greenbug (Homoptera: Aphididae) dynamics and damage to sorghum. Journal of Economic Entomology, 75: 695–698.

Ba-Angood, S.A. and Stewart, R.K., 1980. Economic thresholds and economic injury levels of cereal aphids on barley in Southwestern Quebec. Canadian Entomologist, 112: 759–764.

Bergstrom, G.C., Knavel, D.E. and Kuć, J., 1982. Role of insect injury and powdery mildew in the epidemiology of the gummy stem blight disease of cucurbits. Plant Disease, 66: 683–686.

Bernabe, C.M., 1972. Effects of aphid infestation (*Aphis craccivora* Koch) on the yield of Los Banos bush sitao. The Philippine Entomologist, 2: 209–212.

Bhatia, S.K. and Singh, V.S., 1977. Effect of corn leaf aphid infestation on the yield of barley varieties. Entomon, 2: 63–66.

Bishop, A.L., Walters, P.J., Holtkamp, R.H. and Dominiak, B.C., 1982. Relationships between *Acyrthosiphon kondoi* and damage in three varieties of alfalfa. Journal of Economic Entomology, 75: 118–122.

Bouchery, Y., 1977. Les pucerons *Aphis fabae* Scop. et *Acyrthosiphon pisum* (Harris) (Homoptères, Aphididae) déprédateurs de la féverole de printemps (*Vicia faba* L.) dans le Nord-Est de la France: influence sur le rendement des cultures. Méchanisme de la déprédation. Annales de Zoologie – Ecologie Animale, 9: 99–109.

Capinera, J.L., 1981. Some effects of infestation by bean aphid, *Aphis fabae* Scopoli, on carbohydrate and protein levels in sugarbeet plants, and procedures for estimating economic injury levels. Zeitschift für Angewandte Entomologie, 92: 374–384.

Carlson, P.S. (Editor), 1980. The Biology of Crop Productivity. Academic Press, New York, 471 pp.

Carter, N., Dixon, A.F.G. and Rabbinge, R., 1982. Cereal aphid populations: biology, simulation and prediction. Centre for Agricultural Publishing and Documentation, Wageningen, 97 pp.

Chiarappa, L. (Editor), 1971. Crop loss assessment methods. FAO Manual on the evaluation and prevention of losses by pests, diseases and weeds. Commonwealth Agricultural Bureaux, Farnham, (looseleaved).

Cramer, H.H., 1967. Plant protection and world crop production. Pflanzenschutz-Nachrichten Bayer, 20: 1–524.

Cuperus, G.W., Radcliffe, E.B., Barnes, D.K. and Marten, G.C., 1982. Economic injury levels and economic thresholds for pea aphid, *Acyrthosiphon pisum* (Harris), on alfalfa. Crop Protection, 1: 453–463.

Daebeler, F. and Hinz, B., 1976. Untersuchungen über Saugschäden durch *Aphis fabae* Scop. an Zuckerrüben. Archiv für Phytopathologie und Pflanzenschutz, 12: 105–110.

Daebeler, F. and Hinz, B., 1977. Zum Vergleich von Blattlaussaugschäden an Zuckerrüben durch Zusatzberegnung. Archiv für Phytopathologie und Pflanzenschutz, 13: 199–205.

Dixon, A.F.G, 1971. The role of aphids in wood formation. 1. The effect of the sycamore aphid,

Drepanosiphum platanoidis (Schr.) (Aphididae), on the growth of sycamore, *Acer pseudoplatanus* (L.). Journal of Applied Ecology, 8: 165–179.

Dixon, A.F.G., 1975. Aphids and translocation. In: M.A. Zimmerman and J.A. Milburn (Editors), Encyclopedia of Plant Physiology. Vol 1: Phloem transport. Springer, Berlin, pp 154–170.

Foott, W.H. and Timmins, P.R., 1973. Effects of infestations by the corn leaf aphid, *Rhopalosiphum maidis* (Homoptera: Aphididae), on field corn in Southwestern Ontario. Canadian Entomologist, 105: 449–458.

Forrest, J.M.S., Hussain, A. and Dixon, A.F.G., 1973. Growth and wilting of radish seedlings, *Raphanus sativus*, infested with the aphid, *Myzus persicae*. Annals of Applied Biology, 75: 267–274.

George, K.S. and Gair, R., 1979. Crop loss assessment on winter wheat attacked by the grain aphid, *Sitobion avenae* (F.), 1974–1977. Plant Pathology, 28: 143–149.

Gräpel, H., 1982. Untersuchungen zur wirtschaftlichen Schadensschwelle von Getreideblattläusen. Zeitschrift für Pflanzenkrankheiten und Pflanzenschutz, 89: 1–17.

Harper, A.M. and Freyman, S., 1979. Effect of pea aphid, *Acyrthosiphum pisum* (Homoptera: Aphididae), on cold-hardiness of alfalfa. Canadian Entomologist, 111: 635–636.

Harvey, T.L., Hackerott, H.L. and Sorensen, E.L., 1971. Pea aphid injury to resistant and susceptible alfalfa in the field. Journal of Economic Entomology, 64: 513–517.

Hinz, B. and Daebeler, F., 1981. Schadewirkung der schwarzen Bohnenblattlaus (*Aphis fabae* Scop.) an Ackerbohnen. Nachrichtenblatt für den Pflanzenschutz in der DDR, 35: 175–178.

Hussain, A., Forrest, J.M.S. and Dixon, A.F.G., 1974. Sugar, organic acid, phenolic acid and plant growth regulator content of extracts of honeydew of the aphid, *Myzus persicae* and its host plant, *Raphanus sativus*. Annals of Applied Biology, 78: 65–73.

James, W.C., 1980. Economic, social and political implications of crop loss: a holistic framework for crop loss assessment in agricultural systems. In: P.S. Teng and S.V. Krupa (Editors), Crop Loss Assessment. Miscellaneous Publication, University of Minnesota, Agricultural Experiment Station, no. 7, pp. 10–15.

Johnson, C.B. (Editor), 1981. Physiological Processes Limiting Plant Productivity. Butterworths, London, 395 pp.

Lammerink, J. and Banfield, R.A., 1980. Effect of aphid control by disulfoton on seed yield components and seed quality of oilseed rape. New Zealand Journal of Experimental Agriculture, 8: 45–48.

Lee, G., Stevens, D.J., Stokes, S. and Wratten, S.D., 1981. Duration of cereal aphid populations and the effects on wheat yield and breadmaking quality. Annals of Applied Biology, 98: 169–178.

Lloyd, D.L., Gramshaw, D., Hilder, T.B., Ludke, D.H. and Turner, J.W., 1985. Performance of North American and Australian lucernes in the Queensland subtropics. 3. Yield, plant survival and aphid populations in raingrown stands. Australian Journal of Experimental Agriculture, 25: 91–99.

Maiteki, G.A. and Lamb, R.J., 1985a. Growth stages of field peas sensitive to damage by the pea aphid, *Acyrthosiphon pisum* (Homoptera: Aphididae). Journal of Economic Entomology, 78: 1442–1448.

Maiteki, G.A. and Lamb, R.J., 1985b. Spray timing and economic threshold for pea aphid, *Acyrthosiphon pisum* (Homoptera: Aphididae), on field peas in Manitoba. Journal of Economic Entomology, 78: 1449–1454.

Mallott, P.G. and Davy, A.J., 1978. Analysis of effects of the bird cherry–oat aphid on the growth of barley: unrestricted infestation. New Phytologist, 80: 209–218.

Nuorteva, P., 1956. Studies on the effects of the salivary secretions of some Heteroptera and Homoptera on plant growth. Annales Entomologici Fennici, 22: 108–117.

Ogunlana, M.O. and Pedigo, L.P., 1974. Economic injury levels of the potato leafhopper on soybeans in Iowa. Journal of Economic Entomology, 67: 29–32.

Onstad, D.W., 1987. Calculation of economic-injury levels and economic thresholds for pest management. Journal of Economic Entomology, 80: 297–303.

Parry, W.H., 1974. Damage caused by the green spruce aphid to Norway and Sitka spruce. Annals of Applied Biology, 77: 113–120.

Pearson, C.J. (Editor), 1984. Control of Crop Productivity. Academic Press, Sydney, 315 pp.

Penning de Vries, F.W.T. and Van Laar, H.H. (Editors), 1982. Simulation of Plant Growth and Crop Production. Centre for Agricultural Publishing and Documentation, Wageningen, 308 pp.

Plant, R.E., 1986. Uncertainty and the economic threshold. Journal of Economic Entomology, 78: 1–6.

Putrich, G.S., 1971. Water permeability of the wood of grand fir (*Abies grandis* (Doug.) Lindl.) in relation to infestation by the balsam woolly aphid, *Adelges piceae* (Ratz.). Journal of Experimental Botany, 23: 936–945.

Putritch, G.S. and Talmon-de l'Armee, M., 1971. Effect of balsam woolly aphid, *Adelges piceae*, infestation on the food reserves of the grand fir, *Abies grandis*. Canadian Journal of Botany, 49: 1219–1223.

Rabbinge, R. and Mantel, W.P., 1981. Monitoring for cereal aphids in winter wheat. Netherlands Journal of Plant Pathology, 87: 25–29.

Rabbinge, R. and Vereijken, P.H., 1980. The effect of diseases or pests upon the host. Zeitschrift für Pflanzenkrankheiten und Pflanzenschutz, 87: 409–422.

Rabbinge, R., Ankersmit, G.W. and Pak, G.A., 1979. Epidemiology and simulation of population development of *Sitobion avenae* in winter wheat. Netherlands Journal of Plant Pathology, 85: 197–220.

Rabbinge, R., Drees, E.M., Van der Graaf, M., Verberne, F.C.M. and Wesselo, A., 1981. Damage effects of cereal aphids in wheat. Netherlands Journal of Plant Pathology, 87: 217–232.

Rabbinge, R., Sinke, C. and Mantel, W.P., 1983. Yield loss due to cereal aphids and powdery mildew in winter wheat. Mededelingen van de Faculteit Landbouwwetenschappen, Rijksuniversiteit Gent, 48: 1159-1168.

Rohitha, B.H. and Penman, D.R., 1983. Analysis of damage to lucerne plants (cv. Wairau) by bluegreen lucerne aphid. New Zealand Journal of Agricultural Research, 26: 147–149.

Stern, V.M., 1967. Control of aphids attacking barley and analysis of yield increases in the Imperial Valley, California. Journal of Economic Entomology, 60: 485–490.

Stern, V.M., 1973. Economic thresholds. Annual Review of Entomology, 18: 259–280.

Stone, J.D. and Pedigo, L.P., 1972. Development and economic-injury level of the green cloverworm on soybean in Iowa. Journal of Economic Entomology, 65: 197–201.

Teetes, G.L. and Johnson, J.W., 1974. Assessment of damage by the greenbug in grain sorghum hybrids of different maturities. Journal of Economic Entomology, 67: 514–516.

Thompson, S., 1977. The effect of an attack by the aphid *Schizolachnus pineti* Fabricius on the growth of young Scots pine trees. Scottish Forestry, 3: 161–164.

Van Roermund, H.J.W., Groot, J.J.R., Rossing, W.A.H. and Rabbinge, R., 1987. Calculation of aphid damage in winter wheat, using a simulation model. Mededelingen Faculteit Landbouwwetenschappen, Rijksuniversiteit Gent, 51 (3a): 1125–1130.

Watt, A.D., 1983. The influence of forecasting on cereal aphid control strategies. Crop Protection, 2: 417–429.

Watt, A.D., Vickerman, G.P. and Wratten, S.D., 1984. The effect of the grain aphid, *Sitobion avenae* (F.), on winter wheat in England: an analysis of the economics of control practice and forecasting systems. Crop Protection, 3: 209–222.

Way, M.J. and Cammell, M.E., 1980. Constraints in establishing meaningful economic thresholds in relation to decision-making on pesticide application. EPPO Bulletin, 10: 201–206.

Wellings, P.W. and Dixon, A.F.G., 1987. The role of weather and natural enemies in determining aphid outbreaks. In: P. Barbosa and J.C. Schultz (Editors), Insect Outbreaks. Academic Press, New York, pp. 313–346.

Wilde, G. and Ohiagu, C., 1976. Relation of corn leaf aphid to sorghum yield. Journal of Economic Entomology, 69: 195–197.

Wratten, S.D., 1975. The nature of the effects of the aphids *Sitobion avenae* and *Metopolophium dirhodum* on the growth of wheat. Annals of Applied Biology, 79: 27–34.

Wratten, S.D., 1978. Effects of feeding position of the aphids *Sitobion avenae* and *Metopolophium dirhodum* on wheat yield and quality. Annals of Applied Biology, 90: 11–20.

Zadoks, J.C., 1980. Yields, losses, and costs of crop protection – three views, with special reference to wheat growing in the Netherlands. In: P.S. Teng and S.V. Krupa (Editors), Crop Loss Assessment. Miscellaneous Publication, University of Minnesota, Agricultural Experiment Station, no. 7, pp. 17–22.

APPENDIX – Production loss attributable to aphids

In this appendix we attempt to estimate the effect of aphids on agricultural production. Such exercises are fraught with difficulties and these figures should be viewed with caution. However, they represent a conservative view of the general scale of losses due to aphids. For this study we have used the FAO figures for the world-wide production of thirteen categories of agricultural commodities in 1984 (Table 10.2.4). Cramer (1967), in a general study of crop losses, estimated those losses attributable to insects, diseases and weeds. In discussing crop losses due to insects he frequently attributed some losses to particular types of insects. For example, he calculated that insect pests on wheat in North and Central America cause an average loss in production of 4.9% and attributed 23.3% of this loss to *Schizaphis graminum* (Rondani). Our estimates of crop losses use this type of general information and are tabulated

TABLE 10.2.4

World production of selected agricultural commodities ($\times 10^6$ kg) in 1984 (based on Anonymus, 1985)

Commodity	Africa	North and Central America	South America	Asia	Europe	Oceania	U.S.S.R.
Wheat	9267	96 131	16 625	176 219	128 565	18 875	76 000
Barley	3665	23 876	599	15 985	79 452	6059	42 000
Maize	22 201	218 437	34 804	100 157	60 264	392	13 000
Rice	8582	8521	14 542	433 197	1959	658	2500
Other cereals	19 388	42 213	10 271	47 878	36 712	3682	29 008
Potatoes	5830	20 353	10 355	82 135	107 106	1130	85 300
Other tubers	85 818	3008	28 968	161 435	149	1656	0
Pulses	5561	3400	3466	23 732	4083	586	7079
Vegetables[a]	26 201	35 665	10 081	206 961	69 050	1945	35 187
Fruit[b]	33 939	41 487	44 365	83 939	73 653	3725	18 409
Tree nuts	296	806	127	1659	892	7	85
Sugar cane	68 511	180 363	307 534	348 827	349	30 185	0
Sugar beet	3212	21 078	2463	29 153	152 272	0	85 300

[a]Including melons.
[b]Excluding melons.

TABLE 10.2.5

Approximate estimates of world production losses ($\times 10^6$ kg per year) attributable to aphids for selected agricultural commodities; based on Anonymous (1985) and Cramer (1967); the tabulated figures are estimates of losses in the absence of specific control strategies

Commodity	Africa	North and Central America	South America	Asia	Europe	Oceania	U.S.S.R.
Wheat	61	1377	10	112	723		471
Barley	6	466		10	112	?	109
Maize	?	867	24	69	68		43
Rice				?			
Other cereals	?	266	?	?	66		72
Potatoes	?	263	108	?	845	?	658
Other tubers	?			?			
Pulses				?			
Vegetables	?	257	?	?	159	?	231
Fruit		108		?	142	?	175
Tree nuts		?		?	?		
Sugar cane		1745					
Sugar beet		34		?	1927		881

? = aphid losses thought to be likely but insufficient information is available to estimate the scale of loss.

in Table 10.2.5. Insufficient data exists for many crops grown in much of the world and thus this exercise is not comprehensive. Nevertheless, the average yearly total crop loss due to aphids still amounts to $12\,465 \times 10^6$ kg. This represents about 2.0% of all losses attributable to insect pests on these commodities, and this figure is likely to be grossly underestimated.

10.3 Viruses Transmitted by Aphids

EDWARD S. SYLVESTER

INTRODUCTION

Aphids are major vectors of plant viruses, and the number of transmission patterns that have evolved are unique in their variety compared to those found with other vector groups. While much information is available concerning the characterization of specific aphid-borne viruses, it is convenient to discuss these viruses in the familiar non-persistent, semi-persistent, and persistent modes of transmission. Although these terms refer to transmission patterns, frequently one finds that the adjectives are used in reference to the viruses themselves. The term stylet-borne (Kennedy et al., 1962) has been used for non-persistent, and it was believed to more adequately describe how certain aphid-borne viruses are carried. Similarly, the term circulative (Black, 1959) was suggested as an alternative to the term persistent, again to more precisely indicate how the virus is transmitted by the insect, viz. it is ingested, moved from the gut to the haemocoel and thence to the salivary glands for discharge. Current evidence suggests that viruses transmitted in the persistent mode fall into two categories, non-propagative and propagative (Huff, 1931). In both these sub-groups the viruses are circulative, but for many there is little evidence that they multiply in their aphid vectors. Propagative viruses replicate within tissues of the vector, and this amplification stage appears to be necessary before transmission will occur.

Stylet-borne as a possible mechanism of transmission has been repeatedly questioned by Harris (1977), who prefers an "ingestion–egestion" foregut contamination mechanism. The mechanism of transmission of the semi-persistent viruses is obscure, and circulative, like persistent, is noncommittal as a possible replicative process in the vector. An overview of the different groups of viruses is given in Fig. 10.3.1; the groups themselves are discussed below.

NON-PERSISTENT TRANSMISSION

Non-persistent transmission involves the most transient of vector relationships. The essential mechanism of transmission is thought by some to involve virus that is attached or absorbed to the tip of the stylet–labial complex, or on the inner surfaces of the distal $20\,\mu m$ of the maxillary canal (Taylor and Robertson, 1974), hence the so-called stylet-borne virus (Kennedy et al., 1962). Others argue that transmission may involve ingestion and adsorption of virus to stylet, buccal, or oesophageal surfaces with subsequent release by regurgitation (Harris and Bath, 1973). Definitive work on how these viruses are acquired, and how they are transported to and inoculated into a plant has yet to

---

Section 10.3 references, p. 83

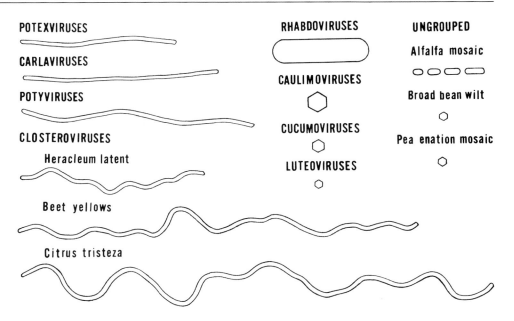

POTEXVIRUSES

CARLAVIRUSES

POTYVIRUSES

CLOSTEROVIRUSES

Heracleum latent

Beet yellows

Citrus tristeza

RHABDOVIRUSES

CAULIMOVIRUSES

CUCUMOVIRUSES

LUTEOVIRUSES

UNGROUPED

Alfalfa mosaic

Broad bean wilt

Pea enation mosaic

Fig. 10.3.1. Diagrammatic representation of the relative size and shape of the major groups of viruses transmitted by aphids. When vectored by aphids, the potexviruses, carlaviruses, potyviruses, caulimoviruses (in part), cucumoviruses, alfalfa mosaic, and broad bean wilt are transmitted in the non-persistent mode. The semi-persistent pattern occurs with the closteroviruses and caulimoviruses (in part). The persistent mode (circulative type) occurs with luteoviruses and pea enation mosaic, while the rhabdoviruses, which multiply in their vectors, have a persistent (propagative type) mode of transmission.

be done. Viruses transmitted in the non-persistent mode can be acquired and inoculated during limited probes into plant tissue (see for description of probing etc. section 4.2). Acquisition appears to be somewhat more influenced by natural termination of the probe than is inoculation (Bradley, 1952). Under conditions of experimental manipulation the complete act of transmission can be accomplished in less than a minute (Sylvester, 1949). Preliminary fasting of candidate vectors has been shown to improve acquisition efficiency (Watson, 1938), but a completely satisfactory explanation of how fasting serves to improve acquisition is lacking. Part of the explanation, accepted by most authors, is that fasting increases the propensity of the insects to probe (Bradley, 1952). Maximum levels of inoculativity can be achieved in a single probe (Sylvester, 1949), but once the virus is acquired, infectivity is rapidly lost in most cases (Hoggan, 1933). The rate of loss of infectivity is greater if the aphids are fed, than if fasted (Watson and Roberts, 1940), and although the rate of loss tends to be exponential in fasted insects, it is positively correlated with temperature (Kassanis, 1941). In general, one finds that many species of aphids will transmit each of these viruses, although tests show that the efficiency of transmission varies among species (Drake et al., 1933); when single acquisition probes are used, it can range from an experimental zero to rates as high as 60–80% (Simons and Sylvester, 1953). Some viruses need a helper factor (Clinch et al, 1936; Kassanis and Govier, 1971) if they are to be acquired. The helper factor of cauliflower mosaic virus, apparently a labile protein with a molecular weight of 18×10^3 daltons, presumably aids in attachment of virus to the surfaces from which it is released. But the problem remains, if this is true, where and how is the helper factor carried?

The cauliflower mosaic virus codes for the development of helper factor during the infection cycle (Armour et al., 1983). The helper factors of potyviruses, which have a molecular weight of $5–6 \times 10^4$ daltons, are not absolutely

Potyviruses (Hollings and Brunt, 1981)

Shape	flexuous rod; 11×680–$900\,nm$
Symmetry	helical, pitch $3.4\,nm$
Genome	ssRNA
Percent	5
Molecular weight	3.0–3.5×10^6 daltons
Capsid proteins	1
Subunit mol. wt.	32–34×10^3 daltons
Inclusions	cytoplasmic; banded bodies

specific, i.e. infection with related, but not identical, viruses can provide a helper factor for other viruses (Pirone, 1977).

Viruses typically transmitted in the non-persistent mode are found in several groups, viz., the flexuous rod-shaped potyviruses, potexviruses, carlaviruses, the isometric cucumoviruses and caulimoviruses, as well as the morphologically unique alfalfa mosaic virus.

Potyviruses

There are 85 viruses, infecting monocotyledons and dicotyledons, herbaceous and woody hosts, recognized as belonging, or possibly belonging, to the potato virus Y (potyvirus) group. Those which are aphid-borne form the largest and most economically important group of viruses transmitted by these insects.

Replication occurs in the cytoplasm, and the viruses typically occur in moderately high concentrations, are juice inoculable, fairly stable, good antigens, and serologically interrelated. Some are seed-borne. Vector acquisition may require the presence of a helper factor (Pirone, 1977).

Symptoms include vein clearing, mosaic, mottling, puckering, rugosity, at times with distortion, stunting, and necrosis. Colour breaking in flowers can occur. Some of the viruses have extensive host ranges; most, however, are quite restricted to a few plant species. Many of the viruses exist in a series of strains.

Most potyviruses have aphids as natural vectors, although current classification puts the potyviruses into four sub-groups, depending upon whether they can be transmitted by aphids, mites, whiteflies or fungi. There also is some experimental evidence reported for transmission by leafminer flies (Zitter and Tsai, 1977) and nematodes (Bond and Pirone, 1970).

The lack of specificity, frequently claimed in the aphid transmission of the potyviruses, is not entirely true. Many reports demonstrate a wide range in the transmission efficiency among tested aphid species, clones, or colonies, as well as a failure to transmit even strains of the same virus (Simons, 1969, Karl and Schmelzer, 1971).

Potexviruses

There are at least 12 viruses, with another 21 possible members in this group. The viruses infect both monocotyledons and dicotyledons, herbaceous and woody hosts, are juice inoculable, and most have no natural vectors. Symptoms of infection include mosaic, mottle, vein clearing, vein banding, stunting, necrosis, and ringspots. In many cases symptoms can be mild or inapparent. Most members have narrow host ranges, but the type member, potato virus X, infects 240 species in 16 families.

The aphid transmission of white clover mosaic (Pratt, 1961), may be in doubt (Goth, 1962), but centrosema mosaic is reported to be vectored by four aphid species (Van Velsen and Crowley, 1962) and parsley virus 5 was found to be

Section 10.3 references, p. 83

Potexviruses (Purcifull and Edwardson, 1981)

Shape	slightly flexuous rod; 11–13 × 480–580 nm
Symmetry	helical, pitch 3.2–3.7 nm
Genome	ssRNA
Percent	5–7
Molecular weight	2.1–2.6×10^6 daltons
Capsid proteins	1
Subunit mol. wt.	18–26×10^3 daltons
Inclusions	cytoplasmic; banded bodies

inefficiently transmitted by *Cavariella pastinacae* (Linnaeus) (Frowd and Tomlinson, 1972). Potato aucuba mosaic apparently can be aphid transmitted, but only in the presence of the helper factor associated with potato virus Y (Kassanis and Govier, 1971).

Carlaviruses

Most of the more than 20 viruses, with the exception of those affecting legumes and the poplar mosaic virus, belonging to the carnation latent virus group are not serious in economic terms. In fact, carnation latent virus and potato virus S were discovered by chance during electron microscopy and serology studies.

The name would imply that latency is characteristic. However, several of the viruses, e.g. cowpea mild mottle, hop mosaic, pea streak, poplar mosaic, potato M, and clover vein mosaic, cause distinct symptoms in the host from which they were named. The viruses are juice inoculable and seed transmission, while rare, does occur in some legumes. So far aphids are the only known natural vectors, and the transmission pattern appears to be non-persistent. Transmission efficiency varies considerably, both among viruses and within strains of the same virus (Hagedorn and Hanson, 1951; Kassanis, 1955; Hollings, 1957; Wetter and Völk, 1960; Kassanis, 1961; Schmidt et al., 1966; Bode and Weidemann, 1971; MacKinnon, 1974; Brunt, 1977; Weber and Hampton, 1980).

Carlaviruses (Wetter and Milne, 1981)

Shape	slightly flexuous rod; 12 × 200–690 nm; often curved to one side
Symmetry	helical, pitch 3–3.45 nm
Genome	ssRNA
Percent	5–6
Molecular weight	2.3–2.59×10^6 daltons
Capsid proteins	1
Subunit mol. wt.	31–36.7×10^3 daltons
Inclusions	cytoplasmic; banded bodies; not highly diagnostic

Broad bean wilt (Taylor and Stubbs, 1972)

Shape	isometric; 27 nm in diameter
Components	3; two contain RNA
Genome	ssRNA; divided
Percent	22 and 33
Molecular weight	1.5 and 2.0×10^6 daltons
Capsid proteins	2
Subunit mol. wt.	26 and 42×10^3 daltons, respectively
Inclusions	cytoplasmic; amorphous and crystalline, depending on strain; not diagnostic

Multicomponent RNA viruses

Broad bean wilt

This virus is mechanically inoculable and has a wide host range. Although structurally and chemically similar to the nepo (nematode-transmitted polyhedral) viruses and to the comoviruses (beetle-transmitted cowpea mosaic virus group), it is not serologically related to either group, and it is unique in that it is aphid-transmitted (Taylor and Stubbs, 1972).

Cucumoviruses

Cucumber mosaic, tomato aspermy, and peanut stunt are three serologically distinct entities in this group of juice inoculable, aphid-transmissible viruses. RNA preparations from infected plants often contain a satellite RNA as a fifth species.

Cucumoviruses affect monocotyledons and dicotyledons, including both herbaceous and woody plants. Cucumber mosaic is ubiquitous, and is a virus with several strains and a variety of symptom types. Tomato aspermy is the most important disease of chrysanthemums in Europe and vegetative propagation assures its presence wherever chrysanthemums are grown. Peanut stunt, a disease of legumes and tobacco, is found in the U.S., Europe, Morocco, and Japan. Strains of some of the viruses are distantly serologically related to the other viruses; e.g. *Robinia* mosaic, although serologically related to cucumber mosaic and to peanut stunt viruses, is considered to be a strain of the latter. Purified virus can be acquired through membrane feeding, and a helper factor is not needed (Pirone, 1977). Seed transmission of cucumber mosaic has been recorded in at least 18 species of plants, and both seed transmission and vegetative propagation are frequently important in primary spread. The principal natural vectors are believed to be *Myzus persicae* (Sulzer) and *Aphis gossypii* Glover, although 60–75 aphid species have been implicated, at least experimentally. *Aphis craccivora* Koch is the major vector of peanut stunt.

Alfalfa mosaic

Alfalfa mosaic is world-wide in its distribution, and various strains have been found to infect more than 300 plant species in 47 plant families. In many hosts it is symptomless. In some hosts the virus can be seed-borne. Alfalfa,

Cucumoviruses (Kaper and Waterworth, 1981)

Shape	isometric; 30 nm in diameter
Genome	divided; 4 ssRNA species
Percent	18
Molecular weight	1.1, 1.0, 0.7, and 0.3×10^6 daltons; first 3 needed for replication
Capsid proteins	1
Subunit mol. wt.	24.5×10^3 daltons
Inclusions	cytoplasmic; crystalline; rare; not diagnostic

Alfalfa mosaic (Van Regenmortel and Pinck, 1981)

Shape	4 particle types (EM); all 18 nm wide, and 29, 38, 49, and 58 nm long
Genome	divided; 4 ssRNA species, three largest, genomic; fourth sub-genomic coat protein messenger
Percent	16
Molecular weight	1.0, 0.7, 0.6, 0.3×10^6 daltons
Capside proteins	1
Subunit mol. wt.	24.28×10^3 daltons
Inclusions	amorphous cytoplasmic, granular and transient in tobacco

Section 10.3 references, p. 83

TABLE 10.3.1

Summarization of the transmission characteristics of aphid-borne viruses

Transmission mode	Threshold periods			Latent period	Retention period	Transstadial passage	Transovarial passage	Specificity	Important vectors[a]	Examples
	Acquisition[b]	Inoculation	Transmission							
Non-persistent	≤15 sec	≤15 sec	≤1 min	none	minutes to hours	none	none	low	*Myzus persicae* (Sulzer), *Macrosiphum euphorbiae* (Thomas), *Aphis gossypii* Glover, *Aphis fabae* Scopoli, *Aulacorthum solani* (Kaltenbach) *Acyrthosiphon pisum* (Harris), *Neomyzus circumflexus* (Buckton), *Brevicoryne brassicae* (Linnaeus), *Aphis craccivora* Koch, *Aphis nasturtii* (Kaltenbach), *Myzus ornatus* Laing, *Hyadaphis foeniculi* (Passerini)	Many potyviruses, e.g. bean yellow mosaic, lettuce mosaic, potato A, Y, tobacco etch, tulip breaking, turnip mosaic, watermelon mosaic 1 and 2; carlaviruses[c], e.g. hop mosaic, potato M and S; cucumoviruses, e.g. cucumber mosaic; alfalfa mosaic
Semi-persistent	≤30 min	≤15 min	≤1 h	none	hours to a few days	none	none	moderate to high	*Myzus persicae*, *Aulacorthum solani*, *Aphis fabae*, *Brevicoryne brassicae*, *Chaetosiphon fragaefolii* (Cockerell), *Toxoptera aurantii* (Boyer de Fonscolombe)	Caulimoviruses[d], e.g. cauliflower mosaic, strawberry vein banding; closteroviruses, e.g. beet yellows, *Citrus* tristeza
Persistent Circulative	≤30 min	≤15 min	≤24 h	a few hours	days	present	none (?)[e]	moderate to high	*Myzus persicae*, *Macrosiphum euphorbiae*, *Aulacorthum solani*, *Acyrthosiphon pisum*, *Metopolophium dirhodum* (Walker), *Aphis gossypii*, *Rhopalosiphum maidis* (Fitch), *Rhopalosiphum padi* (Linnaeus), *Aphis craccivora*, *Aphis fabae*, *Sitobion avenae* (Fabricius), *Cavariella aegopodii* (Scopoli), *Chaetosiphon fragaefolii*, *Schizaphis graminum* (Rondani)	Luteoviruses, e.g. barley yellow dwarf, bean leaf roll, beet western yellows, carrot red leaf, groundnut rosette assistor virus, potato leafroll, soybean stunt; strawberry yellow edge; pea enation mosiac
Propagative	≤30 min	≥15 min	days	days	days	present	present to a low degree	high	*Aphis idaei* (Van der Goot), *Chaetosiphon fragaefolii*, *Hyperomyzus lactucae* (Linnaeus)	raspberry vein chlorosis, strawberry crinkle, lettuce necrotic yellows

[a] Listed approximately in descending order of importance, but all in the non-persistent mode transmit ten or more viruses.

[b] In the non-persistent mode acquisition and inoculation follow stylet probing, in the other modes acquisition is presumed to follow phloem ingestion, while inoculation occurs with salivation during a feeding penetration.

[c] Carlaviruses tend to exhibit a high degree of vector specificity.

[d] The mode of transmission of the caulimoviruses (and of one potyvirus, viz. pea seed-borne mosaic) apparently varies from non-persistent to semi-persistent, depending upon the specific vector.

[e] Potato leafroll, a luteovirus with a circulative mode of transmission, has been reported to be transovarially passed to a limited extent.

clover, pea, potato, tobacco, pepper, tomatoes, and celery all can be economically affected by infection. Symptoms vary with strain, host and environmental
conditions, and they can be latent, transitory, green or yellow mosaic, or with
local or systemic necrosis (Jaspars and Bos, 1980).

Experimentally at least 13 out of 20 species of aphids tested have transmitted
one or more strains (Kennedy et al., 1962), and the transmission pattern is
non-persistent (Swenson, 1952). Aphids can acquire purified virus by membrane
feeding, and as with cucumber mosaic virus, no helper factor is needed (Pirone,
1977).

SEMI-PERSISTENT TRANSMISSION

A small group of viruses are transmitted by aphids in the semi-persistent
mode. In this mode preliminary fasting does not noticeably increase the
efficiency of acquisition. Unlike the non-persistent pattern, acquisition appears
to be a function of feeding, rather than probing. Thus there is a positive
correlation between the length of the acquisition access period and infectivity
(Table 10.3.1). Given that feeding may be required for efficient acquisition, a
greater degree of vector specificity occurs in those cases where transmission is
restricted to the semi-persistent mode. However, there is no latent period and
so the maximum inoculativity occurs at the termination of an acquisition
feeding. Usually inoculation threshold periods are shorter than the corresponding acquisition threshold periods. Inoculativity declines over a period of
hours, and the rate of decline is temperature sensitive, but unlike the non-
persistent mode, retention is not strongly influenced by whether the vector is
feeding or fasting. The evidence to date indicates that infective larvae lose
their inoculativity by moulting. The mechanism of transmission is not known,
but the simplest hypothesis suggests that adsorption and release occur over a
much longer time span in the semi-persistent mode than in the non-persistent
one. Virus-like particles were found associated with the foregut of *Cavariella
aegopodii* (Scopoli), the aphid vector of the *Anthriscus* yellows/parsnip yellow
fleck complex (Murant et al., 1976).

Bimodal transmission

There is evidence that some viruses can be transmitted in either the non-
persistent or the semi-persistent mode, depending upon the species, or the
biotype of the vector species used. The phenomenon has been termed bimodal
transmission (Lim and Hagedorn, 1977). The most extensive published data
involve isometric dsDNA cauliflower mosaic virus, and here the semi-
persistent mode dominates the pattern of transmission by *Brevicoryne brassicae*
(Linnaeus). Transmission by *M. persicae* (Van Hoof 1954; Hamlyn, 1955;
Chalfant and Chapman, 1962) or *Lipaphis erysimi* (Kaltenbach) (Namba and
Sylvester, 1981) tends to be non-persistent. However, the reports indicated that
even with these species, the transmission pattern is variable.

In the other instance, the ssRNA, flexuous rod potyvirus (pea seed-borne
mosaic), typically transmitted in a bimodal (non-persistent/semi-persistent)
pattern by the New London biotype of *Macrosiphum euphorbiae* (Thomas) (Lim
and Hagedorn, 1977).

The mechanism of bimodal transmission has yet to be determined. The rapid
immediate transmission would appear to involve acquisition and release
during probing, as in the non-persistent mode. But during the early stages of
penetration and feeding, in the case of some highly specific vector–virus combinations, virus may accumulate (by being adsorbed to cuticular surfaces

Section 10.3 references, p. 83

Closteroviruses (Lister and Bar-Joseph, 1981)

Shape	flexuous rod; 10–18 nm wide and three classes of length, 600–740, 1250–1450, and 1650–2000 nm long, respectively
Symmetry	helical, pitch 3.7 nm
Genome	ssRNA
Percent	5–6
Molecular weight	2–4.5×10^6 daltons
Capsid proteins	1
Subunit mol. wt.	23–27×10^3 daltons
Inclusions	none

through specific charge relationships?) on the interior surfaces of the mandibles, the stylet food canal, or perhaps the foregut; surfaces from which it is slowly released when additional probes or feeding penetrations are made. In the bimodal transmission of pea seed-borne mosaic virus the immunolatex labelling of the virus occurred on the mandibular stylet tips of the London biotype of the aphid vectors (Lim et al., 1977). The data of Lim et al. also suggested that prolonged acquisition feeding, 5 days or more, depressed the probability of obtaining infective insects. One explanation could be that phloem contains little or no virus and prolonged feeding gradually flushes away any virus that may be absorbed.

Closteroviruses

This is an ill-defined group of viruses responsible for disease with yellowing and necrotic spotting as the predominant symptoms. The viruses are difficult to transmit by juice inoculation, and the only vectors known are aphids. The viruses are divided into three groups. The one with particles that have a modal length ranging from 600 to 740 nm perhaps includes three viruses, only one of which, viz. *Heracleum* latent virus, is aphid-transmissible in a semi-persistent manner. The second group, whose particles range in length from 1250 to 1450 nm, has three aphid-borne viruses, including beet yellows. The third group consists of perhaps eight viruses, with particles of 1650 to 2000 nm in length. Six are considered to be transmitted by aphids, one of which is *Citrus* tristeza, a virus of major concern.

Beet yellows, a disease thought to be European in origin, occurs in most of the major sugar beet areas of the world, with considerable economic consequences. The vector–virus relationships, especially with *M. persicae*, are known in detail (Watson, 1940, 1946; Bennett and Costa, 1954). The lack of prolonged retention, i.e. the estimated half-life for retention of inoculativity at moderate temperatures (ca. 22°C) is about 8 h, was the basic reason for establishing the semi-persistent pattern of aphid transmission as a distinct category (Sylvester, 1956). The main field vectors of the beet yellows virus are considered to be *M. persicae* and *Aphis fabae* Scopoli, although experimentally at least 20 additional species or subspecies have been implicated (Kennedy et al., 1962).

Citrus tristeza virus and its major vector, *Toxoptera citricidus* (Kirkaldy) may have originated in southern Asia (Dickson et al., 1951; Dickson and Flock, 1959). Additional important vectors are *Aphis gossypii* and *Aphis spiraecola* (Patch), and *Toxoptera aurantii* (Boyer de Fonscolombe) (Norman and Grant, 1956). Tristeza now is a worldwide major disease of *Citrus*. Introduction of the disease into other regions began by the distribution of infective propagative material. Spread was facilitated when such introductions were made into areas with efficient vectors, or when an efficient vector was also introduced (McClean, 1957). It also has been proposed that active spread can occur

Cauliflower mosaic virus (Shepherd, 1981; Shepherd and Lawson, 1981)

Shape	spherical; ca. 50 nm in diameter
Genome	dsDNA
Percent	ca. 16
Molecular weight	ca. 5×10^6 daltons
Capsid proteins	1
Subuinit mol. wt.	$44{-}58 \times 10^3$ daltons
Inclusions	cytoplasmic; compact, spheroid, containing two proteins in addition to virus

through the appearance of mutant strains that are efficiently transmitted by local vectors (Bar-Joseph et al., 1978).

Caulimoviruses

This is a relatively small group of viruses with restricted host ranges and, usually, a semi-persistent mode of transmission by aphid vectors. Disease symptoms are mosaic and mottles. At times vein-banding is prominent. The DNA replicates in nuclei, but the virions may be assembled in characteristic cytoplasmic inclusion bodies. There are at least six, possibly nine, members. The vector–virus relationships have been studied in most detail in the case of cauliflower mosaic (see "Bimodal transmission" p. 71), dahlia mosaic, and the strawberry vein-banding virus, but the data are variable. Generally it has been found that acquisition is not particularly influenced by preliminary fasting. Acquisition can occur during probing or feeding, i.e. in contrast to what happens with non-persistent viruses, prolonging the acquisition period does not decrease acquisition efficiency. Inoculation also can occur by probing or feeding. The retention of inoculativity again is variable, negatively correlated with temperature, generally prolonged if the insects are fasted rather than fed, and at times a function of the vector species. Examples of infectivity retention periods: Brierley and Smith (1950) give a maximum of 3 h for dahlia mosaic virus; cauliflower mosaic virus is reported to be retained for three, but not four, days (Day and Venables, 1961); and the half-life of strawberry vein-banding virus in *Chaetosiphon fragaefolii* (Cockerell) and *C. jacobi* Hille Ris Lambers has been estimated to be approximately 10 h, independent of the vector species (Frazier and Sylvester, 1960).

Unclassified viruses

There is a series of other viruses, reported to be aphid-transmitted in an apparently semi-persistent mode, that have not been characterized except in terms of plant hosts, aphid vectors, and symptomatology. Several of these infect small fruits (strawberries and various bushberries), while others remain curiosities. Mild clover mosaic resembles broad bean wilt virus, but is reported to be semi-persistently transmitted by aphids (Gerhardson, 1977). One virus, viz. *Anthriscus* (chervil) yellows is of special interest because of its helper role in the transmission of parsnip yellow fleck. *Anthriscus* yellows was reported originally to have a persistent relationship with its vector *C. aegopodii* (Murant and Goold, 1968), but further studies described it as an isometric virus, ca. 22 nm in diameter, whose vector relationships were semi-persistent (Elnagar and Murant, 1976; Murant and Roberts, 1977). As previously mentioned, virus-like particles have been seen in the foregut of an aphid vector in the *Anthriscus* yellows/parsnip yellow fleck complex (Murant et al., 1976), and acquisition of the parsnip yellow fleck virus apparently is dependent upon the vectors having simultaneous or previous access to *Anthriscus* yellows virus.

PERSISTENT TRANSMISSION

In the persistent mode of transmission, both the acquisition and inoculation efficiency rises with increasing access or feeding time. Fasting does not increase acquisition efficiency, nor does it, in contrast to feeding, affect the period of time over which infectivity will be retained. A latent period of 12 h or more is usually present. The period over which infectivity can be retained may cover the whole life span of the aphids, but the rate of transmission declines with their age. Some data indicate that beginning with first-stage larvae, acquisition efficiency decreases and the length of the latent period increases as a function of age. Infectivity is not affected by moulting, and transovarial passage is extremely rare or non-existent. Again, since feeding is required for acquisition, and usually for inoculation (pea enation mosaic is an exception), the number of species that are vectors is quite limited. There is little evidence that the viruses replicate in their aphid vectors. A possible exception is the unconfirmed experimental evidence for serial passage of potato leafroll virus in *M. persicae* (Stegwee and Ponsen, 1958).

The luteoviruses form a large group of aphid-borne viruses that are transmitted in the persistent mode. These viruses are serologically related and infection causes yellowing symptoms in herbaceous plants. Other aphid-transmitted viruses that are persistently borne appear to be sufficiently distinct from the luteoviruses so as form other groups of viruses and have to be treated separately.

Luteoviruses

The luteoviruses are only just beginning to be studied on a molecular level. Many of those listed by Rochow and Duffus (1981) as being members, or probable members, of the luteovirus group have not been characterized physically. However, most of the diseases induced in plants are symptomatologically similar, transmitted with the persistent pattern, and to a greater or lesser degree serologically related. At least four of the luteoviruses, i.e. barley yellow dwarf (and isolates), beet western yellows, soybean yellow dwarf, and potato leafroll, have been isolated and partially characterized. The viruses are quite stable and good antigens.

Three major complexes of the luteoviruses can be distinguished. They are those primarily infecting graminaceous hosts (the barley yellow dwarf complex), those that infect chenopodiaceous, solanaceous, and cruciferous hosts (the beet western yellows complex), and those that mainly infect legumes (the soybean yellow dwarf group).

Symptoms are not specific, but typically include stunting, yellowing, as well as reddening, rolling, and increased brittleness of the foliage. Older leaves usually are the ones showing symptoms. High light intensity and lower temperatures are needed to maximize symptom displays.

The host ranges are narrow to wide, depending on the virus. For example,

Luteoviruses (Rochow and Duffus, 1981)

Shape	isometric; 20–30 nm in diameter, depending on the preparation
Genome	ssRNA
Percent	28
Molecular weight	$1.9–2.0 \times 10^6$ daltons
Capsid protein	1
Subunit mol. wt.	$22–24.4 \times 10^3$ daltons
Inclusions	none

barley yellow dwarf is confined to the Gramineae, yet within that family some 110 species of annual and perennial grasses, including the agronomically important cereals, viz. barley, oat, wheat, rye, corn, and rice (giallume disease), have been infected with various isolates of the virus. In contrast to being limited to a single plant family, beet western yellows virus has been shown to infect 150 species in 23 dicotyledonous families, and also oat (*Avena byzantina*) has been infected (Duffus and Rochow, 1978). But even within this group, great differences are often found in host ranges and virulence, and variants or isolates can be host specific, as with the beet and turnip isolates of the beet western yellows virus in England and on the European continent. *M. persicae* is the major vector of beet western yellows and related viruses affecting chenopodiaceous, cruciferous and solanaceous hosts.

Barley yellow dwarf is worldwide in distribution, and isolates vary in their degree of vector specificity and serological relatedness. Research has emphasized five isolates that are placed in two serological groups. One group contains two serologically related isolates (RPV and RMV) that tend to be specifically transmitted by *Rhopalosiphum padi* (Linnaeus) and *R. maidis* (Fitch), respectively. The other group of three, serologically interrelated, isolates (SAV, SGV, and PAV) tend to be specifically transmitted by *Sitobion avenae* (Fabricius), *Schizaphis graminum* (Rondani), and relatively non-specifically by both *R. padi* and *S. avenae*, respectively. In mixed infections in plants the protein capsids of either the RPV or the RMV isolate can encapsulate the genome of the SAV isolate. When this occurs, the two *Rhopalosiphum* species can serve as temporary vectors of the genome of vector-specific SAV (Rochow, 1977). In the field the abundance and mixture of isolates varies both seasonally and regionally.

Potato leafroll virus continues to be of major concern to the potato industry. The virions are isometric, ca. 25 nm in diameter, and contain about 28% nucleic acid. Although Sarkar (1976) presented evidence that the nucleic acid was DNA, perhaps double stranded, it is now widely accepted that the nucleic acid is of the RNA type. It has some serological relationship with other luteoviruses (Kubo and Takanami, 1979; Rowhani and Stace-Smith, 1979). Serial passage has been reported with at least one isolate of potato leafroll in *M. persicae* (Stegwee and Ponsen, 1958), and transovarial passage has also been claimed (Miyamoto and Miyamoto, 1966, 1971). However, other research has not shown the virus to multiply in this vector species (Harrison, 1958; Eskandari et al., 1979).

Unclassified viruses

Pea enation mosaic

This virus commonly infects and overwinters in a variety of legumes in the northern temperate zone. Pea enation mosaic virus can be juice inoculated, which is unusual for viruses with the persistent relationship.

The minor protein associated with aphid-transmissible strains is assumed to allow passage of the virus into the salivary glands (Hull, 1977). The coat protein is associated with RNA 1. It has been demonstrated that by mechanical inoculation isolates can cease to be aphid transmissible. This is accompanied by the loss of the minor protein and a reduction in the size of the particles in the middle and bottom components. Mixtures of aphid-transmissible and non-aphid-transmissible isolates permit insect transmission of the non-transmissible isolate. There is a considerable degree of vector specificity. Five aphid species have been reported to transmit the virus, but *Acyrthosiphon pisum* (Harris) probably is the major field vector. Economic losses in peas and broadbeans occur, but severity of the damage depends upon cultivar susceptibility and the age of the plants when infection occurs.

Section 10.3 references, p. 83

Pea enation mosaic (Hull, 1981)

Shape	isometric; 24–27.5 and 29–34 nm in diameter, proportion varying with strain (larger — aphid transmissible)
Genome	divided; two or three ssRNA species
Percent	27–28, in each particle
Molecular weight	1.4–2.3, 1.1–1.68, and 0.11–0.36 $\times$ 10^6 daltons, for RNA 1, 2, and 3 (may be a cleavage product of RNA 1), respectively; RNA 1 and 2 needed for replication
Capsid proteins	2
Subunit mol. wt.	major, 22 $\times$ 10^3; minor (aphid transmission factor), 56 $\times$ 10^3 daltons, respectively.
Inclusions	none

PROPAGATIVE TRANSMISSION

The transmission pattern found in the propagative mode of transmission is similar to that found in the circulative mode. Acquisition and inoculation are both enhanced by extending the feeding time, there is a latent period, transstadial passage, and a relatively prolonged period during which infectivity is retained.

The latent period is prolonged, a matter of days, and there is no evidence that the latent period is shorter in the immature stages. The viruses replicate in the vector's body, and a limited degree of transovarial passage of the viruses occurs in those instances where this phenomenon has been examined. When inoculated with virus by injection, infection of *Hyperomyzus lactucae* (Linnaeus) with the sowthistle yellow vein rhabdovirus decreases the longevity of the vector. A similar phenomenon occurs when *Chaetosiphon jacobi* or *C. fragaefolii* are injected with strawberry crinkle virus. With the possible exception of certain isolates of potato leafroll virus, all of the currently known propagative aphid-borne viruses are rhabdoviruses.

Plant rhabdoviruses

This is a relatively large group of bacilliform viruses. The proteins vary somewhat with the virus, but generally include a glycoprotein (G), one or two matrix proteins (M or M_1 and M_2) associated with the envelope, and a protein (N) with which the nucleic acid forms the nucleocapsid. Some of the viruses have an additional protein (L), that possibly is associated with the nucleocapsid, as well as a nonstructural protein (Ns). The RNA itself is not infectious, and at least two of the plant rhabdoviruses have an RNA-dependent RNA

Plant rhabdoviruses (Francki et al., 1981, Peters, 1981)

Shape	bacilliform; 45–95 $\times$ 180–400 nm
Nucleocapsid	tubular, helical, pitch 4.0–4.5 nm; enclosed in a peplomer-studded, lipoprotein membrane derived from the host cell
Genome	ssRNA, negative polarity
Percent	1
Molecular weight	4 $\times$ 10^6 daltons
Proteins	at least 5; N (nucleocapsid) and G (glycoprotein, outside the membrane) proteins are found in all plant rhabdoviruses tested; other proteins can include L (large), M (matrix) and Ns (non-structural), or M_1 and M_2 (matrix)
Subunit mol. wt.	55–64, 71–91, 145, 22–25, 40, 27–44, and 22–39 $\times$ 10^3 daltons, respectively
Inclusions	none

transcriptase, needed to transcribe copies of the viral RNA. The viruses mature by budding off host cell membranes, and two groups have been proposed, depending on whether budding occurs off the inner nuclear membrane or from the endoplasmic reticulum (Peters, 1981).

Because of their unique size and shape, many plant rhabdoviruses have been described simply because particles have been found in negatively stained leaf-dip preparations or in thin sections. Francki et al. (1981) indicate that only 25 of the 49 rhabdoviruses described from plants have been experimentally transmitted. These authors list nine of the viruses as being aphid-borne (other vectors include leafhoppers, planthoppers, a true bug in the family Piesmidae, and perhaps a mite). With the nine aphid-transmitted rhabdoviruses, evidence of their propagative nature is substantial only in the case of four, viz. lettuce necrotic yellows, sowthistle yellow vein, strawberry crinkle, and coriander feathery red vein (Misari and Sylvester, 1983).

Only two of the aphid-borne rhabdoviruses are reported to have economic significance. Lettuce necrotic yellows of South Australia has caused up to 50% of the lettuce crop being lost in some fields, with the result that some growers have abandoned lettuce production. The other virus, strawberry crinkle, is one of several viruses affecting commercial strawberries whose presence has necessitated the development of intensive certification programs to assure a supply of virus-free plant material, needed to maintain commercial production of strawberries.

EPIDEMIOLOGY AND MANAGEMENT OF APHID-BORNE VIRUSES

It would be difficult to manage a recurring virus disease that involves an aphid transmission cycle without some understanding of the nature of the endemic and epidemic patterns of maintenance and spread. Epidemiological research normally treats both these forms of the disease. Epidemics, while perhaps devastating in their effects, usually are short-lasting. Some follow the introduction of a favourable and susceptible host into an existing endemic situation. Others occur when environmental changes (frequently meteorological or agronomic) permit the endemic transmission cycle to exceed economically tolerable limits.

Basic to understanding the transmission cycle is knowledge of the hosts, both target and natural, the vectors, and the pathogen. Once these have been assessed, then begins the long process of establishing the various interactions and relationships among these fundamental components, as well as their distribution in space and time. Knowledge is needed of the "when and how" of acquisition and inoculation, and "when, how, and where" the cycle is maintained over time. In final analysis, although epidemics cease because the number of suscepts are reduced or the transmission cycle is constrained by the physical environment, they are prevented by keeping the frequency of transmission below some critical level (Swenson, 1968; Carter, 1973).

The spread of plant viruses in the field varies with the type of vector–virus relationship (Zitter, 1977). Since infectivity is rapidly lost, viruses transmitted in the non-persistent mode are extremely sensitive to the distance between successive host plants. In some cases the virus is seed-borne, or carried by vegetatively reproduced planting stock, and the virus sources are introduced at random in the field at planting time. In these cases, spread from these sources, chiefly by winged aphids, can result in a gradual increase in the number of diseased plants around the earlier diseased plants. At times the crop plant is not colonized by the vectors, such as in lettuce or soybean mosaic (Irwin and Goodman, 1981).

Section 10.3 references, p. 83

Frequently the endemic cycle is maintained in weed hosts and the virus must be carried from the weeds into the crop of concern. In such situations, there may be a pronounced, rather steeply decreasing exponential, gradient of infection from the edge of the field downwind from the weed source of virus (Gregory and Read, 1949).

When viruses are transmitted in the semi-persistent or persistent mode, in which infectivity is retained for longer periods of time than in the non-persistent mode, colonizing infective alates can, in addition to an edge effect, inoculate plants further into the field. As colonization advances and more and more plants become infested, isolated patches of disease may appear. These may gradually increase in size and coalesce. If the crop plant serves as its own source of virus, such as in potatoes and small fruit crops, the sources again are randomly distributed through the field and colonizing aphid vectors can soon cause the disease to radiate out from these loci.

Much of our knowledge on the pattern of virus spread by aphids has come from trapping alates. In some cases trapping has been used to monitor aphid flight activity as an aid to disease management and the prediction of disease risk. Trap catches (Taylor and Palmer, 1972), whether from sticky cylinders, vertical yellow boards, shallow yellow pans, or suction devices, more often are better correlated with the spread of persistent and semi-persistent viruses than the non-persistent ones. But even here, predictions of risk, based on averaging aphid numbers from local trap catches, have more reliability when applied to the immediate general area than to individual fields.

The most abundantly caught species may not be the most important vector, nor is a vector species that is important in one area necessarily the most important in all areas. If the predominant barley yellow dwarf isolates in an area are those with a high vector specificity for *Rhopalosiphum* spp., then an outbreak of *S. avenae*, the specific vector of the SAV isolates, will not be followed by a virus epidemic (Rochow, 1967). *M. persicae* is considered to be the major vector of potato viruses in most areas. However, Gabriel (1958) reported *Aphis nasturtii* (Kaltenbach) and *A. frangulae* (Kaltenbach) to be the major vectors to potato virus Y in Poland, and Van Hoof (1977) suggested that other species may contribute considerably to the early season spread of potato virus Y in the Netherlands.

Unexceptionally, aphids are responsive to climatic conditions, and much research has been done in sampling and monitoring vector populations, with the purposes of protecting certain crops from excessive damage due to virus spread. In temperate zones, low winter temperatures severely limit the over-wintering of viviparae of such species as *M. persicae*, and in the absence of snow cover there can be substantial mortality of diapausing eggs. The host plants of aphids have different optimum temperatures for growth, and a greater temperature range over which they can survive, than do the aphids. This frequently means that plants can be grown in low-risk areas for virus spread, such as areas of high rainfall, altitude and winds, or areas that are hot and dry. In temperate zones, aphid flight, favoured by still air near the ground can occur when temperatures rise above 13°C. Flight activity increases up to a peak at ca. 20°C, following which it then decreases until ca. 30°C, when activity ceases (Broadbent, 1953). It has been reported (Van der Plank, 1944) that the spread of aphid-borne potato viruses in central South Africa ceases when the daily mean maximum temperature reaches 32°C.

Most management programs for aphid-borne virus diseases in agricultural crops are based on agrotechnical methods. Although the overall strategy might be to find and interrupt the weakest link in the chain, the tactics of control or management vary somewhat. The major approaches are: (1) elimination of virus sources, (2) isolation from virus sources, (3) manipulation of the crop, and a series of methods directed at reducing numbers of active vectors, either regionally or locally (Harrewijn, 1983; for details, see sections 11.5.1 and 11.5.3).

Elimination of virus sources

Quarantine and certification programs. Quarantines (Kahn, 1982) are used in an attempt to prevent the introduction of infectious material into a region that is free of the disease of concern. Normally quarantines either prohibit the importation of certain materials, or they require that the imported material be certified as being disease-free. Without the existence of natural barriers, long distances or climatic differences, quarantines against aphid-borne viruses are of doubtful effectiveness. However, they are visible reminders of an existing danger, and frequently can serve as political and educational tools.

If the virus of concern has a restricted distribution, such as the non-persistent aphid-borne plum pox, which is currently restricted to Europe, coupled with natural barriers, the expense and inconvenience of a strict quarantine is worthwhile to prevent introduction into new areas, such as the U.S.A. The movement of diseased nursery stock, rather than by vectors, is considered the most likely means of long-distance spread.

Certification schemes (Quiot and Labonne, 1982) are not preventive in nature, but are attempts to eliminate the pathogen from a certain amount of source material so as to assure that commercially available planting material has a low incidence of infection. Certification is commonly used where the commercial crop is the source of its own infection. Aphid-borne viruses primarily restricted to a vegetatively reproduced commercial crop host that serves as the reservoir, overwintering, and primary virus source, are obvious targets for management by certification programs. To be successful, such programs not only are dependent upon growers' support, but they also must be managed by legislative action.

Crops typically involved include fruits and small fruits, citrus, potatoes, bulbs, and many vegetatively reproduced ornamentals, such as carnation and chrysanthemum. A similar problem exists when the primary source is a seed-borne virus, such as in many of the non-persistent legume viruses and lettuce mosaic. In the Salinas Valley of California, the problem of lettuce mosaic essentially was eliminated through the use of a certification program for the production of seed (Grogan et al., 1952; Kimble et al., 1975). In Europe, seed control may not be the only answer, since in certain areas 4–20% of groundsel, *Senecio vulgaris*, weeds have been found to be infected with lettuce mosaic virus (Kemper, 1962).

Elaborate programs to insure adequate supplies of certified seed potatoes have been used in several countries throughout the world in the management of the potato virus diseases (Hille Ris Lambers, 1955, 1972).

Weed control. Weed control (Bos, 1981) may be directed at the elimination of the sources of the virus, or at elimination of vector populations. Normally weed control is indicated when the spread is short-range, incoming, and can be rather specifically targeted (Doolittle and Walker, 1926; Wellman, 1937). On a local level, weed control has been used successfully in the management of some pepper and celery virus problems in Florida (Orsenigo and Zitter, 1971). The viruses involved were non-persistent and the incidence decreased rapidly as one moved from the edges of the fields. Elimination of the weed host from the immediate border of susceptible plantings helped to control the incidence of viruses. Weed control is expensive, and in areas of extremely high incidence of both weed hosts and vectors of certain viruses, the planting of susceptible crops may not be practical. Within-the-field weed control generally is more effective by using herbicides prior to planting than by attempting to eliminate weeds by cultivation. In some cases, the use of cultivation to cut the weeds, and then allowing them to dry in the field, may aggravate the problem by forcing the vectors off the preferred (weed) host and on to the non-preferred crop hosts.

In the U.S.A., important reservoirs of beet western yellows are cruciferous weeds such as mustards, wild radish, and rockets. In Great Britain, weeds apparently are not so important, and instead the overwintering sources are believed to be sugar beet and mangold clamps (Duffus, 1971; Rochow and Duffus, 1981). In Australia, the *Hyperomyzus* aphid vectors of lettuce necrotic yellows virus do not colonize lettuce; rather, they are found on the herbaceous weed host, *Sonchus oleraceus*. Spread apparently occurs when transient vectors temporarily feed on lettuce as the alates search for new host plants (Boakye and Randles, 1974). The elimination of sowthistle from the immediate vicinity of commercial lettuce fields is obviously indicated (Stubbs et al., 1963).

Roguing. The practice of removing diseased material from the field is common. In general the benefits realized from roguing have been variable, and usually it is not effective, since by the time visible symptoms have appeared and roguing can be done, secondary spread can have taken place. Roguing was part of a successful program used to eliminate an extremely destructive strain of cucumber mosaic in the plantation banana crop owned by a large fruit company in Honduras. Plantings were inspected three times a year. If a diseased plant was found, the surrounding 10 ha were sprayed with nicotine, and the infected plant and its nearest 80 neighbours were pulled and chopped. The cleared area was then sprayed with an organophosphate (malathion) and the spraying repeated after 20 and 40 days. The area was continually inspected for any regrowth. All *Plantago* (an alternate host of the virus) was removed, and replanting delayed for 6 months. Using this protocol, the incidence in 5700 ha was reduced 200-fold, from 21 to 0.1 plants per 400 ha, over a 4-year period and the introduction of the disease into new farms was prevented (Adam, 1962).

Isolation from virus sources

Time. Normally time is used as a control tactic in the form of crop-free periods, a practice that depends upon the lack of a significant endemic cycle among the weed hosts. Crop-free periods have been an effective control method on several crops, especially with viruses that have a non-persistent or/and semi-persistent relationship with their vectors. Celery mosaic in southern California was managed by a crop-free period (Milbrath, 1936), and crop-free periods are now standard practice with the beet yellowing complex of viruses.

Space. The use of space usually involves some form of physical separation, especially when the same crop species is used for dual purposes, for example the use of sugar beets both for the production of sugar and seed. Again, in the management of the beet yellowing complex of virus diseases in Great Britain, the steckling crop, the second-year crop of the biennial beet which is used to produce the seed, is isolated from the annually harvested sugar beet crop. In the cultivation of potatoes, it generally is advised to grow the foundation and certified seed-potato crops in areas isolated from commercial potato crops. This may be difficult to bring about, but in any event the commercial crop obviously should be downwind from the seed-potato crop. In Great Britain the seed-potatoes are grown in Scotland, in areas that traditionally have a low incidence of aphid vectors (Ebbels, 1979) (for calculation of distances, see section 11.5.3).

Space also can involve the isolation from weed sources of the virus, but usually this is only effective with non-persistent viruses, where retention of inoculativity is brief.

Another form of isolation that has been used is that of protective cropping. At least two forms of this practice have been successfully used to a limited extent. One involves the technique of mixed plantings. The beet steckling crop can be protected by sowing a barley cover crop (Hull, 1952) and a similar practice has been used in Japan to protect the radish crop (Shirahama, 1957). Here a rice-cover crop is grown. In Africa, delaying the removal of weeds from the fields has been shown to reduce the incidence of aphid-borne groundnut rosette virus. It appears that the continuous ground cover presented by both the weeds and peanuts reduces the chances of infection (Smartt, 1961).

Another scheme that attempts to isolate the crop from virus sources, or from active vectors, is the use of barriers. The protection of *Brassica* seed beds (crucifers in many situations are transplanted) from infection by the cauliflower mosaic virus has been achieved by planting every 12th row with barley (Broadbent, 1957). Again the breaking of the bare ground-cover and row pattern may affect the landing rate of the aphid vectors or increase the landing rate on the most immediate target of opportunity, in this case the highest plant, which is the barley. Experimental reduction in the spread of some of the non-persistent aphid-borne viruses, such as potato virus Y on pepper in Florida, has been demonstrated by the use of a 15-m barrier of a non-susceptible crop such as corn, beans, or cucumber (Simons, 1957). This was true whether or not the barrier crop was subsequently treated with a pesticide. However, to be practical, the barrier crop also must have a potential for economic return. Barriers also have been used in combination with other methods in the management of the spread of some viruses in lilies and tulips in the Netherlands.

Crop manipulation

Alternations in planting and harvesting dates are practices that, when agronomically possible, can make a difference in the amount of virus spread. In Australia, the recommendation for the control of carrot motley dwarf virus is to plant after the spring flight of the aphid vector, *C. aegopodii* (Stubbs, 1948). In potato, resistance to infection increases with age, and this can be used to reduce virus spread by planting presprouted potatoes to advance the growth cycle (Birecki et al., 1964). In Denmark it was found that a difference of one month in delaying the planting of beets could result in a 40% increase in the incidence of the beet yellowing viruses in these younger plants (Hansen, 1950).

Part of the management of the disease in the seed-potato crop in the Netherlands involves early lifting (harvesting) of the potatoes, which is preceded by destruction of the tops (Hille Ris Lambers, 1955) (see also section 11.5.1).

Another facet of crop manipulation is control over the size of the field and the density of the planting. If the number of vectors is constant and limited and there is no colonization and secondary build-up of vectors, then the proportion of diseased plants decreases as the size of the field increases. This generally has been found to be true when comparing the rates of infection in field crops, where large areas have been planted, versus small plantings of horticultural crops. A corollary of this is that a similar type of protection can be achieved by increasing the density of the planting. Increased density of planting has been correlated with the decreased incidence of virus disease in the groundnut crops in Africa (Van der Plank, 1948).

A third general category of crop manipulation is the use of resistant varieties. This is the major approach to the resolution of vectored pathogens and is about the only approach under tropical conditions, where the endemic cycle is continously active and the crop plant is a readily acceptable host for the

Section 10.3 references, p. 83

pathogen and the vectors. Although immunity is generally sought for in breeding programs, it usually is not possible to achieve. However, varying levels of resistance frequently are available (see also section 11.4).

Resistance quite frequently increases with age (Bennett, 1960; Cadman and Chambers, 1960; Smartt, 1961), and so it must be combined with other practices of crop manipulation, such as time of planting, in order to maximize its effectiveness. One example is the practice of planting presprouted potatoes in Poland.

Miscellaneous methods

Most of these are directed at the vector and involve attempts to reduce the general population or to reduce that portion of the population that is directly impacting on the immediate crop.

Some studies have indicated that high nitrogen levels increase crop susceptibility and the rate of vector build-up (Janssen, 1929). However, fertilization measures are not always feasible, because high levels of nitrogen may already be present in the soil (see further section 11.5.3).

Alates of some aphid species are attracted to yellow (Moericke, 1950), and attempts have been made to demonstrate the possible usefulness of crop colour. Lettuce varieties in which reddish colours predominate are visited by fewer aphids than are the green varieties (Müller, 1964), but again it is doubtful if colour preferences of aphids will dictate those of the public, nor has the degree of reduced attractiveness been shown to be a major deterrent to virus spread.

Chemical control is always an intriguing possibility, but the returns frequently are not cost-effective, even though reduction in disease incidence can be demonstrated (Broadbent, 1969). Mineral oils have been found to have some use in reducing the spread of certain non-persistent aphid-borne viruses (Bradley et al., 1962; Simons, 1982). In the most successful instances, the crops involved occurred in small acreages. The use of reflective surfaces, either aluminum (Smith et al., 1964) or white polyethylene (Adlerz and Everrett, 1968; Smith and Webb, 1969), is a relatively new approach. Apparently such surfaces serve to repel the landing of incoming alate aphids. Again their use would be limited to crops with a rather high cash value that are intensively cultivated under limited conditions and acreages (Harpaz, 1982).

Finally there are attempts to reduce the general vector load in an area. In the Netherlands, bird-cherry, *Prunus serotina*, an overwintering host of *M. persicae*, was planted extensively in a reforestation program to protect coniferous forest seedlings (Hille Ris Lambers, 1955). This practice has been discontinued. The species has been spread by birds, occurs everywhere (Hille Ris Lambers, 1971) and is now a weed in the Netherlands. Harpaz (1982) specifically mentions the relatively recent severe outbreak of barley yellow dwarf on cereal crops in Chile vectored by the introduced species of *Metopolophium dirhodum* (Walker) and *S. avenae*. The population levels of these two species far exceeded those found in their native area, the Mediterranean region, and an extensive international project was undertaken to introduce natural enemies. The occurrence of carrot motley dwarf in Australia, a persistent virus transmitted by *C. aegopodii*, an aphid introduced into Australia, was significantly reduced following the introduction of parasites as part of a biological control program directed at this vector species (L.L. Stubbs, personal communication, 1985).

It is an established doctrine that parasites and predators have regulatory effects on the overall population oscillations of vectors, and therefore must contribute to the limitations that exist on virus transmission that occurs in both endemic and epidemic cycles. However, the early-season delay in the build-up of parasite and predator populations usually provides aphids with a sufficient "transmission window" for virus spread to occur before vector populations "crash". It has been commonly found that parasites and predators do not effectively regulate virus spread (see section 11.2), particularly those that are transmitted in the non-persistent mode. However, virus spread that depends upon colonizing aphids undoubtedly is subject to reduction or limitation by natural control factors.

REFERENCES

Adam, A.V., 1962. An effective program for the control of banana mosaic. Plant Disease Reporter, 46: 366–370.

Adlerz, W.C. and Everett, P.H., 1968. Aluminum foil and white polyethylene mulches to repel aphids and control watermelon mosaic. Journal of Economic Entomology, 61: 1276–1279.

Armour, S.L., Melcher, U., Pirone, T.P., Lyttle, D.J. and Essenberg, R.C., 1983. Helper component for aphid transmission encoded by region II of cauliflower mosaic virus DNA. Virology, 129: 25–30.

Bar-Joseph, M., Sacks, J.M. and Garnsey, S.M., 1978. Detection and estimation of citrus tristeza virus infection rates based on ELISA assays of packing house fruit samples. Phytoparasitica, 6: 145–149.

Bennett, C.W., 1960. Sugar beet yellows disease in the United States. Technical Bulletin USDA, 1218, 63 pp.

Bennett, C.W. and Costa, A.S., 1954. Observation and studies of virus yellows of sugar beet in California. Proceedings of the American Society of Sugar Beet Technology, 8 (part 1): 230–235.

Birecki, M., Gabriel, W. and Osínska, J., 1964. Influence of some treatments on the seed value of potatoes. Part I. Influence of some treatments and of their interactions on the virus infections in two potato varieties. Roczniki Nauk Rolniczych, Seria A, 88: 235–258 (in Polish, with English summary).

Black, L.M., 1959. Biological cycles of plant viruses in insect vectors. In: F.M. Burnet and W.M. Stanley (Editors), The Viruses. Academic Press, New York and London, Vol. 2, pp. 157–185.

Boakye, D.B. and Randles, J.W., 1974. Epidemiology of lettuce necrotic yellows virus in South Australia III. Virus transmission parameters, and vector feeding behaviour on host and non-host plants. Australian Journal of Agricultural Research, 25: 791–802.

Bode, O. and Weidemann, H.L., 1971. Untersuchungen zur Blattlausübertragbarkeit von Kartoffel-M- und -S-virus. Potato Research, 14: 119–129.

Bond, W.P. and Pirone, T.P., 1970. Evidence for soil transmission of sugarcane mosaic virus. Phytopathology, 60: 437–440.

Bos, L., 1981. Wild plants in the ecology of virus diseases. In: K. Maramorosch and K.F. Harris (Editors), Plant Diseases and Vectors: Ecology and Epidemiology. Academic Press, New York and London, pp. 1–33.

Bradley, R.H.E., 1952. Studies on the aphid transmission of a strain of henbane mosaic virus. Annals of Applied Biology, 39: 78–97.

Bradley, R.H.E., Wade, C.V. and Wood, F.A., 1962. Aphid transmission of potato virus Y inhibited by oils. Virology, 18: 327–329.

Brierley, P. and Smith, F.F., 1950. Some vectors, hosts, and properties of dahlia mosaic virus. Plant Disease Reporter, 34: 363–370.

Broadbent, L., 1953. Aphids and virus diseases in potato crops. Biological Review, 28: 350–380.

Broadbent, L., 1957. Investigation of virus diseases of brassica crops. Agricultural Research Council Report Series, No. 14. Cambridge University Press, London, 94 pp.

Broadbent, L., 1969. Disease control through vector control. In: K. Maramorosch (Editor), Viruses, Vectors, and Vegetation. Wiley-Interscience, New York, pp. 593–630.

Brunt, A.A., 1977. Some hosts and properties of narcissus latent virus, a carlavirus commonly infecting narcissus and bulbous iris. Annals of Applied Biology, 87: 355–364.

Cadman, C.H. and Chambers, J., 1960. Factors affecting the spread of aphid-borne viruses in potato in eastern Scotland. III. Effects of planting date, roguing and age of crop on the spread of potato leaf-roll and Y viruses. Annals of Applied Biology, 48: 729–738.

Carter, W., 1973. Insects in Relation to Plant Disease. Wiley-Interscience, New York, 759 pp.

Chalfant, R.B. and Chapman, R.K., 1962. Transmission of cabbage viruses A and B by the cabbage aphid and the green peach aphid. Journal of Economic Entomology, 55: 584–590.

Clinch, P.E.M., Loughnane, J.B. and Murphy, P.A., 1936. A study of the aucuba or yellow mosaics of the potato. Scientific Proceedings of the Royal Dublin Society, 21: 431–448.

Day, M.F. and Venables, D.G., 1961. The transmission of cauliflower mosaic virus by aphids. Australian Journal of Biological Sciences, 14: 187–197.

Dickson, R.C. and Flock, R.A., 1959. Insect vectors of tristeza virus. In: J.M. Wallace (Editor), Citrus Virus Diseases. University of California, Division of Agricultural Sciences, pp. 97–100.

Dickson, R.C., Flock, R.A. and Johnson, M.McD., 1951. Insect transmission of citrus quick-decline virus. Journal of Economic Entomology, 44: 172–176.

Doolittle, S.P. and Walker, M.N., 1926. Control of cucumber mosaic by eradication of wild host plants. Department Bulletin USDA, No. 1461, 14 pp.

Drake, C.J., Tate, H.D. and Harris, H.M., 1933. The relationship of aphids to the transmission of yellow dwarf of onions. Journal of Economic Entomology, 26: 841–846.

Duffus, J.E., 1971. Role of weeds in the incidence of virus diseases. Annual Review of Phytopathology, 9: 319–340.

Duffus, J.E. and Rochow, W.F., 1978. Neutralization of beet western yellows virus by antisera against barley yellow dwarf virus. Phytopathology, 68: 45–49.

Ebbels, D.L., 1979. A historical review of certification schemes for vegetatively-propagated crops in England and Wales. Agricultural Development and Advisory Service Quarterly Review, 32: 21–58.

Elnagar, S. and Murant, A.F., 1976. Relations of the semi-persistent viruses, parsnip yellow fleck and anthriscus yellows, with their vector, *Cavariella aegopodii*. Annals of Applied Biology, 84: 153–167.

Eskandari, F., Sylvester, E.S. and Richardson, J., 1979. Evidence for lack of propagation of potato leaf roll virus in its aphid vector, *Myzus persicae*. Phytopathology, 69: 45–47.

Francki, R.I.B., Kitajima, E.W. and Peters, D., 1981. Rhabdoviruses. In: E. Kurstak (Editor), Handbook of Plant Virus Infections and Comparative Diagnosis. Elsevier/North-Holland Biomedical Press, Amsterdam, pp. 455–489.

Frazier, N.W. and Sylvester, E.S., 1960. Half-lives of transmissibility of two aphid-borne viruses. Virology, 12: 233–244.

Frowd, J.A. and Tomlinson, J.A., 1972. The isolation and identification of parsley viruses occurring in Britain. Annals of Applied Biology, 72: 177–188.

Gabriel, W., 1958. Etudes sur les vecteurs des maladies à virus de la pomme de terre en Pologne. Parasitica, 14: 119–134.

Gerhardson, B., 1977. Some properties of a new legume virus inducing mild mosaic in red clover, *Trifolium pratense*. Phytopathologische Zeitschrift, 89: 116–127.

Goth, R.W., 1962. Aphid transmission of white clover mosaic virus. Phytopathology, 52: 1228.

Gregory, P.H. and Read, D.R., 1949. The spatial distribution of insect-borne plant-virus diseases. Annals of Applied Biology, 36: 475–482.

Grogan, R.G., Welch, J.E. and Bardin, R., 1952. Common lettuce mosaic and its control by the use of mosaic-free seed. Phytopathology, 42: 573–578.

Hagedorn, D.J. and Hanson, E.W., 1951. A comparative study of the viruses causing Wisconsin pea stunt and red clover vein mosaic. Phytopathology, 41: 813–819.

Hamlyn, B.M.G., 1955. Aphid transmission of cauliflower mosaic on turnips. Plant Pathology, 4: 13–16.

Hansen, H.P., 1950. Investigations on virus yellows of beets in Denmark. Transactions of the Danish Academy of Technical Sciences, 1: 1–68.

Harpaz, I., 1982. Nonpesticidal control of vector-borne viruses. In: K.F. Harris and K. Maramorosch (Editors), Pathogens, Vectors, and Plant Diseases: Approaches to Control. Academic Press, New York and London, pp. 1–21.

Harrewijn, P., 1983. The effect of cultural measures on behaviour and population development of potato aphids and transmission of viruses. Mededelingen Faculteit van de Landbouwwetenschappen, Rijksuniversiteit Gent, 48: 791–799.

Harris, K.F., 1977. An ingestion-egestion hypothesis of noncirculative virus transmission. In: K.F. Harris and K. Maramorosch (Editors), Aphids as Virus Vectors. Academic Press, New York and London, pp. 165–220.

Harris, K.F. and Bath, J.E., 1973. Regurgitation by *Myzus persicae* during membrane feeding: its likely function in transmission of nonpersistent plant viruses. Annals of the Entomological Society of America, 66: 793–796.

Harrison, B.D., 1958. Studies on the behavior of potato leaf roll and other viruses in the body of their aphid vector *Myzus persicae* (Sulz.). Virology, 6: 265–277.

Hille Ris Lambers, D., 1955. Potato aphids and virus diseases in the Netherlands. Annals of Applied Biology, 42: 355–360.

Hille Ris Lambers, D., 1971. *Prunus serotina* (American bird cherry) as a host plant of Aphididae in the Netherlands. Netherlands Journal of Plant Pathology, 77: 140–143.

Hille Ris Lambers, D., 1972. Aphids: their life cycles and their role as virus vectors. In: J.A. de Bokx (Editor), Viruses of Potatoes and Seedpotato Production. Centre for Agricultural Publishing and Documentation, Wageningen, pp. 36–56.

Hoggan, I.A., 1933. Some factors involved in aphid transmission of the cucumber-mosaic virus to tobacco. Journal of Agricultural Research, 47: 689–704.

Hollings, M., 1957. Investigation of chrysanthemum viruses II. Virus *B* (mild mosaic) and chrysanthemum latent virus. Annals of Applied Biology, 45: 589–602.

Hollings, M. and Brunt, A.A., 1981. Potyviruses. In: E. Kurstak (Editor), Handbook of Plant Virus Infections and Comparative Diagnosis. Elsevier/North-Holland Biomedical Press, Amsterdam, pp. 731–807.

Huff, C.G., 1931. A proposed classification of disease transmission by arthropods. Science, 74: 456–457.

Hull, R., 1952. Control of virus yellows in sugar beet seed crops. Journal of the Royal Agricultural Society, 113: 86–102.

Hull, R., 1977. Particle differences related to aphid-transmissibility of a plant virus. Journal of General Virology, 34: 183–187.

Hull, R., 1981. Pea enation mosaic virus. In: E. Kurstak (Editor), Handbook of Plant Virus Infections and Comparative Diagnosis. Elsevier/North-Holland Biomedical Press, Amsterdam, pp. 239–256.

Irwin, M.E. and Goodman, R.M., 1981. Ecology and control of soybean mosaic virus. In: K. Maramorosch and K.F. Harris (Editors), Plant Diseases and Vectors: Ecology and Epidemiology. Academic Press, New York and London, pp. 1–33.

Janssen, J.J., 1929. Effect of fertilization on the health condition of potatoes. Tijdschrift over Plantenziekten, 35: 119–151 (in Dutch).

Jaspars, E.M.J. and Bos, L., 1980. Alfalfa Mosaic Virus. C.M.I./A.A.B. Descriptions of Plant Viruses, No. 229, 7 pp.

Kahn, R.P., 1982. The host as a vector: exclusion as a control. In: K.F. Harris and K. Maramorosch (Editors), Pathogens, Vectors, and Plant Diseases: Approaches to Control. Academic Press, New York and London, pp. 123–149.

Kaper, J.M. and Waterworth, H.E., 1981. Cucumoviruses. In: E. Kurstak (Editor), Handbook of Plant Virus Infections and Comparative Diagnosis. Elsevier/North-Holland Biomedical Press, Amsterdam, pp. 257–332.

Karl, E. and Schmelzer, K., 1971. Untersuchungen zur Übertragbarkeit von Wassermelonen-mosaik-virus durch Blattlausarten. Archiv für Pflanzenschutz, 7: 3–11.

Kassanis, B., 1941. Transmission of tobacco etch viruses by aphids. Annals of Applied Biology, 28: 238–243.

Kassanis, B., 1955. Some properties of four viruses isolated from carnation plants. Annals of Applied Biology, 43: 103–113.

Kassanis, B., 1961. Potato paracrinkle virus. European Potato Journal, 4: 14–25.

Kassanis, B. and Govier, D.A., 1971. New evidence on the mechanism of aphid transmission of potato C and potato aucuba mosaic viruses. Journal of General Virology, 10: 99–101.

Kemper, A., 1962. Zum Auftreten von Salatmosaikvirus an Salat (*Lactuca sativa* L.) und an Kreuzkraut (*Senecio vulgaris* L.). Zeitschrift für Pflanzenkrankheiten, 69: 653–663.

Kennedy, J.S., Day, M.F. and Eastop, V.F., 1962. A conspectus of aphids as vectors of plant viruses. Commonwealth Institute of Entomology, London, 114 pp.

Kimble, K.A. Grogan, R.G. Greathead, A.S., Paulus, A.O. and House, J.K., 1975. Development, application, and comparison of methods for indexing lettuce seed for mosaic virus in California. Plant Disease Reporter, 59: 461–464.

Kubo, S. and Takanami, Y., 1979. Infection of tobacco mesophyll protoplasts with tobacco necrotic dwarf virus, a phloem-limited virus. Journal of General Virology, 42: 387–398.

Lim, W.L. and Hagedorn, D.J., 1977. Bimodal transmission of plant viruses. In: K.F. Harris and K. Maramorosch (Editors), Aphids as Virus Vectors. Academic Press, New York and London, pp. 237–251.

Lim, W.L., De Zoeten, G.A. and Hagedorn, D.J., 1977. Scanning electronmicroscopic evidence for attachment of a nonpersistently transmitted virus to its vector's stylets. Virology, 79: 121–128.

Lister, R.M. and Bar-Joseph, M., 1981. Closteroviruses. In: E. Kurstak (Editor), Handbook of Plant Virus Infections and Comparative Diagnosis. Elsevier/North-Holland Biomedical Press, Amsterdam, pp. 809–844.

MacKinnon, J.P., 1974. Detection, spread, and aphid transmission of potato virus S. Canadian Journal of Botany, 52: 461–465.

McClean, A.P.D., 1957. Virus infections in citrus trees. Plant Protection Bulletin FAO, 5: 133–141.

Milbrath, D.G., 1936. Celery mosaic. California Department of Agriculture Bulletin 25, 578 pp.

Misari, S.M. and Sylvester, E.S., 1983. Coriander feathery red-vein virus, a propagative plant rhabdovirus, and its transmission by the aphid *Hyadaphis foeniculi* Passerini. Hilgardia, 51: 1–38.

Miyamoto, S. and Miyamoto, Y., 1966. Notes on aphid-transmission of potato leafroll virus. Science Report of the Hyogo University, Faculty of Agriculture, 7: 51–66.

Miyamoto, S. and Miyamoto, Y., 1971. Notes on aphid-transmission of potato leafroll virus. 2. Transference of the virus to nymphs from viruliferous adults of *Myzus persicae* Sulz. Science Report of the Faculty of Agriculture, Kobe University, 9: 59–70 (in Japanese, with English summary).

Moericke, V., 1950. Über das Farbsehen der Pfirsichblattlaus (*Myzodes persicae* Sulz.). Zertschrift für Tierpsychologie, 7: 265–274.

Müller, H.J., 1964. Über die Anflugdichte von Aphiden auf farbige Salatpflanzen. Entomologia Experimentalis et Applicata, 7: 85–104.

Murant, A.F. and Goold, R.A., 1968. Purification, properties and transmission of parsnip yellow fleck, a semi-persistent, aphid-borne virus. Annals of Applied Biology, 62: 123–137.

Murant, A.F. and Roberts, I.M., 1977. Virus-like particles in phloem tissue of chervil (*Anthriscus cerefolium*) infected with anthriscus yellows virus. Annals of Applied Biology 85: 403–406.

Murant, A.F., Roberts, I.M. and Elnagar, S., 1976. Association of virus-like particles with the foregut of the aphid *Cavariella aegopodii* transmitting the semi-persistent viruses anthriscus yellows and parsnip yellow fleck. Journal of General Virology, 31: 47–57.

Namba, R. and Sylvester, E.S., 1981. Transmission of cauliflower mosaic virus by the green peach, turnip, cabbage, and pea aphids. Journal of Economic Entomology, 74: 546–551.

Norman, P.A. and Grant, T.J., 1956. Transmission of tristeza virus by aphids in Florida. Proceedings of the Florida Station of the Horticultural Society, 69: 38–42.

Orsenigo, J.R. and Zitter, T.A., 1971. Vegetable virus problems in South Florida as related to weed science. Proceedings of the Florida Station of the Horticultural Society, 84: 168–171.

Peters, D., 1981. Plant Rhabdovirus Group. C.M.I./A.A.B. Descriptions of Plant Viruses, No. 244, 6 pp.

Pirone, T.P., 1977. Accessory factors in nonpersistent virus transmission. In: K.F. Harris and K. Maramorosch (Editors), Aphids as Virus Vectors. Academic Press, New York and London, pp. 221–235.

Pratt, M.J., 1961. Studies on clover yellow mosaic and white clover mosaic viruses. Canadian Journal of Botany, 39: 655–665.

Purcifull, D.E. and Edwardson, J.R., 1981. Potexviruses. In: E. Kurstak (Editor), Handbook of Plant Virus Infections and Comparative Diagnosis. Elsevier/North-Holland Biomedical Press, Amsterdam, pp. 627–693.

Quiot, J.B. and Labonne, G., 1982. Controlling seed and insect-borne viruses. In: K.F. Harris and K. Maramorosch (Editors), Pathogens, Vectors, and Plant Diseases: Approaches to Control. Academic Press, New York and London, pp. 95–122.

Rochow, W.F., 1967. Predominating strains of barley yellow dwarf virus in New York: changes during ten years. Plant Disease Reporter, 51: 195–199.

Rochow, W.F., 1977. Dependent virus transmissions from mixed infections. In: K.F. Harris and K. Maramorosch (Editors), Aphids as Virus Vectors. Academic Press, New York and London, pp. 253–273.

Rochow, W.F. and Duffus, J.E., 1981. Luteoviruses and yellows diseases. In: E. Kurstak (Editor), Handbook of Plant Virus Infections and Comparative Diagnosis. Elsevier/North-Holland Biomedical Press, Amsterdam, pp. 147–170.

Rowhani, A. and Stace-Smith, R., 1979. Purification and characterization of potato leafroll virus. Virology, 98: 45–54.

Sarkar, S., 1976. Potato leafroll virus contains a double-stranded DNA. Virology, 70: 265–273.

Schmidt, H.E., Karl, E. and Schmidt, H.B., 1966. Die Übertragung eines stäbchenförmigen Hopfenvirus durch *Aphis fabae* Scop. Naturwissenschaften, 53: 537.

Shepherd, R.J., 1981. Cauliflower Mosaic Virus. C.M.I./A.A.B. Descriptions of Plant Viruses, No. 243. 6 pp.

Shepherd, R.J. and Lawson, R.H., 1981. Caulimoviruses. In: E. Kurstak (Editor), Handbook of Plant Virus Infections and Comparative Diagnosis. Elsevier/North-Holland Biomedical Press, Amsterdam, pp. 847–878.

Shirahama, K., 1957. Studies on radish mosaic disease and its controlling. Agricultural Improvement Extension Work Conference, Tokyo, 107 pp. (in Japanese, with English summary).

Simons, J.N., 1957. Effects of insecticides and physical barriers on field spread of pepper veinbanding mosaic virus. Phytopathology, 47: 139–145.

Simons, J.N., 1969. Differential transmission of closely related strains of potato virus Y by the green peach aphid and the potato aphid. Journal of Economic Entomology, 62: 1088–1096.

Simons, J.N., 1982. Use of oil sprays and reflective surfaces for control of insect-transmitted plant viruses. In: K.F. Harris and K. Maramorosch (Editors), Pathogens, Vectors, and Plant Diseases: Approaches to Control. Academic Press, New York and London, pp. 71–93.

Simons, J.N. and Sylvester, E.S., 1953. Acquisition threshold period of western celery mosaic virus for four species of aphids. Phytopathology, 43: 29–31.

Smartt, J., 1961. The diseases of groundnuts in Northern Rhodesia. Empire Journal of Experimental Agriculture, 29: 79–87.

Smith, F.F. and Webb, R.E., 1969. Repelling aphids by reflective surfaces, a new approach to the control of insect-transmitted viruses. In: K. Maramorosch (Editor), Viruses, Vectors and Vegetation. Wiley-Interscience, New York, pp. 631–639.

Smith, F.F., Johnson, G.V., Kahn, R.P. and Bing, A., 1964. Repellency of reflective aluminum to transient aphid virus-vectors. Phytopathology, 54: 748 (Abstract).

Stegwee, D. and Ponsen, M.B., 1958. Multiplication of potato leaf roll virus in the aphid *Myzus persicae* (Sulz.). Entomologia Experimentalis et Applicata, 1: 291–300.

Stubbs, L.L., 1948. A new virus disease of carrots: its transmission, host range, and control. Australian Journal of Scientific Research, 1: 303–332.

Stubbs, L.L., Guy, J.A.D. and Stubbs, K.J., 1963. Control of lettuce necrotic yellows virus disease by the destruction of common sowthistle (*Sonchus oleraceus*). Australian Journal of Experimental Agriculture and Animal Husbandry, 3: 215–218.

Swenson, K.G., 1952. Aphid transmission of a strain of alfalfa mosaic virus. Phytopathology, 42: 261–262.

Swenson, K.G., 1968. Role of aphids in the ecology of plant viruses. Annual Review of Phytopathology, 6: 351–374.

Sylvester, E.S., 1949. Beet-mosaic virus–green peach aphid relationships. Phytopathology, 39: 417–424.

Sylvester, E.S., 1956. Beet yellows virus transmission by the green peach aphid. Journal of Economic Entomology, 49: 789–800.

Taylor, C.E. and Robertson, W.M., 1974. Electron microscopy evidence for the association of tobacco severe etch virus with the maxillae in *Myzus persicae* (Sulz.). Phytopathologische Zeitschrift, 80: 257–266.

Taylor, L.R. and Palmer, J.M.P., 1972. Aerial Sampling. In: H.F. van Emden (Editor), Aphid Technology. Academic Press, London and New York, pp. 189–234.

Taylor, R.H. and Stubbs, L.L., 1972. Broad Bean Wilt Virus. C.M.I./A.A.B. Descriptions of Plant Viruses, No. 81, 4 pp.

Van der Plank, J.E., 1944. Production of seed potatoes in a hot, dry climate. Nature, 153: 589–590.

Van der Plank, J.E., 1948. The relation between the size of fields and the spread of plant-disease into them. Part 1. Crowd diseases. Empire Journal of Experimental Agriculture, 16: 134–142.

Van Hoof, H.A., 1954. Differences in the transmission of the cauliflower mosaic virus by *Myzus persicae* Sulzer and *Brevicoryne brassicae* L. Tijdschrift over Plantenziekten, 60: 267–272.

Van Hoof, H.A., 1977. Determination of the infection pressure of potato virus Y^N. Netherlands Journal of Plant Pathology, 83: 123–127.

Van Regenmortel, M.H.V. and Pinck, L., 1981. Alfalfa mosaic virus. In: E. Kurstak (Editor), Handbook of Plant Virus Infections and Comparative Diagnosis. Elsevier/North-Holland Biomedical Press, Amsterdam, pp. 415–421.

Van Velsen, R.J. and Crowley, N.C., 1962. *Centrosema* mosaic: a new virus disease of *Crotalaria* spp. in Papua and New Guinea. Australian Journal of Agricultural Research, 13: 220–232.

Watson, M.A., 1938. Further studies on the relationship between Hyoscyamus virus 3, and the aphis *Myzus persicae* (Sulz.) with special reference to the effects of fasting. Proceedings of the Royal Society of London, Series B, 125: 144–170.

Watson, M.A., 1940. Studies on the transmission of sugar-beet yellows virus by the aphis, *Myzus persicae* (Sulz.). Proceedings of the Royal Society of London, Series B, 128: 535–552.

Watson, M.A., 1946. The transmission of beet mosaic and beet yellows viruses by aphides; a comparative study of a non-persistent and a persistent virus having host plants and vectors in common. Proceedings of the Royal Society of London, Series B, 133: 200–219.

Watson, M.A. and Roberts, F.M., 1940. Evidence against the hypothesis that certain plant viruses are transmitted mechanically by aphides. Annals of Applied Biology, 27: 227–233.

Weber, K.A. and Hampton, R.O., 1980. Transmission of two purified carlaviruses by the pea aphid. Phytopathology, 70: 631–633.

Wellman, F.L., 1937. Control of southern celery mosaic in Florida by removing weeds that serve as sources of mosaic infection. Technical Bulletin USDA, No. 548, 16 pp.

Wetter, C. and Milne, R.G., 1981. Carlaviruses. In: E. Kurstak (Editor), Handbook of Plant Virus Infections and Comparative Diagnosis. Elsevier/North-Holland Biomedical Press, Amsterdam, pp. 695–730.

Wetter, C. and Völk, J., 1960. Versuche zur Übertragung der Kartoffelviren M und S durch *Myzus persicae* Sulz. European Potato Journal, 3: 158–163.

Zitter, T.A., 1977. Epidemiology of aphid-borne viruses. In: K.F. Harris and K. Maramorosch (Editors), Aphids as Virus Vectors. Academic Press, New York and London, pp. 385–412.

Zitter, T.A. and Tsai, J.H., 1977. Transmission of three potyviruses by the leafminer, *Liriomyza sativae* (Diptera: Agromyzidae). Plant Disease Reporter, 61: 1025–1029

Chapter 11 Control of Aphids

11.1 Chemical Control

A. SCHEPERS

INTRODUCTION

Sucking insects like aphids can affect plant growth and crop production in different ways:

(1) Nutrient drain, which – if many aphids are present – can cause direct reduction of plant productivity.

(2) Transmission of virus diseases, which indirectly hamper the growth and productivity of plants. If plants are in a susceptible (young) stage of growth and sources of virus are in the neighbourhood, only few aphids are needed to cause severe virus infections. Generally the effects of virus infections are most pronounced in vegetatively propagated plants, e.g. potatoes, because the disease is transferred with the seed tubers to the progeny, in which it can cause considerable loss of yield. If many plant species, e.g. weeds, act as a host of the virus in question, crops generatively grown from healthy seed can already be heavily attacked in an early growth stage, also resulting in crop failure.

(3) Toxins present in the saliva of certain aphid species, brought into the plant by sucking aphids, can cause leaf deformation or can lead to the development of galls on the leaves and therefore can negatively influence growth and productivity of the crop.

(4) Excretion products of aphids deposited on the leaves of plants will attract saprophytic fungi, which cover the leaf surface and accelerate the ageing of leaves and therefore will lead to reduction in photosynthesis capacity (see Chapter 10 for more details on damage).

Prevention of crop damage means protecting the crop against attack, and therefore, ever since these negative effects of aphids have been known attempts have been made to prevent or to diminish these by controlling the aphids by chemical, biological (see section 11.2) or integrated control methods (see section 11.5).

APHIDS AND APHICIDES

Before World War II, chemical control of aphids was mainly restricted to the application of nicotine and some arsenical products. Sprayed on a crop they kill the aphids that come into contact with these compounds, but they have neither systemic nor residual effects. Since the war, chemical control of insects has made rapid progress, starting with the development of DDT formulations and other chlorinated hydrocarbons such as lindane, which showed a residual effect but had no systemic properties. However, the persistence of residues of the chlorinated hydrocarbons causes accumulation in the foodchain, and since this phenomenon has become known unrestrained application of those products no longer seems justified and is already prohibited in many countries.

The development of systemic aphicides, starting with the formulation of organophosphorous compounds, opened new perspectives in aphid control, particularly with regard to virus reduction in crops, like potatoes, because of the longer residual activity of these compounds. Later on the carbamate insecticides/nematicides added more possibilities, as did the development of synthetic pyrethroids. All these products play an invaluable role in the control of plant damage and plant diseases caused by aphids. The difference in chemical composition and the alternate use of the various types of aphicides makes it possible to prevent the development of aphid resistance.

Desirable properties of a good aphicide are: selective toxicity to aphids; systemic activity; a reasonable residual activity; rapid action; and low phytotoxicity.

Selectivity

It is very important that the aphicide does not kill parasites and predators of aphids. If it does, treatments which start early in crop development have to be continued throughout the vegetation period, because there is no build-up of a predator or parasite population, and therefore interruption of the treatments would cause severe crop damage by aphids. However, frequent treatment with the same aphicides will favour the development of resistance in aphids, particularly when they are not applied in alternation with different formulations.

Selectivity is also important in saving pollinating insects; this is especially the case when insect pollination is the only possibility for propagation and yield development, e.g. in fruit growing. In the group of organophosphorous and carbamate compounds there are products with a high rate of selectivity. Pyrethroids, however, are not or only little selective.

Aphicides should be no more than slightly toxic to animals and birds.

Systemic activity

Because aphids are mainly phloem-feeding insects and can land and colonize on the crop almost continuously, systemic action of an aphicide is essential, in order to reduce the number of treatments. This is particularly important when reduction in virus infection is the aim of the application. If the population has to be reduced only to prevent damage by nutrient drain a non-systemic aphicide will often be effective enough. Systemic activity means that the aphicide can be applied not only to the foliage, but also to the soil or to the seed. In cases of soil application, sufficient water is needed for transport of the active ingredient to the roots and for its uptake into the vascular system of the plants. In addition, soil aphicides should have sufficient stability to prevent quick dissolution and rinsing out.

Residual activity and persistence

A certain degree of residual activity can be favourable when treated crops are reinfested by aphids, just as it is with systemic compounds. However, in food and fodder products the aphicide should have decomposed into harmless compounds before harvest.

Rapid action

In order to prevent transmission of non-persistent viruses in particular, aphicides should act rapidly. This means that an aphicide should, besides having systemic action, also kill the aphids or reduce their activity quickly by contact or by vapour action. Some aphicides, particularly those from the group of synthetic pyrethroids, show a certain degree of repellent action, which deters the aphids from landing or from inserting stylets into the leaves.

Low phytotoxicity

Aphicides themselves should not harm the crop on which they are used, in order to prevent yield reduction.

APPLICATION OF APHICIDES

The most efficient way of applying an aphicide is determined by a number of conditions. Important is application at the right moment. This means that when prevention of virus infections is the aim and aphids are already present before crop emergence, soil or seed application is preferable to crop spraying or dusting. As has been mentioned above, soil application can only give good results if the soil contains sufficient water. In dense crops, dusting may be better than spraying on the leaf canopy. When spraying on warm clear days, care should be taken to use sufficient water in order to prevent quick evaporation. In such circumstances application in the evening or early morning is preferred. The aphicide should always be well distributed over the crop or over the seed. In the case of soil application the chemical should be given in the cultivation rows, so that the active ingredient can easily reach the developing roots of the seeds.

The duration of protection of the crop against aphids after a treatment differs with the type of aphicide used and depends on the rate of residual activity and persistence. Many organophosphorus and carbamate spray compounds have an effective action period of 10–14 days. Other exhibit shorter action periods and are particularly suitable for application in crops that are intended to be consumed soon. For each situation the right treatment is available. Application of ingredients with a high persistence is never justified in consumption or fodder crops.

Aphicides applied to the soil are released slowly and have an average action period of two months or more. They are very suitable for protecting plants from aphid infestation immediately after their emergence. Much caution should be taken when using them in consumption or fodder crops.

CHEMICAL CONTROL TO REDUCE DIRECT DAMAGE

Solanaceae

Although it is now generally known that damage to potatoes by aphids is mainly caused by virus transmission and virus attack, direct damage by nutrient drain has already been noticed long ago. Quanjer noted in 1939: "Direct damage by aphids is hard to estimate, in addition comes the undoubtedly substantial damage which they can cause as virus transmitters".

Many early examples of yield increase in **potato** (*Solanum tuberosum*) obtained by insect control can be found in the American literature. Hill (1948) gave a compilation of many investigations, in which DDT was frequently used as an insecticide. However, it appeared that the elimination of other insects of which the potato leafhopper, *Empoasca fabae* (Harrington), was the most

Section 11.1 references, p. 112

important, was the main reason for the yield increases. Trials of Harding (1962) involving few aphids but many leafhoppers confirmed these conclusions. Pond (1962) described the effect of some insecticides on aphid development and on tuber yield in experiments in New Brunswick, Canada, from 1948–1960. Although aphid infestation was rather high in certain years and the aphid numbers were mostly reduced by 90% or more, only once a significant increase in yield by aphid control was found. In other trials, from 1963–1966, Pond (1967) obtained good control of the aphids *Myzus persicae* (Sulzer), *Macrosiphum euphorbiae* (Thomas) and *Aphis nasturtii* (Kaltenbach) and increase of tuber yields with either soil application of aldicarb or with three sprays of oxydemeton-methyl.

Kolbe (1978) mentioned feeding damages of 6–25% in experiments in Germany, depending on the potato variety. About 90% of the aphids in his experiments were *M. persicae*. In England in 1965, which was a year of heavy aphid infestations, Hails (1967) found a yield increase of 9% with the variety King Edward when there was aphid control throughout the season. In 1964, 1966 and 1967, however, aphid numbers were low and yields were not influenced by aphid control. De Mey and Pauwels (1978) obtained higher yields when they applied different types of granular systemic insecticides to the soil, resulting in an effective aphid control for at least eleven weeks. Southall and Sly (1976) found a yield increase of at least 10% when they took control measures at a level of more than 500 aphids (*M. persicae, M. euphorbiae, A. nasturtii* and *Aulacorthum solani* (Kaltenbach)) per 100 leaves. Cancelado and Radcliffe (1979), referring to experiments in Minnesota, mentioned an action threshold of 30 apterous aphids (*M. persicae*) per 105 leaves for the spraying of ware potatoes. Starting spraying at a lower threshold meant more applications and proved to be adverse because of the higher risk of development of resistance (see section 11.1.1).

Yield improvement was also reported from Egypt by Abdel Salam et al. (1975), who used various insecticides as sprays or soil-applied granules, and from India by Awate and Pokharkar (1977). In both cases *M. persicae* was the major aphid species, but other insects were mentioned and could also have played a role.

In the Netherlands, if aphid density is below 50 aphids per (compound) potato leaf in a closed crop of ware potatoes, control is assumed to be not economical. This is based on the amount of the nutrient drain effected by such a density, which is calculated to be not more than 2–4 kg dry matter per ha per day (P. Harrewijn, personal communication, 1982; J. Lampe, personal communication, 1972). Predators can control further increase of the population or even will decrease it. It should be kept in mind that frequent chemical control from early crop growth on will favour the development of aphid resistance (Cancelado and Radcliffe, 1979) and in addition will inhibit predator development.

Nevertheless, in most regions of the Netherlands, one spray with an aphicide, about mid-June, is mostly advised in ware potato growing to prevent attack of the crop by toproll. This phenomenon is caused by the aphid *M. euphorbiae* (Gibson, 1974); in the Netherlands particularly by the subspecies *rosae* of this aphid species. In small numbers, this subspecies can already induce toproll; a heavy attack can negatively influence the photosynthetic rate of the potato leaves and therefore the tuber yield (Gibson et al., 1976; Schepers, 1983).

In **tobacco** (*Nicotiana* spp.), several pests can affect growth and development; aphids, mostly *M. persicae*, are of minor importance as far as direct damage concerns, but they are harmful as virus transmitters, together with *Thrips tabaci* (Lindeman). Several predator species are active in controlling aphids and other pests in tobacco (Dirimanov and Dimitrov, 1975). Hence much attention should be paid to the use of selective insecticides in chemical control of aphids. Stefanov and Georgiev (1973) found pirimicarb to be the least toxic to mammals. The residues left in and on the product should also determine the choice of insecticide. Tappan et al. (1974) found no residues of acephate after curing and fermentation, probably because of the high temperatures during processing. This was in contrast with the pyrethroids fenvalerate and permethrin, which left residues (Tappan et al., 1982).

On **tomato** (*Lycopersicon esculentum*), several aphid species are responsible for crop damage, besides other insects such as thrips, leafhoppers and others. Muminov et al. (1976) found a yield reduction of up to 40% by insect attack in glasshouses. Control can be done with several insecticides, but, as is the case with all freshly consumed vegetables, safety terms must be maintained. From flowering onwards only compounds with low persistence can be used.

On **egg plant** (*Solanum melongena*), aphids, mostly *Aphis gossypii* Glover, are of minor importance in direct damage of the crop. Control measures to eradicate other pests, such as *Amrasca* and *Leucimodes* spp., will also kill aphids when present. Increased yields and better quality of the fruits are reported after good insect control (Joshi and Sharma, 1973).

M. persicae is the most harmful aphid in **sweet pepper** (*Capsicum annuum*). It can cause feeding injury, and the quality of the fruits is negatively influenced by the honeydew, which attracts black sooty-mold fungi. Fruits of heavily attacked plants remain smaller. Both granular-soil and foliar-spray applications of insecticides can reduce damage (Burbutis et al., 1972; Cruz, 1974).

Gramineae

Cereal crops are known to be attacked by aphids all over the world. Aphids are transmitters of virus diseases in cereals, of which barley yellow dwarf virus seems to be the most harmful one. However, important damage is also caused through drain of nutrients and contamination of the leaves with honeydew. This will stimulate the growth of saprophytic and possibly necrotrophic fungi on the leaves and will accelerate the ageing of the plants. Both indirect effects decrease the photosynthesis capacity of the crop (Rabbinge et al., 1981). These authors found yield losses of up to 700 kg per ha ($\pm$ 11%) in wheat (*Triticum*) compared with treatments in which aphids were kept under complete control; 72% of the yield loss was found to be due to direct sucking damage, the remaining 28% was caused by indirect effects. Mantel et al. (1982) showed that with an increasing yield level the proportion of yield loss caused by direct sucking decreases and that of indirect damage increases, due to accelerated ageing caused by the honeydew.

In Western Europe, mainly three aphid species are responsible for crop damage in cereals. *Sitobion avenae* (Fabricius) is the most harmful species, because it prefers the ears or panicles as a feeding site. Therefore it becomes most numerous after flowering and causes much damage by direct sucking on the bases of the spikelets. Two other species, *Metopolophium dirhodum* (Walker) and *Rhopalosiphum padi* (Linnaeus), develop earlier and stay mainly on the leaves and stems. In Scandinavia, *R. padi* seems to be the most harmful aphid species, causing serious yield decreases in barley (*Hordeum vulgare*), oats (*Avena sativa*) and wheat (Rautapää and Uoti, 1976; Andersson, 1977). Rautapää and Uoti found maximum yield losses by aphid attack of 999 kg (ca. 30%) in barley, 967 kg (ca. 24%) in oats and 345 kg (ca. 7%) in wheat in Finland from 1968–1975.

In regions with a milder climate, *Schizaphis graminum* (Rondani) and *Rhopalosiphum maidis* (Fitch) also play an important role in cereal crops. The

Section 11.1 references, p. 112

latter species can be particularly harmful in maize (*Zea mays*) (Rajagopal and Channa Basavanna, 1977; Ganguli and Raychaudhuri, 1980), causing yield loss and virus transmission, but is also reported as a pest in barley in India (Verma et al., 1979; Chillar and Verma, 1982). *S. graminum* is important in wheat growing in Eastern Europe (Babenko, 1980) and in North and South America (Piek and Glazer, 1980; De Pew, 1982), and particularly attacks sorghum crops in those regions (Daniels, 1969; Popov, 1972; Cate et al., 1973a; De Pew, 1974; Barbulescu, 1975; Feese and Wilde, 1975; Kindler and Staples, 1981). Crops are attacked in an early stage of growth, probably because of the anholocyclic overwintering of this aphid. *S. graminum* particularly causes great damage to the infested plants by inducing leaf necrosis and growth deformations by the toxins in its saliva (Wood, 1965).

An excellent review on crop damage by aphids was given by Kolbe (1969). As has been confirmed in more recent literature, yield losses varied highly in the described investigations. Aphid damage and aphid control have received continuous attention in agricultural research. However, aphid damage in cereal crops has been increasing in the last two decades and control measures have had to be intensified, certainly in the West.

There are certain reasons for these developments: (1) Changes in cultivation methods, in particular the application of higher quantities of nitrogen, divided in two or three separate gifts, have led to higher nitrogen levels in the plants during a longer period of the season. This favours the reproduction rate of aphids and suppresses wing formation, both resulting in higher aphid densities (Vereijken, 1979) (see sections 11.5.2 and 11.5.3). (2) Chemical control of weeds and fungi has an unfavourable influence on aphid predators (Vickerman and Sunderland, 1975; Hellpap, 1982).

Control measures against aphids are only sensible if the expected gain through yield increase is higher than the cost of the treatments. Kolbe and Linke (1974) found an economic control threshold of 20–30 aphids per ear in wheat, based on the results of many experiments in Western Germany between 1968 and 1972 (Fig. 11.1.1). If the aphids were not killed at this stage of

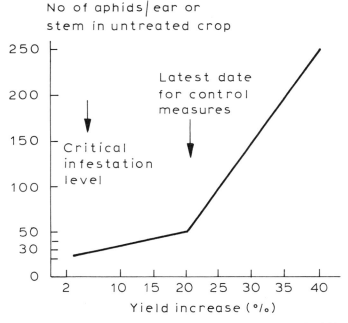

Fig. 11.1.1. Yield increase of wheat, *Triticum*, by control of aphids at various infestation levels (after Kolbe and Linke, 1974).

TABLE 11.1.1

Action threshold for aphid control in wheat in relation to yield levels in untreated fields

Yield (kg per ha)	Acceptable maximum number of aphids per culm	Acceptable level of infested culms (%)
≤ 5500	14.2	95
5500–6500	12.4	90
6500–7000	8.2	85
7000–7500	5.3	80
≥ 7500	3.8	70

Reference: Mantel et al., 1982.

population development, infestation could grow and reached 150 aphids per ear, causing a yield reduction of 30%. The same threshold was mentioned by Andersson (1977) in Sweden for wheat and oats. Kontev (1975) reported an economic threshold of 30–40 aphids per ear until the milky-ripe stage. However, George and Gair (1979) found, in about 50 trials in England and Wales, an average yield increase of 12.5% if one pirimicarb spray was given at the onset of flowering and if five or more *S. avenae* were present per ear. Spraying later in the season gave a much lower yield increase. For *M. dirhodum* a threshold of 30 aphids per flag-leaf was recommended (Mantel et al., 1982). Vereijken (1979) pointed out the problems of setting a threshold because further population development is hard to predict; much depends on variable factors such as weather conditions and predator development.

The number of *S. avenae* individuals in wheat is strictly correlated with the number of infested ears. Therefore, where this aphid species is most abundant in wheat, the level of aphid infestation can be expressed as percentage of infested ears. A control threshold of 80% of infested ears would be safe for the period before milky-ripeness. When this growth stage of the crop has already been reached before the aphids become numerous, 90% infestation can be tolerated (Vereijken, 1979). Control at an early stage, when only few aphids are present, can have the opposite effect; the population development of aphids is favoured by killing the predators. Basedow (1980) also observed re-colonization after control in the flowering stage and recommended spraying with aphicides after this stage. Mantel et al. (1982) worked out that the action threshold is flexible and should depend on the expected yield level. Table 11.1.1 shows that at higher yield levels fewer aphids can be tolerated. Some authors obtained good results by spraying 50-m-wide border strips of large wheat fields (Babenko, 1980; Basedow, 1980), since aphid infestations often start there. This system saves 75–80% of aphicide compared with full-field spraying and also allows complete utilization of the potential of benificial insects present in the rest of the field.

In south-eastern Europe, South America and the southern states of the U.S.A. good aphid control and often high yield increases have been obtained in wheat and sorghum after application of granular insecticides to the soil during planting time (Cate et al., 1973; Barbulescu, 1975; Castillo and Acevedo, 1976). The most important aphid species were *S. graminum*, (particularly in sorghum) *M. dirhodum* and *R. padi*. These species can attack cereal crops in an early growth stage by infesting leaves and stems. In the regions where the experiments were done, anholocyclic overwintering can lead to early migration to important culture crops.

Root aphids can cause serious damage in **rice** (*Oryza sativa*) especially in the dry (upland) situation. Although several aphid species can be involved, three

Section 11.1 references, p. 112

of them seem to be most harmful: *Anoecia fulviabdominalis* (Sasaki),
Rhopalosiphum rufiabdominalis (Sasaki) and *Tetraneura nigriabdominalis*
(Sasaki). Ants play an important role in the infestation of the subterranean
parts of the rice plants by root aphids. Attacks during the early growth stages
of rice cause the most serious damage, and the yield may be reduced by up to
50% (Yano et al., 1983). Chemical control with soil insecticides at the time of
sowing of the rice can be very effective by killing both aphids and associated
ants (Tanaka, 1961).

In the chemical control of aphids, selective aphicides should be used as much
as possible. Hellpap (1982), using parathion, oxydemeton-methyl and pirimicarb against aphids in wheat, found that pirimicarb was the most selective
compound. It had no effect on the ladybird beetle *Coccinella septempunctata*
(Linnaeus) and on *Chrysoperla (Chrysopa) carnea* Stephens, and it was less
toxic to the syrphid fly *Metasyrphus corollae* (Fabricius) than the other insecticides. Parathion caused heavy mortality of all predators, while oxy-demeton-
methyl showed a high toxicity to both Coccinellidae and Syrphidae. Ba-Angood
and Stewart (1980) compared granular treatments of thiofanox and carbofuran
with sprays of dimethoate and pirimicarb against *R. maidis* and *S. avenae* in
barley. All agents gave good control of the aphids, and the granular treatment
did not significantly reduce the number of parasites (*Aphidius* spp.) and predators (Coccinellidae and Chrysopidae). However, both sprays reduced the
populations of benificial insects, pirimicarb less so than dimethoate. Basedow
(1984) found pirimicarb and oxy-demeton-methyl to be relatively harmless to
high numbers of natural enemies of aphids in cereal fields.

Chenopodiaceae

Many organisms threaten the growth and development of **sugar beet** (*Beta
vulgaris*). Since the seeds of this crop are precision-drilled, attack by arthropods, nematodes and fungi can cause much damage, particularly during
germination and early development of the plants. In the Netherlands, aphids
are of minor importance as far as direct damage concerns, in comparison with
other pests (W. Heijbroek, personal communication, 1985). Certain aphid species can be very harmful while transmitting virus diseases (see pp. 107–109), e.g.
M. persicae and in Eastern Europe also *Aphis fabae* Scopoli (Dubnik, 1977;
Fritzsche et al., 1984). Large numbers of the latter aphid species can also cause
direct damage; in the Netherlands (Anonymous, 1984), soil application of granular insecticides is recommended in regions were damage is expected to occur.

In Sweden, Möllerström (1963) observed 16% loss of root yield and 0.13%
decrease of sugar content caused by *A. fabae* in plots not treated with insecticides. Here, beet yellows virus was almost absent. Similarly Fritzsche et al.
(1974) reported a possible yield damage of 15% by this aphid. In some regions
in England control of *A. fabae* is also frequently necessary (Hull and Heathcote, 1967).

Van Steyvoort (1983) in Belgium could ascribe about 5% of the yield loss in
sugar beet to physical feeding damage by aphids. Control of early feeding may
be done by one insecticidal spray in anticipation of later control of virus
transmission. In this case it is important to apply a selective aphicide in order
to spare the aphid predators, so that they will be able to control aphid activity
later in the season. Besides foliage attack, plant damage has been described for
the root aphid, *Pemphigus fuscicornis* (Koch), from Eastern Europe (Gapanova,
1977; Grigorov, 1982). Culture methods including destruction of weeds (in
particular Chenopodiaceae), crop rotation and precise fertilization are good
possibilities to control this type of damage. In addition, chemical control can
be applied by regular sprays with lindane.

Leguminosae

In **alfalfa** (*Medicago sativa*), an important forage crop in many parts of the world, three aphid species are mainly responsible for direct damage: *Acyrthosiphon kondoi* Shinji, *Acyrthosiphon pisum* (Harris) and *Therioaphis trifolii* f. *maculata* (Buckton) (see also 11.4). Infestation of *A. kondoi* at a density of only 20 aphids per stem on regrowing shoots in spring causes severe damage, such as stunting of the plants and curling, yellowing and premature fall of the leaves. This will result in the loss of at least one cutting. But if infestation takes place when the plants are already half-grown or taller still (25–30 cm) even 40–50 aphids per stem hardly damage the crop (Sharma and Stern, 1980). *A. pisum* and *T. trifolii* f. *maculata* can also cause serious injury to alfalfa grown for forage and for seed production (Caballero, 1972; Garcia, 1974; Machain, 1978; Cuperus and Radcliffe, 1983). Stokovskaya et al. (1977) recommended not using the first cutting of alfalfa for seed production, but only the later ones.

The development of aphid-resistant varieties looks promising (Harper, 1983), but many of these varieties still lack persistence and are poorly adapted to grazing (Brieze-Stegeman, 1979) (see section 11.4). That is the reason why chemical control of aphids will often be necessary. In the case of early infestation during the first regrowth in spring, Sharma and Stern (1980) found that control should already be done at a density of 10–12 aphids per stem in order to avoid crop damage by the aphid population, growing rapidly while the predators are still absent. Economic injury levels for *A. pisum* were calculated by Cuperus and Radcliffe (1983) and were expressed as aphid-days (number of aphids × number of days). The highest tolerable level of aphid-days was 3948 per pendulum sweep (with a sweep net of 38 cm diameter) or 3850 per vacuum sample (of 0.48 cm^2) or 114 aphids per stem.

Through the application of only one insecticidal spray, either with acephate or with thiometon, in the spring, Cliffe (1977) reduced the infestation of *A. kondoi* for at least 42 days and obtained yield increases of 60 and 84%, respectively. O'Connor and Hart (1977) controlled *A. kondoi* for five weeks with only one pirimicarb spray with doses varying from 50 to 250 g a.i. per ha. Yields increased by 45–70%, with the highest yield increase being found at the lowest rate of application. East et al. (1977) tested several insecticides and obtained a significant reduction of *A. kondoi* infestation for 13–20 days. Surface-broadcast applications with granular systemic insecticides required much higher rates of active ingredient to achieve a control similar to that achieved with foliage sprays. Moore (1977) reached 100% control for 42 days with several organophosphorous compounds.

In general, aphids in **alfalfa can** quite easily be controlled with a wide variety of insecticides (Summers, 1975). However, Walters and Forrester (1979) found resistance of *T. trifolii* f. *maculata* against a number of insecticides of the organophosphate and carbamate groups. As in other crops, aphid predators can play an important role in the control of aphids. The selectivity of the chemical compound to be used should therefore be taken into consideration. In trials of Summers et al. (1975) and of Syrett and Penman (1980) pirimicarb was found to be the most selective control agent when compared with a series of organophosphorous compounds or with the pyrethroid fenvalerate.

In **beans**, in the moderate climate zone, the major aphid species is *A. fabae*. It acts together with other aphids (e.g. *A. pisum* and *M. persicae* and in warmer climates also with *Aphis craccivora* Koch and *A. gossypii*) as a virus transmitter, but it also causes physical feeding damage. For instance, infestation with *A. pisum* during flowering led to extreme yield losses: 60% or more (Hinz and Daebeler, 1984). Chemical control against aphids is necessary and must

Section 11.1 references, p. 112

often be combined with the control of other pests, e.g. *Liriomyza, Bruchus, Apion, Lixus* and *Sitona* spp. in Mediterranean countries (Fam, 1983; Hariri and Tahhan, 1983).

Bean flowers are intensively visited by honey bees and other pollinating insects, which means that chemical control measures should be taken before or after flowering. If control is needed during flowering, insecticides with low toxicity for honey bees must be used (Andersson, 1975). From experiments in Sweden, Andersson concluded that damage by *A. fabae* mostly occurs at a rather late stage of colonization. Consequently, an action threshold of 25 aphids per plant on at least 25% of the plants can be used (see also section 11.4).

In England most bean crops sown in spring are treated with insecticides to control *A. fabae* (Bardner et al., 1978). In the experiments of Bardner et al., treatments before flowering were the most effective; phorate granules gave the best results and minimized the danger for bees. At least 5% of the plants should be infested before control is economically sensible. Gould and Graham (1977) found phorate granules to be less effective in dry springs. Cammel and Way (1977) and Way et al. (1977) described different control schemes for *A. fabae* in England. They found, over a period of 5 years, that the best economic results were achieved when insecticide sprays were applied according to a forecasting system based on counts of eggs during the winter and on spring counts of active stages of *A. fabae* on its host, the spindle tree (*Euonymus europaeus*). Romankov and Zarzycka (1978) obtained significantly higher pod numbers and seed

Fig. 11.1.2. *Vicia faba* heavily infested with *Aphis fabae* Scopoli, showing leaf deformation (see arrow).

yields with only one spray of pirimicarb in the period that *A. fabae* was most abundant. Dosages of 250 and 500 g per ha were equally effective.

In the Netherlands, **field beans** (*Vicia faba*) are often heavily infested with *A. fabae*. This aphid prefers to colonize the top of the stem (Fig. 11.1.2) causing growth inhibition, leaf deformation and lower pod numbers (Verhoeven, 1968). Heavy infestation can also induce dropping or shrivelling of immature pods (Banks and Macaulay, 1967). However, no exact data for physical damage caused by feeding exist, as it is difficult to separate these effects from those of virus infections, mostly caused by *A. pisum* and *M. persicae*. Here chemical control is done when *A. fabae* becomes numerous, in which case pirimicarb is mostly used.

In **pea** (*Pisum* spp.), *A. pisum* is the most important aphid. Hinz and Daebeler (1984) found yield losses of up to 30% when the crop was infested in the five-leaf stage by this aphid. In the Ukraine, *A. pisum* destroyed 50–70% of the crops in years with outbreaks of the aphid population (Sanin et al., 1975). Sprays with methylparathion or formothion gave effective control. Posylaeva (1976), in a search for insecticides which are less harmful to beneficial insects, found 0.1% pirimicarb a.i. to be very effective, with a mortality of natural enemies of only 10%. Dimethoate, at the same concentration, controlled aphids less effectively; moreover, it killed all natural enemies. Here, too, treatments of 50-m-wide border strips of large fields during the period of mass immigration of aphids gave satisfactory control.

In India, application of carbofuran granules (1 kg a.i. per ha) controlled *A. pisum* and *Macrosiphum pisi* (Kaltenbach) for up to 63 days, while phorate and aldicarb granules were less effective (Mote, 1979).

Combined control of aphids and other pests (e.g. thrips, *Sitona* spp., *Phytomyza* spp.) is often necessary. In Denmark, Thygesen (1971) obtained yield increases of 6–24% with a single insecticide application in June. In India, Srinivasan and Shukla (1980) obtained good control of pests with granular insecticides applied at sowing time followed by a second application 40 days after plant germination.

Cruciferae

Although many aphid species can occur on **brassica** crops (*Brassica* spp.) *Brevicoryne brassicae* (Linnaeus) is the most harmful species all over the world. It damages the crop by direct feeding but also acts as a virus transmitter. In India and some other Asian countries *Lipaphis erysimi* (Kaltenbach) is also of major significance on cabbage, rape and mustard (Lee, 1973; Nath, 1975; Prasad, 1978a, b; Mohan et al., 1981; Kalra et al., 1983). *B. brassicae* has a high reproduction rate and forms large colonies on the underside of leaves, which turn lumpy with curling edges and soon show yellow spots. As a consequence photosynthesis will decrease. Considerably higher yields of cabbage can be obtained by good aphid control (French et al., 1977; Gloria, 1978). Most aphid damage, however, is caused by heavy contamination of the cabbage heart, reducing food quality. It is therefore that in the Netherlands chemical control starts as soon as aphids are observed on the crop. Systemic insecticides with long residual activity are sprayed, but from four weeks before harvest onwards, only treatments with low persistence are tolerated. If caterpillars are an additional pest in the crop and aphids are not too abundant, pyrethroids must be recommended instead of longer acting aphicides (Buishand and Snoek, 1985). Graham (1984) also points to the necessity of early aphid control on cabbage, because if the crop becomes heavily infested, serious injury has already occurred.

Different insecticides should be used alternatively to prevent aphid

Section 11.1 references, p. 112

resistance, and selectivity should be taken into account. However, even selective aphicides will indirectly affect the population development of their predators by taking away their food supply (Alexandrescu and Hondru, 1980). The persistence of insecticides differs much, so safety terms should always be kept in mind. However, washing and cooking of vegetables remove much of the possible residues (Sarode and Lal, 1981).

Frequent application of insecticides can result in some phytotoxicity in **Brussels sprouts** (*B. oleracea gemmifera*). For this and economical reasons, insecticide sprays should be determined by careful monitoring of the pests (Jackson, 1982). Granular insecticides applied to the soil at planting time give long protection against aphid attack Narkiewicz-Jodko, 1976; Tandon and Bhalla, 1976; Van Dijk and Krause, 1978). However, the effectivity of the compound varies with soil humidity and is lower under dry conditions (Suett, 1977).

Lammerink and Banfield (1980) provided good control in **rape seed** (*B. napus*) until the full flowering stage by applying a granular insecticide together with the seed in the furrow. A considerable higher pod number, number of seeds per pod and hence seed yield was obtained. Carlson (1973) obtained good control of *B. brassicae* and *L. erysimi* and increased the seed yield of rape with insecticide sprays. *L. erysimi* caused less damage than *B. brassicae*.

As was mentioned above, *L. erysimi* was found to be very harmful to **rape** and **mustard** (*B. juncea*) in India. Control of this aphid by application of granular systemic insecticides to the soil or by regular sprays after the onset of flowering led to higher seed yield (Nath, 1975; Prasad, 1978a, b, 1979; Bhadoria et al., 1982; Kalra et al., 1983). However, Sharma and Kathri (1978), although getting good aphid control with demeton-S-methyl, did not obtain any yield increase, probably because of adverse effects on pollinating insects, thus confirming the importance of the use of selective control agents.

Compositae

The "salad" crops like **lettuce** (*Lactuca sativa*) and **endive** (*Cichorium endivia*) can seriously by injured by the root aphid *Pemphigus bursarius* (Linnaeus). It colonizes the roots, which may lead to the death of the plants and hence to large yield losses. The aphid usually hibernates during the egg stage on its host, *Populus nigra*, but it can also overwinter in the soil as radicicolae (Hauri et al., 1973). Although insecticidal treatments of the winter host gave quite good results, the wide spread of *P. nigra* in private gardens and public parks makes complete control unfeasible (Toscano et al., 1977). Therefore it is better to control the aphids on the (summer) crop. Good results can be obtained with soil application of granular systemic insecticides before sowing (Gentile and Vaughan, 1974; Thompson and Percivall, 1981). However, Toscano et al. (1977) found inconclusive results with side-dressing of certain granular insecticides in commercial lettuce fields.

The major aphid species feeding on the leaves of lettuce are *Nasonovia ribis-nigri* (Mosley), *M. persicae* and *M. euphorbiae*. They also act as virus transmitters, and in large numbers they cause losses by nutrient drain from the leaves. High loss of quality is caused by heavy contamination of the crop through honeydew production. The above-mentioned treatment just before sowing against root aphids will also protect the crop against leaf aphids, at least for part of the growth period. If this is not the case, extra sprays have to be applied to keep the crop clean. The safety term should be watched. Acephate and pirimicarb showed no phytotoxic effects on lettuce and had only low mammalian toxicity (Simonet, 1980). Heptenophos has much effect on aphids and a very low persistence.

Safflower (*Carthamus tinctorius*) is a crop, particularly grown in South Asian countries, that can seriously be damaged by aphids. The aphid *Uroleucon (Uromelan) compositae* (Theobald) mainly occurs on this crop. Grain losses by aphid attack were found to be reduced by 67% when the crop was sprayed with phosphamidon (Basavanagoud et al., 1981). In field trials, Grewal et al. (1982) achieved nearly 100% control of *M. persicae* for two weeks after spraying with oxy-demeton-methyl or monocrotophos. Eight insecticides, including phosphamidon and pirimicarb 0.05% a.i. left acceptable residues in the seeds of treated plants (Goud et al., 1981).

M. persicae is the most important species attacking **chrysanthemum**, (*Chrysanthemum spp.*). Control is needed to harvest clean, marketable plants and is done by sprays or by soil drenching (Webb and Argauer, 1974; Oetting, 1982). Phytotoxicity is undesirable in this crop but often occurs, although the degree varies with the chrysanthemum variety. Most phytotoxic were dimethoate (Baranowski, 1976), monocrotophos (Lindquist and Spadafora, 1975) and carbofuran (Oetting et al., 1977). Tank-mixes of insecticides with fungicides increased phytotoxicity (Schuster and Engelhard, 1979).

Rosaceae

On **fruit trees** aphids can cause much harm to growth, yield and fruit quality. Most noxious is the species *Dysaphis plantaginae* (Passerini) (Pasqualini et al., 1982; Van Frankenhuyzen and Gruys, 1983). It mainly occurs on apple and its infestations causes the fruits to remain very small; they will be malformed and do not ripen. Yield losses ranging from 50 to 70% have been reported (Hull and Starner, 1983). Other important aphids on fruit trees are *Eriosoma lanigerum* (Hausmann), *Aphis pomi* de Geer, *Dysaphis devecta* (Walker) and *Rhopalosiphum insertum* (Walker) on apple and pear, *Brachycaudus helichrysi* (Kaltenbach) and *Hyalopterus pruni* (Geoffroy) on plum and cherry, *M. persicae* and *Appelia tragopogonis* Kaltenbach on peach.

Although virus diseases can be transmitted by some species, most harm by

TABLE 11.1.2

Toxicity of insecticides to aphid enemies

Active ingredients	Hymenoptera	*Aphelinus mali*	Coccinellidae	*Prospaltella perniciosi*	Syrphidae	Neuroptera	Predatory mites
Azinphos methyl	+	+ +	+ +	+ +	+ +	+ +	+ (+)
Bromophos	+	+	0	0	0	0	+ +
Diazinon	+	+ +	+	+ +	+ +	+	+
Dimethoate	+ +	+ +	+ +	+ (+)	+ +	+ +	+ (+)
Endosulfan	+	+	0(+)		0(+)	+	0(+)
Fenthion		+ +	+				+ +
Heptenophos						+	
Malathion	+ +	+ +	+ +	0(+)	+ +	+ +	+ +
Methidathion	+		+ +	+		0	+ +
Methomyl							+ +
Phosalone	+ (0)	+	+ (0)	+ (0)	+	+ (0)	+ (0)
Phosmet			0(+)			+ (+)	+ +
Phosphamidon	+	+	+ +	+ +		+	+ (+)
Pirimicarb	0	+	0		+	0	0
Propoxur	+		+		+	+	+ (+)
Thiometon			+ +			+ +	+
Vamidothion	+	0(+)	0(+)	0		0	+ (+)

+ +, High toxicity; +, moderate toxicity; 0, no toxicity; (0), (+) indicate values between that of the two symbols.
Reference: Fischer-Colbrie (1982).

Section 11.1 references, p. 112

aphids is done by direct damage. The heavy nutrient drain imposed by the aphids causes growth inhibition and yield loss. The aphids further cause the development of gall-like swellings and curling and necrosis of the leaves, leading to poor fruit setting and early leaf fall. Excreted honeydew on leaves and fruits stimulates the development of molds, which inhibit photosynthesis and give quality loss. Young trees in nurseries and young shoots on trees in orchards are malformed by heavy aphid attack. Wounds made by aphids, for example *E. lanigerum*, on branches and twigs can act as gateways for infection with cancer.

Natural enemies of aphids play an important role in their control, but in many cases chemicals are indispensable for fruit protection. Although many insecticides are available, control measures should be chosen in such a way that natural enemies are spared. In Table 11.1.2 some information is given on the toxicity of insecticides to aphid enemies.

In chemical control of aphids it is important not to eradicate pollinating insects. This means that treatments should be done before or after flowering. If, in exceptional cases, chemical treatments are necessary during flowering, only insecticides that are harmless in that respect should be used. In the Netherlands pirimicarb is the only recommended substance in such cases (Anonymous, 1985).

Vamidothion appeared to give excellent control against aphids in fruit trees (Hameed et al., 1974; Gloria, 1978; Pasqualini et al., 1982). However, it has a high persistence and application in water-yielding regions is not allowed in the Netherlands (Anonymous, 1985). Penman and Chapman (1980) found that the synthetic pyrethroids fenvalerate and permethrin were not effective. When they replaced fenvalerate with azinphos-methyl in their spraying programme, they observed an outbreak of *E. lanigerum* within one season, due to the lower toxicity of the former compound against aphids. However, Hall (1979) obtained good control of insect and mite pests with fenvalerate, permethrin and fen-propathrin in doses ranging from 1.5 to 12 g a.i. per 100 l water at a rate of 15–18 l per tree. Cooke et al. (1976), through field experiments over a period of three years, concluded that to obtain an equal level of control of *R. insertum*, the doses of the insecticides used could be reduced to one-tenth of normal (2250 l per ha reduced to 225, 435 or 22.5 l per ha, respectively).

Most aphid species overwinter in the egg stage on their host. *E. lanigerum* hibernates as nymph on the trees, partly in crevices on the trunk and branches, partly on the roots. During a rather long period in spring migration takes place from the roots to the upper parts. A spray of insecticides on the leaves will not control the root forms of the aphid, and so rapid reinfestation of the trees can occur. In this case, Attri and Sharma (1971) obtained good results with soil application of granular insecticides. Chander and Dogra (1977) state that the upward migration in April and downward migration in October can effectively be controlled by a soil application of aldicarb or phorate. One extra spray in May or in October/November will give full control.

Roses (*Rosa*) can be seriously affected by the aphid *Macrosiphum rosae* (Linnaeus). Chumak and Petrov (1975) obtained good control by the application of an insecticide–fungicide mixture including dimethoate and zineb. Natskova (1974) found dimethoate the least and pirimicarb the most selective to aphid predators.

Strawberries (*Fragaria*) and **raspberries** (*Rubus*) can suffer from direct aphid damage. Many species can be involved; *Chaetosiphon fragaefolii* (Cockerell) is the most common one (Kennedy et al., 1976). Mother plants of strawberry must be kept free of aphids by periodical sprays of systemic insecticides (Kacharmazov et al., 1976). During the period of flowering no treatments should be applied that are harmful to bees. Especially fruit-bearing crops should receive particular attention with regard to safety terms.

Malvaceae

Cotton (*Gossypium*) is attacked by several sucking insects, e.g. the cotton aphid *A. gossypii*. Other important pests are white flies, thrips, spider mites, cicadellids, jassids and bollworms. A complex of these insect pests can cause considerable yield loss (Katarki and Thimniah, 1969; Sivaprakasam and Balasubramanian, 1981). In Burundi, Autrique (1980) reported yield losses ranging from 20 to 40%, mainly due to aphids. Similarly, in Thailand, *A. gossypii* has been recorded as being the most important pest on cotton and a possible vector of leafroll disease in this crop. The aphids prefer to feed on young leaves and stems, and a heavy infestation can kill the young plants or at least can reduce the quality of the crop (Mabbett, 1982). Cauquil et al. (1982) studied the distribution of *A. gossypii* on cotton in the Central African Repub- lic, and came to the conclusion that the colonies of this aphid can be found mostly on the lower surfaces of the leaves at the base of the plant. Therefore control by spraying with a horizontal boom, particularly with non-systemic insecticides, appeared to be insufficient.

Chemical control of aphids and other sucking pests in cotton will often be necessary. Bartsch (1978) concluded that chemical pest control on cotton in the Gezira region of Sudan is highly profitable, even when the costs are sharply rising. In the U.S.S.R. damage thresholds have been established; for *A. gossypii* it is 20–25 individuals per 100 leaves (Alimukhamedov and Mirpulatova, 1978).

Various insecticides can be used against aphids in cotton. Application of granular systemic insecticides to the soil at sowing time gave early control and reduced insect attack for several weeks. It led to increased plant vigour and yield (Khalil and Rizk, 1972; Bindra et al., 1973; Borle et al., 1980). However, soaking of the seeds in a 1% solution of aldicarb or of its metabolites sulfoxide and sulfone gave serious inhibition of germination, whereas coating of the seeds with those toxicants at 2.5 and 5% w/w gave no harmful effects (Regupathy and Subramaniam, 1982). Application of cattle dung significantly reduced the action of aldicarb when applied to the soil (Regupathy and Subramaniam, 1979).

For insecticidal sprays several compounds can be selected. Systemic insec- ticides are preferred, as they kill the aphids on all parts of the plants. In order to prevent the build-up of adverse side effects from intensive spraying, com- pounds should be used alternately. Low-volume and ultra-low-volume appli- cation of insecticides – the latter in undiluted form, e.g. by air craft – gave results just as good as those obtained with conventional spray methods (Stefanov, 1972; Harcharan Singh and Vijay, 1975; Sokhta and Rasulev, 1981).

Miscellaneous

Urticaceae. Hop (*Humulus lupulus*) is infested by the damson-hop aphid, *Phorodon humuli* (Schrank) and can cause serious reduction of yield and quality. Sometimes the crop can even be completely destroyed (Perju et al., 1979). Aphid control is particularly important during the development of the cones, because it is almost impossible to kill the aphids that have already penetrated (Tröster and Griesel, 1983). Infestation of hop can therefore be reduced most effectively by spraying other host plants in spring (Taran and Semenyuk, 1976). Because of the rapid growth of the hop plant, control meas- ures must be applied frequently and care should be taken to wet all plant parts (Tröster and Griesel, 1983). Good control results were obtained when air-craft application was added to ground control (John and Seefried, 1977). Granular insecticides applied to the soil are effective for long-term aphid control, with aldicarb being preferable to phorate because it better controls another pest of hop, the two-spotted spider mite (Cone, 1975). Allen and Easterbrook (1981) found drench treatments with the persistent insecticide mephosfolan greatly effective, because of immediate uptake by the plant roots. Caldicott (1974) obtained satisfactory control of aphids on the foliage as well as in the cones by 3–4 insecticidal sprays (e.g. phosphamidon, dimethoate, pirimicarb or propoxur) early in the season, followed by a mephosfolan drench in early July. Meier and Jossi (1983) obtained good aphid control with insecticide treatments in an experiment where the untreated plots had an infestation level of almost 1000 aphids per leaf and no yield at all. Highest yields were obtained with the synthetic pyrethroids deltamethrin and cypermethrin.

Cucurbitaceae. Mostly aphids are not so important in causing direct damage to crops like cucumber, cucurbit, melon, squash, courgette or gherkin; occasionally the cotton aphid *A. gossypii* has been reported from southern countries. Control can be achieved by seed treatment at sowing time or by spraying after emergence of the crop. Several insecticides are suitable (Abdel Shaheed et al., 1972). Soil application of aldicarb gives long-term control (Chahal and Mann, 1977), without adverse effect on syrphid predators (Abdel Salam et al., 1972).

Rutaceae. Citrus trees can suffer from aphid attack (Ebeling, 1959). This can lead to nutrient drain and defoliation of young shoots as well as to virus transmission, and growth vigour of the trees is unfavourably influenced. The main aphid species on citrus is *Aphis citricola* van der Goot. Spraying of various insecticides (e.g. dimethoate, demeton-methyl, pirimicarb or propoxur) is suitable for control (Rivero and Jaen, 1972; Koli et al., 1981; Maheswari, 1981) Trunk treatment can also give excellent results (Bullock, 1972). This method is of particular value in areas where water is scarce (Tao and Wu, 1968). Tao and Wu also recommend the destruction of ants' nests in the vicinity of the trees, since ants have a positive effect on the development of the aphids. Application of systemic insecticides to the soil can also be done. The residual activity in the leaves after soil application of ethiofencarb (18 g a.i. per tree) was sufficient for aphid control and lasted for more than 42 days (Aharonson et al., 1979).

Coniferae. *Chermes abietis* (Linnaeus) is a common pest of conifer species. Other noxious aphids are *Cinara atlantica* (Wilson), *Essigella pini* (Wilson) and *Mindarus abietinus* (Koch). They cause gall formation, the stunting and death of twigs and growth inhibition of young trees. Damage often leads to loss of esthetic quality, e.g. the trees become unmarketable as Christmas trees (Maksymov, 1975). Chemical control of aphids can prevent this, using soil treatments of aldicarb, carbofuran and disulfoton (Nielsen and Balderston, 1977; Hood and Fox, 1980). Dusts and sprays of insecticides are also effective, particularly if treatments start early in the growth season (Schread, 1971; Puritch et al., 1980).

CHEMICAL CONTROL TO REDUCE VIRUS DISEASES

Solanaceae

Viruses can lead to excessive yield loss in **potato** because they cause infectious diseases which are transferred with the seed tubers to the progeny crop. Yield loss can rise to more than 80%, depending on the type of virus, the potato variety, the rate of infection and growth conditions, (Arenz and Hunnius, 1959;

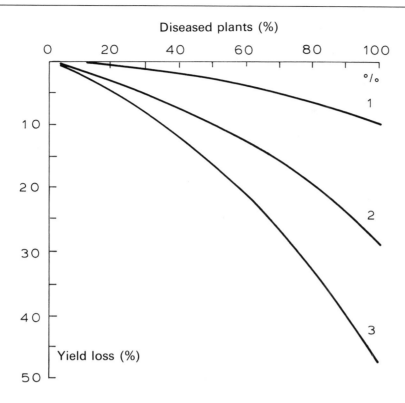

Fig. 11.1.3. Relation between the percentage of diseased plants and yield loss of potatoes, *Solanum tuberosum*, for different viruses: (1) PVX (potato virus X) and PVS (potato virus S); (2) PVA (potato virus A) and PVY^N (potato virus new Y); (3) PLRV (potato leaf roll virus) and PVY^O (potato virus old Y).

Borghardt et al., 1964; Jotoff, 1971; Dédić, 1975; Kolbe, 1981). The yield of partly diseased potato fields is less affected, because the healthy plants will compensate for the yield loss of the neighbouring diseased ones. Van der Zaag (1977) constructed curves which show an average yield loss of potato crops plotted against the percentage of plants diseased by different viruses (Fig. 11.1.3).

Since it became known that potato viruses are transmitted by aphids, chemical control measures against these insects were studied to prevent virus spread. The effect of aphid control in a seed-potato field depends on the type of virus, on the healthiness of surrounding potato fields and on the number and flight activity of winged aphids. Wingless aphids contribute little to virus spread, unless they are dispersed with the rogueing of diseased plants (Hille Ris Lambers et al., 1953). The use of (systemic) insecticides effectively inhibits the spread of potato leafroll virus (PLRV) from infected plants within the field, as is shown in Table 11.1.3 (Schepers et al., 1954). The persistent mode of transmission of PLRV implies that the aphid vector, *M. persicae*, is killed after feeding and acquisition of the virus from a treated, infected plant before other plants can be inoculated. However, infection by viruliferous aphids immigrating from surrounding untreated fields cannot be prevented (Hille Ris Lambers et al., 1953; Broadbent et al., 1956, 1958; Rönnebeck, 1957).

Regular foliage sprays of systemic insecticides after plant emergence, application of granules at planting time, or side-dressing before plant emergence give good inhibition of PLRV spread (Burt et al., 1960; Powell and Mondor, 1973; Bauernfeind and Chapman, 1978). Marco (1980) obtained good control and very few infections with a programme in which soil-applied granules of aldicarb or ethiofencarb at planting time (early May) were alternated

TABLE 11.1.3

Distribution of potato leaf roll virus infection (in %) in rows parallel to a row of secondarily diseased plants after three or eight treatments with a systemic insecticide

Treatment[a]	Row number											Average percentage of leafroll
	0	1	2	3	4	5	6	7	8	9	10	
Water	100	74	37	22	19	16	16	14	6	6	4	21.4
Systox 3×	100	7	1	3	3	1	3	2	3	3	3	2.9
Systox 8×	100	1	1	0	2	1	0	0	0	2	1	1.0

[a]Systox applied at 0.1% a.i., 1300 l per ha.
Reference: Schepers et al. (1954).

with foliage sprays with pirimicarb, methamidofos or ethiofencarb from the end of July until the harvest in September.

Due to the non-persistent mode of transmission of aphid-borne mosaic viruses of potato, insecticide treatments to the crop have no or little inhibiting effect on the spread of those viruses (Gersdorf, 1960; Burt and Heathcote, 1964; Schepers, 1972). They are transmitted by many aphid species (Van Hoof, 1980), among them species which, in search of a suitable host, pay only short visits to many plants in the crop. These viruses can be transmitted within seconds, and after probing an infected plant the aphid can visit any number of plants, thereby possibly transmitting mosaic viruses. However, a reduction in the spread of potato virus Y (PVY) by the use of insecticides has sometimes been reported (Broadbent et al., 1956, 1958; Van der Wolf, 1961; Hornig, 1963; Nirula and Kumar, 1969). Good control of non-persistent viruses can also be obtained with regular sprays of mineral oil (Schepers et al., 1978).

Tobacco is infected by potato virus Y (PVY), by tobacco vein mottling virus (TVMV) and by alfalfa mosaic virus (AMV). However, until 1976, when a widespread outbreak of PVY and TVMV occurred, no serious losses were known from virus attack in U.S.A. Although tobacco is colonized by *M. persicae*, this aphid does not seem to be a key vector of the viruses. Other aphid species, including *A. citricola* and *A. gossypii* are thought to be the main vectors. Application of insecticides failed to delay or to reduce the incidence of those three non-persistent viruses (Reagan et al., 1979). In Russia another virus disease (unnamed by the author), causing dwarfing of plants and transmitted by *Thrips tabaci* is known in tobacco. Transmission was reduced by periodical sprays of insecticides, e.g. dimethoate, fenitrothion, malathion or phosalone (Tkatch, 1979). The best results were obtained by rotation of these insecticides.

Tomato mosaic virus (TMV) is one of the mosaic causing viruses in **tomato**. These viruses are non-persistently transmitted by several aphid species. Control of aphids by insecticides will be of little value. Tomato varieties resistant against TMV have been developed (Buishand, 1979).

Aphids, in particular *M. persicae*, are transmitters of PVY and cucumber mosaic virus (CMV) to **sweet peppers**. The transmission is non-persistent and insecticides, although reducing aphid attack and feeding damage, are of little use. Instead of this, mineral oil sprays can reduce virus infections to a large extent (Loebenstein et al., 1970; Gracia and Boninsegna, 1977; Su, 1982).

Gramineae

The most important virus in **cereals** is barley yellow dwarf virus (BYDV).

Although all cereals are susceptible, barley and oats are more severely attacked than wheat, and the risk of infection increases when sowing in early autumn and late spring. Changing the sowing time can give reasonable control (Plumb, 1977). The virus is persistent and the main vector is *R. padi* (Plumb, 1977; Horellou and Evans, 1980; Port, 1983). There are indications that the virus is transmitted to cereal crops sown in the autumn by aphids migrating from maturing maize fields, which often act as a reservoir of BYDV. Yield losses due to BYDV attack of 3.5 tonne per ha have been reported, and chemical control of aphids can be highly effective: significant reduction in disease incidence and yield increases up to 90% were found after one single insecticide treatment (Horellou and Evans, 1980). Control must be done in the two- or three-leaf stage (Lechapt, 1979), and in any case within 7 days after the first aphids appear in that stage (Horellou and Evans, 1980). Kendall et al. (1983) also obtained better results with an early treatment – in October or November – than with one in December or March.

Synthetic pyrethroids give better aphid control in comparison with organophosphates (Horellou and Evans, 1980; Kendall and Smith, 1982). Horellou and Evans, using permethrin, mentioned the following advantages of this compound in BYDV control in their experiments on winter barley: (1) good persistence, which is useful to control aphid migration into the crop over a reasonable period of time; (2) a negative temperature coefficient, i.e. it is more toxic to aphids at lower temperatures – this may be important for treating cereals in autumn; (3) antifeeding/repellent activity against a range of insect species, possibly thus preventing aphid alatae from landing and settling.

It was concluded that 5% yellowing of the crop can cause significant yield losses in barley, and by this criterion a permethrin spray of 40 g a.i. per ha was the optimum dose for control.

One of the viruses attacking **maize** is maize dwarf mosaic virus (MDMV). A weed, namely Johnson grass (*Sorghum halepense*) functions as a virus reservoir (Onazi and Wilde, 1974; Milinko et al., 1979). Because this virus is non-persistent, chemical control measures of the aphid vectors did not reduce virus infection (Onazi and Wilde, 1974; Rains and Christensen, 1983). However, mineral oil sprays on the crop were effective in that respect (Ferro et al., 1980; Szatmari-Goodman and Nault, 1983). Spread of another virus in maize, the persistent maize chlorotic dwarf virus (MCDV), could indeed be reduced via aphid control (Rains and Christensen, 1983).

Chenopodiaceae

Sugar beet can be infected by several aphid-borne viruses: beet yellows virus (BYV), beet mild yellowing virus (BMYV), beet mosaic virus (BMV) and beet western yellows virus (BWYV). The beet-colonizing aphid species *M. persicae* and – to a lesser extent – *A. fabae* transmit those viruses into the crop from other sources, such as fodderbeet clamps, certain weed species and volunteer beet plants in other crops. Aphid species that do not colonize but frequently visit the crop, e.g. *B. helichrysi* and the cereal aphids *R. padi, M. dirhodum* and *S. avenae*, although not very effective virus transmitters, can still be important vectors of the non-persistently transmitted BMV (Katis and Gibson, 1984).

Virus attack can lead to serious yield decreases in beets grown for sugar as well as for seed production. Losses of up to 30% of sugar and more than 50% of seed have been reported (Fritzsche et al., 1974; Steudel, 1976; Wiesner, 1983). The degree of loss is related to the time of virus infection during the growth period. Average yield losses due to virus yellows are assumed to be 3% of the potential yield per week of infection (Heathcote, in Mumford, 1977). If

Section 11.1 references, p. 112

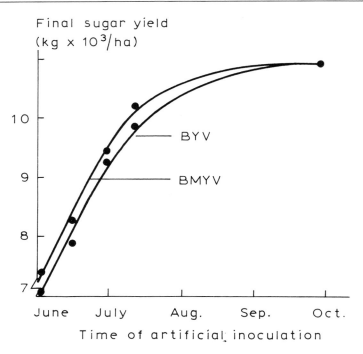

Fig. 11.1.4. Effect of the time of inoculation with two beet yellows viruses on the final yield of sugar beet, *Beta vulgaris* var. *altissima* (after Heijbroek, 1984).

symptoms of yellowing appear as late as early August, no significant yield depression will occur (Heijbroek, 1984). The damage can vary strongly from year to year; only if 20% or more of the beet plants show symptoms of virus yellows by the end of August will crop yield be diminished (Heathcote, 1978). Final yield losses induced by artificial inoculations with the yellowing viruses BYV and BMYV at different times during the growth period were calculated from an experiment in the Netherlands in 1982; these are shown in Fig. 11.1.4 (Heijbroek, 1984).

Virus incidence and yield loss can be decreased by the use of insecticides. Fritzsche et al. (1974) obtained 53% less infection by BMV after two or three insecticide sprays in beets grown for seed production. Wiesner (1983) mentioned 79% less BMV infection and even 91% less BYV infection by repeated chemical control of aphids in beets sown in autumn. In the experiments of Proeseler et al. (1976), the effect of insecticides on BMV spread (53%) was increased by another 25% if citol oil was added to the treatment. Fritzsche et al. (1984) observed lower numbers of aphids and less spread of BYV and BMYV if certain pigment-carrier mixtures were added to the insecticide sprays. Crop colour became more intensive. The pigments used were iron oxide, permanent red extra, umbra, ultramarine and Berlin blue. Carriers are dispersals of polymers without organic solvents.

Hull and Heathcote (1967) found a much-reduced incidence of virus yellows, expressed as "infected plant weeks" (IPW) and an increased yield of sugar beets with one spray of insecticides in experiments from 1954–1966. The best results were obtained when the treatment was done at the moment that a spray warning was given by the British Sugar Corporation. For example, in 1962–1966 sprayings at warning time reduced IPW by 37%, spraying two weeks earlier or two weeks later reduced IPW by 24% and 25%, respectively. A warning service operates in many countries, informing the farmers about the right spraying time. The internationally agreed threshold is one *M. persicae* per four beet plants (Heijbroek, 1984). One insecticidal treatment may be sufficient, if given at the right moment. Tamaki et al. (1979), in the U.S.A., reported that with adequate timing the three currently recommended applications could be reduced to only one.

However, it should be remarked that fields near drainage ditches must sometimes receive an early insecticide treatment after mild winters, because aphids are then able to overwinter on weeds in these ditches. Nesterenko (1976) also recommended border treatments to delay virus spread.

Granular insecticides applied during sowing may give protection to early virus infections. When viruses attack later or when the weather is dry, sprays are preferred (Steudel, 1962). Mixing of granular insecticides with the seed can give reduced germination (Lewartowski and Trojanowski, 1972).

In **spinach** (*Spinacea oleracea*), mosaic viruses (cucumber mosaic, beet mosaic) can occur. Both are non-persistently transmitted by several species. Growth of infested plants can stop completely (J. De Kraker, personal communication, 1984). The persistent beet yellows virus can also attack spinach. Aphid control can reduce the spread of those viruses to a certain extent but is difficult because of safety problems in this fast-growing crop. Mosaic-resistant varieties are available, and therefore chemical control is not applied in spinach in the Netherlands.

Leguminosae

In leguminous crops virus diseases are not so important as direct damage (see p. 97–99). In many cases the spread of virus diseases is also reduced by control of the aphids. In **alfalfa** the non-persistent alfalfa mosaic virus is cosmopolitan. The virus is transmitted by at least 13 aphid species (Ashby et al., 1979), among them many species that do not colonize the crop. Alfalfa is also the winter host plant of the pea top yellowing virus, a virus mainly transmitted by the aphid *A. pisum* to beans and peas. In the Netherlands, farmers are advised not to grow peas and beans adjacent to alfalfa (Anonymus, 1985). Insecticidal treatments can partly reduce the spread of this persistent virus, which can be particularly harmful in peas. However, many pea varieties are resistant to this virus (Anonymus, 1985). Good control of BLV could be achieved by multiple insecticide sprays or by a granular insecticide applied at sowing time. However, when alfalfa was grown adjacent to the pea crop, results were much less favourable (Stoltz and Forster, 1984).

Furthermore, **pea** can be attacked by mosaic viruses. The pea enation mosaic virus (PEMV) is persistent, the bean common mosaic virus (BCMV) and the pea seed-borne mosaic virus (PSbMV) are non-persistent, so that insecticide treatments will have limited success. PEMV, BCMV and PSbMV and the non-persistent bean yellows mosaic virus (BYMV) also occur in (faba) beans. The most important vectors in **bean** are *A. fabae*, *A. pisum* and *M. persicae*. Generally insecticide sprays are hardly effective, but favourable results can be obtained by control measures over large areas applied at one time (Schmidt et al., 1977). Oil sprays can reduce the spread of these viruses (Walkey and Dance, 1979). Subterranean clover red leaf virus (SCRLV) causes important yield reductions of bean in Tasmania (Johnstone, 1978). Its spread by *A. solani* was strongly inhibited by regular sprays up till pod set (Johnstone and Rapley, 1981).

Bean (pea) leaf-roll virus (BLRV) can cause considerable yield loss in bean. Early infected plants often die prematurely. *A. fabae* is the most important vector of this virus. Two applications of a systemic insecticide, the first soon after emergence and the second 10–15 days later, protected the crop very well (Schmutterer and Thottappilly, 1972).

Ground nuts (*Arachis hypogaea*) can be attacked by the rosette disease, a

Section 11.1 references, p. 112

virus transmitted by the aphid *A. craccivora*. Reduced incidence and better yields could be obtained by regular crop spraying with insecticides, e.g. dimethoate, thiometon, menazon or trichlorphon (Gangadharan et al., 1972; Davies, 1975a, b).

Cruciferae

B. brassicae and *M. persicae* are the main virus vectors in **brassica** crops. Important viruses are cauliflower mosaic virus (CaMV), turnip mosaic virus (TuMV) and broccoli necrotic yellows virus. Another virus, the turnip yellow mosaic virus, is transmitted by flea beetles such as *Phyllotetra* and *Psylliodes* spp. (Hill, 1983). CaMV is known to reduce yield in winter-hard cauliflower, while TuMV causes internal necrosis of cold-stored white cabbage (Whitwell, 1977), thus reducing quality. The yield of swedes can seriously be affected by TuMV (Tomlinson et al., 1976). Remnants of cabbage crops left in the field after harvesting are important sources of infection, from which aphids transmit viruses to newly-grown crops in spring. Therefore, in addition to the aphid control by insecticides that is done to prevent direct damage – which only partly inhibits the spread of these non-persistent viruses – timely ploughing after the harvest in autumn or spring will be important in minimizing the risk of virus infection. Moreover, there should be a distance of at least 100 m between seedbeds with young plants and cabbage crops or plants kept for seed (Whitwell, 1977).

Compositae

Several aphid species can transmit virus diseases to **lettuce** and **endive**. In susceptible varieties lettuce mosaic virus (LMV) can cause large yield losses in cabbage-lettuce, as growth is inhibited and the head will not close. Other host plants of this virus are Compositae, such as *Aster*, *Tagetes*, *Zinnia* and *Senecio*. The virus is also transmitted by pollen and by seed (Schmidt et al., 1981; Crüger, 1983). Modern varieties of cabbage-lettuce are tolerant against this virus disease, while varieties of crisphead lettuce are not.

Other virus diseases are lettuce necrotic yellows virus and cucumber mosaic virus. Like LMV they are non-persistent, so that insecticide treatments will not have much value in preventing infection.

In floriculture there are also a number of non-persistent viruses. As insecticide treatments also are not very effective here, Hakkaart (1968) tested the application of mineral oil on **chrysanthemum** to prevent spread of tomato aspermy virus, with good results. Application of vegetable (olive) oil was not successful.

Rosaceae

Although aphid transmitted virus diseases do occur in **fruit trees**, direct damage by aphids is more important in this crop (see pp. 101–102). Aphids transmitting viruses are, for example, *M. persicae* and *A. pomi*. In addition to the use of insecticides, control has also been attempted by the use of chemosterilants. Compounds such as 5-fluorouracil, thiotepa, tepa, metepa and apholate have been shown to be effective in inducing sterility and reducing the reproductive potential of aphids. However, beneficial insects such as Coccinellidae and Chrysopidae are also destroyed (Sharma, 1980).

Viruses can have serious detrimental effects on the production of both runners and fruits of **strawberries** (Becker and Rich, 1956; Craig, 1957; McGrew and Scott, 1959; Freeman and Mellor, 1962). The viruses involved are mottle, mild yellow edge and crinkle viruses. They are all non-persistent. Several aphid species can transmit those viruses, an important one being *C. fragaefolii*, a widespread aphid species in different parts of the world. The aphid *Rhodobium porosum* (Sanderson) is also a vector of strawberry viruses and can be abundant in Nova Scotia, Canada (Craig and Stultz, 1964). To inhibit damage by virus diseases in strawberry crops, it is important to control infection of the mother plants in nursery stocks by growing those plants in an isoláted area of at least a few miles' diameter under a strict regime of insecticide treatments. After the young plants have been transplanted, frequent sprays are necessary (Kacharmazov et al., 1976); alternatively, granular insecticides can be applied at planting time (Chisholm and Specht, 1978; Shanks, 1981).

Malvaceae

A. gossypii is the transmitter of the agent causing the "blue disease" in **cotton** in Central Africa. This disease causes serious damage to the crop and is believed to be of viral origin. Either seed coating with insecticides, with or without an additional spray or soil application, or two sprays with a 15-day interval after crop emergence could prevent aphid infestation for at least two months of the growing season; this also increased the height of the plants and the quality of the seed. Treatments were only economical for fields that have a potential yield of at least 1 tonne of seed cotton per ha (Cauquil et al., 1982, 1983).

From Thailand a leafroll disease has been reported in cotton (Mabbett, 1982). Studies pointed to a persistent virus, transmitted by *A. gossypii*. Levels of diseased plants have reached 10%, particularly in late-sown cotton (Sachs, 1979). The effects of insecticides on the disease are not clear, since Jones and Wangboonkong (1964) observed the symptoms on plants in spite of complete insecticidal protection.

Miscellaneous

Liliaceae. Loss of weight and quality of tulips (*Tulipa*) and lilies (*Lilium*) is caused by the tulip breaking virus (TBV). Hyacinth mosaic virus (HMV) can reduce growth and yield of hyacinths (*Hyacinthus*). Irises (*Iris*) can be infected by two different strains of mosaic virus, the iris mild mosaic virus and the iris mild yellow mosaic virus. All these viruses are non-persistently transmitted by several aphid species, colonizing or just visiting the crop. Insecticidal treatments have limited effects on virus spread. In the Netherlands, Asjes (1981, 1985) found a consistent decrease in the spread of non-persistent viruses in flower bulbs sprayed weekly with permethrin, cypermethrin, deltamethrin and fenvalerate. The effect of pirimicarb was very variable. The most effective prevention of virus spread, in all crops, was obtained with mineral oil sprays at 20 or 15 l per ha, but these treatments had a negative effect on bulb yields. Sutton and Garrett (1978), in Australia, did not obtain any positive results from insecticide or oil sprays in attempts to control TBV in tulips. Infection with cucumber mosaic virus can reduce flower yield of *Gladiolus*. In the experiments of Adams et al. (1976), insecticidal treatments did not increase the proportion of flowering plants, although aphid numbers were reduced.

Umbelliferae. *Cavariella aegopodii* (Scopoli) is the main vector of carrot mottle virus in England. Early migration from its winter hosts, *Salix* sp., to **carrots** (*Daucus carota*) can have an extremely negative effect on that crop (Watson and Heathcote, 1965). However, as the virus is persistent, timely protection of the crop by (systemic) insecticides will prevent yield depression. Usually, as in the Netherlands, aphids do not play an important rôle in carrot growing, since the crop had to be treated early against the carrot fly, which will also control the aphids.

Several aphid species can transmit celery mosaic virus (CeMV) to **celery** (*Apium graveolens*) crops. This virus hibernates in several umbelliferous weeds and can cause crop failure, particularly in regions where celery is cultivated for many years. CeMV is non-persistent and therefore insecticides are of little value. Mineral oil sprays (Matthieu and Verhoyen, 1983) and the use of resistant or tolerant varieties (Walkey and Ward, 1984) show better prospects.

Cucurbitaceae. Several aphid species are transmitters of cucumber mosaic virus (CMV) and related mosaic virus in cucurbitaceous crops such as cucumber, melon, courgette and gherkin. Kibata (1979) suggested that other insects are also able to transmit these mosaic viruses. Transmission is non-persistent and CMV has many host plants: more than 200 according to Crüger (1983).

Insecticides have only little effect. However, Singh (1969) obtained good control of the spread of cucumis virus 2 in experiments with bottlegourd (*Lagenaria siceraria*) by six weekly applications of insecticides against *A. gossypii*. But insecticides, and also repellents, had no effect on the spread of cantaloupe mosaic viruses (Dickson et al., 1949). CMV is the limiting factor for growing cucumber in Israel in spring, because of the wide virus spread by aphids. Excellent control of CMV spread and significantly higher yields were obtained by the use of mineral oil (Loebenstein et al., 1964, 1966).

REFERENCES

Solanaceae

Abdel Salam, A.M., Abassy, A.M., Assem, M.A. and El-Minshawy, A.M., 1975. Studies on potato pests in Egypt. The 8th International Plant Protection Congress, Moscow, Reports and Informations Section III, Chemical Control, Part I, pp. 13–21.

Arenz, B. and Hunnius, W., 1959. Der Einfluss verschiedener Virusarten auf die Ertragsbildung bei der Kartoffel. Bayerischer Landwirtschaftlicher Jahrbuch, 36: 163–173.

Awate, B.G. and Pokharkar, R.N. 1977. Chemical control of potato pest complex. Pesticides, 11: 40–42.

Bauernfeind, R.J. and Chapman, R.K., 1978. Green peach aphid control in relation to the incidence of potato leafroll virus. Abstract, Proceedings of the North Central Branch of the Entomological Society of America, 75th Annual Conference, 33: 54–55.

Borghardt, G., Bode, O., Bartels, R. and Holz, W., 1964. Untersuchungen über die Minderungen des Ertrages von Kartoffelpflanzen durch Virusinfektionen. Nachrichtenblatt der Deutschen Pflanzenschutzdienstes (Braunschweig), 16: 150–154.

Broadbent, L., Burt, P.E. and Heathcote, G.D. 1956. The control of potato virus diseases by insecticides. Annals of Applied Biology, 44: 256–273.

Broadbent, L., Heathcote, G.D. and Mason, E.C., 1958. An Essex Farm trial on the insecticidal control of potato virus spread. Plant Pathology, 7: 53–55.

Burbutis, P.P., Davies, C.P., Kelsey, L.P. and Martin, C.E., 1972. Control of green peach aphid on sweet peppers in Delaware. Journal of Economic Entomology, 65: 1436–1438.

Burt, P.E. and Heathcote, G.D., 1964. The use of insecticides to find when leafroll and Y-viruses spread within potato crops. Annals of Applied Biology, 54: 13–22.

Burt, P.E., Broadbent, L. and Heathcote, G.D., 1960. The use of soil insecticides to control potato aphids and virus diseases. Annals of Applied Biology, 48: 580–590.

Cancelado, R.E. and Radcliffe, E.B., 1979. Action thresholds for green peach aphid on potatoes in Minnesota. Journal of Economic Entomology, 72: 606–609.

Cruz, C., 1974. Evaluation of new insecticides for control of the green peach aphid, *Myzus persicae* (Sulzer) on peppers. Journal of Agriculture of the University of Puerto Rico 58: 489–490.

Dědič, P. , 1975. The effect of Virus A (PVA) yield in some potato varieties. Sbornik ÚVTI-Ochrony Rostlin, 11: 127–133 (in Czech, with English summary).

De Mey, W. and Pauwels, W.J., 1978. Field trials in potatoes made with thiofanox in the United Kingdom in 1977. Mededelingen van de Fakulteit der Landbouwwetenschappen, Rijks Universiteit Gent, 43: 641–648.

Dirimanov, M. and Dimitrov, A., 1975. Rôle of useful insects in the control of *Thrips tabaci* Lind. and *Myzodes persicae* Sulz. on tobacco. 8th International Plant Protection Congress, Moscow, Section V, Biological and Genetic Control, pp. 71–72.

Gersdorf, E., 1960. Sind Spritzungen gegen das "Y-virus" der Kartoffel aussichtsreich? Kartoffelbau, 11: 70–71.

Gibson, R.W., 1974. The induction of top-roll symptoms on potato plants by the aphid *Macrosiphum euphorbiae*. Annals of Applied Biology, 76: 19–26.

Gibson, R.W., Whitehead, D., Austin, D.J. and Simkins, J., 1976. Prevention of potato top-roll by aphicide and its effect on leaf area and photosynthesis. Annals of Applied Biology, 82: 151–153.

Gracia, O. and Boninsegna, J.A., 1977. Application of mineral oils to control virus in peppers. Idia, 349–354: 35–40 (in Spanish).

Hails, M.R., 1967. The effect of aphids on ware yield (of potatoes). Annual Review, Terrington Experimental Husbandry Farm, 8: 18–20.

Harding, J.A., 1962. Systemic insecticide effects on Irish potato fields during 1962. Journal of the Rio Grande Valley Horticultural Society, 16: 94–97.

Hill, R.E., 1948. Research on potato insect problems. A review of recent literature. American Potato Journal, 25: 107–128.

Hille Ris Lambers, D., Reestman, A.J. and Schepers, A., 1953. Insecticides against aphid vectors of potato viruses. Netherlands Journal of Agricultural Science, 1: 188–201.

Hornig, 1963. Auftreten und Bekämpfung der Blattläuse im Pflanzkartoffelbau. Kartoffelbau, 14: 56–58.

Joshi, M.L. and Sharma, J.C., 1973. Insecticides for the control of brinjal aphid (*Aphis gossypii* Glov.), jassid (*Amrasca devestans* Dist.) and fruit-borer (*Leucinodes orbonalis* Guen.) in Rajasthan. Indian Journal of Agricultural Science, 43: 436–438.

Jotoff, L., 1971. Untersuchungen über Ertragsdepressionen bei Kartoffeln, die durch verschiedene Viruskrankungen und Klimatbedingungen in Bulgariën verursacht wurden. Potato Research, 14: 161–614.

Kolbe, W., 1978. Untersuchungen über die Blattlausbekämpfung im Pflanzkartoffelbau (1958–1978) unter besorderer Berücksichtigung von Tamaron und Bedeutung Pflanzengutwechsels für den Kartoffelertrag. Pflanzenschutz Nachrichten, Bayer, 31: 283–297.

Kolbe, W., 1981. Ergebnisse eines Kartoffel-Nachbauversuches. Kartoffelbau, 31: 352–353.

Loebenstein, G., Alper, M. and Levy, S., 1970. Field tests with oil sprays for the prevention of aphid-spread viruses in peppers. Phytopathology, 60: 212–215.

Marco, S., 1980. The use of insecticides to control potato leafroll virus in seed-potato crops on the Golan Heights. Phytoparasitica, 8: 61–71.

Muminov, A., Askaraliev, Kh. and Oripov, Kh., 1976. The control of sucking pests of vegetables in glasshouses. Zashchita Rastenii, 2: 27–28 (in Russian).

Nirula, K.K. and Kumar, R., 1969. Soil application of systemic insecticides for control of aphid vectors and leaf-roll and "Y"-viruses in potato. Indian Journal of Agricultural Science, 39: 699–702.

Pond, D.D., 1962. Insecticidal field trials for the control of potato aphids in New Brunswick, 1948–1960. Journal of Economic Entomology, 56: 227–230.

Pond, D.D., 1967. Field evaluation of insecticides for the control of aphids on potatoes. Journal of Economic Entomology, 60: 1203–1205.

Powell, D.M. and Mondor, T.W., 1973. Control of the green peach aphid and suppression of leaf-roll on potatoes by systemic soil insecticides and multiple foliar sprays. Journal of Economic Entomology, 66: 170–177.

Quanjer, H.M., 1939. The green peach aphid, a threat to the culture of potato and other crops. Tijdschrift over Plantenziekten, 5: 1–8 (in Dutch).

Reagan, T.E., Gooding, G.V., Jr. and Kennedy, G.G., 1979. Evaluation of insecticides and oil for suppression of aphid-borne viruses in tobacco. Journal of Economic Entomology, 72: 538–540.

Rönnebeck, W., 1957. Fortschritte bei der Bekämpfung insektenübertragbarer Pflanzenseuchen. Naturwissenschaftlicher Rundschau, 8: 283–286.

Schepers, A., 1972. Control of aphid vectors in the Netherlands. In: J.A. de Bokx (Editor), Viruses of Potatoes and Seed Potato Production. Center for Agricultural Publishing and Documentation, Wageningen, pp. 174–187.

Schepers, A., 1983. Toproll in potatoes. Bedrijfsontwikkeling, 14: 735–738 (in Dutch).

Schepers, A., Reestman, A.J. and Hille Ris Lambers, D., 1954. Some experiments with Systox. Proceedings of the 2nd Conference on Potato Virus Diseases, Lisse–Wageningen, pp. 75–83.

Schepers, A., Bus, C.B. and Styszko, L., 1978. Effects of mineral oil on seed potatoes. Abstracts of Papers, 7th Triannual Conference EAPR, Warsaw, pp. 269–270.

Southall, D.R. and Sly, J.M.A., 1976. Routine spraying of potatoes to control aphids and potato blight during 1969–1973. Plant Pathology, 25: 89–98.

Stefanov, D. and Georgiev, G., 1973. Possibilities for controlling resistant forms of *Myzus persicae* in tobacco. Rastitelna Zashchita, 21: 25–28 (in Bulgarian, with English summary).

Su, T.H., 1982. Aphid vectors of sweet pepper viruses. Plant Protection Bulletin of Taiwan, 24: 257–264.

Tappan, W.B., Wheeler, W.B. and Lundi, H.W., 1974. Insect control and chemical residues after applying acephate on cigar-wrapper and flue-cured tobaccos in Florida. Journal of Economic Entomology, 67: 648–650.

Tappan, W.B., Wheeler, W.B., Johnson, J.T. and Rich, J.R., 1982. Insect control and chemical residues from organophosphates and synthetic pyrethroids applied alone or in tank mixes on flue-cured tobacco. Journal of Economic Entomology, 75: 1143–1146.

Tkatch, M.T., 1979. Protecting of tobacco. Zashchita Rastenii, 8: 34–36 (in Russian).

Van Hoof, H.A., 1980. Aphid vectors of potato virus Y. Netherlands Journal of Plant Pathology, 86: 159–162.

Van der Wolf, J.P.M., 1961. Virus transmission and vector control in seed potatoes. Höfchenbriefe Bayer Pflanzenschutz Nachrichten, 17: 113–184.

Van der Zaag, D.E., 1977. Significance of infection of seed potatoes with virus for the yield of tubers. Landbouwkundig Tijdschrift, 89: 154–157 (in Dutch).

Gramineae

Andersson, K., 1977. Why does control of aphids on cereals not succeed? Vaxtskyddsnotiser, 41: 67–76 (in Swedish, with English summary).

Ba-Angood, S.A. and Stewart, R.K., 1980. Effect of granular and foliar insecticides on cereal aphids (Hemiptera) and their natural enemies on field barley in southwestern Quebec. Canadian Entomologist, 112: 1309–1313.

Babenko, V.A., 1980. A test on the rational control of aphids. Zashchita Rastenii, 6: 14–15 (in Russian).

Barbulescu, A., 1975. Effectiveness of some systemic organophosphorous insecticides in the control of the cereal greenbug (*Schizaphis graminum* Rond.). Probleme de Protectia Plantelor, 3: 155–174 (in Rumanian, with English summary).

Basedow, T., 1980. Studies on the ecology and control of the cereal aphids (Hom. Aphididae) in northern Germany. WPRS Bulletin 1980/III/4: 67–84.

Basedow, T., 1984. Die Getreideblattläuse. Nachrichtenblatt des Deutschen Pflanzenschutzdienstes (Braunschweig), 36: 61.

Castillo, B.D. and Acevedo, A.J., 1976. Protection with aphicides during various phenological stages of winter wheat (*Triticum aestivum* L.) cultivar Melifen. Agricultura Tecnica 36: 93–98 (in Spanish, with English summary).

Cate, J.R., Bottrell, D.G. and Teetes, G.R., 1973a. Management of the greenbug on grain sorghum. 1. Testing foliar treatments of insecticides against greenbug and corn leaf aphids. Journal of Economic Entomology, 66: 945–952.

Cate, J.R., Bottrell, D.G. and Teetes, G.R., 1973b. Management of the greenbug on grain sorghum. 2. Testing seed and soil treatment for greenbug and corn leaf control. Journal of Economic Entomology, 66: 953–959.

Chillar, B.S. and Verma, A.N., 1982. Yield losses caused by the aphid *Rhopalosiphum maidis* (Fitch) in different varieties/strains of barley crop. Haryana Agricultural Univeristy Journal of Research, 12: 298–300.

Daniels, N.E., 1969. Greenbug control in grain sorghum. Progress Report Texas Agricultural Experiment Station, 2669, 2 pp.

De Pew, L.J., 1974. Controlling greenbugs in grain sorghum with foliar and soil insecticides. Journal of Economic Entomology, 67: 553–555.

De Pew, L.J., 1982. Chemical control of greenbugs in winter wheat under cool temperatures in Kansas. Journal of the Kansas Entomological Society, 55: 562 (Abstract).

Feese, H. and Wilde, G., 1975. Planting time applications of systemic insecticides on grain sorghum for greenbug control; interactions with herbicides and effect on predators (Hemiptera, Homoptera: Aphididae). Journal of the Kansas Entomological Society, 48: 396–402.

Ferro, D.H., Mackenzie, J.D. and Margolies, D.C., 1980. Effect of mineral oil and a systemic insecticide on field spread of aphid-borne maize dwarf mosaic virus in sweet corn. Journal of Economic Entomology, 73: 730–735.

Ganguli, R.N. and Raychaudhuri, D.N., 1980. Studies on *Rhopalosiphum maidis* Fitch (Aphididae Homoptera) – a formidable pest of *Zea mays* (maize) in Tripura. Science and Culture 46: 259–261.

George, K.S. and Gair, R., 1979. Crop loss assessment on winter wheat attacked by the grain aphid, *Sitobion avenae* (F.), 1974–1977. Plant Pathology, 28: 143–149.

Hellpap, C., 1982. Untersuchungen zur Wirkung verschiedener Insektizide auf Predatoren von Getreideblattläusen unter Freilandbedingungen. Anzeiger für Schädlingskunde, Pflanzenschutz und Umweltschutz, 55: 129–131.

Horrellou, A. and Evans, D.D., 1980. Control of barley yellow dwarf virus with permethrin on winter barley in France. Proceedings, 1979 British Crop Protection Conference – Pests and Diseases, pp. 9–15.

Kendall, D.A. and Smith, B.D., 1982. Evaluation of insecticides to control aphid vectors of barley yellow dwarf virus in winter barley. Annals of Applied Biology, 100 (Supplement): 24–25.

Kendall, D.A., Smith, B.D., Burchill, R.G. and Chinn, N.E., 1983. Comparison of insecticides in relation to application date for the control of barley yellow dwarf virus in winter barley. Annals of Applied Biology, 100 (Supplement): 22–23.

Kindler, S.D. and Staples, R., 1981. *Schizaphis graminum*: effect on grain sorghum exposed to severe drought stress. Environmental Entomology, 10: 247–248.

Kolbe, W., 1969. Untersuchungen über das Auftreten verschiedener Blattlausarten als Ursache von Ertrags- und Qualitätsminderungen im Getreidebau. Pflanzenschutz Nachrichten Bayer, 22: 177–211.

Kolbe, W. and Linke, W., 1974. Studies of cereal aphids; their occurrence, effect on yield in relation to density levels and their control. Annals of Applied Biology, 77: 85–87.

Kontev, Kh., 1975. The control of aphids on wheat. Rastitelna Zashchita, 23: 15–17 (in Bulgarian).

Lechapt, G., 1979. Mise au point sur la jaunisse nanisante de l'orge – Perspectives de lutte raisonnée. Phytiatrie – Phytopharmacie, 28: 59–68.

Mantel, W.P., Rabbinge, R. and Sinke, J., 1982. Effects of aphids on the yield of winter wheat. Gewasbescherming, 13: 115–124 (in Dutch).

Milinko, I., Peti, J. and Papp, I., 1979. Problems and possibilities for control of maize dwarf mosaic virus in Hungary. Acta Phytopathologica Academiae Scientiarum Hungariae, 14: 127–131.

Onazi, O.C. and Wilde, G., 1974. Factors affecting transmission of maize dwarf mosaic (MDMV) and its control. Nigerian Journal of Entomology, 1: 51–55.

Pike, K.S. and Glazer, M., 1980. Compatibility of insecticide–fungicide wheat seed treatments with respect to germination, seedling emergence, and greenbug control. Journal of Economic Entomology, 73: 759–761.

Plumb, R.T., 1977. Aphids and virus control on cereals. Proceedings, 1977 British Crop Protection Conference – Pests and Diseases, pp. 903–913.

Popov, P., 1972. An efficacious method for the control of aphids on sorghum. Rastitelna Zashchita, 20: 7–9 (in Bulgarian).

Port, C.M., 1983. Timing of aphicides to control aphid vectors of barley yellow dwarf virus in winter barley. Annals of Applied Biology, 102 (Supplement): 18–19.

Rabbinge, R., Drees, E.M., Van der Graaf, M., Verberne, F.C.M. and Wesselo, A., 1981. Damage effects of cereal aphids in wheat. Netherlands Journal of Plant Pathology, 87: 217–232.

Rains, B.D. and Christensen, C.M., 1983. Effect of soil-applied carbofuran on transmission of maize chlorotic dwarf virus and maize dwarf mosaic virus to susceptible field corn hybrid. Journal of Economic Entomology, 76: 290–293.

Rajagopal, D. and Channa Basavanna, G.P., 1977. Preliminary studies of the chemical control of maize insects. Mysore Journal of Agricultural Science, 11: 73–76.

Rautapää, J. and Uoti, J., 1976. Control of *Rhopalosiphum padi* (L.) (Hom., Aphididea) on cereals. Annales Agriculturae Fenniae, 15: 101–110.

Szatmari-Goodman, G. and Nault, L.R., 1983. Tests of oil sprays for suppression of aphid-borne maize dwarf mosaic virus in Ohio sweet corn. Journal of Economic Entomology, 76: 144–149.

Tanaka, T., 1961. The rice root aphids, their ecology and control. Special Bulletin, College of Agriculture Utsonomiya, no. 10, 83 pp.

Vereijken, P.H., 1979. Feeding and multiplication of three cereal aphid species and their effect on yield of winter wheat. Centre for Agricultural Publishing and Documentation, Wageningen, Agricultural Research Report no. 888, 58 pp.

Verma, G.D., Matho, D.N. and Haque, M.N., 1979. The loss in grain yield due to aphids and other pests on certain varieties of barley grown in North Bihar. Science and Culture, 45: 370–371.

Vickerman, G.P. and Sunderland, K.D., 1975. Arthropods in cereal crops: nocturnal activity, vertical distribution and aphid predation. Journal of Applied Ecology, 12: 755–765.

Wood, E.A., 1965. Effect of foliage infestation of the English grain aphid on yield of Triumph wheat. Journal of Economic Entomology, 58: 778–779.

Yano, K., Miyake, T. and Eastop, V.F., 1983. The biology and economic importance of rice aphids (Hemiptera: Aphididae): a review. Bulletin of Entomological Research, 73: 539–566.

Chenopodiaceae

Anonymus, 1984. Manual for the Control of Diseases, Pests and Weeds in Agricultural Crops. Netherlands Ministry of Agriculture and Fisheries, 72 pp. (in Dutch).

Dubnik, H., 1977. Beobachtungen über das Auftreten von *Myzus persicae* auf Beta-rüben, den Blattlausbefall und die Blattlausbekämpfung. Nachrichtenblatt für den Pfanzenschutz in der DDR, 31: 84–87.

Fritzsche, R., Amme, M., Sass, O. and Proeseler, G.,. 1974. Vectorbekämpfung im Zucker- und Futterrübenstecklingsanbau. Nachrichtenblatt für die Pflanzenschutz in der DDR, 28: 182–185.

Fritzsche, R., Thiele, S., Kramer, W., Thust, U., Geissler, K. and Kochmann, W., 1984. Wirkung von Insektizid-Farbstoffgemischen auf den Befall von Zuckerrüben mit Vergilbungsviren und deren Vektoren *Myzus persicae* (Sulz.) sowie *Aphis fabae* Scop. Archive für Phytopathologie und Pflanzenschutz, Berlin, 20: 33–38.

Gapanova, A.F., 1977. Trophic links of the beet root aphid *Pemphigus fuscicornis* Koch (Homoptera, Aphidoidea) and measures for controlling it. Entomologicheskoe Obozrenie, 56: 292–298 (in Russian, with English summary).

Grigorov, S., 1982. The protection of sugar beet against the sugar beet aphid. Rastitelna Zashchita, 30: 29–34 (in Bulgarian).

Heathcote, G.D., 1978. Review of losses caused by virus yellows in English sugar beet crops and the cost of partial control with insecticides. Plant Pathology, 27: 12–17.

Heijbroek, W., 1984. The beet yellows disease warning system: 25 years of integrated control. Publications of the Sugar Beet Research Institute, Bergen op Zoom, pp. 1–9 (in Dutch).

Hull, R. and Heathcote, G.D., 1967. Experiments on the time of application of insecticide to decrease the spread of yellowing viruses of sugar beet 1954–1966. Annals of Applied Biology, 60: 469–478.

Katis, N. and Gibson, R.W., 1984. Transmission of beet mosaic virus by cereal aphids. Plant Pathology, 33: 425–427.

Lewartowski, R. and Trojanowski, H., 1972. Investigations on the effectiveness of granular insecticides in the control of bean aphid (*Aphis fabae* Scop.) on sugar beet. Biuletyn Instytutu Ochrony Roslin, 54: 505–519, (in Polish, with English summary).

Möllerström, G., 1963. Different kinds of injury to leaves of the sugar beets and their effect on yield. Swedish National Institute for Plant Protection Contributions, 12: 95: 299–309.

Mumford, J.D., 1977. Farmer attitudes towards the control of aphids on sugar beet. Proceedings, 1977 British Crop Protection Conference – Pests and Diseases, pp. 263–270.

Nesterenko, N.I., 1976. Lebaycid for the protection of sugar beet. Zashchita Rastenii, 7: 35–37.

Proeseler, G., Fritzsche, E., Karl, E., Geissler, K. and Zschiegner, H.J., 1976. Bekämpfung der Blattläuse als Virusüberträger der Beta-Rübe in Parzellen-und Grossversuchen durch Systeminsektizide und Citolöl. Archive für Phytopathologie und Pflanzenschutz, Berlin, 12: 127–134.

Steudel, W., 1962. Vergleichende Untersuchungen zur Wirkung von Disyston-Granulat auf Zuckerrübenschädlinge bei Zugabe zur Saat oder nach dem Auflauf zur Hacke. Nachrichtenblatt des Deutschen Pflanzenschutzdienstes (Braunschweig), 14: 104–107.

Steudel, W., 1976. Der Einfluss einer Bodenbehandlung mit Saatschutzmitteln auf den Aufgang und die durch Vergilbungsviren verursachten Ertragsverluste bei Zuckerrüben. Nachrichtenblatt des Deutschen Pflanzenschutzdienstes (Braunschweig), 28: 117–121.

Tamaki, G., Fox, L. and Butt, B.A., 1979. Ecology of the green peach aphid as a vector of beet western yellows virus of sugar beets. Technical Bulletin, Science and Education Administration, USDA, 1599, 16 pp.

Van Steyvoort, L., 1983. Evolution de la lutte contre la jaunisse de la betterave en Belgique. Publication Trimestrielle, Institut Royal Belge pour l'Amélioration de la Betterave, 51: 49–61.

Wiesner, K., 1983. Blattläuse und Virosen in Bestanden der direkten und indirekten Vermehrung von Beta-rüben. Nachrichtenblatt für den Pflanzenschutz in der DDR, 37: 59–62.

Leguminosae

Anonymus, 1985. Crop Protection; Arable Farming Central No. 6. Publication Netherlands Advisory Service for Arable Farming in the IJsselmeerpolders and North Holland, Lelystad, 26 pp. (in Dutch).

Andersson, S., 1975. Field experiments with chemical control of the black bean aphid in field beans in 1969–1970. Meddelanden Statens Vaxtskyddsantstalt, 16: 101–113 (in Swedish, with English summary).

Ashby, J.W., Forster, R.L.S., Fletcher, J.D. and Teh, P.B., 1979. A survey of sap-transmissible viruses of lucerne in New Zealand. New Zealand Journal of Agricultural Research, 22: 637–640.

Banks, C.J. and Macaulay, E.D.M., 1967. Effects of *Aphis fabae* Scop. and of its attendant ants and insect predators on yields of field beans (*Vicia faba* L.). Annals of Applied Biology, 60: 445–453.

Bardner, R., Fletcher, K.E. and Stevenson, J.H., 1978. Pre-flowering and post-flowering insecticide applications to control *Aphis fabae* on field beans: their biological and economic effectiveness. Annals of Applied Biology, 88: 265–271.

Brieze-Stegeman, R., 1979. Blue-green aphid. Journal of Agriculture of Tasmania, 50: 82–84.

Caballero, V.C., 1972. Survey, bionomics and control of the principal pests that attack crops of lucerne and red clover in Chile. Proceedings of the first Latin-American Congress of Entomology, Curzo, Peru, 1971; 15: 201–214 (in Spanish with German summary).

Cammel, M.E. and Way, M.J., 1977, Economics of forecasting for chemical control of the black bean aphid, *Aphis fabae*, on the field bean, *Vicia faba*. Annals of Applied Biology, 85: 333–343.

Cliffe, A.M., 1977. Insecticidal control of blue-green lucerne aphid. In: M.J. Hartley (Editor), Proceedings 30th New Zealand Weed and Pest Control Conference, pp. 168–169.

Cuperus, G.W. and Radcliffe, E.B., 1983. Thresholds for pea aphid and potato leafhopper on alfalfa. Proceedings, 10th International Congress on Plant Protection, Brighton, 1983, Vol. 1, p. 106.

Davies, J.C., 1975a. Use of menazon insecticide for control of rosette disease of ground-nuts in Uganda. Tropical Agriculture, 52: 359–367.

Davies, J.C., 1975b. Insects for the control of the spread of groundnut rosette disease in Uganda. PANS, 21: 1–8.

East, R., Pottinger, R.P. and Newman, L.M., 1977. Chemical control of blue-green lucerne aphid in northern North Island. In: M.J. Hartley (Editor), Proceedings, 30th New Zealand Weed and Pest Control Conference, pp. 160–164.

Fam, E.Z., 1983. Reducing losses due to insects in faba beans in Egypt. Fabis Newsletter, 7: 48–49.

Gangadharan, K., Ayyavoo, R., Kandaswamy, T.K. and Krishnamurthy, C.S., 1972. Efficacy of various insecticides in the control of rosette disease of groundnut. Madras Agricultural Journal, 59: 657–659.

Garcia, A.C., 1974. Chemical control of the green lucerne aphid, *Acyrthosiphon pisum* (Harris) and its effect on the predator *Hippodamia convergens* G. Revista Peruana de Entomologia, 17: 92–94 (in Spanish, with English summary).

Gould, H.J. and Graham, C.W., 1977. The incidence of *Aphis fabae* Scop. on spring-sown field beans in south-east England and the efficiency of control measures. Plant Pathology, 26: 189–194.

Hariri, G. and Tahhan, O., 1983. Insect damage and grain yield of faba bean, lentil and chickpea. Proceedings, 10th International Congress on Plant Protection, Brighton, 1983, Vol. 1, p. 127.

Harper, A.M., 1983. Spotted alfalfa aphid. Agrifax, Alberta Agriculture, No. Agdex 622–17: 2 pp.

Hinz, B. and Daebeler, F., 1984. Zur Schadewirkung der Erbsenblattlaus (*Acyrthosiphon pisum* (Harris)) an grosskörnigen Leguminosen. Nachrichtenblatt des Pflanzenschutzdienstes in der DDR, 38: 1:79–180.

Johnstone, G.R., 1978. Diseases of broad bean (*Vicia fabae* L.) and green pea (*Pisum sativum* L.) in Tasmania caused by subterranean clover red leaf virus. Australian Journal of Agricultural Research, 29: 1003–1010.

Johnstone, G.R. and Rapley, P.E.L., 1981. Control of subterranean clover red leaf virus in broad bean crops with aphicides. Annals of Applied Biology, 99: 135–141.

Machain, L.M., 1978. Chemical control of the spotted aphid, *Therioaphis maculata* Buckton, on seed lucerne with new and commercial products in the Mexican Valley, Baja California. Informe Tecnico de la Coordinacion Nacional del Apoyo Entomologico; Instituti Nacional de Investigaciones Agricolas, Mexico, Technical Report 1978, 3: 10–18 (in Spanish).

Moore, M.S., 1977. Insecticide screening for blue-green lucerne aphid control. In: M.J. Hartley (Editor), Proceedings, 30th New Zealand Weed and Pest Control Conference, pp. 165–167.

Mote, U.N., 1979. Efficacy of different granulated systemic insecticides against pea aphid. Indian Journal of Plant Protection, 7: 221–222.

O'Connor, B.P. and Hart, R.T.K., 1977. Pirimicarb for the control of blue-green lucerne aphid. In: M.J. Hartley (Editor), Proceedings, 30th New Zealand Weed and Pest Control Conference, pp. 170–172.

Posylaeva, G.A., 1976. Pirimor for protection from the pea aphid. Zashchita Rastenii, 32 (in Russian).

Romankov, W. and Zarzycka, K., 1978. Control of the bean aphid (*Aphis fabae* Scop.) on bean using Pirimor 50 DP. Prace Naukowe Instytutu Ochrony Roslin, 20: 183–187 (in Polish, with English summary).

Sanin, V.A., Khukhrii, O.V., Kushnerik, V.M., Lep'okhin, M.S., Voityuk, O.G. and Gorbach, V.Y., 1975. Effectiveness of low- and ultra- low volume sprays for the control of the pea aphid. Zakhist Roslin, 21: 32–38 (in Ukranian).

Schmidt, H.E., Dubnik, H., Karl, E., Schmidt, H.B. and Kamann, H., 1977. Verminderung von Virusinfektionen der Ackerbone (*Vicia fabae* L.) im Rahmen der Blattlausbekämpfung auf Grossflächen. Nachrichtenblatt für den Pflanzenschutz in der DDR, 31: 247–250.

Schmutterer, H. and Thottappilly, G., 1972. Zur wirtschaftlichen Bedeutung und Ausbreitung des Erbsenblattrollvirus in Ackerbohnenbestand sowie zur chemischen Bekämpfung der Vektoren. Zeitschrift für Pflanzenkrankheiten und Pflanzenschutz, 79: 478–484.

Sharma, R. and Stern, V., 1980. Blue alfalfa aphid: Economic threshold levels in southern California. California Agriculture, 34(2): 16–17.

Srinivasan, K. and Shukla, R.P., 1980. Control of pea leaf miner (*Phytomyza atricornis* Meigen) and aphids (*Acyrthosiphon pisum* Harris) with certain granular insecticides. Pesticides, 14: 19–20.

Stokovskaya, T.M., Servetnik, L.G. and Boiko, L.G., 1977. Test on the control of pests of seed lucerne. Zashchita Rastenii, 12: 24–25 (in Russian).

Stoltz, R.L. and Forster, R.L., 1984. Reduction of pea leaf roll of peas (*Pisum sativum*) with systemic insecticides to control the pea aphid (Homoptera: Aphididae) vector. Journal of Economic Entomology, 77: 1537–1541.

Summers, C.G., 1975. Efficacy of insecticides and dosage rates applied for control of the Egyptian alfalfa weevil and pea aphid. Journal of Economic Entomology, 68: 864–866.

Summers, C.G., Coviello, R.L. and Cothran, W.R., 1975. The effect on selected entomophagous insects of insecticides applied for pea aphid control in alfalfa. Environmental Entomology, 4: 612–614.

Syrett, P. and Penman, D.R., 1980. Studies of insecticide toxicity to lucerne aphids and their predators. New Zealand Journal of Agricultural Research, 23: 575 580.

Thygesen, T., 1971. The pea midge (*Contarinia pisi* Winn.) and other insect pests in pea cultivation. Biology and control. Tidskrift for Planteavl, 75: 825–842 (in Danish, with English summary).

Verhoeven, W.B.L., 1968. Diseases and Damages of Agricultural Crops and their Control. H. Veenman and Sons, Wageningen, 296 pp. (in Dutch).

Walkey, D.G.A. and Dance, M.C., 1979. The effect of oil sprays on aphid transmission of turnip mosaic, beet yellows, bean common mosaic and bean yellow mosaic viruses. Plant Disease Reporter, 63: 877–881.

Walters, P.J. and Forrester, N., 1979. Resistant lucerne of aphids at Tamworth. Agricultural Gazette of New South Wales, 90: 5, 7.

Way, M.J., Cammel, M.E., Alford, D.V., Gould, H.J., Graham, C.W., Lane, A., Light, W.I.St.G., Rayner, J.M., Heathcote, G.D., Fletcher, K.E. and Seal, 1977. Use of forecasting in chemical control of black bean aphid, *Aphis fabae* Scop., on spring-sown field beans, *Vicia faba* L. Plant Pathology, 26: 1–7.

Cruciferae

Alexandrescu, S. and Hondru, N., 1980. Selectivity of some insecticides used for the control of the cabbage aphid (*Brevicoryne brassicae* L.) on cabbage crops. Analele Institutului de Cercetari pentru Protectia Plantelor, 16: 375–383 (in Rumanian, with English summary).

Bhadoria, U.S., Deole, J.Y. and Dhamdhere, S.V., 1982. Efficacy of some foliar insecticides against mustard aphid, *Lipaphis erysimi* (Kalt.). Pesticides, 16: 23–25.

Buishand, Tj. and Snoek, N.J., 1985. The Growth of Red and White Cabbage. PAGV, Lelystad, 85 pp. (in Dutch).

Carlsson, E.C., 9173. Cabbage and turnip aphids and their control and damage on rape and mustard. Journal of Economic Entomology, 66: 1303–1304.

French, N., Nichols, D.B.R., Wilson, W.R., Rogerson, J.P. and Wright, A.J., 1977. Control of pests and mildew and their effect on yield of swedes in northern England. In: R.A. Fox (Editor), Proceedings, Symposium on Problems of Pest and Disease Control in Northern Britain, University of Dundee, pp. 64–66.

Gloria, B.R., 1978. Chemical control of aphids *Brevicoryne brassicae* (L.) and *Myzus persicae* (Sulz.) on cabbage. Revista Peruana de Entomologia, 21: 115–117 (in Spanish).

Graham, C.W., 1984. Cabbage aphid. Leaflet, Ministry of Agriculture, Fisheries and Food, London, no. 269, 8 pp.

Hill, S.A., 1983. Viruses of brassica crops. Leaflet, Ministry of Agriculture, Fisheries and Food, London, no. 370, 8 pp.

Jackson, T., 1982. Rationalisation of spray programmes for control of insect pests in Brussels sprouts. In: M.J. Hartley (Editor), Proceedings, 35th New Zealand Weed and Pest Control Conference, pp. 308–311.

Kalra, V.K., Gupta, D.S. and Yadava, T.P., 1983. Effect of cultural practices and aphid infestation on seed yield and its comparant traits in *Brassica juncea* (L). Czern and Coss. Haryana Agricultural University Journal of Research, 13: 115–120.

Lammerink, J. and Banfield, R.A., 1980. Effect of aphid control by disulfoton on seed yield components and seed quality of oilseed rape. New Zealand Journal of Experimental Agriculture, 8: 45–48.

Lee, H.S., 1973. Field evaluation of insecticides for control of aphids and diamondback moth on cabbages. Plant Protection Bulletin, Taiwan, 15: 130–133.

Mohan, N.J., Krishnaiah, K. and Kumar, N.K.K., 1981. Chemical control of mustard aphid, *Lipaphis erysimi* (Kalt.) and leaf webber, *Crocidolomia binotalis* (Zell.) on cabbage. Pesticides, 5: 29–32.

Narkiewicz-Jodko, J., 1976. Dynamics of development and control of cabbage aphid (*Brevicoryne brassicae* L.). Materialy z XVI Sesji Naukowej Instytutu Ochrony Roslin, 12–14 Cuty 1976: 185–200 (in Polish, with English summary).

Nath, D.K., 1975. Control of mustard aphid, *Lipaphis erysimi* (Kalt.) by soil application of insecticides. Science and Culture, 41: 428–429.

Prasad, S.K., 1978a. Chemical control of mustard aphid, *Lipaphis erysimi* (Kalt.). Indian Journal of Entomology, 40: 328–332.

Prasad, S.K., 1978b. Control of mustard aphid, *Lipaphis erysimi* (Kaltenbach). Indian Journal of Entomology, 40: 401–404.

Prasad, S.K., 1979. Control of mustard aphid, *Lipaphis erysimi* (Kaltenbach) by granular systemic insecticides. Indian Journal of Agriculture, 41: 39–42.

Sarode, S.V. and Lal, R., 1981. Persistence of lindane on cauliflower. Indian Journal of Entomology, 43: 408–412.

Sharma, K.C. and Kathri, N.K., 1978. Studies on biology and control of the mustard aphid *Lipaphis erysimi* Kalt. on mustard at Khulmaltar. Nepalese Journal of Agricultura, 13/14: 63–71.

Suett, D.L., 1977. Influence of placement depth on the behaviour of disulfoton in soil and brussels sprouts and on cabbage aphid control. Proceedings, 1977 British Crop Protection Conference – Pests and Diseases, pp. 427–442.

Tandon, P.L. and Bhalla, O.P., 1976. Control of cabbage aphid with granular systemic insecticides through soil. Indian Journal of Entomology, 38: 181–186.

Tomlinson, J.A., Ward, C.M. and Webb, M.J.W., 1976. Brassica virus diseases. Annual Report, National Vegetable Research Station, p. 107.

Van Dijk, L.P. and Krause, M., 1978. Persistence and efficacy of disulfoton on cabbages. Phytophylactica, 10: 53–55.

Whitwell, J.D., 1977. The relevance of pest and disease control in systems of modern vegetable production. Proceedings, 1977 British Crop Protection Conference – Pests and Diseases, pp. 915–919.

Compositae

Baranowski, T., 1976. Control of the two-spotted spider mite (*Tetranychus urticae* Koch) and the peach aphid (*Myzus persicae* Sulz.) on chrysanthemum with pesticides of Polish and foreign origin and considerations of their phytotoxic effects on different varieties. Roczniki Nauk Rolniczych E; 6; 1: 165–183 (in Polish, with English summary).

Basavanagoud, K., Kulkarni, K.A. and Thontadarya, T.S., 1981. Estimation of crop loss in safflower due to the aphid, *Dactynotus compositae* Theobald (Hemiptera; Aphididae). Mysore Journal of Agricultural Science, 15: 279–282.

Crüger, G., 1983. Pflanzenschutz im Gemüsebau. Verlag Eugen Ulmer, Stuttgart, 482 pp.

Gentile, A.G. and Vaughan, A.W., 1974. Control of the lettuce root aphid in Massachusetts. Journal of Economic Entomology, 67: 556.

Goud, K.B., Kulkarni, K.A. and Thontadarya, T.S., 1981. Residues of insecticides in safflower seed. Pesticides, 15: 12–13.

Grewal, G.S., Singh, G. and Sandhu, S.S., 1982. Efficacy of some new insecticides in controlling *Myzus persicae* (Sulz.) on safflower. Journal of Research, Punjab Agricultural University: 351–354.

Hakkaart, F.A., 1968. Prevention of the spread of non-persistent plant viruses by oil spraying. Mededelingen van de Directie Tuinbouw, 31: 262–269 (in Dutch, with English summary).

Hauri, P., Freuler, J., Bertuchoz, P. and Hugi, H., 1973. Premières observations sur *Pemphigus bursarius* L., puceron des racines nuisible aux composées cultivées, et essai de lutte préliminaire. Revue Suisse de Viticulture, Arboriculture et Horticulture, 5: 179–182.

Lindquist, R.K. and Spadafora, R.R., 1975. Summary of experiments for green peach aphid control on greenhouse chrysanthemums – soil drench applications. Ohio Florists Association, Bulletin no. 550, pp. 3–5.

Oetting, R.D., 1982. Systemic activity of acephate, butoxycarboxim and butocarboxim for control of *Myzus persicae* on ornamentals. Journal of the Georgia Entomological Society, 17: 433–438.

Oetting, R.D., Morishita, F.S., Jefferson, R.N., Humphrey, W.A. and Besemer, S.T., 1977. Aphid control on chrysanthemums and carnations. California Agriculture, 31(12): 7–9.

Schmidt, H.E., Weber, I. and Kegler, H., 1981. Wirtschaftlich wichtige Virus-krankheiten bei Gemüse und ihre Bekämpfung. Nachrichtenblatt für den Pflanzenschutz in der DDR, 35: 220–225.

Schuster, D.J. and Engelhard, A.W., 1979. Insecticide – fungicide combinations for control of arthropods and *Ascochyta* blight on chrysanthemum. Phytoprotection, 60: 125–135.

Simonet, D.E., 1980. Strategies for aphid control in field-grown leaf lettuce. Ohio Report on Research and Development, 65(3): 44–46.

Thompson, A.R. and Percivall, A.L., 1981. Protection of field-sown outdoor lettuce against top and root aphids with soil applications of granular insecticide products. Annals of Applied Biology, 97 (Supplement 2): 10–11.

Toscano, N.C., Kido, K., Snyder, M.J., Koehler, C.S., Kennedy, G.C. and Sevacherian, V., 1977. Insecticides evaluated for lettuce root aphid control. California Agriculture, 31(4): 4–5.

Webb, R.E. and Argauer, R.J., 1974. Uptake of monocrotophos by chrysanthemum cultivars and resulting control of melon aphid. Journal of Economic Entomology, 67: 251–252.

Rosaceae

Anonymus, 1985. Recommendations for 1985, Crop Protection and Growth Regulators. Joint Publication, Netherlands Fruitgrowers' Association and Horticultural Advisory Service for Crop Protection, Wageningen, 36 pp. (in Dutch).

Attri, B.S. and Sharma, P.L., 1971. Granules systemic insecticides for the control of woolly aphid,

Eriosoma lanigerum (Hausm.) on apple (*Malus pumila* Mill.). Indian Journal of Agricultural Science, 41: 627–631.

Becker, R.F. and Rich, A.E., 1956. Increased runner production and fruit yield of virus-free strawberry plants over commercial stocks in New Hampshire. Plant Disease Reporter, 40: 947–951.

Chander, R. and Dogra, G.S., 1977. A technique of granule application for woolly apple aphid control. Entomologists' Newsletter, 7: 42.

Chisholm, D. and Specht, H.B., 1978. Residues and control of aphids on strawberries with banded surface applications of disulfoton. Journal of Economic Entomology, 71: 469–472.

Chumak, N.O. and Petrov, A.S., 1975. On the effectiveness of a combination of chemical measures for the control of pests and diseases of oilbearing rose. Zakhist Roslin, 21: 73–77 (in Ukranian).

Cooke, B.K., Herrington, P.J., Jones, K.G. and Morgan, N.G., 1976: Pest and disease control with low doses of pesticides in low and ultralow volumes applied to intensive apple trees. Pesticide Science, 7: 30–34.

Craig, D.L., 1957. A two-year comparison of virus-free and common stock strawberry plants. Plant Disease Reporter, 41: 79–82.

Craig, D.L. and Stultz, H.T., 1964. Aphid dissemination of strawberry viruses in Nova Scotia. Canadian Journal of Plant Science, 44: 235–239.

Fischer-Colbrie, P., 1982. Die Blutlaus – ein neuer alter Schädling in Österreichischen Apfelanlagen. Pflanzenarzt, 35: 26–28.

Freeman, J.A. and Mellor, F.C., 1962. Influence of latent virus on vigor, yield and quality of British Sovereign strawberries. Canadian Journal of Plant Sciences, 42: 602–610.

Gloria, B.R., 1978. Chemical control of the "woolly aphis" *Eriosoma lanigerum* (Hausm.) on apple. Revista Peruana de Entomologia, 21: 118–120 (in Spanish).

Hall, F.R., 1979. Effects of synthetic pyrethroids on major insect and mite pests of apple. Journal of Economic Entomology, 72: 441–446.

Hameed, S.F., Adlakha, R.L. and Sud, V.K., 1974. Control of *Eriosoma lanigerum* (Hausm.), by systemic insecticides. Indian Journal of Agricultural Science, 44: 301–303.

Hull, R.A. and Starner, V.R., 1983. Effectiveness of insecticide application times to correspond with the development of rosy-apple aphid (Hemiptera: Aphididae) in apple. Journal of Economic Entomology, 76: 594–598.

Kacharmazov, V., Zamfirov, T.S. and Choleva, B., 1976. Protection of strawberry mother plants from pests and diseases. Rastitelna Zashchita, 24: 19–21 (in Bulgarian).

Kennedy, G.G., Oatman, E.R. and Voth, V., 1976. Suitability of Plictran and Pirimor for use in a pest management program on strawberries in Southern California. Journal of Economic Entomology, 69: 269–272.

McGrew, J.R. and Scott, D.H., 1959. Effect of two virus complexes on the responses of two strawberry varieties. Plant Disease Reporter, 43: 385–389.

Natskova, V., 1974. The possibility of combining chemical control of the rose-leaf aphid, *Macrosiphum rosae* (L.) with control by its natural enemies. Rastitelna Zashchita, 22: 10–13 (in Bulgarian).

Pasqualini, E., Briolini, G., Memmi, M. and Monari, S., 1982. Guided control tests against apple aphids. Bollettino dell' Istituto di Entomologia della Universita degli Studi di Bologna, 36: 159–171 (in Italian, with English summary).

Penman, D.R. and Chapman, R.B., 1980. Woolly apple aphid outbreak following use of fenvalerate in apple in Canterbury, New Zealand. Journal of Economic Entomology, 73: 49–51.

Shanks, C., 1981. Strawberry aphids and strawberry viruses. Extension Bulletin of the Cooperative Extension Service, College of Agriculture, Washington State University, Pullman, WA, EB 1012, 2 pp.

Sharma, M.L., 1980. Chemical sterilants, virus vectors, insect control. Acta Phytopathologica, 11th International Symposium, Fruit Tree Virus Diseases, 15: 403–405.

Van Frankenhuyzen, A. and Gruys, P., 1983. Guided Control of Diseases and Pests on Apple and Pear. Plant Protection Service, Wageningen, 4th edn., 128 pp. (in Dutch).

Malvaceae

Alimukhamedov, S.N. and Mirpulatova, N.S., 1978. Fundamental problems in the protection of cotton. Zashchita Rastenii, 6: 24–25 (in Russian).

Autrique, A., 1980. Control of insect pests of cotton in Burundi. African Journal of Plant Protection, 2: 131–138.

Bartsch, R., 1978. Economic problems of pest control. Examined for the case of the Gezira/Sudan. Weltforum Verlag, München, 136 pp.

Bindra, O.S., Sidhu, A.S., Singh, G. and Brar, K.S., 1973. Control of sucking pests of cotton by soil application of granular systemic insecticides. Indian Journal of Agricultural Science, 43: 352–356.

Borle, M.N., Ramarao, B. and Deshmukh, S.D., 1980. Residual toxicity of some granular systemic insecticides as soil application in the control of cotton aphid, *Aphis gossypii* Glover. Indian Journal of Entomology, 42: 142–147.

Cauquil, J., Vincens, P., Denechere, M. and Mianze, T., 1982. Nouvelle contribution sur la lutte chimique contre *Aphis gossypii* Glover, ravageur du cotonnier en Centrafrique. Coton et Fibres Tropicales, 37: 333–350.

Cauquil, J., Vincens, M. and Girardot, M., 1983. La lutte contre le puceron du cotonnier (*Aphis gossypii* Glover) en Centrafrique. Mededelingen van de Fakulteit Landbouwwetenschappen, Rijksuniversiteit Gent, 48: 341–347.

Harcharan Singh and Vijay, N.S., 1975. Comparison of LVC- and EC-formulations of pesticides for the control of cotton pests. Journal of Research, Punjab Agricultural University, 12: 145–151.

Jones, A.J. and Wangboonkong, S., 1964. Progress Reports from Experiment Stations – Thailand. Season 1962/1963. Cotton Research Corporation, London (cited in Mabbett, 1982).

Katarki, H. and Thimniah, G., 1969. Insecticidal trials for the control of major insect pests on laxmi cotton in Mysore State. Mysore Journal of Agricultural Science, 3: 174–186.

Khalil, F.M. and Rizk, G.A., 1972. Field evaluation of new insecticides for control of some early season cotton insects in Upper Egypt. Bulletin of the Entomological Society of Egypt, Economic Series, 6: 111–116.

Mabbett, T.H., 1982. The economic insect pests of cotton in Thailand. II. The cotton aphid (*Aphis gossypii* Glover). Thai Journal of Agricultural Science, 15: 67–68.

Regupathy, A. and Subramaniam, T.R., 1979. Effect of organic matter in the soil on the performance of aldicarb to cotton. Indian Journal of Agricultural Science, 49: 613–616.

Regupathy, A. and Subramaniam, T.R., 1982. Performance of aldicarb and its metabolites against cotton aphid, *Aphis gossypii* Glov. Indian Journal of Agricultural Science, 52: 130–134.

Sachs, Y., 1979. Cotton pest scouting schemes in Thailand. Report of the National Agricultural Extension Project, Department of Agricultural Extension, Bangkok, no. 2 (cited in Mabbett, 1982).

Sivaprakasam, K. and Balasubramanian, G., 1981. Impact of plant protection on cotton yield. Pesticides, 15: 11–13.

Sokhta, A.A. and Rasulev, F.K., 1981. Ultra-low-volume sprays in the control of pests of cotton. Zashchita Rastenii, 11: 14 (in Russian).

Stefanov, S., 1972. Ultra-low-volume spraying for the control of pests of cotton. Rastitelna Zashchita, 20: 16–18 (in Bulgarian).

Miscellaneous

Abdel Shaheed, G.A., Abdel-Salam, A.M., Assem, M.A. and Amin, S.M., 1972. Control of the pests of snake-cucumber (*Cucumis melo* L. var. *flexuosus* L.) and cucumber (*C. sativus* L.) in the Arabic Republic of Egypt. Indian Journal of Agricultural Science, 42: 95–99.

Abdel-Salam, A.M., Assem, M.A., Abdel Shaheed, G.A., Hammad, S.M. and Ragab, F.Y., 1972. Chemical control of some squash pests in U.A.R. Zeitschrift für Angewandte Entomologie, 70: 169–174.

Adams, R.G., Lilly, J.H. and Gentile, A.G., 1976. Evaluation of some insecticides in controlling aphids and reducing flower losses from virus disease in gladiolus-plantings. Journal of Economic Entomology, 69: 171–172.

Aharonson, N., Neubauer, I., Ishaaya, I. and Raccah, B., 1979. Residues of croneton and its sulfoxide and sulfone metabolites in citrus (Clementine trees) following a soil treatment for the control of *Aphis spiraecola*. Journal of Agriculture and Food Chemistry, 27: 265–268.

Allen, J.G. and Easterbrook, M.A., 1981. Mephosfolan concentrations in hop (*Humulus lupulus* L.) leaves and flowers after systemic uptake and the toxicity to caged aphids (*Phorodon humuli*). Journal of Horticultural Science, 56: 81–87.

Asjes, C.J., 1981. Control of stylet-borne spread of aphids of tulip breaking virus in lilies and tulips, and hyacinth mosaic virus in hyacinths by pirimicarb and permethrin sprays versus mineral oil sprays. Mededelingen van de Fakulteit Landbouwwetenschappen, Rijksuniversiteit Gent, 46: 1073–1077.

Asjes, C.J., 1985. Control of field spread of non-persistent viruses in flowerbulb crops by synthetic pyrethroid and pirimicarb insecticides, and mineral oils. Crop Protection, 4: 485–493.

Bullock, R.C., 1972. Trunk treatment with systemics for aphid control on Florida citrus. Florida Entomologist, 55: 165–172.

Caldicott, J.J.B., 1974. The control of the damson-hop aphid with mephosfolan. Proceedings 7th British Insecticide and Fungicide Conference, 1973, pp. 457–464.

Chahal, B.S. and Mann, G.S., 1977. Chemical control of seed-maggot, *Hylemia* sp. and aphid, *Aphis gossypii* Glover, on seedlings of musk-melon (*Cucumis melo* L.) var. *hara madhu*. Journal of Research, Punjab Agricultural University, 14: 238–240.

Cone, W.W., 1975. Crown-applied systemic acaricides for control of the two-spotted spider mite and hop aphid on hops. Journal of Economic Entomology, 68: 684–686.

Crüger, G., 1983. Pflanzenschutz im Gemüsebau. Verlag Eugen Ulmer, Stuttgart, 482 pp.

Dickson, R.C., Swift, J.E., Anderson, L.D. and Middleton, J.T., 1949. Insect vectors of cantaloupe mosaic in California's Desert Valleys. Journal of Economic Entomology, 42: 770–774.

Ebeling, W., 1959. Subtropical Fruit Pests. University of California Press, Berkeley, 226 pp.

Hood, W.M. and Fox, R.C., 1980. Control of aphids on loblolly pine in north-western south Carolina. Journal of the Georgia Entomological Society, 15: 105–108.

John, F. and Seefried, W., 1977. Gegenwartiger Stand des Pflanzenschutzes mit Bodenmaschinen und Luftfahrzeugen in Hopfenbau der DDR. Nachrichtenblatt für den Pflanzenschutz in der DDR, 31: 150–152.

Kibata, G.N., 1979. Preliminary attempts to control the insect transmitted water-melon mosaic virus (WMV-K) on baby marrows. Kenya Entomologist's Newsletter, 9: 8–9.

Koli, S.Z., Makar, P.V. and Choudhary, K.G., 1981. Seasonal abundance of citrus pests and their control. Indian Journal of Entomology, 43: 1834–187.

Loebenstein, G., Alper, M. and Deutsch, M., 1964. Preventing aphid-spread cucumber mosaic virus with oils. Phytopathology, 54: 960–962.

Loebenstein, G., Deutsch, M., Frankel, H. and Sabar, Z., 1966. Field tests with oil sprays for the prevention of cucumber mosaic virus in cucumbers. Phytopathology, 56: 512–516.

Maheswari, A.S., 1981. Seasonal occurrence and chemical control of *Myzus persicae* Sulz. on orange (var. Nagpur). Pesticides, 15: 23–24.

Maksymov, J.K., 1975. Zur Bekämpfung von Gallenläuse aus der Gattung *Sacchiphantes* (Adelgidae, Homoptera) an der Fichte (*Picea abies* (L.) Karsten). Anzeiger für Schädlingskunde, Pflanzenschutz und Umweltschutz, 48: 113–118.

Matthieu, J.L. and Verhoyen, M., 1983. Efficacité inhibitrice des huiles minerales sur la transmission du virus de la mòsaique du celeri. I. Inhibition de la transmission aphidienne en relation avec la viscosité des huiles. Mededelingen van de Fakulteit Landbouwwetenschappen, Rijksuniversiteit Gent, 48: 823–828.

Meier, W. and Jossi, W., 1983. Einsatz von Pyrethroid-Insektiziden in Feldkulturen. Mitteilungen für die Schweizerische Landwirtschaft, 31: 29–42.

Nielsen, D.G. and Balderston, C.P., 1977. Control of eastern spruce and Cooley spruce gall aphids with soil-applied systemic insecticides. Journal of Economic Entomology, 70: 205–208.

Perju, T., Ghizdavu, I., Marusca, C., Gherman Isuto, P., Paal, G., Ongerth, M. and Balint, I., 1979. Investigations on the biology, ecology and control of the green hop aphid (*Phorodon humuli* Schrank, Aphididae–Homoptera). Probleme de Protectia Plantelor, 7: 97–110 (in Rumanian, with English summary).

Puritch, G.S., Nigam, P.C. and Carrow, J.R., 1980. Chemical control of balsam woolly aphid (Homoptera: Adelgidae) on seedlings of *Abies amabilis*. Journal of the Economic Entomological Society of British Columbia, 77: 15–18.

Rivero, J.M. del and Jaen, A., 1972. Notes on control tests against the aphid *Aphis gossypii* Glover on "Satsuma" mandarin. Anales del Instituto Nacional de Investigaciones Agrarias, Serie: Proteccion Vegetal, 2: 233–235 (in Spanish).

Schread, J.C., 1971. Control of the eastern spruce gall aphid. Circular, Connecticut Agricultural Experiment Station, no 242, 6 pp.

Singh, B.R., 1969. Control of a mosaic disease of bottle-gourd (*Lagenaria siceraria* Standl.) caused by a strain of *Cucumis* virus 2 at Kanpur, India. Labdev Journal of Science and Technology, B (Life Sciences), 7: 239–240.

Sutton, J. and Garrett, R.G., 1978. The epidemiology and control of tulip breaking virus in Victoria. Australian Journal of Agricultural Research, 29: 555–563.

Tao, C.C. and Wu, K.C., 1968. Report on citrus insect control study by chemicals in Taiwan. Plant Protection Bulletin, Taiwan, 10: 57–64.

Taran, F.I. and Semenyuk, V.I., 1976. The protection of hops "Polesskoe gold". Zashchita Rastenii, 6: 21–22 (in Russian).

Tröster, H. and Griesel, A., 1983. Pflanzenschutzmassnahmen als Bestandteil intensiver Hopfenproduktion. Nachrichtenblatt für den Pflanzenschutz in der DDR, 37: 107–109.

Walkey, D.G.A. and Ward, C.M., 1984. The reaction of celery (Apium graveolens L.) to infection by celery mosaic virus. Journal of Agricultural Science, 103: 415–419.

Watson, M.A. and Heathcote, G.D., 1965. The use of sticky traps and the relation of their catches of aphids to the spread of viruses in crops. Annual Report, Rothamsted Experiment Station, 1965; 292–300.

11.1.1 Resistance of Aphids to Insecticides

A.L. DEVONSHIRE

OCCURRENCE OF INSECTICIDE-RESISTANT APHIDS

Extensive use of insecticides has led to widespread development of resistance in many insect species, not the least in aphids, and this represents one of the major threats to the future success of chemical pest control. Resistance occurs when a population can no longer be controlled adequately by the recommended application rates of pesticides that give good control of other populations of the same species. It should not be confused with natural tolerance, where the "normal" susceptible variants of a given species are unaffected by application rates known to be generally effective on other pests. Before resistance becomes apparent as significant control failure, incipient resistance develops by selection and build-up of resistance genes in the population. When such resistance is suspected under practical conditions, it is important to be able to confirm it rapidly and discount other factors that might have also adversely affected insecticide performance. To this end, various laboratory bioassays have been developed, some of which are recommended as standardised methods by the World Health Organization and Food and Agriculture Organization of the United Nations; these are discussed in section 8.10, p. 120). Once identified, detailed studies of resistance can provide guidelines for the most appropriate alternative chemicals to use.

In a survey of the worldwide incidence of resistance up to 1980, Georghiou (1981) listed over 400 arthropods reported to have developed resistance to one or more classes of insecticide, and this included 18 aphid species. Table 11.1.1.1 shows only those resistant aphids for which published bioassay data are now available. The other examples represent isolated reports, often in the form of a review (Knipling, 1954; Wiessman, 1955; Asakawa, 1975), for which corroborative evidence seems not to have been published. In some of these cases, as well as in some of the isolated examples in Table 11.1.1.1, changing to different types of insecticide or to different crop varieties appears to have by-passed the resistance problem, perhaps explaining to some extent the absence of confirmatory publications. In addition to those listed in Table 11.1.1.1 *Aphis nasturtii* (Kaltenbach) has also proved difficult to control with pirimicarb. However, it was not clear whether this arose from resistance or from the inherent insensitivity to pirimicarb of this particular species (Laska, 1981).

Overall then, although there are now reports of 20 or so resistant aphid species, the problem is most persistent, widespread and serious in a few, notably *Myzus persicae* (Sulzer), *Phorodon humuli* (Schrank) and *Aphis gossypii* Glover, and this section is therefore concerned primarily with these three. These cases seem to be associated with situations where a large proportion of the population is subjected to strong selection pressure. For example, since the

---

Section 11.1.1 references, p. 135

TABLE 11.1.1.1

Examples of insecticide-resistant aphid species for which published bioassay data are available

Species	Insecticide class[a]	References[b]
Aphis fabae Scopoli	OP	Hlinakova and Hurkova (1976)
Aphis gossypii Glover	OP	Kung et al. (1964)
	carb	Furk et al. (1980), Silver (1984)
	pyr	Zil'bermints and Zhuravleva (1984)
Brevicoryne brassicae (Linnaeus)	OP	Ripper (1961)
Chaetosiphon fragaefolii (Cockerell)	OC	Shanks (1967)
Chromaphis juglandicola (Kaltenbach)	OP	Michelbacher et al. (1954)
Hyalopterus pruni (Geoffroy)	OP	Hurkova (1973a)
Myzocallis coryli (Goeze)	carb	Aliniazee (1983)
Myzus persicae (Sulzer)	OP and carb	many (e.g. Hurkova (1973b), Sawicki and Rice (1978), Baker (1978b), Attia et al. (1979))
	pyr	Sawicki and Rice (1978), Attia and Hamilton (1978), Otto (1980)
	OC	Bauernfeind and Chapman (1985)
Phorodon humuli (Schrank)	OP	many (e.g. Hrdy and Zeleny (1968), Muir (1979))
	carb	Muir (1979), Lewis and Madge (1984)
Schizaphis graminum (Rondani)	OP	Teetes et al. (1974), Chang et al. (1980)
Therioaphis trifolii (Monell)	OP	Stern (1962), Walters and Forrester (1979)
	carb	Walters and Forrester (1979)

[a]OP, organophosphorus; OC, organochlorine; carb, carbamate; pyr, pyrethroid.
[b]For a comprehensive list of references, see Georghiou (1981).

hop is the only summer host for *P. humuli*, the majority of the population of this aphid in a hop-growing region will come into contact with insecticides on the crop (Muir, 1979).

Similarly, in some peach-growing areas of Europe where holocyclic overwintering of *M. persicae* is believed to predominate (Blackman, 1974), a large proportion of the regional population may be treated with insecticides in the spring on the primary host (Baranyovits, 1973). In glasshouses, *M. persicae* and *A. gossypii* reproduce continuously by parthenogenesis and are pests throughout the year; therefore these partially isolated anholocyclic populations are often very heavily selected with insecticides. Some are not simply resistant, but virtually immune to some chemicals. This intensive selection of isolated populations of *M. persicae* under glass contrasts with the field situation, where the insecticide-treated portion of the outdoor population is regularly diluted by untreated aphids during migration, probably explaining why resistance took longer to develop outdoors. Although alatae of strongly resistant *M. persicae* could migrate from glasshouses and colonise field crops, and are known to survive such a transfer as apterae on chitted potatoes (Dunn and Kempton, 1977), so far they appear not to contribute significantly to the outdoor population; distinct resistant variants predominate in the two environments (Sawicki et al., 1980). However, with few such exceptions, our understanding of these ecological aspects of aphids is generally inadequate to understand why resistance develops more readily in some situations than others (Blackman, 1975).

Of the cases of insecticide-resistant aphids confirmed in laboratory studies, most are based on bioassays of field populations, either as collected or after mass-rearing. Whilst of value for *detecting* resistance, such tests give only limited insight into its nature, since they measure the composite response of what is likely to be a heterogeneous population. However, as discussed in section 8.10 (p. 124), a better understanding of the problem can be gained by establishing and studying clonal cultures from these populations, perhaps after selection with an insecticide in the laboratory. Although this inevitably means studying a large number of clones to ensure they are typical, such an approach has been successful in characterising resistance in those species discussed below.

THE BIOCHEMICAL NATURE OF RESISTANCE IN APHIDS

Of the 20 or so aphid species recorded as being insecticide resistant, evidence of its biochemical nature is available for only *M. persicae*, the most intensively studied, *P. humuli* and *A. gossypii*.

Attempts to infer the biochemical cause of resistance in aphids indirectly from bioassays with and without synergists have had limited success. Sesamex, a mixed-function oxidase inhibitor, virtually abolished dimethoate resistance in *M. persicae* but enhanced disulfoton resistance to almost immunity (Needham and Sawicki, 1971). A similar synergist, piperonyl butoxide, reduced carbamate resistance from about 10-fold to 2-fold, suggesting oxidative metabolism of these compounds (Attia et al., 1979). It also synergised parathion against *M. persicae* larvae, but for unknown reasons not against adults (Kirknel and Reitzel, 1973). Sesamex has been shown to decrease dramatically the rate of evaporation of topically-applied insecticides from aphid cuticle, an effect probably involved with its influence on insecticide toxicity (Devonshire and Needham, 1974). Mixed function oxidases could not be detected *in vitro* in *M. persicae* (Devonshire, 1973; Oppenoorth and Voerman, 1975), but this was because endogenous inhibitors of these enzymes were present in the aphid homogenates (Devonshire, 1973). Esterase-inhibiting synergists also affected insecticidal activity, but the synergist doses required were close to those killing aphids directly (Oppenoorth and Voerman, 1975; Devonshire and Moores, 1982), and in one case they were as high as 500 times the dose of insecticide required to kill susceptible aphids.

It has been suggested (Von Amiressami and Petzold, 1977) that symbionts are involved in resistance, based on differences in the number and size of the symbionts between a susceptible and a resistant strain of *M. persicae* after treatment with parathion. However, it was not clear whether this difference was a cause or a consequence of resistance. Furthermore, since these differences between strains are not found universally (Ball and Bailey, 1978), their possible involvement remains uncertain.

Those species for which there is direct biochemical evidence for the nature of the resistance are discussed below.

Myzus persicae

The discovery of a positive correlation in some clones between resistance and the activity of enzymes (esterases) hydrolysing naphth-1-yl acetate (Needham and Sawicki, 1971) was a major milestone, not only in understanding the biochemical cause of resistance in *M. persicae*, but also in gaining detailed knowledge of the distribution of resistant aphids in field populations (see below: "Detailed studies"). Precedents for such a correlation were the known associations in houseflies and leafhoppers between insecticide resistance and non-specific esterase activity. These initial observations with aphids were supplemented by more extensive studies of *M. persicae* collected from glasshouses and the field, which confirmed the association between resistance and esterase activity (Beranek, 1974a; Devonshire, 1975; Sawicki et al., 1978; Wachendorff and Klingauf, 1978; Hamilton et al.; 1981).

Since esterases are known to vary greatly in activity and type between individuals of many organisms, such apparent correlations with resistance do not prove unequivocally a causal link between the two characters; this link was established later for *M. persicae* by more detailed biochemical analysis. The first step was the resolution by electrophoresis of the various esterases present in aphid homogenates, and demonstration that the increased enzyme activity arose predominantly from changes in one (carboxylesterase E4) of several naphthyl acetate-hydrolysing enzymes (Beranek, 1974a; Devonshire, 1975). Preliminary evidence (Bunting and Van Emden, 1981) that esterase band patterns and intensities on polyacrylamide gels were affected by aphid diet has since been discounted (White, 1983).

Independently of these esterase studies with model substrates, Oppenoorth and Voerman (1975) demonstrated that the most resistant of three aphid clones, differing by up to 30-fold in sensitivity to parathion, hydrolysed paraoxon, methyl paraoxon and malaoxon between 2.5 and 9 times faster than the least resistant clone. Inability to resolve, by starch gel electrophoresis, the proteins hydrolysing paraoxon and naphthyl acetate provided preliminary evidence that a single enzyme might be acting both on model substrates and insecticides (Beranek and Oppenoorth, 1977). However, unequivocal evidence that only one enzyme was responsible came from kinetic studies of the hydrolysis of the two substrates by purified E4 (Devonshire, 1977). Furthermore, whether purified from susceptible or resistant aphids, the enzyme had identical catalytic properties, i.e. resistant aphids have more molecules of an enzyme present in susceptible aphids, rather than having mutant more-efficient enzymes. This was subsequently confirmed by direct measurements of the amount of E4 protein in susceptible and resistant aphids (Devonshire and Moores, 1982), and also by immunological detection methods (Devonshire et al., 1986a). A genetic regulatory mechanism was thus implicated as the cause of resistance in *M. persicae*.

Although of little toxicological significance, a mutant form of E4 is present in resistant aphids of normal karyotype (see below) (Devonshire et al., 1983; ffrench-Constant et al., 1988). It is very closely related to E4 and, like E4, relies for its effect on being produced in large quantities. It is characterised by its slightly higher M_r, 66 000 compared to 65 000 for E4, and slightly higher (approx. 1.5-fold) catalytic centre activity with some substrates. Both forms are glycosylated, and differ from their nascent forms by M_r approx. 8000 (Devonshire et al., 1986b).

Preliminary attempts to understand the genetic mechanism for increased E4 production were made by crossing susceptible and resistant aphids, but were hampered by the difficulty of breeding *M. persicae* sexually. No clearcut interpretation could be made from measurements of the distribution of E4 activity in the small number of F_1, F_2 and backcross clonal progenies obtained (Blackman et al., 1977; Blackman and Devonshire, 1978). However, an indication of the genetic mechanism responsible for "overproduction" of E4 came from measurements of the molar amount of this enzyme in aphids from a series of seven clones with progressively greater insecticide resistance (Devonshire and Sawicki, 1979). These studies showed that the amount of E4 doubled between each successive variant (Table 11.1.1.2). The most likely cause of such a regular stepwise increase is gene amplification, i.e. the presence of multiple copies of the structural gene for E4 arising from a series of tandem gene duplications. In the most resistant clones, E4 accounts for 1–2% of total aphid protein, and this arises from their increased content of mRNA encoding E4 (Devonshire et al., 1986b). Recent molecular studies have confirmed that gene amplification causes increased esterase production (Field et al., 1988). Thus, the gene for E4 was cloned (cDNA) and used to determine esterase gene copy number in susceptible and resistant insects; not only did resistant aphids have

TABLE 11.1.1.2

Amount of carboxylesterase E4[a] and resistance[b] in clones of *Myzus persicae* (Sulzer)

Clone	Origin[c]	Approx. resistance factor to dimethoate	E4 (pmol per mg aphid $\pm$ s.d.)	Putative number of E4 genes per aphid
US1L(S)	F	1	0.37 $\pm$ 0.20	1
240N	F	3	0.85 $\pm$ 0.18	2
MS1G(R_1)	F	8	1.78 $\pm$ 0.75	4
French R	F	15	4.8 $\pm$ 0.6	8
T1V(R_2)	F	100	6.7 $\pm$ 0.7	16
PirR	G	250	11.8 $\pm$ 1.2	32
G6	G	500	24.7 $\pm$ 0.2	64

[a] See Devonshire and Sawicki (1979).
[b] See Devonshire (1977).
[c] F, field; G, glasshouse; (S, R_1 and R_2 variants, of which clones US1L, MS1G and T1V are characteristic, are commonly found in the field in the U.K.)

TABLE 11.1.1.3

The rate of hydrolysis (k_3) of different groups of insecticide by purified carboxylesterase E4 from *M. persicae*, and the calculated maximum amounts (ng) of each that could be removed by sequestration and hydrolysis in one hour by susceptible (US1L) and resistant (G6) aphids[a]

Insecticide class	k_3 (h^{-1} $\pm$ s.d.)	Amount detoxified in one hour per aphid (ng)	
		US1L	G6
O,O-dimethylphosphate	3.12 $\pm$ 0.24	0.16	10.3
O,O-diethylphosphate	0.33 $\pm$ 0.03	0.05	3.3
N,N-dimethylcarbamate	0.10 $\pm$ 0.01	0.04	2.7
N-monomethylcarbamate	0.09 $\pm$ 0.01	0.04	2.7

[a] See Devonshire and Moores (1982).

more esterase genes according to their degree of resistance, but restriction patterns varied with the karyotype of the aphid (see below). There are ample precedents for the phenomenon of gene amplification in microorganisms and mammalian cell culture lines resistant to cytotoxic drugs (see Devonshire and Sawicki, 1979), but *M. persicae* is one of the first intact higher organisms identified to use this mechanism as a means of developing resistance to a toxicant.

Increased production of E4 is the only resistance mechanism identified in *M. persicae*; it causes cross-resistance in varying degrees to carbamate and organophosphorous insecticides, and probably also to pyrethroids (Devonshire and Moores, 1982; ffrench-Constant et al., 1987). The catalytic efficiency of the enzyme with insecticidal esters as substrates is extremely low, with each molecule of enzyme hydrolysing (k_3) only 0.1–3 molecules of insecticide per hour (Table 11.1.1.3). However, the molar amount of E4 in resistant aphids is so great that a substantial proportion of a lethal dose of insecticide is detoxified simply by binding on an equimolar basis to the catalytic centre of the enzyme. Furthermore, since so many E4 molecules are present, even a slow turnover of this bound insecticidal substrate (i.e. hydrolysis) can remove a further significant amount of toxicant. The speed at which different insecticidal classes are hydrolysed (k_3, Table 11.1.1.3) is generally reflected in the degree of resistance the enzyme causes.

These biochemical studies have significant practical implications. Having established that this one enzyme confers resistance to a wide range of insecticides, it is clear that there is no value in adopting the strategy of mixing or alternating insecticides as a means of overcoming this type of resistance. However, the biochemical studies, coupled with toxicological assessment of insecticides from different classes, provide guidelines for choosing the most appropriate chemicals to control resistant aphids and to minimise selection pressure on this mechanism in field populations (Devonshire and Moores, 1982).

Phorodon humuli

In *P. humuli* there appears to be a correlation, similar to that in *M. persicae*, between resistance and esterase activity. However, this is based on data from a small number of *P. humuli* clones and no detailed biochemical studies have been done to establish a causal relationship between the two characters.

Von Beck and Büchi (1980), studying four clones from wild hops and four taken from cultured hops after treatment with acephate and methomyl, found increased esterase activity in those clones resistant to acephate and oxydemeton-methyl in laboratory bioassays. As with *M. persicae*, the difference in total esterase activity arose from changes in a single esterase band identified by electrophoresis. The possible involvement of an esterase in resistance indicated by this work on aphids from northern Switzerland was supported by studies of a susceptible clone from northern England and three resistant clones, two from different hop-growing areas of England and one from Czechoslovakia (Lewis and Madge, 1984). In this later work, the more-intensely staining region on polyacrylamide gels, characteristic of resistant aphids, could be resolved into a group of four bands, all of which were more active in resistant clones. The levels of resistance to several insecticides were 10–20-fold, whether measured by bioassay using artificial diets containing insecticide (Von Beck and Büchi, 1980) or by spraying aphids and then transferring them to clean leaves (Lewis and Madge, 1984).

The concomitant loss of both resistance and esterase activity in clones reared under conditions designed to preclude contamination by stray aphids provides further support for the involvement of these enzymes in resistance (Lewis and Madge, 1984).

Although synthetic pyrethroids are being used increasingly to control *P. humuli*, there is as yet no published evidence for substantial resistance to them or for cross-resistance from organophosphorus and carbamate insecticides (Muir, 1979; Herdeg, 1982; Lewis and Madge, 1984). However, Büchi (1981) found that of the aphids present on hops in the field after treatment with deltamethrin, only a small proportion had high esterase activity; this contrasts with a population present after treatment with methomyl, in which high-esterase variants predominated. He concluded that the pyrethroids did not preferentially kill the aphids with low esterase activity and assumed that those present after treatment were resistant to this insecticide as the result of another mechanism. However, in the absence of bioassay data to confirm resistance or measurements of the frequency of high-esterase aphids before treatment with deltamethrin or methomyl, these conclusions should be considered provisional.

The more active esterases in resistant *P. humuli* have been shown to cross-react with the antiserum to E4 from *M. persicae* (Devonshire et al., 1986a). This established not only a degree of homology between the proteins, but also that the *Phorodon* esterases are produced in larger molar amounts by the resistant strains, as occurs in *M. persicae*.

Aphis gossypii

Many insect and acarine species have become resistant to organophosphorus and/or carbamate insecticides by developing mutant forms of acetylcholinesterase less sensitive to inhibition by these chemicals. However, although the possibility of this type of resistance occurring in aphids has been examined (Smissaert, 1976; Devonshire and Moores, 1982), it was found only recently as the mechanism responsible for pirimicarb resistance in *A. gossypii* (Silver, 1984). Whilst organophosphorus resistance of unknown biochemical nature was identified in this aphid 20 years ago, pirimicarb resistance was first reported by Furk et al. (1980), and this particular form of resistance is now known to be specific for pirimicarb, with resistance factors of the order of 1000, compared with a maximum of 20-fold to a series of other carbamates and organophosphorus insecticides, and virtually none to pyrethroids (Silver, 1984).

The clones of *A. gossypii* studied were host specific, although this was not associated with resistance; the same susceptible clone was used by both groups (Furk et al., 1980; Silver, 1984) and would only survive on cucumber, as would a resistant clone from Japan, whereas the resistant clones from glasshouses in the U.K. and the Netherlands would only infest chrysanthemums. These host specificities were associated with qualitative differences in electrophoretic non-specific esterase patterns; the Japanese resistant clone gave the same esterase mobility variants as the susceptible, whereas the Dutch and British resistant clones gave a different pattern. However, any influence of the host plant on the esterases could not be evaluated, since none of the clones could be transferred between hosts. Superimposed on these qualitative differences in esterases was a quantitative difference, with all resistant clones having slightly less activity in the two major esterases (Furk et al., 1980; Silver, 1984). However, these minor differences were not thought to be involved in resistance; and, furthermore, Silver (1984) found no evidence for reduced penetration or enhanced metabolism of pirimicarb in resistant compared to susceptible clones.

The only biochemical difference associated with pirimicarb resistance in these *A. gossypii* clones was in the properties of their acetylcholinesterase (Silver, 1984). These enzymes were characteristic of those from other aphid species (Zahavi et al., 1972; Smissaert, 1976; Manulis et al., 1981) in being sensitive to inhibition by DTNB (5,5'-dithiobis(2-nitrobenzoic acid)), the colorimetric thiol reagent commonly used for their assay, indicating the involvement of this group in their catalytic activity. The radiometric method of Johnson and Russell (1975) with ^{3}H acetylcholine was therefore used to characterise the enzymes. The acetylcholinesterase from both Japanese and Dutch resistant clones was approx. 700-fold less sensitive to inhibition by pirimicarb than the susceptible enzyme, as judged from their bimolecular rate constants (k_i); this arose primarily from a 350-fold decrease in affinity (K_a), with a further slight decrease in carbamylation rate (k_2); see Devonshire (1977) for a definition of these kinetic constants.

In common with the cross-resistance spectra of the aphids, acetylcholinesterase-insensitivity was virtually specific to pirimicarb, with less than a 5-fold difference between enzymes in rate of inhibition by ethiofencarb or aldicarb. Resistance was also associated with a slight (approx. 2-fold) decrease in sensitivity to DTNB inhibition, and a 2–4-fold increase in efficiency of the enzyme (Silver, 1984; assessed by measuring the catalytic centre activity).

The specificity of the insensitivity for one insecticide is unusual, and Silver (1984) suggested that inhibition by pirimicarb of acetylcholinesterase from *A. gossypii* might involve its binding to a site (allosteric) distinct from the substrate-binding site. A recent report of reciprocal cross-resistance when *A.*

Section 11.1.1 references, p. 135

gossypii was selected with either pirimiphos-methyl or permethrin (Zil'ber-mints and Zhuravleva, 1984) clearly implicates some other biochemical mechanism in this particular case.

KARYOTYPE AND INSTABILITY OF RESISTANCE IN *M. PERSICAE*

A common feature of much of the work on resistance in *M. persicae*, and to a smaller extent in *P. humuli* (Lewis and Madge, 1984), has been the numerous reports of spontaneous loss of resistance in clonal lineages (Dunn and Kempton, 1966; Hurkova, 1971; Needham and Sawicki, 1971; Beranek, 1974b; Boness and Von Unterstenhofer, 1974); a recent report of instability (Bauernfeind and Chapman, 1985) is not directly comparable with the foregoing since it was only observed in non-clonal cultures. These observations were based primarily on monitoring the changes by bioassay of mass clonal cultures, and some might well have arisen from contamination of the cultures, as discussed by Blackman (1979). However, by using a sensitive enzyme marker (E4) for resistance, it became possible to monitor changes in individual offspring at every generation, and this, in conjunction with strict precautions to preclude contamination, has established unequivocally that changes in resistance, both decreasing and increasing. (Sawicki et al., 1980; Bunting and Van Emden, 1980), do occur in certain clones at an extremely high frequency. The exact genetic mechanism remains unknown; genetic recombination (endomeiosis) seems to be the likely cause and this has often been invoked, although Blackman (1970) has argued against this. However, in all cases so studied loss or gain of resistance occurred only in clones heterozygous for a particular chromosomal translocation, suggesting that this is in some way involved with resistance and its instability.

The translocation was first recognised by Blackman and Takada (1975) and attributed, on the basis of biometric characterisation, to either a simple or reciprocal translocation involving autosomes 1 and 3. It was also studied by Lauritzen (1982) using a chromosome banding technique, and this work implicated autosomes 1 and 2; however, the two reports are believed (R.L. Blackman, personal communication, 1984) to refer to the same translocation. It occurs throughout the world (Blackman et al., 1978), but has been studied most intensively in U.K. glasshouse populations and Japanese field populations.

The translocation is common in the former environment, where it is associated with high E4 activity and strong insecticide resistance; the few translocated susceptible clones reported from glasshouses may have been derived from very resistant aphids by loss of E4 activity (Blackman and Takada, 1975), as discussed above. It should be stressed, however, that not all strongly resistant translocated clones are unstable; some originating both under glass (Blackman et al., 1978) and in the field (Devonshire and Sawicki, 1979), have been reared for many generations with no loss of resistance. This variable effect of the translocation on stability of resistance is also illustrated by studies in which sexual females from a karyotypically normal, insecticide-susceptible clone were crossed with males from a translocated very resistant clone (Blackman et al., 1978). Of the 100 clones established from the progeny, 65 were susceptible (in terms of E4 activity) and lacked the translocation; the remainder had the translocation, and all of these produced at least some aphids with high E4 activity. The two characters, resistance and translocation, thus showed complete linkage. However, whilst some of the translocated clones had stable resistance like the parent, others produced a mixture of susceptible and resistant individuals. This intraclonal variation was suggested (Blackman et al., 1978) to arise from a so-called variegated (V-type) position effect by inactivation of the E4 locus by repositioned heterochromatin. However, the phenomenon of instability is known to be a feature of protein overproduction when this relies on gene amplification, the mechanism responsible for resistance in *M. persicae* (Devonshire and Sawicki, 1979; Field et al., 1988).

All the Japanese clones of normal karyotype were susceptible, and the translocated ones resistant to some degree (Blackman et al., 1978). However, with one exception, resistance was low (3–15-fold) in contrast to translocated clones from British glasshouse populations. It is clear, therefore, that whilst *very* strong resistance has always been associated with the translocation, the reverse is not generally true, i.e. the translocation does not invariably cause strong resistance. When discussing the same bioassay data on these clones, Takada (1979) could not correlate resistance with E4 activity. However, heterogeneity within the clones in respect of E4 activity might have caused difficulty in classification both in bioassays and electrophoresis, perhaps accentuated since each clone appears to have been bioassayed only once, and electrophoresis was done on a "mass" homogenate of one to five aphids. The aphids were collected in Japan between 1971 and 1978 and reared without insecticidal selection; it is possible that when collected, they were very resistant, but that this resistance was unstable and disappeared in the absence of selection pressure.

Aphids heterozygous for the translocation are probably selected against during sexual reproduction, so that in areas where this is important, aphids of normal karyotype will have a reproductive advantage during this phase of their life cycle. However, translocation heterozygotes are successful in warm climates or glasshouses, where they can continuously reproduce parthenogenetically, and the use of insecticides on mixed populations favours the build-up of translocated aphids as a consequence of their resistance.

DETAILED STUDIES OF RESISTANT POPULATIONS OF *M. PERSICAE*

Although resistant aphids are known to occur throughout the world (Georghiou, 1981), the reports cannot give a full picture of the distribution and development of resistance. Even for *M. persicae*, for which the most extensive information is available, the individual occurrences are generally difficult to compare because different criteria are used to assess resistance. However, in this species the known biochemical resistance mechanism, carboxylesterase E4, has been exploited to elucidate in detail the structure of some resistant populations in the U.K. This was possible because the enzyme can be characterised in single insects, thus allowing their individual resistance and the heterogeneity of populations to be determined.

The most widely-used method has been to resolve E4 from other esterases by electrophoresis, stain for esterase activity, and then classify the aphid according to the intensity of the E4 band. Such electrophoretic techniques are commonly used by geneticists to study populations when classification is normally based on qualitative differences between enzymes, as opposed to the quantitative assessment necessary for E4. However, whilst the *identification* of resistance by this method is quite straightforward since even slightly resistant aphids differ clearly in their E4 content from susceptible insects, assessing their *degree* of resistance is generally somewhat subjective. Although broad groupings can be made (R_1 and R_2, see Sawicki et al., 1978), aided by the presence in R_1 aphids and absence in R_2 aphids (translocated) of another esterase (E5), further subdivision cannot be done unequivocally from a single assay on an individual insect. However, by cloning the aphids and making quantitative measurements of E4 in a large number of progeny (Sawicki et al., 1980) more accurate classification is possible, and seven variants have been identified in this way. The need for cloning can be avoided when studying field populations if the recently-developed immunoassay for E4 is used; the large amount of quantitative data this provides can be presented as histograms superimposed on the E4 distributions curves of standard clones for comparison (ffrench-Constant and Devonshire, 1988).

Different systems of nomenclature have been used by various authors for the electrophoretically-resolved esterases from *M. persicae*, and these qualitative definitions have been correlated by Takada (1979); one of these (Devonshire, 1977) is now most-widely accepted. In the quantitative classification used by Baker (1977), it appears from both the definition in the test and the photograph, that RAE($-$) and RAE($+ + +$) correspond to S and R_1 respectively, and that RAE($+$) represents a form with intermediate E4 activity; R_2 aphids were apparently not detected. This correlation is supported by a study (Devonshire and Needham, 1975) of populations from the same area and over the same period as that of Baker (1977), which showed clearly by quantitative esterase determination that only S and R_1 aphids were present. Furthermore in later work (Baker, 1978a), RAE($+$) and RAE($+ + +$) were not classified separately.

Quantitative determination of esterase activity is particularly useful for distinguishing between different resistant variants in which E4 contributes most of the esterase activity; it is not as sensitive as the electrophoretic technique for distinguishing susceptible and slightly resistant (R_1) aphids in which the small increase in E4 activity is superimposed on a comparatively high "background" contributed by other esterases. A simplified semiquantitative method, the "tile test" (Sawicki et al., 1978), is useful as a preliminary screen of field populations for the presence of very resistant *M. persicae*.

Since the application of such techniques to the study of field populations has generally involved the characterisation of a relatively small number of insects, it is important that the sample be collected in such a way as to avoid clonal groups and be as representative as possible of the population as a whole. Despite such precautions, the significance of some of the differences reported between individual samples should not be overestimated, especially when these are based on the analysis of few insects.

This limitation of sample size has been overcome by a recently introduced immunoassay (Devonshire et al., 1986a) which greatly facilitates the study of resistant populations. Antiserum specific for E4 is used to trap this enzyme in 96-well microtitration plates, and the "background" esterases removed by washing. Bound E4 is measured by its own esterase activity with naphthyl butyrate as substrate. This simplified procedure was found to be much faster and more reproducible than enzyme-linked immunoassay (ELISA). It gives a quantitative measure of E4, enabling the reliable discrimination between individual S, R_1, R_2 and the strongly resistant aphids common in glasshouse populations. Furthermore, because one person can now characterise 1000–2000 aphids per day compared with a maximum of 200 per day by electrophoresis or 100 per day by total esterase assay, the limiting factor is no longer the analysis, but the collection of insects.

Electrophoretic characterisation of E4 in over 4000 aphids from 250 field samples (Sawicki et al., 1978), coupled with bioassays using demeton-S-methyl and pirimicarb, established the broad distribution of resistant aphids throughout the U.K. in 1976. Resistance was mainly of the R_1 type; this variant accounted for the majority of the aphids collected in eastern England, but was not so predominant in the less-heavily sprayed central and western regions. Surprisingly, not only was resistance (R_1) common in northern England and central and western Scotland, where insecticidal selection would be expected to be less than in East Anglia, but the more resistant variant (R_2) was detected only in these areas (see also Sykes, 1977; Devonshire et al., 1977); for example, of 171 aphids collected at six sites in Scotland, only 7 were S, 111 were R_1 and 53 were R_2. It is not clear why the R_2 variant was so common in these samples from the north, nor why susceptible aphids predominated in the east of Scotland (see below).

Even though the proportion of R_1 aphids in a field was found to increase immediately after spraying (Sawicki et al., 1978), an overall effect of insecticide use on regional populations throughout the summer was not apparent; thus, although two of the fields from which the Derbyshire samples were collected were treated in June with demeton-S-methyl, the proportions of R_1 aphids in these samples remained fairly constant at 77/158 (49%) in May, 138/305 (45%) in June and 46/126 (37%) in July. Over the same period, samples from fields in East Anglia (Broom's Barn), all of which were treated between mid-June and early July, had 170/205 (83%), 355/425 (84%) and 68/71 (96%) R_1 aphids.

More detailed conclusions were drawn by Baker (1977, 1978a) from similar studies of *M. persicae* collected mainly from the east of England and Scotland. Only five of the sixty aphids from Scotland were resistant, whereas, in agreement with Sawicki et al. (1978), susceptible aphids were much less common in East Anglia, especially in those fields where insecticides were applied shortly before sampling (Baker, 1977). In this area, there appeared to be a higher proportion of susceptible aphids on the windward sides of four fields and this was interpreted as indicating that, since this "is where one would expect to find the highest proportion of windborne new arrivals", the aerial population at the time of sampling was largely composed of susceptible aphids, and by implication, that resistant aphids were more abundant on the crops even though two of them were untreated. This work also emphasised the association between the activity of the esterase causing resistance (E4) and the mobility of another esterase, ESE (equivalent to E1/E2 of Devonshire, 1975, 1977), susceptible aphids usually having the "fast" form and resistant aphids the "slow" form. More than 90% of the aphids characterised had one of these combinations.

The Scottish populations were studied more intensively the following year (Baker, 1978a) with emphasis on the different forms of ESE which were shown to comprise homozygous and heterozygous forms of a dimeric enzyme, the (s) and (f) forms being renamed (ss) and (sf), and the very fast form, found only in Scotland, as (ff). The work showed that none of the 122 aphids collected in March from anholocyclic overwintering sites in E. Scotland (Edinburgh–E. Lothian) was resistant. In the samples collected in mid-June from the potato crop, resistant aphids accounted for 3 of the 80 insects collected (i.e. 3/80) in this area, 0/13 near Dundee and 4/29 in Ayrshire. In early July, the corresponding proportions on untreated crops were 7/65, 1/49 and 22/91. Taken together with the characteristic variants at the ESE locus, these findings suggested that only susceptible aphids overwinter in E. Scotland and the resistant variants found later in the season were believed to be immigrants from the west of Scotland and northern England.

In an electrophoretic survey of 2700 aphids from U.K. populations during 1980–1984, the nationwide distribution of resistance remained broadly similar to that during the 1970s (Furk, 1986). However, by 1985/1986, aphids more resistant than R_1 were common in samples collected from unsprayed sites in southern and eastern England, and those from insecticide-treated crops were predominantly R_2 and R_3 (ffrench-Constant and Devonshire, 1988).

In a similar study, Von Büchi and Häni (1984) also used electrophoresis of esterases to examine resistance in twelve samples (1100 aphids in all) of *M. persicae* populations collected in Switzerland between 1977 and 1982. Very resistant aphids (R_2) were common in the two samples from peach, whereas the R_1 variant was the most abundant on potatoes and sugar beet. Although regional variation was found in the proportions of the variants, with R_2 aphids

---

Section 11.1.1 references, p. 135

absent in the samples from western Switzerland, there was no apparent systematic change with time.

The E4 immunoassay has made possible detailed studies on the selection of resistance in *M. persicae* populations by different insecticide classes. Carbamate, organophosphorus and pyrethroid insecticides all selected variants with high esterase activity (R$_2$), but this occurred most rapidly after repeatedly spraying a pyrethroid (ffrench-Constant et al., 1987) as expected from laboratory bioassays (Sawicki and Rice, 1987).

Although the technique of measuring esterase activity has been used by many other workers, the studies mentioned above represent the only use of the method for detailed characterisation of field populations. In all this work, it is important to remember that, although increased E4 activity is a positive indicator of resistance, its absence can only show that this particular resistance mechanism is absent. However, all cases of organophosphorus resistance so studied have been associated with increased E4 activity.

The most detailed survey of resistance based solely on bioassay is that by Weber (1985), who examined the susceptibility of more than 1000 clones established from field populations (seven samples averaging approx. 150 clones) collected in the lower Rhine Valley during 1980–1982. The toxicity of parathion to these clones varied over a range of 3000-fold. However, the susceptible reference clone also varied in its response by as much as 9-fold when bioassayed at different times during the survey, and it seems likely that much of the variation in clones collected from the field is attributable to this variability in the bioassay. Whilst it is not possible to draw unequivocal comparisons with other work, it appears that most of the clones examined fell within the normal distribution ranges expected for S and R$_1$ aphids, but with a significant proportion equivalent to R$_2$ or higher. The repeated grouping in the bioassay results referred to by the author, and attributed to the gene doubling phenomenon, is not immediately apparent in the data presented, and the "13 duplication levels" suggested are unlikely. The extended range almost certainly has two origins: the superimposition of the bioassay variability on the intrinsic differences in sensitivity between clones, and the lack of direct proportionality between E4 content and resistance (Devonshire, 1977; Pedersen, 1984). Despite these reservations about some of the conclusions, this was a thorough and significant contribution to our knowledge of resistance in aphids.

THE FUTURE

As in the past, resistance problems will probably continue to be overcome by changing to alternative chemicals, but this will lead to more and more restriction on the range of suitable aphicides. In some cases, e.g. *P. humuli* and glasshouse populations of *M. persicae*, this is already a serious problem. One of the consequences of resistance is that aphids, although killed by the chemical, die more slowly. In situations where insecticides are used to control aphid-borne viruses, this is important since the aphids might then have the opportunity to transmit the virus before dying (see section 11.1, p. 90 and section 11.5.1, p. 270). In such cases, careful monitoring for resistance is particularly important if the virus is to be controlled as well as the aphid infestation.

The development of resistance emphasises the need for control strategies to prolong the useful life of insecticides. These might involve the use of integrated control in situations where this is practicable, improved forecasting systems to minimise insecticide use, and the development of novel control chemicals such as antifeedants or other behaviour-controlling chemicals. It might be expected that the performance of such compounds lacking an ester group would be unaffected by the known resistance mechanisms in aphids, but some clones of organophosphorus-resistant *M. persicae* are insensitive to their own alarm pheromone, (E)-β-farnesene (Dawson et al., 1983).

It is commonly assumed that resistant variants in general are less "fit" and therefore at a disadvantage in the absence of insecticidal selection when in a mixed population with susceptible insects. This assumption has been made for aphid populations (e.g. Baker, 1977), and, indeed, resistant aphids appear not to overwinter asexually in eastern Scotland. However, this generalisation does not seem applicable to *M. persicae*. A preliminary study (Banks and Needham, 1970) showed that one resistant clone reproduced faster than a susceptible clone when reared on turnip plants in a glasshouse. Similarly, susceptible and slightly resistant clones of *M. persicae* reared on sugar beet "developed" to a given stage in 4–5 weeks whilst the more resistant clones gave similar infestations after only 3 weeks (Hurkova, 1971). A more detailed study under carefully controlled conditions was made by Eggers-Schumacher (1983). He assessed reproductive potential of *M. persicae* from the number of embryos in aphids, their rate of birth and their development time. By these criteria, the three resistant clones studied were more "fit" on preferred hosts than four susceptible clones, whether reared at high (22–29°C) or low (7–10°C) temperature. However, the susceptible clones transferred more readily to less suitable hosts. In a recent study, Weber (1985) was unable to find any correlation between resistance and reproduction rate, longevity or larval mortality.

Although even less information on fecundity is available for *P. humuli*, one resistant strain was found to reproduce at a higher rate and produce larger offspring than susceptible aphids (Lorriman and Llewellyn, 1983).

Furthermore, there is evidence that aphids surviving treatment with organophosphorus insecticides (Gordon and McEwan, 1984; Lowery and Sears, 1986a,b) or pyrethroids (A.D. Rice, personal communication, 1985; ffrench-Constant et al., 1987) produce nymphs more rapidly than untreated aphids; this phenomenon, coupled with the higher survival rate of resistant insects, could encourage the build-up of resistant populations. Thus, although these independent studies characterised a limited number of clones, they are in agreement that resistant aphids were, if anything, slightly more "fit", so contradicting the generally-held belief about relative fitness, and emphasising that, as far as *M. persicae* is concerned, resistant forms will remain with us and become the "normal" population against which future aphicides must be effective.

REFERENCES

Aliniazee, M.T., 1983. Carbaryl resistance in the filbert aphid (Homoptera: Aphididae). Journal of Economic Entomology, 76: 1002–1004.

Asakawa, M., 1975. Insecticide resistance in agricultural insect pests of Japan. Japan Pesticide Information, 23: 5–8.

Attia, F.I. and Hamilton, J.T., 1978. Insecticide resistance in *Myzus persicae* in Australia. Journal of Economic Entomology, 71: 851–853.

Attia, F.I., Hamilton, J.T. and Franzmann, B.A., 1979. Carbamate resistance in a field strain of *Myzus persicae* (Sulzer) (Hemiptera: Aphididae). General and Applied Entomology, 11: 24–26.

Baker, J.P., 1977. Assessment of the potential for and development of organophosphorus resistance in field populations of *Myzus persicae*. Annals of Applied Biology, 86: 1–9.

Baker, J.P., 1978a. Electrophoretic studies on populations of *Myzus persicae* in Scotland from March to July, 1976. Annals of Applied Biology, 88: 1–11.

Baker, R.T., 1978b. Insecticide resistance in the green peach-potato aphid *Myzus persicae* (Sulz.) (Hemiptera: Aphididae). New Zealand Journal of Experimental Agriculture, 6: 77–82.

Ball, B.V. and Bailey, L., 1978. The symbiotes of *Myzus persicae* (Sulz.) in strains resistant and susceptible to demeton-S-methyl. Pesticide Science, 9: 522–524.

Banks, C.J. and Needham, P.H., 1970. Comparison of the biology of *Myzus persicae* Sulz. resistant and susceptible to dimethoate. Annals of Applied Biology, 66: 465–468.

Baranyovits, F., 1973. The increasing problem of aphids in agriculture and horticulture. Outlook on Agriculture, 7: 102–108.

Bauernfeind, R.J. and Chapman, R.K., 1985. Nonstable parathion and endosulfan resistance in green peach aphids (Homoptera: Aphididae). Journal of Economic Entomology, 78: 516–522.

Beranek, A.P., 1974a. Esterase variation and organophosphate resistance in populations of *Aphis fabae* and *Myzus persicae*. Entomologia Experimentalis et Applicata, 17: 129–142.

Beranek, A.P., 1974b. Stable and non-stable resistance to dimethoate in the peach-potato aphid (*Myzus persicae*). Entomologia Experimentalis et Applicata, 17: 381–390.

Beranek, A.P. and Oppenoorth, F.J., 1977. Evidence that the elevated carboxylesterase (esterase 2) in organophosphorus-resistant *Myzus persicae* (Sulz.) is identical with the organophosphate-hydrolyzing enzyme. Pesticide Biochemistry and Physiology, 7: 16–20.

Blackman, R.L., 1974. Life-cycle variation of *Myzus persicae* (Sulz.) (Hom.: Aphididae) in different parts of the world, in relation to genotype and environment. Bulletin of Entomological Research, 63: 595–607.

Blackman, R.L., 1975. A preliminary report on the genetics of resistance to organophosphates in *Myzus persicae*. Proceedings 8th British Insecticide and Fungicide Conference, 1: 75–78.

Blackman, R.L., 1979. Stability and variation in aphid clonal lineages. Biological Journal of the Linnean Society, 11: 259–277.

Blackman, R.L. and Devonshire, A.L., 1978. Further studies on the genetics of the carboxylesterase regulatory system involved in resistance to organophosphorus insecticides in *Myzus persicae* (Sulzer). Pesticide Science, 9: 517–521.

Blackman, R.L. and Takada, H., 1975. A naturally occurring chromosomal translocation in *Myzus persicae* (Sulzer). Journal of Entomology, 50: 147–156.

Blackman, R.L., Devonshire, A.L. and Sawicki, R.M., 1977. Co-inheritance of increased carboxylesterase activity and resistance to organophosphorus insecticides in *Myzus persicae* (Sulzer). Pesticide Science, 8: 163–166.

Blackman, R.L., Takada, H. and Kawakami, K., 1978. Chromosomal rearrangement involved in insecticide resistance of *Myzus persicae*. Nature, 271: 450–452.

Boness, M. and Von Unterstenhofer, G., 1974. Insektizidresistenz bei Blattlausen. Zeitschrift für Angewandte Entomologie, 77: 1–19.

Büchi, R., 1981. Evidence that resistance against pyrethroids in aphids *Myzus persicae* and *Phorodon humuli* is not correlated with high carboxylesterase activity. Zeitschrift für Pflanzenkrankheiten und Pflanzenschutz, 88: 631–634.

Büchi, R. and Häni, A., 1984. Das Resistenzspektrum in Blattlauspopulationen von *Myzus persicae* (Sulz.) in der Schweiz in den Jahren 1977/8 und 1982. Zeitschrift für Angewandte Entomologie, 98: 239–246.

Bunting, S. and Van Emden, H.F., 1980. Rapid response to selection for increased esterase activity on small populations of an apomictic clone of *Myzus persicae*. Nature, 285: 502–503.

Bunting, S. and Van Emden, H.F., 1981. The effect of conventional and artificial diet on esterase band pattern in *Myzus persicae* (Sulzer). Experientia, 37: 220–221.

Chang, R., Ward, C.R. and Ashdown, D., 1980. Spread and characterization of chemical resistance of biotype D greenbugs in Texas. Journal of Economic Entomology, 73: 458–461.

Dawson, G.W., Griffiths, D.C., Pickett, J.A. and Woodcock, C.M., 1983. Decreased response to alarm pheromone by insecticide-resistant aphids. Naturwissenschaften, 70: 254–255.

Devonshire, A.L., 1973. The biochemical mechanisms of resistance to insecticides with especial reference to the housefly, *Musca domestica* and aphid, *Myzus persicae*. Pesticide Science, 4: 521–529.

Devonshire, A.L., 1975. Studies of the carboxylesterases of *Myzus persicae* resistant and susceptible to organophosphorus insecticides. Proceedings 8th British Insecticide and Fungicide Conference, 1: 67–74.

Devonshire, A.L., 1977. The properties of a carboxylesterase from the peach-potato aphid, *Myzus persicae* (Sulz.), and its role in conferring insecticide resistance. Biochemical Journal, 167: 675–683.

Devonshire, A.L. and Moores, G.D., 1982. A carboxylesterase with broad substrate specificity causes organophosphorus, carbamate and pyrethroid resistance in peach-potato aphids (*Myzus persicae*). Pesticide Biochemistry and Physiology, 18: 235–246.

Devonshire, A.L. and Needham, P.H., 1974. The fate of some organophosphorus compounds applied topically to peach-potato aphids (*Myzus persicae* [Sulz.]) resistant and susceptible to insecticides. Pesticide Science, 5: 161–169.

Devonshire, A.L. and Needham, P.H., 1975. Resistance to organophosphorus insecticides of peach-potato aphid (*Myzus persicae*) from sugar beet in 1975. Proceedings 8th British Insecticide and Fungicide Conference, 1: 15–19.

Devonshire, A.L. and Sawicki, R.M., 1979. Insecticide-resistant *Myzus persicae* as an example of evolution by gene duplication. Nature, 280: 140–141.

Devonshire, A.L., Foster, G.N. and Sawicki, R.M., 1977. Peach-potato aphid, *Myzus persicae* (Sulz.), resistant to organophosphorus and carbamate insecticides on potatoes in Scotland. Plant Pathology, 26: 60–62.

Devonshire, A.L., Moores, G.D. and Chiang, C., 1983. The biochemistry of insecticide resistance in the peach-potato aphid, *Myzus persicae*. In: J. Miyamoto et al. (Editors), IUPAC Pesticide Chemistry, Human Welfare and The Environment, pp. 191–196.

Devonshire, A.L., Moores, G.D. and ffrench-Constant, R.H., 1986a. Detection of insecticide resistance by immunological estimation of carboxylesterase activity in *Myzus persicae* (Sulzer) and cross reaction of the antiserum with *Phorodon humuli* (Schrank) (Hemiptera: Aphididae). Bulletin of Entomological Research, 76: 97–107.

Devonshire, A.L., Searle, L.M. and Moores, G.D., 1986b. Quantitative and qualitative variation in the mRNA for carboxylesterases in insecticide-susceptible and resistant *Myzus persicae* (Sulz.). Insect Biochemistry, 16: 659–665.

Dunn, J.A. and Kempton, D.P.H., 1966. Non-stable resistance to demeton-methyl in a strain of *Myzus persicae*. Entomologia Experimentalis et Applicata, 9: 67–73.

Dunn, J.A. and Kempton, D.P.H., 1977. The development on potatoes of a glasshouse strain of *Myzus persicae* resistant to organophosphorus insecticides. Annals of Applied Biology, 85: 175–179.

Eggers-Schumacher, H.A., 1983. A comparison of the reproductive performance of insecticide-resistant and susceptible clones of *Myzus persicae*. Entomologia Experimentalis et Applicata, 34: 301–307.

ffrench-Constant, R.H. and Devonshire, A.L., 1988. Monitoring frequencies of insecticide resistance in *Myzus persicae* (Sulzer) (Hemiptera: Aphididae) in England during 1985–86 by immunoassay. Bulletin of Entomological Research, 78: 163–171.

ffrench-Constant, R.H., Devonshire, A.L. and Clark, S.J., 1987. Differential rate of selection for resistance by carbamate, organophosphorus and combined pyrethroid and organophosphorus insecticides in *Myzus persicae* (Sulzer) (Hemiptera: Aphididae). Bulletin of Entomological Research, 77: 227–238.

ffrench-Constant, R.H., Devonshire, A.L. and White, R.P., 1988. Spontaneous loss and re-selection of resistance in extremely resistant *Myzus persicae* (Sulzer). Pesticide Biochemistry and Physiology, 30: 1–10.

Field, L.M., Devonshire, A.L. and Forde, B.G., 1988. Molecular evidence that insecticide resistance in peach-potato aphids (*Myzus persicae*, Sulz.) results from amplification of an esterase gene. Biochemical Journal, 251: 209–212.

Furk, C., 1986. Incidence and distribution of resistant strains of *Myzus persicae* (Sulzer) (Hemiptera: Aphididae) in England and Wales 1980–84. Bulletin of Entomological Research, 76: 53–58.

Furk, C., Powell, D.F. and Heyd, S., 1980. Pirimicarb resistance in the melon and cotton aphid, *Aphis gossypii* Glover. Plant Pathology, 29: 191–196.

Georghiou, G.P., 1981. The occurrence of resistance to pesticides in arthropods; an index of cases reported through 1980. FAO Plant Production and Protection Series, Rome, 192 pp.

Gordon, P.L. and McEwen, F.L., 1984. Insecticide-stimulated reproduction of *Myzus persicae*, the green peach aphid (Homoptera: Aphididae). Canadian Entomologist, 116: 783–784.

Hamilton, J.T., Attia, F.I. and Hughes, P.B., 1981. Multiple resistance in *Myzus persicae* (Sulzer) in Australia. General and Applied Entomology, 13: 65–68.

Herdeg, G., 1982. Untersuchungen zur Insektizidresistenz der Hopfenblattlaus (*Phorodon humuli* Schrank) im Hopfenanbaugebiet Tettnang. Zeitschrift für Pflanzenkrankheiten und Pflanzenschutz, 89: 468–474.

Hlinakova, M. and Hurkova, J., 1976. Black bean aphid (*Aphis fabae*) resistant to thiometon (Homoptera, Aphididae). Acta Entomologica Bohemoslovaca, 73: 286–288.

Hrdy, I. and Zeleny, J., 1968. Hop aphid (*Phorodon humuli*) resistant to thiometon. Acta Entomologica Bohemoslovaca, 65: 183–187.

Hurkova, J., 1970. Resistance of greenhouse populations of *Myzus persicae* (Sulz.) to some organophosphorus insecticides (Homoptera, Aphididae). Acta Entomologica Bohemoslovaca, 67: 211–217.

Hurkova, J., 1971. "Clonal" variability of response to thiometon in *Myzus persicae* (Sulz.) (Homoptera, Aphididae). Acta Entomologica Bohemoslovaca, 68: 372–376.

Hurkova, J., 1973a. Resistance of the plum aphid in peach orchards. Agrochemia, 13: 205–206 (in Czech).

Hurkova, J., 1973b. Multiple resistance to insecticides in greenhouse samples of *Myzus persicae* (Sulz.) (Homoptera, Aphididae). Acta Entomologica Bohemoslovaca, 70: 13–19.

Johnson, C.D. and Russell, R.L., 1975. A rapid simple radiometric assay for cholinesterase, suitable for multiple determinations. Analytical Biochemistry, 64: 229–238.

Kirknel, E. and Reitzel, J., 1973. Insecticide resistance in the green peach aphid (*Myzus persicae* Sulz.) Tidsskrift for Planteavl, 77: 191–199 (in Danish, with English summary).

Knipling, E.F., 1954. On the insecticide resistance problem. Agricultural Chemicals, 9: 46–47.

Kung, K.Y., Chang, K.L. and Chai, K.Y., 1964. Detecting and measuring the resistance of cotton aphis to systox. Acta Entomologica Sinica, 13: 1–9.

Laska, P., 1981. Low pirimicarb-susceptibility of buckthorn-potato aphid, *Aphis nasturtii* (Homoptera, Aphididae). Acta Entomologica Bohemoslovaca, 78: 61–62.

Lauritzen, M., 1982. Q- and G-band identification of two chromosomal rearrangements in peach-potato aphids, *Myzus persicae* (Sulzer), resistant to insecticides. Hereditas, 97: 95–102.

Lewis, G.A. and Madge, D.S., 1984. Esterase activity and associated insecticide resistance in the damson-hop aphid, *Phorodon humuli* (Schrank) (Hemiptera: Aphididae). Bulletin of Entomological Research, 74: 227–238.

Lorriman, F. and Llewellyn, M., 1983. The growth and reproduction of hop aphid (*Phorodon humuli*) biotypes resistant and susceptible to insecticides. Acta Entomologica Bohemoslovaca, 80: 87–95.

Lowery, D.T. and Sears, M.K., 1986a. Stimulation of reproduction of the green peach aphid, *Myzus persicae* (Sulzer), by azinphosmethyl applied to potatoes. Journal of Economic Entomology, 79: 1530–1533.

Lowery, D.T. and Sears, M.K., 1986b. Effect of exposure to the insecticide azinphosmethyl on reproduction of green peach aphid (Homoptera: Aphididae). Journal of Economic Entomology, 79: 1534–1538.

Manulis, S., Ishaaya, I. and Perry, A.S., 1981. Acetylcholinesterase of *Aphis citricola*: properties and significance in determining toxicity of systemic organophosphorus and carbamate compounds. Pesticide Biochemistry and Physiology, 15: 267–274.

Michelbacher, A.E., Fullmer, O.H., Cassil, C.C. and Davis, C.S., 1954. Walnut aphid resistant to parathion in northern California. Journal of Economic Entomology, 47: 366–367.

Muir, R.C., 1979. Insecticide resistance in damson-hop aphid, *Phorodon humuli* in commercial hop gardens in Kent. Annals of Applied Biology, 92: 1–9.

Needham, P.H. and Sawicki, R.M., 1971. Diagnosis of resistance to organophosphorus insecticides in *Myzus persicae* (Sulz.) Nature, 230: 125–126.

Oppenoorth, F.J. and Voerman, S., 1975. Hydrolysis of paraoxon and malaoxon in three strains of *Myzus persicae* with different degrees of parathion resistance. Pesticide Biochemistry and Physiology, 5: 431–443.

Otto, D., 1980. Nachweis von Kreuzresistenz eines Dimethoat-resistenten *Myzus persicae*-Sulz. -Stammes gegenüber Pyrethroiden. Archiv für Phytopathologie und Pflanzenschutz, Berlin, 16: 283–285.

Pedersen, O.C., 1984. An insecticide bioassay for *Myzus persicae* validated by probit analysis and esterase measurement. Ecological Bulletins, 36: 50–56.

Ripper, W.E., 1961. Selective insect control and its application to the resistance problem. Miscellaneous Publications of the Entomological Society of America, 2: 153–156.

Sawicki, R.M. and Rice, A.D., 1978. Response of susceptible and resistant peach-potato aphids *Myzus persicae* (Sulz.) to insecticides in leaf-dip bioassays. Pesticide Science, 9: 513–516.

Sawicki, R.M., Devonshire, A.L., Rice, A.D., Moores, G.D., Petzing, S.M. and Cameron, A., 1978. The detection and distribution of organophosphorus and carbamate insecticide-resistant *Myzus persicae* (Sulz.) in Britain in 1976. Pesticide Science, 9: 189–201.

Sawicki, R.M., Devonshire, A.L., Payne, R.W. and Petzing, S.M., 1980. Stability of insecticide resistance in the peach-potato aphid, *Myzus persicae* (Sulzer). Pesticide Science, 11: 33–42.

Shanks, C.H., 1967. Resistance of the strawberry aphid to endosulfan in Southwestern Washington. Journal of Economic Entomology, 60: 968–970.

Silver, A.R.J., 1984. The biochemical nature of pirimicarb resistance in two glasshouse strains of *Aphis gossypii* (Glover). Ph.D. Thesis. University of Reading, U.K., 236 pp.

Smissaert, H.R., 1976. Reactivity of a critical sulfhydryl group of the acetylcholinesterase from aphids (*Myzus persicae*). Pesticide Biochemistry and Physiology, 6: 215–222.

Stern, V.M., 1962. Increased resistance to organophosphorus insecticides in the parthenogenetic spotted alfalfa aphid, *Therioaphis maculata*, in California. Journal of Economic Entomology, 55: 900–904.

Sykes, G.B., 1977. Resistance in the peach-potato aphid (*Myzus persicae* (Sulz.)) to organophosphorus insecticides in Yorkshire and Lancashire. Plant Pathology, 26: 91–93.

Takada, H., 1979. Esterase variation in Japanese populations of *Myzus persicae* (Sulzer) (Homoptera: Aphididae), with special reference to resistance to organophosphorus insecticides. Applied Entomology and Zoology 14: 245–255.

Teetes, G.L., Schaefer, C.A., Gipson, J.R., McIntyre, R.C. and Latham, E.E., 1975. Greenbug resistance to organophosphorus insecticides on the Texas High Plains. Journal of Economic Entomology, 68: 214–216.

Von Amiressami, M. and Petzold, H., 1977. Symbioseforschung und Insektizidresistenz. Ein licht- und elektronenmikroskopischer Beitrag zur Klärung der Insektizidresistenz von Aphiden unter Berücksichtigung der Mycetom-Symbionten bei *Myzus persicae* Sulz. Zeitschrift für Angewandte Entomologie, 82: 252–259.

Von Beck, A.K. and Büchi, R., 1980. Esterasetest zum Nachweis der Insektizidresistenz bei der Hopfenblattlaus, *Phorodon humuli* Schrk. Zeitschrift für Angewandte Entomologie, 89: 113–121.

Wachendorff, U. and Klingauf, F., 1978. An esterase assay for the diagnosis of resistance in aphids. Zeitschrift für Pflanzenkrankheiten und Pflanzenschutz, 85: 218–227.

Walters, P.J. and Forrester, N., 1979. Resistant lucerne aphids at Tamworth. The Agricultural Gazette of New South Wales, 90: 7.

Weber, G., 1985. Population genetics of insecticide resistance in the green peach aphid, *Myzus persicae* (Sulz.) (Homoptera, Aphididae). Zeitschrift fur Angewandte Entomologie, 99: 408–421.

White, T.C.R., 1983. The effect of diet on esterase band pattern in *Myzus persicae* (Sulzer) – a disclaimer. Experientia, 39: 884–885.

Wiesmann, R., 1955. Der heutige Stand des Insektizid-Resistenzproblems. Mitteilungen aus der Biologischen Zentralanstalt für Land- und Forstwirtschaft. 83: 17–37.

Zahavi, M., Tahori, A.S. and Klimer, F., 1972. An acetylcholinesterase sensitive to sulfhydryl inhibitors. Biochimica et biophysica Acta, 276: 577–583.

Zil'bermints, I.V. and Zhuravleva, L.M., 1984. Response of melon and greenhouse aphids to Ambush and Actellic. Khimiya v Sel'skom Khozyaistva, 3: 37–40 (in Russian).

11.2 Biological Control of Aphids

MARY CARVER

DEFINITION AND SCOPE

Biological control can be described as the limitation of the abundance of living organisms and their products by other living organisms. The term is difficult to define and any comprehensive definition is probably too broad to be meaningful. The term encompasses both the biotic component of natural control, that is, the naturally occurring regulation of the numbers of a species by the action of its natural enemies, and, also, applied or manipulative biological control, which is the use by man of biological means to control a plant or animal (or its products) that has become a pest.

Natural control is considered under various guises in other sections of this book. This section will for the most part consider some aspects of biological control as applied by man.

The classical example is the application of biological control to an immigrant species which has attained pest status in an invaded habitat, due, it is presumed, to the absence of the natural enemies that regulate its numbers in its native habitat. Natural enemies of one or more species from its original habitat are therefore selected and introduced into the new habitat in the hope that, on establishment, they will create a new and humanly more tolerable population balance of the pest.

So-called "classical" biological control may be applied not only to exotic species using natural enemies from the area of origin of the pest but also to pests in their native habitat using controlling agents either, more rarely, from the native habitat or, more likely, from elsewhere, or from other hosts (Wilson, 1974).

In usual practice, the selected agent is reared in large numbers, and inoculative releases are made in the target area. If establishment occurs, and an acceptable reduction in pest status of the host is achieved, the effect can be considered permanent and no further input may be required.

Biological control may also be achieved by periodical inundation. In comparison with inoculative releases, the beneficial agent is repetitively or periodically released in such large numbers as to overwhelm the host population at a critical period (Mackauer, 1972), e.g. at a susceptible time in the host's life cycle. It acts as a biological pesticide and prompt reduction of the pest population rather than permanent establishment of the agent is the aim.

Other factors may contribute to the attainment of pest status of an invading species, e.g. a preferred or more susceptible or more plentiful host, or a more favourable climate. These may require other means of control, such as the breeding or induction of resistance or tolerance in crop plants, and the

Section 11.2 references, p. 161

adjustment of cultural practices so as to favour a low incidence of pests and disease or a high incidence of natural enemies.

Other means of control also considered by some under the term biological control include other forms of habitat manipulation, such as field placement of artificial food for natural enemies, as either a supplement or attractant; genetical manipulation of pest populations, e.g. by inundative releases of sterile males; and the use of pheromones and insect growth regulators.

The advantages of classical biological control are numerous, especially when compared with chemical control. The method is non-polluting, non-toxic, and self-perpetuating. Though the initial costs may be high, no subsequent expense is incurred, and the results are generally permanent. Biological control is sometimes the only possible means of control, for instance, when chemical control measures would be more costly than the crop loss sustained, in cases of development of pesticide resistance in the target pest, or in the absence of or pending studies on host plant resistance.

Effective pest control, however, may sometimes be possible only by integrated control, that is, by the selective and integrated use of two or more individually effective methods, e.g. biological control in combination with chemical control. The concept of integrated control was first defined by Stern et al. (1959) and has since developed to embrace the wider concepts of pest management and integrated pest management (IPM) (Huffaker, 1980).

These methods and those concerned with host resistance and habitat and genetical manipulation will be considered in detail in sections 11.4 and 11.5. This section will concentrate on biological control by the manipulation of natural enemies.

SOME PROCEDURES AND PRINCIPLES OF BIOLOGICAL CONTROL

The literature may be consulted for detailed information on currently and generally accepted procedures in applied biological control (e.g. DeBach, 1964; Huffaker and Messenger, 1976). Briefly: correct identification of the pest is a necessary prerequisite, and prior knowledge of its biology and basic ecology in the target area desirable. In the case of exotic pests, the area of indigenity is ascertained, if possible, and biological control agents of the pest species are selected from there, and/or from areas in which they have become effectively established. Ideally, importations are made of large stocks of recent field origin from climatic and ecological environments matching those in the target area. The stocks can thus be presumed to be genetically diverse. The potential and host-specificity of the agents are assessed, and secondary parasites and predators are excluded, preferably before, or in quarantine after, introduction. Correct identification of the agents is essential. Voucher specimens are retained. For preference, the introduced agents are mass-reared on the target host and host plant. The agents are released when and where conditions are optimal, and as soon as possible so as to minimise genetic attenuation and adaptation to laboratory conditions. It may be convenient or desirable to make use of an outdoor "nursery site" while awaiting optimal conditions or in the hope of adapting the agents to the new environment prior to release. The introduced agent should be well established and have gained a stable population balance before an attempt is made to establish its success.

In the experience of the author and colleagues, in practising biological control of aphids, selection of potentially suitable agents is possible using available knowledge and, in the interests of efficiency and economy, all the procedures and testing required to assess their suitability and effect establishment are best done after introduction, and before or during mass-rearing and releasing. Procedures are discussed in more detail and with special reference to aphids in section 11.2.1.

Experience has shown that biological control agents often perform better in their new habitat than in the original one, presumably because of the absence of their own natural enemies. A generally accepted tenet is that biological control is more likely to succeed in a permanent, continuous habitat such as on a perennial or woody host than in a temporary, disjunct one such as an annual crop or a herbaceous host, but there is no agreement as to what effects better biological control – a complex of introduced natural enemies, which could lead to competitive displacement, or a single promising agent (Ehler and Hall, 1982, Keller, 1984); highly specific or generalised agents; parasites or predators. The empirical records provide little evidence that multiple introductions are detrimental, and favour specific rather than generalised parasites, and parasites rather than predators (Huffaker et al., 1976). Hokkanen and Pimentel (1984) proposed, with questionable evidence, the preferred selection of biological control agents allopatric with their hosts on the grounds that the co-adaptation of the co-evolved host and enemy may prevent the latter from being effective as a control agent. The records, however, rather point to the sympatric, presumably co-adapted, agent as being of paramount importance in biological control.

HISTORY OF BIOLOGICAL CONTROL, WITH SPECIAL REFERENCE TO APHIDS

The application of biological control has always been an appealing and popular endeavour, and literature on the subject abounds (Sweetman, 1958; DeBach, 1964, 1974; Huffaker and Messenger, 1976; Delucchi, 1976; Ridgway and Vinson, 1977; Van den Bosch et al., 1982). The validity of the method was first demonstrated by the dramatic control in the U.S.A. of *Icerya purchasi* Maskell, the cottony cushion scale, a coccoid pest of citrus of Australian origin, following the introduction of the coccinellid *Rodolia cardinalis* (Mulsant) and the agromyzid fly *Cryptochaetum iceryae* (Williston) from Australia into the U.S.A. in 1888 (Wilson, 1963; Clausen, 1978). Their success together with that of the pyralid moth, *Cactoblastis cactorum* (Berg), from Argentina, in controlling prickly pear cactus (*Opuntia* spp.) in Australia (1926–33) (Wilson, 1960; Clausen, 1978) provide two of the best and most quoted examples of successful biological control by insects. Undoubtedly the most successful endeavour of all, however, has been the biological control of a mammal by an insect-borne virus, namely the control of the European rabbit in Australia following the introduction there in the 1950s of the myxoma (myxomatosis) poxvirus, and the dissemination of the virus by naturally occurring mosquito vectors and imported fleas (Fenner, 1983).

The popularity of biological control waned after the introduction and widespread use of organophosphorous pesticides in the 1940s but was again renewed with the recognition of the severe limitations and dangers of chemical control. Today it is acknowledged that biological control is often now only possible within the framework of integrated control (Wilson and Huffaker, 1976), because other control measures are usually also recommended in combination with biological methods against both the target pest and other pest species in the ecosystem.

Laing and Hamai (1976) and Van den Bosch et al. (1982) provided lists of biological control projects throughout the world, and Clausen (1978) gave a review of candidate pests and their agents. Hughes (Table 11.2.1.1) gives synoptic summaries of attempts at biological control of aphids in Australia and elsewhere.

Section 11.2 references, p. 161

Biological control of aphids was advocated as long ago as 1734 by de Réaumur, who recommended the collection of eggs of an aphidivorous fly (Syrphidae?, Chamaemyiidae?) for their control in greenhouses (Simmonds et al., 1976). C.V. Riley was the first to apply biological control to aphids (and the first to apply it internationally) when, in 1873, during the phylloxera scare in Europe, he shipped the predacious mite *Tyroglyphus phylloxerae* Riley from the U.S.A. to France to combat the scourge. The mite became established but was not effective (DeBach, 1964). Later attempts at biological control of aphids mostly involved insect predators, especially coccinellids, which were introduced into several countries, including Hawaii, Australia, New Zealand, and South Africa, from 1874 onwards. Some of these attempts were small-scale and poorly documented, and involved doubtful identifications. Only a small number of the agents became established, among which were the coccinellids, *Coccinella undecimpunctata* Linnaeus in New Zealand and *Coelophora inaequalis* (Fabricius) in Hawaii (Wilson, 1960; Hodek, 1973; Clausen, 1978).

The first acknowledged case of successful biological control of an aphid species was that of the woolly apple aphid, *Eriosoma lanigerum* (Hausmann), a serious pest of apple of North American origin and now worldwide in distribution (see also Table 11.2.1.3A). The aphelinid *Aphelinus mali* Haldeman, a specialised parasite of *Eriosoma*, was imported, originally from the U.S.A., into 51 countries and geographical entities from 1920 onwards, and became established in 42 countries, with excellent to good results in some, and poor in others (Howard, 1929; Clausen, 1978). Assessment of the effectiveness of this parasite is difficult because *E. lanigerum* infests not only the aerial parts but also the roots of susceptible hosts, where it is not parasitised by *A. mali*, and, more importantly, because the extensive use of aphid-resistant rootstocks of apple and of insecticides to control other pests such as the codling moth, has certainly contributed to the decline in the incidence of *E. lanigerum*.

The most spectacular success, however, was scored against the spotted alfalfa aphid, *Therioaphis trifolii* f. *maculata* (Buckton), a species of Old World origin, which invaded south-western U.S.A. in 1953 (see also Table 11.2.1.3C). Aphidiid parasites, *Trioxys complanatus* Quilis and *Praon exsoletum* (Nees), and the aphelinid *Aphelinus asychis* Walker were imported from the Middle East and quickly became established (Van den Bosch et al., 1964) (Fig. 11.2.1). *T. complanatus*, especially, played a very important role in the suppression of the aphid, but again assessment of effectiveness of the parasites was difficult because of the operation of other controlling factors such as aphid-resistant alfalfa varieties and existent indigenous predators, as well as strip-harvesting procedures, etc. This case is also notable because it led to the formulation of the concept of integrated control (Stern et al. 1959).

The spotted alfalfa aphid was first detected in Australia in 1977 and within ten months had dispersed southwards from Queensland to Victoria and westwards to Western Australia, devastating monocultures of a susceptible lucerne (alfalfa and lucerne are synonyms for *Medicago sativa*; both names will be used here). It did not reach Tasmania until 1980. A nationwide, collaborative campaign was launched to combat the aphid (Anonymous, 1978; Lehane, 1982) and use was made of the knowledge gained from the American experience. *T. complanatus*, *P. exsoletum* and *A. asychis* were introduced from North America and the Mediterranean area, and the entomophthorous fungus, *Zoophthora radicans* (Brefeld) Batko, from Israel (Milner et al., 1982). All have become established, *T. complanatus* very successfully so. The susceptible lucerne (cv. Hunter River) is being replaced by more aphid-resistant varieties such as cvs. Siriver, CUF101 and Condura 73, and the role of native arthropod predators is recognised as being significant. The spotted alfalfa aphid is now no longer a major pest in Australia due, it is claimed, to biological control by *T. com-*

Fig. 11.2.1.. Parasites ovipositing in *Therioaphis trifolii* f. *maculata* (Buckton). (A) *Trioxys complanatus* Quilis; (B) *Aphelinus asychis* Walker (courtesy of L.T. Woolcock, CSIRO, Canberra).

planatus (Hughes et al., 1987). The aphid reached New Zealand in 1982, but has not attained pest status there. *T. complanatus* from Australia has been released (see also Table 11.2.1.1F).

Other lucerne-infesting aphids have also been successful subjects of biological control. The pea aphid, *Acyrthosiphon pisum* (Harris), is a palaearctic

Section 11.2 references, p. 161

species which has been known in North America since the end of last century, chiefly as a serious pest of alfalfa and peas (see also Table 11.2.1.3D). Native predators and parasites are plentiful but of variable control value. In 1958, *Aphidius smithi* Sharma and Subba Rao, a specialised parasite of *A. pisum*, was introduced from India (Hagen and Schlinger, 1960) and readily became established, and in 1963 the polyphagous *Aphidius ervi* Haliday was introduced from Europe. *A. smithi*, in particular, was considered to be a highly successful parasite, but in recent years its relative abundance has declined and there is evidence that it is being displaced by *A. ervi* and, to a lesser extent, by *Aphidius pisivorus* Smith (Mackauer and Kambhampati, 1986). *A. smithi* was subsequently successfully introduced into Hawaii, South America and Europe (Clausen, 1978).

A. *pisum* was first detected in New Zealand in 1976 and in Australia in 1979 (an innocuous, shrub-infesting form has been present in Tasmania for many years). *A. smithi* and another monophagous, but palaearctic aphidiid, *Aphidius eadyi* Starý, González and Hall (= *A. urticae* Haliday partim), were introduced into both countries and *A. pisivorus* into Australia. *A. smithi* has become established in Tasmania, but almost certainly not in mainland Australia, where drought conditions prevailed at the time and the aphid, though initially spreading rapidly, did not become a serious pest. *A. smithi* has not become established in New Zealand, but *A. eadyi* is common (Cameron et al. 1981; P.J. Cameron, personal communication, 1986); *A. ervi*, however, which was imported to control *Acyrthosiphon kondoi* Shinji (see below), is an important parasite of *A. pisum* in both countries (see also Table 11.2.1.1H).

In 1975 the blue-green alfalfa/lucerne aphid, *A. kondoi*, an enigmatic aphid of uncertain origin, appeared suddenly as a serious pest of lucerne in California and New Zealand. It was first detected in Australia in 1977, within weeks of *T. trifolii* f. *maculata*, and in South America and South Africa later. It had previously been known only as a rare aphid in Manchuria (its type locality and possible area of origin) and Japan (Takahashi, 1965). Despite searches for its natural enemies (González et al., 1978, 1979), the only effective agent found was *A. ervi*, a widespread, polyphagous aphidiid of palaearctic origin. *A. ervi* was introduced into North America again, and into Australia and New Zealand, in all of which places it is now widespread (Mackauer and Kambhampati, 1986; Milne, 1986; P.J. Cameron, personal communication, 1986); (see also Tables 11.2.1.1G and 11.2.1.3H). The aphidiids *Ephedrus plagiator* (Nees) (introduced into Australia and New Zealand) and *Praon barbatum* Mackauer (into New Zealand) did not become established. *A. ervi* was also successfully introduced into Argentina in 1972 against *A. pisum* and in 1978 against *A. kondoi* (Botto and Hernández, 1982).

The success of the campaign against *T. trifolii* f. *maculata* in the U.S.A. prompted an attempt at the biological control of the palaearctic phyllaphidine *Chromaphis juglandicola* (Kaltenbach), a pest of walnut in California. The aphidiid *Trioxys pallidus* (Haliday), which was imported from France in 1959, did not disperse beyond the coastal areas but a presumably more heat-tolerant import from Iran in 1968 quickly became widespread and so effective that it can now be credited with almost total commercial control of the aphid (Van den Bosch et al., 1982). This case is also important because it represents a case of biological control by a single, specialised agent operating in the absence of other biotic and human controls, and as such is assessable (see also Table 11.2.1.3F). A different type of *T. pallidus* was also successfully introduced from Europe into Oregon, U.S.A. in 1984, as a biological control agent of the originally palaearctic hazel/filbert aphid. *Myzocallis coryli* (Goeze) (Messing and Ali Niazee, 1987).

In the 1970s, species of palaearctic aphidiids and an aphelinid were introduced from Europe to control exotic phyllaphidines (and their honeydew) on ornamental street trees in California. *Trioxys curvicaudus* Mackauer, a monophagous parasite of *Eucallipterus tiliae* (Linnaeus) on *Tilia*, became established at the release sites, as did *Trioxys tenuicaudus* Starý as a parasite of *Tinocallis platani* (Kaltenbach) on *Ulmus* (see also Table 11.2.1.3J). In addition, sticky bands were used on tree trunks to exclude the argentine ant, *Iridomyrmex humilis* (Mayr), from the aphid colonies. As a consequence, the aphid populations were reduced to the point where insecticide treatments became unnecessary (Olkowski et al., 1974, 1982a, b; Zuparko, 1983).

The case of the carrot aphid, *Cavariella aegopodii* (Scopoli), is an example of a small-scale operation for which, nevertheless, a claim of outstanding success has been made (see Table 11.2.1.3E), namely, that the parasite *Aphidius salicis* Haliday, an import from California in 1962, has controlled the carrot aphid in Australia and thereby also suppressed carrot motley dwarf virus disease, of which *C. aegopodii* is the vector. The claim, if substantiated, represents a unique case of control of a serious crop disease by control of its transmitting agent (Van den Bosch et al., 1976).

Cereal-infesting aphids have also been candidates for biological control in several countries. The greenbug, *Schizaphis graminum* (Rondani), known in North America since approximately 1882 as a pest of small grains, in 1968 became a serious pest of sorghum, which had hitherto been only an incidental host. This new sorghum-feeder (biotype C of *S. graminum* (Harvey and Hackerott, 1969)) may have been the result of a new introduction into North America (Mackauer et al., 1976). *Aphelinus asychis* was introduced from Iran in 1970, but apparently has not become established (Archer et al., 1974; see also Table 11.2.1.3I). However, *S. graminum* is a preferred host of the native, polyphagous aphidiid *Lysiphlebus testaceipes* (Cresson), which is co-extensive with the aphid. Native predators are conspicuous, but attempts to translocate the coccinellid *Hippodamia convergens* Guérin-Méneville within the U.S.A. have been unsuccessful (Starks et al., 1975).

The cereal aphids *Metopolophium dirhodum* (Walker) and *Sitobion avenae* (Fabricius) are rather recent (1966) adventives to South America. *Aphidius ervi* and members of the *Aphidius rhopalosiphi* de Stefani Perez/*A. uzbekistanicus* Luzhetzki complex were introduced into southern Chile from mediterranean France in 1976–77 and are considered to be potentially important regulators of these aphids (Norambuena, 1981). They may also have spread from there to Brazil and Argentina (Botto and Hernández, 1982). *M. dirhodum* has recently invaded New Zealand (1981) and Australia (Carver, 1984a). Accordingly, stocks of the *Aphidius rhopalosiphi* group were introduced into New Zealand in 1985 from England and southern France (J. Farrell, personal communication, 1985) and into Australia in 1986 from Chile and New Zealand. Establishment has apparently not been effected in Australia.

The sowthistle aphid, *Hyperomyzus lactucae* (Linnaeus), is the principal vector of lettuce necrotic yellows virus, a serious disease of lettuce in Australia and New Zealand. In 1981, the aphidiid *Aphidius sonchi* Marshall and *Praon volucre* (Haliday) were introduced into Australia in an attempt to control the aphid and thereby the disease. Both species were imported from the mediterranean area, and a consignment of *A. sonchi* was also sent from Japan. The mediterranean stock of *A. sonchi* has become widely established in southern Australia, and *P. volucre* is reported to be established in Tasmania (see also Table 11.2.1.1.I). The principal host plant of the virus and vector, *Sonchus oleraceus*, is a ubiquitous synthanthropic ephemeral, and this case provides a very good example of successful establishment of a control agent in a non-permanent, dispersed habitat. Because of the unusual simplicity of the scenario – one virus, one vector, one effectual aphid host plant, and an absence of

Fig. 11.2.2. Mass-rearing of *Lysiphlebus testaceipes* (Cresson) in *Aphis craccivora* Koch on *Vicia faba*, at CSIRO, Canberra. (A) Early mummification period; (B) post-mummification period.

hymenopterous parasites and chemical or other human controls – the project was considered ideally suitable for post-operative assessment (Carver and Woolcock, 1986).

The following year, the reputedly polyphagous aphidiids, *Lysiphlebus fabarum* (Marshall) and *L. testaceipes*, were introduced into Australia from the Mediterranean area and California, respectively, to control the cowpea aphid, *Aphis craccivora* Koch, principally in its capacity as a pest of beans (Carver, 1984b; see also Table 11.2.1.1.J) (Fig. 11.2.2). *L. testaceipes* has become established in at least one release site but *L. fabarum* has not been recovered. The nearctic *L. testaceipes*, on the other hand, is spreading rapidly in the Mediterranean area following its deliberate recent introduction into France as a biological control agent of *Toxoptera aurantii* (Boyer de Fonscolombe) and other aphid pests of citrus (Tremblay, 1984). *A. craccivora* has in recent years become a pest of forage lupins in Western Australia (but not in the eastern States). Accordingly, in 1986, *Trioxys indicus* Subba Rao and Sharma (Aphidiidae), an Indian parasite of *Aphis* and its allies, was introduced from India and released in Western Australia (J. Sandow, personal communication, 1985), New South Wales and Victoria (L.T. Woolcock, personal communication, 1986).

A North American lachnid, *Cinara cronartii* Tissot and Pepper, has become a widespread pest of *Pinus* spp. in South Africa since its first appearance there in 1974. In 1983, a species of *Pauesia* (Aphidiidae) was introduced from the U.S.A. for its control. Its successful establishment seems likely (Kfir et al., 1985).

Other attempts include the introduction of *L. testaceipes* into Hawaii and Tonga (Starý, 1970; Stechmann, 1987), Brazilian stock of the widely polyphagous *Aphidius colemani* Viereck into southern France (Tardieux and Rabasse, 1986), and *Aphelinus varipes* (Foerster) into U.S.A. (Archer et al., 1974). Starý

(1970, 1987) should be consulted for information on other introductions of aphidiid parasites.

The use of predators as aphid control agents has not been so successful. Many species, mostly coccinellids and chamaemyiids, have been introduced into North America since 1933 to control the adelgid *Adelges piceae* (Ratzeburg) on *Abies balsamea* and have become established, but only the derodontid beetle, *Laricobius erichsoni* (Rosenhauer) from Europe is considered to be effective (Clausen, 1978; see also Table 11.2.1.3Bf). The palaearctic *Coccinella septempunctata* Linnaeus is an exception. Attempts to establish it as a general predator in the U.S.A. between 1957 and 1975 apparently did not succeed until 1973 (Angalet et al., 1979), since when it has spread throughout most of the eastern two-thirds of North America (Schaefer et al., 1984). The North American coccinellid *H. convergens* has been introduced into many areas of the world but has become established in only Peru, Chile, Venezuela and Hawaii (Hagen et al., 1976), and translocations within the U.S.A. have failed. Several other coccinellid species have been transported between continents (Hodek, 1973; Valentine, 1975; Clausen, 1978), and species of Syrphidae, Chamaemyiidae and Chrysopidae have been introduced into New Zealand (Thomas, 1977; Valentine, 1975).

The inundative method of biological control has been sparingly used on aphids. In the U.S.A., inundative releases were made of imported *C. septempunctata* to combat the potato aphid, *Macrosiphum euphorbiae* (Thomas) (Hodek, 1973), and also of chrysopid eggs (New, 1975; Driestadt et al., 1986) and simulated honeydew as a food supplement and attractant (Hagen and Bishop, 1979; see also Table 11.2.1.3G).

The potential of biological control in controlled and circumscribed environments such as greenhouses has been appreciated in recent years (see section 11.2.2). and tried agents include *Aphidius matricariae* Haliday against the green peach aphid, *Myzus persicae* (Sulzer), on chrysanthemums (Scopes and Biggerstaff, 1973), the cecidomyiid *Aphidoletes aphidimyza* (Rondani) in Finland (Markkula et al., 1979), hymenopterous parasites and predators (Lyon, 1986) and the fungus, *Verticillium lecanii* (Zimmerman) Viegas or "Vertalec" (Hall, 1981). Attempts to use coccinellids, however, have been discouraging (Hodek, 1973).

This history must include reference to the collaborative research on "Biological Control of Aphids" undertaken as part of the International Biological Programme (IBP) during the period 1967–1974. The aim was to investigate the feasibility of developing an integrated control programme against the worldwide pest, *M. persicae* and other associated aphids, on different crops (Mackauer et al., 1976). The collaborators concluded that, though improvement of the impact of natural enemies on *M. persicae* was improbable with the present available methods, control is most likely to result from further breeding for host plant resistance to *M. persicae* and from determination of its biotypic characteristics under given situations especially on non-pre-crop hosts.

INCREASED NEED FOR BIOLOGICAL CONTROL

One of the reasons for the increase in the number of attempts at biological control of aphids in recent years is undoubtedly a greater need, as a consequence of the increased incidence of invasion of new areas (countries) by aphid species. Listed below are some evidently itinerant species followed by the areas invaded and the approximated dates (in parentheses) of first detection there: *Acyrthosiphon kondoi*: [India (1971), native or exotic?]; U.S.A. (1974); New
    Zealand (1975); Australia (1977); South America (1977); South Africa (1981).

Acyrthosiphon pisum: South America (1969); New Zealand (1976); Australia (1979).

Aloephagus myersi Essig, African, *Pistacia, Bromelia*: Australia (1986).

Aphis citricola van der Goot, widespread, polyphagous: western Africa (ca. 1974); eastern Africa (1980).

Appendiseta robiniae (Gillette), nearctic, *Robinia pseudoacacia*: Italy (1981); Switzerland (1981); France (1982); England (1982).

Brachycaudus rumexicolens (Patch), holarctic, *Rumex acetosella*: Australia (1985).

Cinara cronartii, nearctic, *Pinus* spp.: South Africa (1974).

Diuraphis noxia (Mordvilko), the Russian wheat aphid, palaearctic, grasses and cereals: South Africa (1978), Mexico (1980), Argentina(?); U.S.A. (1986).

Dysaphis aucupariae (Buckton), the wild service aphid, palaearctic, *Sorbus* and *Plantago* (plantain): Australia (1981); New Zealand (1981).

Dysaphis lappae (Koch), palaearctic, Cynareae (thistles): Australia (1982).

Eulachnus rileyi (Williams), nearctic, *Pinus* spp.: South Africa (1980).

Hyadaphis tataricae (Aizenburg), Asiatic, *Lonicera tatarica* and others: U.S.S.R. (Moscow) (1935); Germany (1965); Canada (1976); U.S.A. (1979).

Hysteroneura setariae (Thomas), the rusty plum aphid, nearctic, *Prunus* and grasses: Philippines (1958); Hong Kong (1962); South Africa (1967); southern India (1967); Australia (1967).

Impatientinum asiaticum Nevsky, Asiatic, *Impatiens parviflora*: U.S.S.R. (Moscow) (1967); Czechoslovakia (1969); Denmark (1972); England (1982).

Macrosiphum albifrons Essig, nearctic, *Lupinus* (lupin): England (1981); the Netherlands (1984); Germany (1986).

Macrosiphum ptericolens Patch, nearctic, *Pteridium aquilinum*: England (1972).

Metopolophium dirhodum, the rose-grain aphid, palaearctic, *Rosa*, grasses and cereals: South America (1966); South Africa (1967); New Zealand (1981); Australia (1984).

Myzus hemerocallis Takahashi, oriental: South Africa (1969); Australia (1987).

Nearctaphis bakeri (Cowen), nearctic, *Pomoidea, Trifolium* and others: Europe (1965); Japan (1967).

Rhopalosiphum insertum (Walker), the apple-grass aphid, holarctic, Pomoidea, grasses and cereals: Australia (1982).

Schizaphis graminum, the greenbug, biotype C: North America (1968).

Sitobion avenae, palaearctic, grasses and cereals: South America (1966).

Therioaphis trifolii f. *maculata*: U.S.A. (1953); Australia (1977); Japan (1980); South Africa (ca 1981); New Zealand (1982).

Toxoptera odinae (van der Goot), oriental, polyphagous: eastern Africa (1983).

Trichosiphonaphis polygoni (van der Goot), oriental, *Polygonum*: U.S.A. (1974).

Aphids commonly travel great distances in air currents, and can be introduced into new areas on imported plants (listed above, perhaps *A. myersi, C. cronartii, E. rileyi, M. hemerocallis*). However, the sharp increase in the amount and rapidity of aircraft traffic could be responsible for some of the recent increase in the numbers of invasions. Aphids can become frozen at high altitudes and (thus?) survive. Granted survival in transit, location of a suitable host soon after landing is all that is required for establishment of a parthenogenetic individual in a new area. The fact that the hosts of the invasive species are mostly legumes, grasses or weeds commonly to be found near airports and on roadsides, or ornamentals in urban areas, may explain why some species become established rather than others. And now that the cut flower industry operates internationally by air, more aphid introductions to new regions may be anticipated.

APHIDS AS CANDIDATE PESTS FOR BIOLOGICAL CONTROL

Population dynamics

Aphids are amenable to the same biological control procedures as other insects, although some features of their behaviour and functioning require or provide the opportunity for different approaches. Parthenogenesis, paedogenesis, viviparity and polymorphism together have provided for a reproductive rate unbeatable by most if not all of their natural enemies (see section 4.5). At the same time, this rapid reproductive rate allows for the early attainment in aphid colonies of overlapping generations, so that preferred developmental stages are quickly available for parasitisation and predation. Aphids generally have the advantage of a temperature threshold of development lower than that of their natural enemies. On the other hand, they are much more environmentally responsive, being critically sensitive to the nutritional status and growth stages of their host plants, and they are more susceptible to extremes of heat and radiation and unanticipated cold, to low humidity, heavy precipitation and uncontrollable crowding than are their enemies. Frequent and drastic population changes are the characteristic outcome of the consequent mortality and dispersal, and these require the selection of biological control agents that can adjust accordingly in the new environment. Failure to do so may be the reason why (polyphagous) predators have not been very successful. Introduced coccinellids, for instance, have dispersed from release areas without settling, a behavioural characteristic which is consistent with their natural tendency to disperse when food supplies run low. In contrast, several coccinellid species (e.g. *R. cardinalis* and *Cryptolaemus montrouzieri* Mulsant) have been successful as control agents of coccoids, which provide more permanent food sources.

Only alate morphs are engaged in long-distance dispersal and colonisation. Consideration should therefore also be given to control agents (hymenopterous parasites) that can suppress wing formation in their aphid hosts.

Limitations of adventives

Exotic aphid species are presumably often the descendants of a small number of accidentally introduced parthenogenetic females, maybe a clone or even a single individual. As such they have low genetic potential or are genotypically identical and may have a more circumscribed phenotype and a more restricted biology than the species in the indigenous habitat. This likelihood illustrates the desirability of matching candidate pest and control agent even below the species level. A species may be introduced into an area more than once, a known example being *T. trifolii* (Monell), an aphid of Old World origin with a wide geographical and leguminous host range. Two morphological extremes of this species are believed to have been introduced into the U.S.A., namely the yellow clover aphid, which infests *Trifolium* and is holocyclic, and the spotted alfalfa aphid, *T. trifolii* f. *maculata*, which infests *Medicago* and is holocyclic only in the cool fringes of its host's geographical range (Carver, 1978).

Heteroecy

Heteroecious (host-alternating) species may require special consideration. Starý (1970) recommended that, in the interests of establishment, control agents of heteroecious aphids should be released in natural or artificial mixed stands of primary and secondary hosts, as were *A. colemani* against

Section 11.2 references, p. 161

Hyalopterus pruni (Geoffroy) on *Prunus* and *Phragmites communis* in Czecho-slovakia (Starý, 1970) and *A. salicis* against *C. aegopodii* on *Salix* and *Foeni-culum vulgare* in Australia (Stubbs et al., 1983). Only stands of a temporary nature are to be recommended, however, so as to avoid the provision of host plant continuity for the pests themselves.

Care should be exercised in the selection of agents because many natural enemies are niche-restricted. For instance, in Australia, *M. persicae* on its primary host, peach, is attacked by the aphidiids, *Ephedrus persicae* Froggatt and *A. colemani* and by the coccinellid *Harmonia conformis* (Boisduval) and other predators. Whereas *E. persicae* restricts itself to this habitat, as a parasite of *Brachycaudus* and *Myzus* spp., *A. colemani*, a widely polyphagous species, is to be commonly found in both of the aphid's primary and secondary niches. Similarly, *H. conformis* prefers an arboreal habitat and is largely replaced on the aphid's herbaceous hosts by *Coccinella repanda* Thunberg.

However, probably everywhere except in northern temperate areas, most heteroecious exotic aphids are predominantly anholocyclic on their secondary hosts and are therefore effectively monoecious and available all the year round, e.g. *Brachycaudus helichrysi* (Kaltenbach), *Rhopalosiphum padi* (Linnaeus), *H. lactucae* and *H. pruni*. Either the autumnal environment is not extreme enough to induce holocycly, or the primary host is absent or rare, or anholocyclic "strains" have been introduced, or the faculty for holocycly has been lost. Biological control would need to be applied in these cases only within the secondary niche.

Virus vectors

Aphids debilitate their hosts both directly as sap feeders, toxifiers and pollutive excreters (see further section 10.1), and indirectly as vectors of plant virus diseases (section 10.3); some species, such as *M. persicae*, may act in both capacities. The two categories call for different control measures (see also section 11.1). Whereas the degree of damage inflicted by sap feeding is usually correlated with aphid numbers, even small numbers of alate aphids can cause heavy viral infection within a crop, merely by probing and without necessarily settling. When this is so, control measures must be directed towards preventing the vectors from reaching the relevant crop (see section 11.5). To this end, Mackauer et al. (1976) recommended the application of biological control to *M. persicae* outside the crop situation, on surrounding herbaceous weeds. And in introducing the aphidiid *A. sonchi* into Australia as a biological control agent against *H. lactucae*, the principal vector of the luteovirus lettuce necrotic yellows (LNYV), complete control was not the aim but rather some degree of suppression of wing production in presumptive alatae plus a reduction of the "crowding effect" on sowthistle, with a consequent reduction in the numbers of flying vectors of LNYV to lettuce (Carver and Woolcock, 1986). *A. sonchi* has become widely established in Australia, but whether or not it will bring about a reduction in the incidence of LNYV is not yet known. This case emphasises the need to have a prior understanding not only of the candidate aphid but of the associated virus(es) too, and the importance of monitoring the vector in the crop concurrently with the virus. *H. lactucae* infests sowthistle, *Sonchus olera-ceus*, a host to LNYV and a ubiquitous weed in and around lettuce crops. In the field, it reputedly probes but does not colonise lettuce and, under certain laboratory conditions, transmits the disease to the crop plant (Boakye and Randles, 1974). *H. lactucae* is very rarely found on lettuce, so how *H. lactucae* can be responsible for the heavy LNYV infections observed is difficult to under-stand. However, the luteovirus beet western yellows (BWYV) has recently been

detected in several economically important hosts in Australia, including lettuce, and could be responsible for a percentage of the viral symptoms previously attributed to LNYV (Johnstone and Duffus, 1984).

Aulacorthum solani (Kaltenbach), which has been proposed as a candidate
aphid pest for biological control in Australia, is an efficient vector of another
luteovirus, soybean dwarf virus (SYD) [= subterranean clover red leaf]. In
mainland Australia *A. solani* is to be found in cool, moist places and at higher
altitudes, and is rare in the open field, clover crops included. Here again, any
study should include the monitoring of both virus and vector and their identification. BWYV is also present in clovers, and several kinds of nutritional
deficiency in the plant produce symptoms similar to those of SYD.

The aphidiid *A. salicis* is credited with the unique distinction of having
virtually eradicated a luteovirus by controlling its aphid vector. In Australia,
before and during the 1940s, carrot root- and seed-yield were severely reduced
as a result of infection by carrot motley dwarf virus complex (CMDV) and/or
epidemic-type infestation by *C. aegopodii*, its aphid vector. In 1962, ten adults
of an *Aphidius* sp. were introduced from California, and their descendants
became established (Stubbs et al., 1983). The species was probably *A. salicis*,
but the parasites were not identified to species. Nowadays, *A. salicis* is apparently co-existent with *C. aegopodii* in south-eastern Australia, probably as a
result of the 1962 introduction. Coincidentally with the latter, and subsequently, the numbers of trapped *C. aegopodii* dropped dramatically and irrevocably
(Hughes et al., 1965; E.J. Martyn, G.T. O'Loughlin, unpublished data) and the
incidence of CMDV declined to such an extent that the disease is now ex-
ceedingly rare (Waterhouse, 1985). Since then, claims have been made (Smith
and Van den Bosch, 1967; Van den Bosch, 1971; Van den Bosch et al., 1976, 1982;
Clausen, 1978) that *A. salicis* has all but eradicated CMDV by controlling *C.
aegopodii*. Nowadays, however, *C. aegopodii* is rare on carrot in Australia,
though of common occurrence on other umbellifers such as fennel, celery and
parsley, and on *Salix*, its primary host, as is its parasite. Strong evidence has
been assembled that the demise of CMDV is due, not to biological control, but
to the replacement in the 1950s and 1960s of virus- and aphid-susceptible
cultivars of carrot, principally cv. Chantenay, by virus-tolerant, less aphid-
susceptible ones such as cvs. Topweight, All Seasons and Western Red (M.
Carver, unpublished).

Dispersive vehicles

Parasitic control agents that can be dispersed within or on their flying hosts
could be greatly advantaged with respect to establishment. But only within the
Homoptera is parasitisation of adults by insects a common phenomenon, and
many sternorrhynchous adults are apterous anyway (e.g. Coccoidea, Aphidoidea). It has been claimed, though no evidence is provided, that dispersing
adult alate aphids parasitised by aphidiids (in egg or larval stages) can contribute significantly to the dispersal of the parasites themselves, namely, that in
Australia 20% of alate *Brevicoryne brassicae* (Linnaeus) alighting on a crop
may be parasitised by immature *Diaeretiella rapae* (M'Intosh) (Gilbert and
Hughes, 1971); that after introduction, *P. exsoletum* dispersed in California
primarily as immatures within alate *T. trifolii* f. *maculata* (Van den Bosch et al.,
1959); and that early stage aphidiids may be carried in early immigrant alatae
of *A. solani* to potato crops (Robert and Rabasse, 1977). Other evidence strongly
indicates that such instances must be rare. Aphidiids generally prefer to parasitise early-instar aphid nymphs, which are killed and mummified before
reaching the adult stage, and in which wing formation in presumptive alatae
is suppressed (Liu, 1983). Only in cases of parasitisation within aphid colonies

Section 11.2 references, p. 161

comprising predominantly late-instar alatoid nymphs are a large number of parasitised alatae likely to be produced. Presumably only those alatae containing early stage parasites can actually fly. These would need to settle successfully after dispersal, stay alive long enough to allow development of the parasites, and settle in sufficiently large numbers close enough to each other to allow establishment by synchronously emerging female and male parasites. Observations over many years have shown that in the field alate mummies are rare, whether formed by Aphidiidae or Aphelinidae, even of those phyllaphidine species such as *T. annulatus* and *Myzocallis castanicola* Baker, of which all adults are alate.

If alate aphids can aid in the dispersal of primary parasites they also can help cynipid hyperparasites to disperse. In Australia, *Alloxysta ancylocera* (Cameron) heavily parasitises *D. rapae* within *B. brassicae*, often to 100% at the end of the season.

Entomophthorous fungi can, however, be dispersed within alate aphids, but to what extent is not known (see also section 9.3).

THE CONTROL AGENTS

In all of the cases of successful biological control of aphids, the control agents have been parasites and, with one exception, members of the family Aphidiidae (Ichneumonoidea). The success of the latter is not surprising in view of their obligatory, co-evolved, conspicuous relationship with aphids, and the extensive study and wide promotion they have received. Species of *Aphelinus* (Chalcidoidea) are generally not common in the field even in their native habitats, and, except for *A. mali*, the impact of introduced *Aphelinus* has been small or secondary to that of aphidiids. They can however be useful in glasshouses (see section 11.2.2) (*Aphelinus abdominalis* (Dalman) and *A*. nr *mariscusae* (Risbec), both thelytokous in Australia, have been common contaminants in aphid cultures there). Whereas coccinellids have been widely researched and used (see section 9.2.1), other predators such as syrphids have received scant attention (section 9.2.3), despite their obvious voracity (as larvae) and apparent efficiency as aphidophages. Chrysopids also appear to have potential (section 9.2.2). The use of fungal pathogens is relatively recent and has some advantages, e.g. production of climate-resistant resting spores, although the need for a humid environment is severely limiting (see section 9.3).

MULTILATERAL CONTROL

In principle, control agents should be selected that are specific to the target pest. Aphids, however, are good candidates for collective control, or what Starý (1967) terms "multilateral control", that is, the use of polyphagy by manipulation of either the agent or the aphid host(s). An example of the former is the use of polyphagous agents that may more easily become established, and can utilise and contribute to the control of alternative hosts in the ecosystem, or in similar or adjacent ecosystems, during periods of scarcity of one or more target species. Needless to say, none of the non-target hosts should be preferred hosts of the agent.

In accordance with this approach, the widely polyphagous aphidiids *L. fabarum* and *L. testaceipes* were introduced into Australia to control *A. craccivora*, a pest of legumes. Laboratory tests demonstrated, however, that the host ranges of both introduced parasites were largely restricted to species of the genera *Aphis* and *Toxoptera*, an indication that both aphidiid species, as currently recognized, are actually complexes of either host-restricted biotypes or of sibling species. In addition, although they readily parasitised and killed *A. citricola, A. nerii* Boyer de Fonscolombe and *Toxoptera citricidus* (Kirkaldy), most individuals tested were unable to complete their development in these species, which therefore act as "egg traps" of the *Lysiphlebus* species (and, incidentally, of other hymenopterous parasites too). The prognosis for the establishment and usefulness of these two parasites in Australia is therefore not good (Carver, 1984b).

L. testaceipes, on the other hand, was introduced into the Mediterranean area in 1973–74 to control *T. aurantii* and other aphids on citrus and has spread so rapidly that it may be displacing the indigenous *L. fabarum* and other parasites (Tremblay, 1984). Here too, *A. citricola* acts as an egg trap (*T. citricidus* is not known from the area). However, suitable hosts (e.g. *T. aurantii*) are more common than in Australia, where unsuitable hosts (*A. citricola, A. nerii* and *T. citricidus*) prevail and where *T. aurantii* is rare.

The aphidiid *A. ervi*, which was introduced into North America, Argentina, New Zealand and Australia as a biological control agent of either *Acyrtosiphon pisum* or *Acyrtosiphon kondoi* on lucerne, has become an important parasite of the other in those areas and, in Australia and Argentina at least, of *Sitobion* and *Metopolophium* also, on grasses and cereals (Botto and Hernández, 1982).

One may manipulate the aphid hosts. For instance, *T. aurantii* and *Aphis citricola* are pests of citrus in Europe and are parasitised by *Lysiphlebus fabarum, Lipolexis gracilis* (Foerster) and *Aphidius matricariae*, which are all polyphagous. Starý (1968) recommends the removal of *Pittosporum tobira*, which is a preferred host of *T. aurantii* in the mediterranean area, from the vicinity of citrus orchards and its replacement by willows and oleanders, the host plants of *Aphis farinosa* Gmelin and *A. nerii*, respectively, in order to provide alternative hosts for the parasites but not for the pests. But *A. nerii* has been shown (with *A. citricola*) to be an unsuitable host for *L. fabarum* and other parasites (Carver, 1984b) and should not be recommended.

In Europe, peach and the reed *Phragmites* are the primary and secondary hosts, respectively, of the mealy plum aphid, *H. pruni*, which is parasitised by *Aphidius colemani*. Starý (1964) recommends removal of *Phragmites* and weedy *Prunus* from the neighbourhood of peach orchards and the promotion instead of another reed, *Arundo donax*, the host plant of *Melanaphis donacis* (Passerini), an alternative host of *A. colemani*.

Innocuous aphid species on wild and cultivated plants can provide valuable reservoirs of parasites common to aphids on nearby crops. Nettle aphids are a good example. *Urtica* species (stinging nettles) are the specialised host plants of *Microlophium carnosum* (Buckton) and *Aphis urticata* Fabricius, which in turn are hosts to a wide range of parasites and predators in Europe, including *Aphidius ervi* and *Aphidius urticae* (parasites of *M. carnosum*) and *L. fabarum* (parasite of *A. urticata*). These aphidiids also parasitise aphids on legumes, cereals, sugar beet and other crops. The promotion and management (e.g. controlled cutting) of nettles in the vicinity of such crops has been proposed (Perrin, 1975; Starý, 1983). Recent evidence indicates, however, that both *A. ervi* and *A. urticae* are complexes of host-restricted entities and may not transfer from *M. carnosum* to the pest hosts (Starý et al., 1980; Cameron et al., 1984). Other alternative aphid hosts that could be similarly exploited are *Acyrthosiphon pisum* subsp. *ononis* (Koch) in Europe on *Ononis* species, which is host to *A. ervi* and *A. eadyi*, valuable parasites of *Acyrthosiphon* species in legume crops (Starý and Lyon, 1980), and *Aphis helianthi* Monell in the U.S.A. on diverse host plants, which hosts the aphidiid *L. testaceipes*, a valuable parasite of the greenbug, *S. graminum* (Eikenbary and Rogers, 1973).

Section 11.2 references, p. 161

The coccinellid *H. conformis* in Australia will breed slowly in winter and multiply in spring on *Casuarina stricta* infested with the native psylloid *Aacanthocnema casuarinae* (Froggatt). Maelzer (1981) recommends the planting of *Casuarina* trees near (orchard?) crops in order to provide a source of predators of crop aphids in spring and a refuge for *H. conformis* when aphids are scarce. An interesting question arises: will *H. conformis* prey on aphids in preference to and in the presence of its native hosts?

MULTIPLE INTRODUCTIONS

The proposition that the operation of more than one biological control agent on a target species could lead to competitive displacement is given further support by evidence suggesting that *A. ervi* (polyphagous) is displacing *A. smithi* (monophagous) in North America (Mackauer and Kambhampati, 1986). Possibly *A. smithi* did not become established in mainland Australia because *A. ervi* was already established in the niche as a parasite of *A. kondoi*. In Tasmania, *A. smithi* is much less common in lucerne than *A. ervi*, even when its host, *A pisum*, is the dominant aphid (S. Suwanbutr, personal communication, 1985) and in New Zealand *A. ervi* may be displacing *A. eadyi*, an earlier established parasite of *A. pisum* (P.J. Cameron, personal communication, 1985).

ANTS

As protectors and predators of aphids and antagonists of their natural enemies, ants play an important role in the population dynamics of those aphid species with which they interact. Interestingly, the same aphid species exhibit the same consistency in their associations with ants in both the native and the exotic environments. Only the ant species are different (except when the ants are themselves exotic, e.g. *Iridomyrmex humilis*). Little is known of the impact of ants on aphid populations and on wing production in aphids in either habitat, and their role is ignored even in intensive aphid studies. Aphid control by means of ant exclusion would appear to be simple, cheap and effective under some circumstances, and has been tried successfully in California against arboreal phyllaphidine aphids by banding tree trunks (Olkowski et al., 1974). However, the same ant species, *I. humilis*, removed 98% of chrysopid eggs released to control *Illinoia liriodendri* (Monell) and its honeydew on tulip tree, also in California (Driestadt et al., 1986).

Ants themselves have been considerably exploited as biological control agents and therefore, indirectly, their aphid mutualists, too. In their role as generalised predators, wood ants of the *Formica rufa* group were extensively translocated, in both deciduous and coniferous forests, especially in Germany and Italy. The incidence of aphids, as honeydew producers, was an important consideration in the selection of target sites. Such promotion of aphids is difficult to justify. However, it is suggested that any consequent injury by aphids to host trees could be offset by increased honey production resulting from increased availability of honeydew for honeybees (Greathead, 1976).

THE EXOTIC ENVIRONMENT

Existent Entomophaga

One of the principal tenets of biological control is that species invading new areas are enabled to multiply unchecked in the absence of the natural enemies

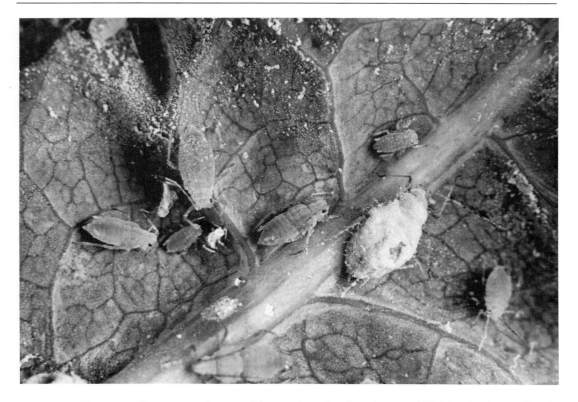

Fig. 11.2.3. *Hyperomyzus lactucae* (Linnaeus) on *Sonchus oleraceus* killed by the fungus *Erynia neoaphidis* Remaudière and Hennebert.

that regulate their numbers in the areas of their origin. Rarely would an invading aphid move into such a void. Predators, native mostly, both predominantly aphidophagous species and non-selective ones (e.g. carabids, staphylinids, spiders), will be found already occupying the niche and, in time, others will move into it in response to the presence of the new prey species. Some suitable parasites, both exotic and native, can also be expected to be present, as well as entomogenous fungi (Milner et al., 1980) (Fig. 11.2.3). For example, when *M. dirhodum* reached Australia (Carver, 1984a) it was quickly subjected to parasitisation by the common aphidophagous fungus *Erynia neoaphidis* Remaudière and Hennebert and by *A. ervi* (introduced to control *A. kondoi*).

Likewise, introduced control agents are expected to perform better than in their original abode because they are themselves freed from the control of their own natural enemies. Again, some aphid hyperparasites and parasites of predators, exotic and native, would be already present in the target niche prior to the introduction of a primary agent.

The exotic entomophages would presumably be, like the immigrant aphid, genetically restricted and possibly possessed of only part of the genetic potential of the species as a whole e.g. the parasites *A. abdominalis* and *E. persicae* are thelytokous (or rarely deuterotokous) in Australia, as are the aphid hyperparasite *Phaenoglyphis villosa* (Hartig) (Cynipoidea) and the cosmopolitan syrphid parasite *Diplazon laetatorius* (Fabricius) (Ichneumonoidea). (Parthenogenetic parasites, like the aphids themselves, presumably have a better chance of becoming established.) In addition, *E. persicae* has a restricted host and niche range in Australia.

Section 11.2 references, p. 161

Invading species

It is as yet unexplained why some insect species invading new environments initially multiply and disperse explosively and then reduce performance to more normal levels before control measures are applied, if applied at all. Both *T. trifolii* f. *maculata* and *A. kondoi* followed this pattern in the U.S.A. and Australia. Possible causes include a dearth of other aphid competitors, an initially slow response by existent entomophaga and operation of natural selection on the host plant by the new aphid; conceivably, the most susceptible lucerne plants succumbed or were rendered less competitive. Alternatively, an unfavourable aspect of climate may not be immediately influential. Australia's hot, dry periods, for instance, might successively reduce the population density of an aphid invader.

An example: Australia

The Australian aphid fauna is meagre, the continent featuring instead a rich, diverse fauna of psylloids and coccoids highly adapted to the dry environment. Of 156 known aphid species, only 20 are indigenous, a similar number are naturally occuring extensions of the northern fauna, and the rest are exotic and mostly pestiferous. The indigenous species are for the greater part phylogenetically, biologically and zoogeographically isolated. The exotic species have hardly invaded the native flora, which is dominated by scleromorphous *Eucalyptus* (Myrtaceae) and *Acacia* (Mimosaceae) and, instead, predominantly infest exotic crop plants, ornamentals and weeds in cultivated areas, the number and diversity of which are understandably restricted. The exotic aphids appear to be largely the result of accidental introduction on plants by early British settlers. Only a small number of species are restricted to exotic weeds, which are of diverse origin. The cultivated areas are concentrated on the eastern, south-eastern and south-western margins of the continent.

The aphid fauna is meagre not only in species but also in numbers, which, for instance, are much lower both on hosts and in the aerial plankton, even during peak periods, than those in northern temperate regions. Aphid pest problems are not as serious either. Imagine, therefore, a fauna that includes only two species of *Macrosiphum*, three species of *Acyrthosiphon*, seven species of *Myzus*, 13 species of *Aphis* (only four of which are common), no *Uroleucon*, relatively few Drepanosiphidae and Lachnidae, and features other notable absences; a fauna that is rare on the native vegetation and not conspicuous for many months of the year. Their paucity is undoubtedly due to a relative dearth of suitable host plants outside of the crop and garden situation and, more importantly, to the shortness of the periods when host plant conditions and climate are favourable for aphid development.

The factors that determine aphid abundance are complex and interacting. The intrinsic physiological phenology of the host plant must be the key determinant. The aphid is a parasite of the contents of the phloem and conditions will be optimal for the aphid when the phloem sap is (nutritionally?) optimal in those parts used by the aphid, be they in the transporting, developing, storage or senescing areas. The plant operates within the constraints imposed by prevailing climatic conditions (temperature, light, moisture) and edaphic conditions. Climatic and edaphic factors also operate on the aphid and its natural enemies, mutualists and competitors, but to differing degrees. In the absence of self-regulating mechanisms, aphid numbers will therefore be highest when host phloem phenology optimal for the aphid is coincidental with climatic conditions also optimal for the aphid but not for its natural enemies and competitors.

In general, in southern Australia, conditions are favourable for plant growth only in relatively restricted periods of spring and autumn because peak light and thermal regimes are out of phase with peak moisture regimes (Fitzpatrick and Nix, 1970; Maelzer, 1981). In the prolonged summer high temperatures, low rainfall and high evaporation rates limit plant growth, and in the short, mild winter low temperatures are the limiting factor. Other characteristics of the climate that are not apparent from meteorological annual averages include long, unpredictable periods of drought conditions and, to a lesser extent, of heavy rainfall, extremes of high but not low temperatures, high light intensities, and a high incidence of warm winter days (above the temperature threshold of development of many aphids and natural enemies).

Aphid numbers in southern Australia generally reflect the growth patterns of their hosts, exhibiting two peaks of seasonal abundance, one in spring (September–November), and another, sometimes lower one in autumn (March–May) (cf. the Middle East; Bodenheimer and Swirski, 1957). In Tasmania, the peaks tend to merge to form one summer peak (Hughes et al., 1964), a situation more closely resembling that in temperate Holarctica. Aphid numbers are extremely low during summer and are also low in winter, when slow, anholocyclic reproduction is usual. One important feature of aphid behaviour is the heavy mortality experienced during hot, dry periods when seemingly whole populations may walk off host plants in a period of hours, never to return. Such drastic reductions are often attributed to enemy action!

The aphid parasite fauna is also restricted. Of only seventeen known species of Aphidiidae and nine of *Aphelinus*, only one is indigenous and seven are recent, deliberate introductions. The most common aphidiid is *A. colemani*, of Indian origin, which has a very wide host range in the Aphidinae and Myzinae (Carver and Starý, 1974). Only one of the *Aphelinus* species is common in the field. The reasons for the parasite paucity are not known. They may not have accompanied the introduced aphids and host plants or, if they did, failed to establish (because of their sexuality perhaps?) or thrive. Building up sufficient numbers during the short periods of host abundance to allow survival at other more unfavourable times (especially summer) could be a prohibitive obstacle. There are only eight known species of hyperparasites, all exotic.

The insect predators are almost exclusively native and not exclusively aphidophagous. Of the many species predacious on the rich, indigenous sternorrhynchous fauna, only two coccinellid, three syrphid, one chrysopid and one hemerobiid species have diversified their diets and habitats to any extent to include exotic aphids. Little is known of the relationship of non-selective predators to aphids in Australia.

The Australian environment then appears to be far from optimal for the exotic aphid. Nevertheless, several important aphid pests occur, and two spectacular invasions have taken place. Biological control has been successfully applied. Possibly the Australian environmental is more favourable to selected, environmentally matched agents than to their aphid hosts.

The gaps in the aphidiid fauna should be filled and selected parasites introduced from environmentally comparable areas. Prime candidate aphid pests include *Macrosiphum euphorbiae* and *M. rosae* (Linnaeus); cereal aphids (*Rhopalosiphum* and *Sitobion*) as vectors of barley yellow dwarf virus; *Pentalonia nigronervosa* Coquerel, the vector of banana bunchy top virus; and the widely polyphagous, multivectorial *M. persicae*. In Australia, expensive measures to combat potato virus diseases transmitted by *M. persicae* take the form of intensive insecticide regimes and seed certification schemes, which can hardly be expected to be abandoned in deference to biological control. *M. persicae* is also the principal vector of beet western yellows virus, which infects

Section 11.2 references, p. 161

a diversity of crop plants and weeds in Australia (Johnstone and Duffus, 1984). Biological control should therefore be applied to *M. persicae*, but on host plants other than potatoes (see also Mackauer et al., 1976).

Introduction into Australia of the nearctic coccinellids *Hippodamia quinquesignata* Kirby and *H. parenthesis* (Say) has been proposed (D.A. Maelzer, personal communication). Importation of predators should, however, be opposed at all costs, not only because of their poor success record but because the scientifically important and little-known but largely innocuous coccoid and psylloid faunas could be put at risk. The argument that the proposed imports are predominantly aphidophagous is unacceptable, because in their area of origin the *Hippodamia* species have a diverse and abundant aphid fauna available as hosts, and it is not possible to predict how they would behave during periods of aphid scarcity in Australia. However, the palaearctic coccinellid *C. undecimpunctata* is a recent adventive to Tasmania (from New Zealand?) which is now well established. It would be interesting to see if it remains strictly aphidophagous.

APHIDS AS CONTROLLERS OF WEEDS

Aphids themselves have been considered as biological control agents of weeds. Wilson (1938) studied the biology of *Aphis chloris* Koch in England as part of an investigation into its potential to control St John's wort, *Hypericum perforatum*, in Australia. In the event, *A. chloris* was not introduced into Australia until 1986 (D.T. Briese, personal communication, 1985), the hope being that it might be effective against new plant growth resulting from summer rains when introduced chrysomelids (*Chrysolina* spp.) are not abundant. The aphid has become established at release sites. How ironic if *A. chloris* proves to be a much-needed reservoir host for the poorly established parasite *L. testaceipes*, imported to control *A. craccivora* (Carver, 1984b). *A. chloris* has recently been introduced into Canada (1979) and South Africa (1982) (D.T. Briese, personal communication, 1985).

The aphids *Dactynotus chondrillae* (Nevsky) and *Chondrillobium blattnyi* (Pintera) were also seriously considered for introduction into Australia for the control of skeleton weed, *Chondrilla juncea*, a cichorine composite of mediterranean origin. The former aphid failed host-specificity tests, the latter was not recommended for introduction because of the possibility of its implication in the transmission of virus diseases of lettuce, another cichorine (Caresche et al., 1974).

The use of aphids as biological control agents is not to be recommended where it could conceivably jeopardise the future chances of applying collective control to aphid pests in the region concerned.

ASSESSMENT

That the application of biological control to aphids is a worthwhile endeavour has been decisively demonstrated by the successful establishment of many introduced agents, especially aphidiids, even in small-scale operations. Success in terms of control effected is extremely difficult to assess. Nevertheless, successful control has been claimed for several programs, as stated previously and in section 11.2.1.

In theory, aphids should be good candidates for biological control – they and their honeydew are attractive food sources for so many entomophages. Nevertheless, it is believed that they are poor prospects because their high reproductive capacities and physiological activity at relatively low temperatures give them an insurmountable advantage over natural enemies (Van den Bosch et al., 1982), especially during the initial build-up of an aphid colony. Furthermore, aphidophages are often left bereft of resources, because of the frequent and rapid changes of aphid populations.

However, it is unrealistic to consider aphids in the light of their potential for increase or to compare them with other insects on the above criteria alone. Aphid prolificity is a response to vulnerability, and their vulnerability to other controlling factors is much greater than that of their natural enemies. When considering biological control, therefore, one should allow for inevitable mortality caused by nutritional and climatic factors. It should be remembered that when these two factors are not controlling, they are promoting, and natural enemies, if present, then provide the only restraint. The acknowledged role of parasites in reducing the numbers of flying colonisers and virus vectors alone justifies their use as control agents.

In the present state of knowledge, suitable biological control agents of aphids can be selected with confidence, but their performance in their new environment cannot be predicted. It cannot yet be explained why one selected agent succeeds and another fails (Hodek, 1973). Assessment is difficult and not only when other factors are also operating. The interrelationships between environment (climatic, edaphic), host plant, aphid, competitor, parasite/predator, hyperparasite and ant are so complex that evaluation of the roles of the components of the system is extremely difficult. Aphid population dynamics, and the non-comparable disparity between a hemimetabolous host and its holometabolous enemy, especially if parasitic, are major complexities, and other factors often inadequately or improperly considered in past studies include: aphid sensitivity to host plant condition; interspecific competition; degrees of parasitisation and hyperparasitisation; effect of parasitisation and ant attendance on alate morph production; effect of ants on parasitisation and predation; significance of ovipositorial feeding by *Aphelinus*; disturbance of hosts by enemies (Ruth et al., (1975) showed that in the laboratory 41% of *S. graminum* may drop from the plant in response to the presence of the aphidiid *L. testaceipes*); identification of aphidophagy in predators; and intra-niche movements by hosts and enemies, e.g. concealment during adverse conditions and non-feeding periods. Some factors cannot as yet be considered. For instance, little is known about the cues for dispersal of adult entomophages or host location and progeny placement. Preliminary observations on the distribution of the newly established *A. sonchi* in the field in Australia have shown that this parasite is widely distributed, with a high percentage of plants bearing mummies but with a low percentage of parasitisation per plant, which raises the possibility that the female parasite disperses after parasitising only a small number of available hosts. Such a dispersal pattern could make the parasite effective at low host densities and could explain the pronounced phototaxis exhibited by gravid females of many aphidiid species in laboratory cultures even in the presence of unparasitised aphids.

So, aphid biological control must continue along the present path, building on the principles and procedures learned. Every project undertaken can provide experience for the next one, whether successful or not. That is, one must continue to "try it and see".

REFERENCES

Angalet, G.W., Tropp, J.M. and Eggert, A.N., 1979. *Coccinella septempunctata* in the United States: Recolonizations and notes on its ecology. Environmental Entomology, 8: 896–901.

Anonymous, 1978. Lucerne Aphid Workshop, Agricultural Research Centre, Department of Agriculture, Tamworth, New South Wales, November 1978, Working Papers, 273 pp.

Archer, T.L., Cate, R.H., Eikenbary, R.D. and Starks, K.J., 1974. Parasitoids collected from greenbugs and corn leaf aphids in Oklahoma in 1972. Annals of the Entomological Society of America, 67: 11–14.

Boakye, D. and Randles, J.W., 1974. Epidemiology of LNYV in South Australia. III. Virus transmission parameters and vector feeding behaviour on host and non-host plants. Australian Journal of Agricultural Research, 25: 791–803.

Bodenheimer, F.S. and Swirski, E., 1957. The Aphidoidea of the Middle East. Weizmann, Israel, 378 pp.

Botto, E.N. and Hernández, M.C., 1982. Parasitism of the aphids *Metopolophium dirhodum* (Walker) and *Sitobion avenae* (F.) by parasitoids of the genus *Aphidius* (Hymenoptera: Aphidiidae) in the Republic of Argentina. IDIA, nos 401–404: 17–19 (in Spanish).

Cameron, P.J., Walker, G.P. and Allan, D.J., 1981. Establishment and dispersal of the introduced parasite *Aphidius eadyi* (Hymenoptera: Aphidiidae) in the North Island of New Zealand, and its initial effect on pea aphid. New Zealand Journal of Zoology, 8: 105–112.

Cameron, P.J., Powell, W. and Loxdale, H.D., 1984. Reservoirs for *Aphidius ervi* Haliday (Hymenoptera: Aphidiidae), a polyphagous parasitoid of cereal aphids (Hemiptera: Aphididae). Bulletin of Entomological Research, 74: 647–656.

Caresche, L.A., Hasan, S. and Wapshere, A.J., 1974. Biology and host specificity of two aphids *Dactynotus chondrillae* (Nevsk.) and *Chondrillobium blattnyi* (Pintera) (Hemiptera) living on *Chondrilla juncea*. Bulletin of Entomological Research, 64: 277–288.

Carver, M., 1984a. *Metopolophium dirhodum* (Walker) newly recorded from Australia. Journal of the Australian Entomological Society, 23: 192.

Carver, M., 1984b. The potential host ranges in Australia of some imported aphid parasites (Hym: Ichneumonoidea: Aphidiidae). Entomophaga, 29: 351–359.

Carver, M. and Starý, P., 1974. A preliminary review of the Aphidiidae (Hymenoptera: Ichneumonoidea) of Australia and New Zealand. Journal of the Australian Entomological Society, 13: 235–240.

Carver, M. and Woolcock, L.T., 1986. The introduction into Australia of biological control agents of *Hyperomyzus lactucae* (L.) (Homoptera: Aphididae). Journal of the Australian Entomological Society, 25: 65–69.

Clausen, C.P. (Editor), 1978. Introduced parasites and predators of arthropod pests and weeds: a world review. United States Department of Agriculture, Agriculture Handbook No. 480, 551 pp.

DeBach, P. (Editor), 1964. Biological Control of Insect Pests and Weeds. Chapman and Hall, London, 844 pp.

DeBach, P., 1974. Biological Control by Natural Enemies. Cambridge University Press, London, 323 pp.

Delucchi, V.L. (Editor), 1976. Studies in Biological Control. International Biological Programme 9. Cambridge University Press, Cambridge, 304 pp.

Driestadt, S.H., Hagen, K.S. and Dahlsten, D.L., 1986. Predation by *Iridomyrmex humilis* (Hym.: Formicidae) on eggs of *Chrysoperla carnea* (Neu.: Chrysopidae) released for inundative control of *Illinoia liriodendri* (Hom.: Aphididae) infesting *Liridendron tulipifera*. Entomophaga, 31: 397–400.

Ehler, L.E. and Hall, R.W., 1982. Evidence for competitive exclusion of introduced natural enemies in biological control. Environmental Entomology, 11: 1–4.

Eikenbary, R.D. and Rogers, C.E., 1973. Importance of alternate hosts in establishment of introduced parasites. Proceedings of the Tall Timbers Conference on Ecological Animal Control by Habitat Management, number 5, Tallahassee, Florida, March 1973, pp. 119–133.

Fenner, F., 1983. Biological control, as exemplified by smallpox eradication and myxomatosis. Proceedings of the Royal Society of London B, 218: 259–285.

Fitzpatrick, E.A. and Nix, H.A., 1970. The climatic factor in Australian grassland ecology. In: R.M. Moore (Editor), Australian Grasslands, A.N.U. Press, Canberra, pp. 3–26.

Gilbert, N. and Hughes, R.D., 1971. A model of an aphid population – three adventures. Journal of Animal Ecology, 40: 525–534.

González, D., White, W., Hall, J.C. and Dickson, R.C., 1978. Aphidiidae imported into California for biological control of *Acyrthosiphon kondoi* and *Acyrthosiphon pisum*. Entomophaga, 23: 239–248.

González, D., Miyazaki, M., White, W., Takada, H., Dickson, R.C. and Hall, J.C., 1979. Geographical distribution of *Acyrthosiphon kondoi* Shinji (Homoptera: Aphididae) and some of its parasites and hyperparasites in Japan. Kontyû, 47: 1–7.

Greathead, D.J. (Editor), 1976, A Review of Biological Control in Western and Southern Europe. Technical Communication of the Commonwealth Institute of Biological Control, No. 7, 182 pp.

Hagen, K.S. and Bishop, G.W., 1979. Use of supplemental foods and behavioural chemicals to increase the effectiveness of natural enemies. In: D.W. Davis, S.C. Hoyt, J.A. McMurtry and M.J. AliNiazee (Editors), Biological Control and Insect Pest Management. University of California Publications in Agricultural Science, No. 4096, pp. 49–60.

Hagen, K.S. and Schlinger, E.I., 1960. Imported Indian parasite of pea aphid established in California. California Agriculture, 14(9): 5–6.

Hagen, K.S., Bombosch, S. and McMurtry, J.A., 1976. The biology and impact of predators. In: C.B. Huffaker and P.S. Messenger (Editors), Theory and Practice of Biological Control. Academic Press, New York, London, pp. 93–142.

Hall, R.A., 1981. The fungus *Verticillium lecanii* as a microbial insecticide against aphids and scale insects. In: H.D. Burges (Editor), Microbial Control of Pests and Plant Diseases 1970–1980. Academic Press, London, pp. 483–498.

Harvey, T.L. and Hackerott, H.L., 1969. Recognition of a greenbug biotype injurious to sorghum. Journal of Economic Entomology, 62: 776–779.

Hodek, I., 1973. Biology of Coccinellidae. W. Junk, The Hague, and Academia, Prague, 260 pp.

Hokkanen, H. and Pimentel, D., 1984. New approach for selecting biological control agents. Canadian Entomologist, 116: 1109–1121.

Howard, L.O., 1929. *Aphelinus mali* and its travels. Annals of the Entomological Society of America, 22: 341–368.

Huffaker, C.B., 1980. New Technology of Pest Control. Wiley-Interscience, New York, 500 pp.

Huffaker, C.B. and Messenger, P.S. (Editors), 1976. Theory and Practice of Biological Control. Academic Press, New York, London, 788 pp.

Huffaker, C.B., Simmonds, F.J. and Laing, J.E., 1976. The theoretical and empirical basis of biological control. In: C.B. Huffaker and P.S. Messenger (Editors), Theory and Practice of Biological Control. Academic Press, New York, London, pp. 41–78.

Hughes, R.D., Casimir, M., O'Loughlin, G.T. and Martyn, E.J., 1964. A survey of aphids flying over eastern Australia in 1961. Australian Journal of Zoology, 12: 174–200.

Hughes, R.D., Carver, M., Casimir, M., Martyn, E.J. and O'Loughlin, G.T., 1965. Comparison of the numbers of aphids flying over eastern Australia in two consecutive years. Australian Journal of Zoology, 13: 823–839.

Hughes, R.D., Woolcock, L.T., Roberts, J.A. and Hughes, M.A., 1987. Biological control of the spotted alfalfa aphid, *Therioaphis trifolii* f. *maculata*, on lucerne crops in Australia, by the introduced parasitic hymenopteran *Trioxys complanatus*. Journal of Applied Ecology, 24: 515–537.

Johnstone, G.R. and Duffus, J.E., 1984. Some luteovirus diseases in Tasmania caused by beet western yellows and subterranean clover red leaf viruses. Journal of Agricultural Research, 35: 821–830.

Keller, M.A., 1984. Reassessing evidence for competitive exclusion of introduced natural enemies. Environmental Entomology, 13: 192–195.

Kfir, R., Kirsten, F. and Van Rensburg, N.J., 1985. *Pauesia* sp. (Hymenoptera: Aphidiidae): a parasite introduced into South Africa for biological control of the black pine aphid, *Cinara cronartii* (Homoptera: Aphidiidae). Environmental Entomology, 14: 597–601.

Laing, J.E. and Hamai, J., 1976. Biological control of insect pests and weeds by imported parasites, predators and pathogens. In: C.B. Huffaker and P.S. Messenger (Editors), Theory and Practice of Biological Control. Academic Press, New York, London, pp. 685–743.

Lehane, L., 1982. Biological control of lucerne aphids. Rural Research, CSIRO, Australia, 114: 4–10.

Liu Shu-sheng, 1983. An investigation of ecological relationships – experimental studies of the interactions between the sowthistle aphid *Hyperomyzus lactucae* and its parasite, *Aphidius sonchi*. Ph.D. thesis, Australian National University, Canberra, 314 pp.

Lyon, J.P., 1986. Use of aphidophagous and polyphagous beneficial insects for biological control of aphids in greenhouses. In: I. Hodek (Editor), Ecology of Aphidophaga. Academia, Prague and W. Junk, Dordrecht, pp. 471–474.

Mackauer, M., 1972. Genetic aspects of insect production. Entomophaga, 17: 27–48.

Mackauer, M. and Kambhampati, S., 1986. Structural changes in the parasite guild attacking the pea aphid in North America. In: I. Hodek (Editor), Ecology of Aphidophaga. Academia, Prague and W. Junk, Dordrecht, pp. 347–356.

Mackauer, M. et al., 1976. *Myzus persicae* Sulz., an aphid of world importance. In: V.L. Delucchi (Editor), Studies in Biological Control. International Biological Programme 9. Cambridge University Press, Cambridge, pp. 51–119.

Maelzer, D.A., 1981. Aphids – introduced pests of man's crops. In: R.L. Kitching and R.E. Jones (Editors), The Ecology of Pests. Some Australian Case Histories. CSIRO, Australia, pp. 88–106.

Markkula, M., Tiittanen, K., Hämäläinen, M. and Forsberg, A., 1979. The aphid midge *Aphidoletes aphidimyza* (Diptera, Cecidomyiidae) and its use in biological control of aphids. Annales Entomologici Fennici, 45: 89–98.

Messing, R.H. and Ali Niazee, M.T., 1987. Introduction and establishment of a distinct biotype of *Trioxys pallidus* in Oregon, U.S.A., for control of filbert (hazelnut) aphids. Poster, Third Symposium, Ecology of Aphidophaga, Teresin, Poland, 31 August–5 September 1987.

Milne, W.M., 1986. The release and establishment of *Aphidius ervi* Haliday (Hymenoptera: Ichneumonoidea) in lucerne aphids in eastern Australia. Journal of the Australian Entomological Society, 25: 123–130.

Milner, R.J., Teakle, R.E., Lutton, G.G. and Dare, F.M., 1980. Pathogens (Phycomycetes: Entomophthoraceae) of the blue-green aphid, *Acyrthosiphon kondoi* Shinji and other aphids in Australia. Australian Journal of Botany, 28: 601–619.

Milner, R.J., Soper, R.S. and Lutton, G.G., 1982. Field release of an Israeli strain of the fungus *Zoophthora radicans* (Brefeld) Batko for biological control of *Therioaphis trifolii* (Monell) f. *maculata*. Journal of the Australian Entomological Society, 21: 113–118.

New, T.R., 1975. The biology of Chrysopidae and Hemerobiidae (Neuroptera) with reference to their usage as biocontrol agents: a review. Transactions of the Royal Entomological Society of London, 127: 115–140.

Norambuena, M.H., 1981. Introduction and establishment of parasitoids (Hymenoptera: Aphidiidae) for biological control of *Metopolophium dirhodum* (Walker) and *Sitobion avenae* (Fabricius) (Homoptera: Aphididae) in southern Chile. Agricultura Tecnica (Chile) 41: 95–102 (in Spanish, with English summary).

Olkowski, W., Pinnock, C., Toney, W., Mosher, G., Neasbitt, W., Van den Bosch, R. and Olkowski, H., 1974. An integrated insect control program for street trees. California Agriculture, 28(1): 3–4.

Olkowski, W., Olkowski, H. and Van den Bosch, R., 1982a. Linden aphid parasite establishment. Environmental Entomology, 11: 1023–1025.

Olkowski, W., Olkowski, H., Van den Bosch, R., Hom, R., Zuparko, R. and Klitz, W., 1982b. The parasitoid *Trioxys tenuicaudus* Starý (Hymenoptera: Aphidiidae) established on the elm aphid *Tinocallis platani* Kaltenbach (Homoptera: Aphididae) in Berkeley, California. Pan-Pacific Entomologist, 58: 59–63.

Perrin, R.M., 1975. The role of the perennial stinging nettle, *Urtica dioica*, as a reservoir of beneficial natural enemies. Annals of Applied Biology, 81: 289–297.

Ridgway, R.L. and Vinson, S.B. (Editors), 1977. Biological Control by Augmentation of Natural Enemies. Plenum Press, New York, London, 480 pp.

Robert, Y. and Rabasse, J.M., 1977. Rôle écologique de *Digitalis purpurea* dans la limitation naturelle des populations du puceron strié de la pomme de terre *Aulacorthum solani* par *Aphidius urticae* dans l'ouest de la France, Entomophaga, 22: 373–382.

Ruth, W.E., McNew, R.W., Caves, D.W. and Eikenbary, R.D., 1975. Greenbugs (Homoptera: Aphididae) force from host plants by *Lysiphlebus testaceipes* (Hymenoptera: Braconidae). Entomophaga, 20: 65–71.

Schaefer, P.W., Dysart, R.J. and Specht, H.B., 1987. North American distribution of *Coccinella septempunctata* (Coleoptera: Coccinellidae) and its mass appearance in coastal Delaware. Environmental Entomology, 16: 368–373.

Scopes, N.E.A. and Biggerstaff, S.M., 1973. Progress towards integrated pest control on year-round chrysanthemums. Proceedings of the 7th British Insecticide and Fungicide Conference Vol. 1: 227–234.

Simmonds, F.J., Franz, J.M. and Sailer, R.I., 1976. History of Biological Control. In: C.B. Huffaker and P.S. Messenger (Editors), Theory and Practice of Biological Control. Academic Press, New York, London, pp. 17–39.

Smith, R.F. and Van den Bosch, R., 1967. Integrated Control. In: W.W. Kilgore and R.L. Doutt (Editors), Pest Control. Academic Press, New York, London, pp. 295–340.

Starks, K.J., Wood, E.A., Jr., Burton, R.L. and Somsen, H.W., 1975. Behavior of convergent lady beetles in relation to greenbug control in grain sorghum. Observations and preliminary tests. United States Department of Agriculture, Agricultural Research Service ARS-S-53, 10 pp.

Starý, P., 1964. Integrated control problems of citrus and peach aphid pests in Italy orchards. Entomophaga, 9: 147–152.

Starý, P., 1967. Multilateral aphid control concept. Annales de la Société Entomologique de France, N.S., 3: 221–225.

Starý, P., 1968. Biological control of aphids – pests of citrus and tea plantations in the Black Sea Coast districts of the U.S.S.R. – Georgia. Bollettino del Laboratorio di Entomologia Agraria "Filippo Silvestri" di Portici, 26: 227–240.

Starý, P., 1970. Biology of Aphid Parasites (Hymenoptera: Aphidiidae) with Respect to Integrated Control. W. Junk, The Hague, 643 pp.

Starý, P., 1983. The perennial stinging nettle (*Urtica dioica*) as a reservoir of aphid parasitoids (Hymenoptera, Aphidiidae). Acta Entomologica Bohemoslovaca, 80: 81–86.

Starý, P., 1987. Subject bibliography of aphid parasitoids (Hymenoptera: Aphidiidae) of the world. 1758–1982. Monographs to Applied Entomology, 25. Paul Parey, Hamburg, 101 pp.

Starý, P. and Lyon, J.P., 1980. *Acyrthosiphon pisum ononis* (Homoptera, Aphididae) and *Ononis* species as reservoirs of aphid parasitoids (Hymenoptera, Aphidiidae). Acta Entomologica Bohemoslovaca, 77: 65–75.

Starý, P., González, D. and Hall, J.C., 1980. *Aphidius eadyi* n.sp. (Hymenoptera: Aphidiidae), a widely distributed parasitoid of the pea aphid, *Acyrthosiphon pisum* (Harris) in the Palearctic. Entomologica Scandinavica, 11: 473–480.

Stechmann, D.-H., 1987. Introduction of *Lysiphlebus testaceipes* (Cresson) (Hymen.: Aphidiidae) into the Kingdom of Tonga, Oceania. Poster, Third Symposium, Ecology of Aphidophages, Teresin, Poland, 31 August–5 September 1987.

Stern, V.M., Smith, R.F., Van den Bosch, R. and Hagen, K.S., 1959. The Integrated Control Concept. Hilgardia, 29: 81–101.

Stubbs, L.L., Smith, P.R. and O'Loughlin, G.T., 1983. Biological control of the carrot willow aphid (*Cavariella aegopodii*) by an introduced microhymenopterous parasite and consequent elimination of the carrot motley dwarf virus disease. Abstracts of the Fourth International Congress of Plant Pathology, Melbourne, Australia, August 1983, p. 123.

Sweetman, H.L., 1958. The Principles of Biological Control. W.C. Brown, Dubuque, Iowa, 560 pp.

Takahashi, R., 1965. Some new and little known Aphididae from Japan (Homoptera). Insecta Matsumurana, 28: 19–61.

Tardieux, I. and Rabasse, J.M., 1986. Host–parasite interrelationships in the case of *Aphidius colemani* (Hymenoptera, Aphidiidae). In: I. Hodek (Editor), Ecology of Aphidophaga. Academia, Prague and W. Junk, Dordrecht, pp. 125–130.

Thomas, W.P., 1977. Biological control of the blue-green lucerne aphid. Proceedings of the Weed and Pest Control Conference, Johnsonville, 1977 pp. 182–187.

Tremblay, E., 1984. The parasitoid complex (Hymenoptera: Ichneumonoidea) of *Toxoptera aurantii* (Homoptera: Aphidoidea) in the mediterranean area. Entomophaga, 29: 203–209.

Valentine, E.W., 1975. Current projects on biological control of insects. Proceedings of the 28th New Zealand Weed and Pest Control Conference, Hastings, pp. 193–195.

Van den Bosch, R., 1971. Biological control of insects. Annual Review of Ecological Systematics, 2: 45–66.

Van den Bosch, R., Schlinger, E.I., Dietrick, E.J., Hagen, K.S. and Holloway, J.K., 1959. The colonization and establishment of imported parasites of the spotted alfalfa aphid in California. Journal of Economic Entomology, 52: 136–141.

Van den Bosch, R., Schlinger, E.I., Hall, J.C. and Puttler, B., 1964. Studies on succession, distribution and phenology of imported parasites of *Therioaphis trifolii* (Monell) in southern California. Ecology, 45: 602–621.

Van den Bosch, R., Beingolea, O.G., Hafez, M. and Falcon, L.A., 1976. Biological control of insect pests of row crops. In: C.B. Huffaker and P.S. Messenger (Editors), Theory and Practice of Biological Control. Academic Press, New York, London, pp. 443–456.

Van den Bosch, R., Messenger, P.S. and Gutierrez, A.P., 1982. An Introduction to Biological Control. Plenum Press, New York, London, 247 pp.

Waterhouse, P.M., 1985. Isolation and identification of carrot red leaf virus from carrot and dill growing in the Australian Capital Territory. Australasian Plant Pathology, 14: 32–34.

Wilson, F., 1938. Some experiments on the influence of environment upon the forms of *Aphis chloris*. Transactions of the Royal Entomological Society of London, 87: 165–180.

Wilson, F., 1960. A Review of the Biological Control of Insects and Weeds in Australia and Australian New Guinea. Technical Communication of the Commonwealth Institute of Biological Control, No. 1, 102 pp.

Wilson, F., 1963. Australia as a Source of Beneficial Insects for Biological Control. Technical Communication of the Commonwealth Institute of Biological Control, No. 3, 28 pp.

Wilson, F., 1974. Biology in Pest Disease Control. 13th Symposium of the British Ecological Society, Oxford, January 1972, pp. 59–72.

Wilson, F. and Huffaker, C.B., 1976. The philosophy, scope, and importance of biological control. In: C.B. Huffaker and P.S. Messenger (Editors), Theory and Practice of Biological Control. Academic Press, New York, London, pp. 3–15.

Zuparko, R., 1983. Biological Control of *Eucallipterus tiliae* (Hom.: Aphididae) in San Jose, Calif., through establishment of *Trioxys curvicaudus* (Hym.: Aphidiidae). Entomophaga, 28: 325–330.

11.2.1 Biological Control in the Open Field

R.D. HUGHES

INTRODUCTION

In this section, biological control is considered only in its sense of a manipulative procedure, instigated by man in response to a perceived pest situation, i.e. setting an exotic natural enemy against a selected aphid pest.

The impact of naturally occurring enemies on aphid populations is dealt with in various sections of Chapter 9. Biological control in closed environments is considered in section 11.2.2 while other uses of biological control agents alone or in combination with other controls are discussed in sections 11.3 and 11.5. Nevertheless, much of what follows is relevant to all biological control projects.

Here we look at what is often called classical biological control as it has been applied to aphid pests in the field. Since 1900 the approach has been used against several aphid pests in Australia (Table 11.2.1.1), where the results have varied from apparent success to apparent complete failure. A similar range of results can be seen in the other biological control attempts against aphids reported in the literature. Only a relatively small number of attempts have been carried through and described in sufficient depth to merit comparative analysis, but if the variability of results is inherent in the method as applied to aphids, any general characteristics of either successes or failures may provide useful leads for further attempts.

My viewpoint is that of a population ecologist, who intuitively feels that biological control should be applied ecology. This position is supported by what seems to be general agreement about many of the ecological attributes of the natural enemies associated with successful biological control attempts, dating back at least to DeBach (1964). However, there is now also general agreement that, even with some knowledge of the biology of a natural enemy and its interaction with the target species, failure of any particular attempt might be more predictable than success. This situation results from the fact that biological control is still more of an art than a science. The causes of failure may be in the preliminary consideration given to an attempt, or in the procedures used. Little specific attention has been given to the practice of biological control as it has been applied to aphid pests.

This section first develops a framework of practice (Table 11.2.1.2), using as a unifying theme Australian experience with an attempt to biologically control the spotted alfalfa aphid, *Therioaphis trifolii* f. *maculata* (Buckton) on lucerne, *Medicago sativa*, using the aphidiid parasite *Trioxys complanatus* Quilis. This framework will then be used in conjunction with a brief review of the ecological attributes of successful natural enemies, in a subsequent analysis of well-documented attempts to biologically control aphid pests (Table 11.2.1.3).

Section 11.2.1 references, p. 194

For the target aphids and natural enemies used in particular attempts at biological control, their full scientific names and authorities will as usual be given only at their first appearance in the text and in all tables. However in this section, where the protagonists are subsequently referred to in the text, the abbreviated name will be accompanied by a reference to the appropriate table. For the example above it would be (1.F.a) referring to Table 11.2.1.1; aphid F and enemy *a*.

THE PRACTICE OF BIOLOGICAL CONTROL OF APHIDS

There is a sequence of seven steps in a biological control project (Table 11.2.1.2). They are grouped in three overlapping phases: the preliminary considerations, the attempt itself and the assessment.

Step 1. Characterisation of target aphid and its environment

A sound knowledge of the target aphid species is as essential to biological control as it is to any other method of control. In the 1930s the assumption that the adelgid on Monterey pine (l.C) in Australia was the same as that occurring on Scots pine resulted in the importation of natural enemies largely from northern Europe rather than from California, and this may partly account for the failure of the predators to establish.

In considering biological control, the nature of the pest problem is particularly important. The worst direct damage caused by the target aphids listed in Tables 11.2.1.1 and 11.2.1.3 is plant death, e.g. by *Acyrthosiphon kondoi* Shinji on lucerne seedling stands, or complete crop loss such as that caused to the aerial parts of lucerne by a heavy *T. trifolii* f. *maculata* infestation (e.g. Lodge, 1980). Most aphids, however, only stunt or distort their host plant, but this still results in severe crop loss. With the stunting of trees, and even perennial herbs, the yield loss may be cumulative, as with *Adelges piceae* (Ratzeburg) on balsam fir (Clark et al., 1971). When *A. kondoi* severely infests mature lucerne, the stunting caused reduces root reserves and the effects are carried over to at least two subsequent crops (Kain et al., 1977). Indirect damage to crops often results from the transmission of virus disease by aphids. In another way, the honeydew produced by aphids can cause crop loss, affect machinery or just be a nuisance (see Chapter 10).

Since both the direct and indirect damage caused by aphids seems to be in proportion to their numbers on host plants, any reduction in numbers caused by a natural enemy would be potentially beneficial. Even in the case of virus transmission, where the feeding of only one infected aphid on a plant may indirectly cause substantial loss, reduction in aphid numbers will reduce the probability of flying aphids landing on a host plant more than proportionally, because reduced crowding of aphids usually results in a lower fraction forming wings.

This may be of particularly value where reservoirs of aphids occur on weeds or other non-crop host plants on which other control methods would be out of the question. Other features of the general environment can have both positive and negative effects on the amenability of a pest problem to biological control. The proximity of flowering weeds, etc., to attract and feed natural enemies may be an advantage. Conversely the regular use of herbicides against such weeds, or of insecticides against the same or other pest problems in the crop, may reduce the chance of biological control. (But see section 11.5 on integrated control.)

The climate in which the target aphid occurs may be particularly important when making a choice of potential sources of a natural enemy.

TABLE 11.2.1.1

Synoptic summary of attempts to biologically control target aphids in Australia, in approximate chronological order

Target aphid and target environment	Selected natural enemy	Type of enemy[a]	Establishment[b]	Reduction of target aphid problem[b]	Verification of role of enemy	Reference
(A) *Eriosoma lanigerum* (Hausmann) Woolly apple aphid	(a) *Exochomus melanocephalus* (Zoubkoff) Coccinellid beetle	Pred	–			Wilson (1960)
Pyrus malus, P. communis Apple, pear Orchard trees	(b) *Aphelinus mali* (Haldemann) Aphelinid wasp	Para	+ + +	+		Wilson (1960)
(B) *Brevicoryne brassicae* (Linnaeus) Cabbage aphid	(a) *Diaeretiella rapae* (M'Intosh) Aphidiid wasp	Para	+ +	+		Jenkins (1948) Wilson (1960) Hughes (1963) Gilbert and Hughes (1971)
Brassica oleracea cvs. Brassica crops/weeds Annual crops						
(C) *Pineus laevis* (Maskell) = *Chermes boerneri* (Annand) Pine adelgid	(a) *Leucopis obscura* Haliday Chamaemyiid fly	Pred	–			Wilson (1960)
Pinus radiata Monterey pine Plantation trees	(b) Four other predators Flies, beetle, lacewing	Pred	–			
(D) *Tuberculatus annulatus* (Hartig) = *Myzocallis annulatus* (Hartig) Oak aphid	(a) *Aphelinus flavus* (Nees) Aphelinid wasp	Para	–			Wilson (1960) Carver and Stary (1974)
Quercus spp. Oaks Street trees						

Section 11.2.1 references, p. 194

TABLE 11.2.1.1 (*Continued*)

Target aphid and target environment	Selected natural enemy	Type of enemy[a]	Establishment[b]	Reduction of target aphid problem[b]	Verification of role of enemy	Reference
(E) *Cavariella aegopodii* (Scopoli) Carrot **aphid** *Daucus carota* Carrot *Salix* spp. Willow	(a) *Aphidius salicis* Haliday Aphidiid wasp	Para	+ +	+ + +		Stubbs (1966) Carver and Stary (1974) Hughes et al. (1964, 1965) Carver (section 11.2)
(F) *Therioaphis trifolii* f. *maculata* (Buckton) Spotted alfalfa aphid	(a) *Trioxys complanatus* Quilis Aphidiid wasp	Para	+ + +	+ + +	+ +	Wilson et al. (1982)
	(b) *Praon exsoletum* (Nees) Aphidiid wasp	Para	+	+ + +	–	Lehane (1982)
Medicago sativa Lucerne Perennial crop	(c) *Aphelinus asychis* Walker Aphelinid wasp	Para	+	+ + +	–	Milner (1984)
	(d) *Zoophthora radicans* Batko Entomophthoran fungus	Path	+ + +	+ + +		Hughes et al. (1987)
(G) *Acyrthosiphon kondoi* (Shinji) Blue-green aphid	(a) *Aphidius ervi* Haliday Aphidiid wasp	Para	+ + +	+ +		Milne (1982, 1986a,b) Snowball and Lukins (1979)
Medicago sativa Lucerne Perennial crop	(b) *Ephedrus plagiator* (Nees) Aphidiid wasp	Para	–			Cordingley et al. (1979) Brieze-Stegeman (1979)
(H) *Acyrthosiphon pisum* (Harris) Pea aphid	(a) *Aphidius smithi* Sharma and Rao Aphidiid wasp	Para	+			Carver (section 11.2)

Host plant / Aphid / Crop	Natural enemy	Type[a]	Symbol[b]		Reference
Medicago sativa / Lucerne / Perennial crop					
	(b) *Aphidius pisivorus* Smith / Aphidiid wasp	Para	–		
	(c) *Aphidius eadyi* Stary / Aphidiid wasp	Para	–		
	(d) *Aphidius ervi* Haliday / Aphidiid wasp	Para	+++	+	Milne (1986a,b)
	(e) *Aphelinus asychis* Walker / Aphelinid wasp	Para	–		
(I) *Hyperomyzus lactucae* (Linnaeus) / Sowthistle aphid	(a) *Aphidius sonchi* Marshall / Aphidiid wasp	Para	++		Carver and Woolcock (1986)
Sonchus oleraceus / Sowthistle / Annual weed	(b) *Praon volucre* Haliday / Aphidiid wasp	Para	+		Brieze-Stegeman (unpubl.)
(J) *Aphis craccivora* Koch / Cowpea aphid	(a) *Lysiphlebus fabarum* (Marshall) / Aphidiid wasp	Para	–		Carver (1984)
Various legumes / Sub-clover, beans, etc. / Annual crops	(b) *Lysiphlebus testaceipes* (Cresson) / Aphidiid wasp	Para	–		Carver (section 11.2)

[a]Para = parasitic Hymenoptera, Path = pathogenic fungus, Pred = predatory insect.
[b]For use of + and – symbols, see p. 192.

Section 11.2.1 references, p. 194

TABLE 11.2.1.2

The seven steps in a biological control project and the overlapping phases in which they are grouped

Step	Preliminary considerations	Attempt	Assessment
(1) Characterisation of target aphid and its environment – Identity of aphid, nature of problem caused – Host plant(s), biotope, climate	×		
(2) Appropriateness of biological control and type of enemy, and specific approach – Crop value, quality requirements, availability of more appropriate control methods, types of enemy, augmentation	×		
(3) Selection of natural enemy (a) Where previous attempt has been made, experience with particular natural enemies, sources and contacts, availability from climatically matched areas (b) Where no previous attempt has been made, surveys in area of origin of the pest, studies of specificity and variation in other desirable characters, searches in climatically matched areas	×	×	
(4) Import, quarantine procedures, identification, examination – Quarantine procedures, insectary rearing, mass-rearing procedures, storage procedures		×	
(5) Release of natural enemies – Direct or after multiplication, release strategies – numbers and sites, manipulative or natural dispersal – Conditions at time of release, repeated release		×	×
(6) Establishment – Monitoring, occurrence after release, initial establishment, unfavourable seasons, prolonged establishment – Competition, resiliance			×
(7) Assessment of impact (a) Decline in pest problem, indirect assessment, reduction of target aphid numbers, before and after monitoring, development of decline over time (b) Verification of role of natural energy, extent of distribution, matching phenology, impact on numbers and population structure, mathematical model or simulation of impact			×

Step 2. Appropriateness of biological control, types of enemy and specific approach

Whilst most aphids would seem prime targets for biological control on the basis of damage they cause, there are several other factors which influence the decision as to the general appropriateness of the method. An indigenous aphid pest is not normally considered as an appropriate target for biological control. This is because such an aphid would already have a suite of indigenous natural enemies adapted to utilize it. The chance of finding an exotic natural enemy sufficiently adapted to an indigenous host so as to biologically control it, seems remote. Another important point is that *some* aphids will be present even when the pest is under good biological control. Thus crop situations where even a small amount of damage is unacceptable (e.g. any cabbage aphids in Brussels sprouts will result in the crop being rejected for freezing) cannot rely on biological control. Aphids can often be readily controlled by the application of insecticides systemic in the host plants. In horticulture this is still the method of choice, because of demands for blemish-free products, the ease of application of the chemicals and because of the high value of the crops relative to the costs involved. However, the situation can be different in other branches of agriculture. This can be seen in the spectrum of lucerne-growing operations common in Australia. These range from highly capitalised irrigated enterprises supplying high-quality baled lucerne for horse feed and for stockfeed milling, through rotational grazing paddocks and wheat–lucerne–fat-lamb rotations, to lucerne-improved pastures and to naturalised lucerne in pastures on a variety of types of country. The high value of the crop and the demand for quality at the top end of the range would favour the use of insecticides against aphids. In the middle range other factors, like avoidance of insecticides in meat producing enterprises, may shift the balance in favour of biological control, and at the bottom of the range it is clearly the method of choice.

Only relatively few groups of organisms are specialised to exploit aphids (see Chapter 9). The three basic sorts are the parasitic Hymenoptera (section 9.1) predatory insects (of various orders; section 9.2) and the pathogenic fungi (section 9.3). For specific target aphids all three sorts of natural enemy may not be available. For example, no parasites of *A. piceae*, the balsam woolly aphid, are known (3.B). Given a choice is available, characteristics of the various sorts of natural enemies may make these more, or less, favourable for use against target aphids in certain biotopes.

The parasitic hymenoptera have life cycles closely associated with those of their aphid hosts. Since only one aphid is consumed to produce each parasite, ultimate body size is limited. The small size of aphid parasites limits their ability to seek out target aphids over long distances, so that their impact on the target aphid depends almost entirely on their relative rates of development and reproduction in the target environment. Winged aphids colonising new host plants often carry immature stages of their parasites with them, e.g. *Diaeretiella rapae* (M'Intosh) in *Brevicoryne brassicae* (Linnaeus) (1.B.a) (Hughes, 1963). Other characteristics and behaviour help maintain interaction at low densities of target aphid. Where free-living stages of aphids disappear from a habitat in unfavourable seasons, the parasite may enter diapause,or more polyphagous species may be able to utilise alternate host aphids. Highly specific parasites may seek out the target species even at very low densities (e.g. *T. complanatus* on *T. trifolii* f. *maculata* (1.F.a)).

Unlike the life cycles of the parasites, those of even specialised aphid predators (sections 9.2.1–9.2.6) are rarely closely linked to those of the aphids preyed upon. Their interaction is maintained by the mobile adult stages of the predators, which seek out concentrations of aphids and lay their eggs close by.

Section 11.2.1 references, p. 194

Some adult predators feed on the target aphid, so that the impact is direct, but with others (e.g. syrphids and chrysopids) only the immature stages are predatory, delaying the impact. Predatory insects consume many aphids during their development so that, unlike the parasites, their body size is not limited to that of the prey aphid. The large adults are usually active fliers and may leave an area where aphid density is low or where conditions are otherwise unfavourable. When away from the target aphids they may utilise other aphids or even alternative prey; in the absence of any prey, they may enter diapause.

The relations between pathogenic fungi (section 9.3) and target aphids are similar to those between parasitic Hymenoptera and aphids, but their life cycles are not so closely associated. The transmission of the disease from an infected cadaver depends very much on the chance of contact between conidiospores and aphids under prolonged conditions of high humidity. The process is most effective at high aphid densities in certain seasons and may result in spectacular population crashes of the target aphid. When target aphid densities are low, the fungus may have to utilise alternative aphid hosts to survive in an area, unless it forms resting spores. When resting spores are formed, as in the case of *Zoophthora radicans* Batko on *T. trifolii* f. *maculata* (1.F.d), the fungus can remain dormant for long periods in the absence of aphids. The weather conditions which cause the re-emergence of host plants and their aphids in an area, are also likely to cause resting spores to germinate and eject conidiospores to restart the life cycle.

The broad characteristics of the basic types of natural enemy suggest some adaptations and some limitations for use in particular biotopes and point to the need for an informed selection. For stable biotopes, which can range from forests or orchards to semi-perennial crops, weak-flying natural enemies, adapted to stay and maintain their hold on target aphids, would seem desirable. For biotopes patchily favourable on the ground, as with rotations of annual crops, more independent natural enemies, such as syrphids or coccinellids, that are able to migrate extensively to seek out the target aphids would seem most appropriate. For biotopes in which local conditions favourable for aphids are widely separated only in time, natural enemies able to maintain themselves in the absence of the target aphids, e.g. parasitic Hymenoptera capable of diapause, or pathogens with resting spores, would perhaps be best.

The last decision in relation to the use of biological control is whether to leave the natural enemy to maintain or re-establish its hold on the target aphid naturally, or to assist the process at crucial times. This can be achieved by modifying the naturally-occurring ratio of the numbers of natural enemy and target aphid. High-cost techniques, like breeding up natural enemies for inundative release, have been tried, e.g. rearing *Aphidius smithi* Sharma and Rao for inundative release against the pea aphid, *Acyrthosiphon pisum* (Harris), (3.D.c.) (Halfhill and Featherstone, 1973). Less costly techniques, like increasing overwinter survival of diapausing parasites by providing shelter, are often more practical, e.g. keeping bundles of apple twigs carrying mummies of *Aphelinus mali* Haldemann among the woolly aphid infestations (3.A.b) in cold storage with the fruit through the winter and then replacing them in the orchards in spring (Greathead and Pschorn-Walcher, 1976). An interesting case of manipulation is the simultaneous exclusion of attendant ant species. Olkowski et al. (1974) used sticky-banding of street trees to reduce interference to introduced *Trioxys* spp. (3.J.a; 3.K.a) by Argentine ants, *Iridomyrmex humilis* (Mayr).

In coming to a decision to attempt biological control and to employ a particular method, general reviews of the field, and papers on specific target aphids, should be consulted.

Step 3. Selection of the natural enemy

Where the pest aphid has been a target for previous biological control attempts, the characteristics of the available natural enemies can often be discovered from the literature. This information should be compared with the characteristics of the target aphid and target environment before making the choice of natural enemy. Particularly important points to consider are: (a) whether the target aphid is a preferred resource of the natural enemy; (b) whether the enemy is accustomed to seeking the target aphid on the host plant concerned, and in the relevant biotope; (c) whether the enemy exists in places where the climate matches that affecting the target aphid; and (d) what other requirements the natural enemy has.

When *T. trifolii* f. *maculata* appeared on lucerne in Australia in 1977, there existed an extensive literature on the attempt to biologically control it in California in the late 1950s (3C.a,c,d), so that the choice of parasitic hymenoptera could be well informed. *T. complanatus* and *Praon exsoletum* (Nees) were both relatively specific parasites of *T. trifolii* f. *maculata*, and was no doubt that the aphid would be the preferred host. This was less certain for the more polyphagous *Aphelinus asychis* Walker. Both *T. complanatus* and *P. exsoletum* occurred on *T. trifolii* f. *maculata* throughout the natural range of the host plant *M. sativa* (Van den Bosch, 1957) and in California operated successfully even in the extremely modified biotope provided by the vast irrigated areas of the alfalfa monocultures. Most parts of Australia where *M. sativa* is grown have climates which can be matched in the natural range of the host plant in the Middle East. The aphid-infested host plants seem to provide all the resources necessary for the parasites.

The literature will also provide information as to possible sources of enemies, both in terms of geographical location and the addresses of potential collaborators who could perhaps collect and send material.

Where no previous attempt at biological control has been recorded, searches for natural enemies of the target aphid may have to be made within its native range. It may be possible for local entomologists to do this and, if a particular natural enemy is an obvious choice, to collect and send material for study, e.g. Aeschlimann (1981). Alternatively, the staff of an international biological control agency may provide this service. Best of all, a visit should be made to appropriate climatic regions to discover, investigate and perhaps collect natural enemies of the target aphid, e.g. Van den Bosch (1957).

If relatively little is known about the natural enemies it is usually worthwhile to study their biology in the country of origin or, less preferably, under quarantine conditions in the country where the control attempt is to be made. Such studies of suitability, specificity and relative rates of increase under a variety of conditions compared with the target aphid, may all help in the initial choice of natural enemy. If the chosen natural enemy is available from more than one source, laboratory studies of this type may help in the selection of one strain for mass rearing, or may indicate greater suitability of a strain for a particular sub-region of the infested area.

Once a natural enemy has been selected and the source of supply of material has been worked out, the preliminaries are over and the second phase – *the attempt* – has already started.

Step 4. Import, quarantine procedures, identification and examination

In some countries where rules concerning the import of certain live organisms do not include quarantine, it may be possible to release imported natural

Section 11.2.1 references, p. 194

enemies directly into the field. However, this should not be done unless certified laboratory-reared material is available, because of the chance of releasing other unwanted organisms at the same time. Although there may be problems in rearing a natural enemy in quarantine, it does give an opportunity for studying the species and making a more informed decision on the release of the species into the field.

Many countries, like Australia, impose severe restrictions on the import of live organisms. Unless permission is sought and the requirements of the applicable regulations fulfilled, natural enemies may be destroyed or be held at the port of entry. Even when material is allowed into the country, other regulations may demand that it be examined and kept under quarantine conditions for a specified period to prevent the arrival of unwanted fellow-travellers. These can always be expected in field-collected insect material but should be looked for even in material from fellow aphid workers and from reputable laboratories. In the first batches of parasites of *T. trifolii* f. *maculata* sent from California to Australia (1.F), *seven* living specimens of *A. pisum*, a pest aphid unknown in Australia until three years later, were found in the packaging – presumably from laboratory cultures at the point of origin. When importing natural enemies, an even more important reason to examine and hold the material in quarantine is to rigorously exclude parasites and diseases of the natural enemies themselves. At CSIRO, aphid parasites are reared in quarantine for at least one generation. None of the original imported material is released, and only fully identified F_1 progeny allowed to leave quarantine. For predators known to be liable to disease, more generations in quarantine may be desirable.

It is very important to be sure of the identity of the imported natural enemy. For aphid parasites (see section 9.1) this is usually a job for a specialist, but even they may not be certain. In all cases it is best, after the F_1 generation has been started, to preserve and lodge all or most of the originally imported material in a properly curated insect collection. This allows problems over the identity of species established to be resolved, as in the case of *Leucopis* spp. introduced to Canada to prey on *A. piceae* (3.B.a) (Clark et al., 1971).

The initial examination of imported material should include making a record of other characteristics of the consignment – particularly the numbers of natural enemies and their sex ratio. Where initial numbers are small or there is an excess of males, problems in establishing the natural enemy may arise. But if only females are present the enemy may be from a uniparental strain. If an adverse male ratio persists in the F_1 generation, this suggests that the strain of natural enemy may not be adapted to the ambient conditions, and that an alternative source of material should perhaps be sought. When a batch of *P. exsoletum* mummies from Quetta in Kashmir was received in Australia (1.F.b) from the Commonwealth Institute of Biological Control, Rawalpindi, it was found to have a sex ratio of $3\male\male:1\female\female$ compared with the ratio of approximately $1\male:1\female$ normal in our cultures on *T. trifolii* f. *maculata*. The history of the material was checked: it had been subjected to a severe temperature change, from 25°C to 40°C, and reared on a non-lucerne biotype of *T. trifolii*. As survival was low in the F_1 generation, the culture was terminated.

Step 5. Release of natural enemies

The release of material is the third and final decision to be made in any biological control programme. It should not be made lightly, and again the decision actually demands answers to several questions: (a) whether to release the particular material at all; (b) what method of release should be used; (c) how many natural enemies should be released at one time?

(a) The release of natural enemies of which the quality is in any way doubtful, could be at best a waste of resources and at worst introduce factors deleterious to further attempts at biological control. The very real pressures to release material just because of efforts already made, should be resisted.

(b) There are many possible ways of releasing natural enemies, which need to be considered. It may be possible to release a species at any stage of its development that is available, i.e. for insects: as adults, as eggs, as larvae, or as pupae; or for diseases, as cadavers or some sort of inoculum. Under particular conditions at the time of release, any one, or a combination of more than one stage could be the most appropriate. The direct release of imported material limits the choice to the development stage in which it was received, but laboratory rearing allows other stages to be released, albeit somewhat later.

With parasitic Hymenoptera and other organisms whose necessary resources all occur in the immediate vicinity of the "crop" infected by the target aphid, the immediate pre-reproductive stage, namely the mated adult female, is the optimum stage to release. Furthermore, as adults can usually be kept alive with minimal resources, e.g. in small cages with sources of water and honey, storage for later release is a practical possibility. However, it is well-known that flying insects kept in cages tend to disperse far and wide rather than respond to other cues when first released. In a large area of target aphid infestation and using mated females this may not be important. Nevertheless, at the time of release the adults may be exposed to unfavourable weather conditions, which their natural phenology and behaviour would allow them to avoid. These problems may largely be circumvented by the release of the material into a cage or muslin sleeve enclosing a portion of the target aphid infestation. This method was used initially in the release of *T. complanatus* and other parasites of *T. trifolii* f. *maculata* in Australia (1.F.a). In South Australia, following the suggestion of Flanders (1959), large $2 \times 1 \times 1$ m muslin cages were placed over infested *M. sativa*, and after the parasites had been present within them for two weeks the cages were opened to allow dispersal of the parasite progeny (Wilson et al., 1982). An analogous technique was used by Milner et al. (1982) to release the fungus disease *Z. radicans* into infestations of *T. trifolii* f. *maculata* (1.F.d). Cadavers of infected *T. trifolii* f. *maculata* were suspended above *M. sativa* infested with the aphid, in a plastic enclosure ensuring the high humidity necessary for conidiospore discharge and germination. The release of an immature stage, e.g. of pupae, allows the complete cycle of adult behaviour to occur under field conditions. Even in cages this method may improve the chances of released natural enemies (Wilson et al., 1982), and for release into the open field the advantages are obvious. Starý (1968) has gone even further, suggesting that "artificial focus units", containing host plants, host aphids and parasites in all stages of development, should be placed among or around infested crops. In that way, adult parasites are released continuously over an extended period of environmental variations. This was the method finally preferred for the release of *T. complanatus* and other parasites of *T. trifolii* f. *maculata* on *M. sativa* (1.F.) in southeastern Australia (Woolcock, 1979). Caged flats ($400 \times 300 \times 100$ mm) of 12-week-old *M. sativa* were inoculated with parasitised and unparasitised aphids and with parasites and held in a glasshouse insectary at 24°C for 16 days before being placed in the field for a 14-day period.

When the adult predator does not attack aphids directly, and/or is involved in behaviour associated with resource requirements other than the target aphids, there is less chance of it settling down at the site of release. In the case of such predators, the optimum stage for release on a target aphid infestation is an early immature stage that attacks the aphids immediately. The more

Section 11.2.1 references, p. 194

complete the contact set up between a released predator and the target aphid, the greater the chance the natural enemy has of becoming adapted to and returning to the target aphid environments in later generations.

(c) In general, the more individuals of a natural enemy that are released at one place, even over a period, the more chance there is of an early establishment there. Greater numbers ensure more and prolonged contacts between the natural enemy and the target aphid. This in turn increases the rate of adaptation of the enemy to the target aphid. However, it is probably equally true that the more sites at which releases are made, the more likely it is that the best environments for early establishment will be found. So the numerical aspects of the release strategy always depends on the resources available to the biological control programme concerned.

Unless the imported consignment of natural enemies is large or import can be repeated many times, the chance of direct release into the field being successful is probably very small. Usually the imported natural enemies have to be reared and allowed to reproduce to multiply their numbers. Putting the imported material into muslin sleeves or cages confining larger than normal populations of the target aphid, and then opening the cages after the F_1 generation has emerged, is the simplest but riskiest method of multiplication. Safer and more informative is to rear the natural enemies under controlled or semi-controlled conditions in an insectary. This allows observations to be made on the life history of the enemy and the development of the best strategies for the stage to be used and the timing of the release. Some multiplication of the original material can usually be achieved without much difficulty during insectary rearing. However, if a release strategy demands large numbers of enemies to be available over a long period, the development of the necessary mass-rearing techniques involves problems of resources and logistics.

To rear large numbers of natural enemies of aphids, it is necessary to acquire or produce the necessary numbers of healthy and clean host plants, to rear stocks of the target aphid on these host plants and finally to rear stocks of the natural enemy on the aphid-infested host plants. Unfortunately, the aphids cause damage and may prevent the growth of their host plants and, of course, the natural enemies hopefully prevent aphid increase. As a result, it is necessary to keep the three rearing operations as separate as possible to avoid contamination. This is best achieved by complete physical separation of facilities, i.e. a separate glasshouse for plants and two separate insectaries for aphids and their enemies. Because plants need to be growing vigorously to tolerate aphid infestation for long periods, insectaries with glass roofs or with lighting adequate for plant growth are necessary. The minimum requirements would be to acquire healthy host plants (uncontaminated with insecticide) from a nursery and to have two insect-proof, walk-in cubicles in an insectary.

The small size of aphids and of some of their natural enemies makes it easy to contaminate host-plant and aphid cultures by accidental transfer of the insects on workers' clothing. The propensity of some target aphids to jump or fall off host plants when disturbed adds to the risk of transfer, and special precautionary measures are necessary. In the CSIRO programme rearing *T. complanatus, T. trifolii* f. *maculata* and *M. sativa* (1.F.a), the spotted alfalfa aphid adults and nymphs often jumped onto clothes, and the ability of the parasite to seek out aphids meant that the chances of contamination were high, even though the rearing operations took place in physically separate buildings. To avoid any risk, work schedules were arranged so that on any one day personnel had to go from plant glasshouse to aphid insectary, and from there to parasite insectary – **never in the reverse direction**. Equally, any other work involving handling of either the parasite or the aphid precluded entry of facilities where they were absent. When more than one stock or strain of the parasite had to be reared, it was not possible to use another separate insectary, but to reduce the risk of cross-contamination insect-proof cubicles were built into the insectary. If necessary, separate insect-proof cages kept stocks separate within cubicles. Again, work schedules were arranged so that manipulation of each stock was carried out on a different day.

Careful cleaning of culture cages is very important. Mummified aphids or predator pupae may remain attached and hidden within the cages and not be killed by routine washing procedures. A special procedure, such as dipping in very hot water, or the use of separate sets of cages clearly labelled "for aphids" or "for natural enemies", should be instituted.

The damage done by aphids to the host plant usually limits both the numbers that can be produced per plant and the time that high aphid densities can be maintained for the multiplication of parasite numbers. In certain cases, optimal procedures have to be evolved by trial and error. For instance, only by waiting until *M. sativa* plants were at least 12 weeks old could they be used to build up and maintain dense infestations of *T. trifolii* f. *maculata* long enough to rear *T. complanatus* through one parasite generation. Some extension of such a time limit can be achieved by the use of oblong "flats" containing many host plants and inoculating only one corner with aphids. The aphid infestation progresses only slowly across the "flat", adding a week or so to the time available for rearing natural enemies (Woolcock, 1979).

The potential production of the mass-rearing operation will depend on the logistics of the procedures involved and on the space in the available facilities. For example, at the height of the CSIRO programme of mass-rearing *T. complanatus* for use against *T. trifolii* f. *maculata*, a very large area of glasshouse space had to be found to grow *M. sativa* from seed and to maintain the lucerne insect-free for 12 weeks before use. If absolutely necessary it may be possible, with the judicious use of insecticides, to clean up host-plant material grown outside before it is brought into the insecticides. The risks involved are significant, however: either of bringing in plants carrying predators and competitors of the target aphid, or of leaving insecticide residues sufficient to affect the insect cultures. While on the subject of chemical contamination of cultures, one of the best natural aphicides, nicotine, is often present in residual amounts from the tobacco-smoking activities of personnel. Smoking should always be banned in insectaries, but more severe restrictions may be necessary, as some aphids are very sensitive. With *T. trifolii* f. *maculata* (1.F), even the clothing of a heavy smoker was found to carry sufficient residue to affect the aphid cultures severely. This is not surprising, as a single puff of tobacco smoke in a room has been shown sufficient to kill a *T. trifolii* f. *maculata* culture (Kircher and Lieberman, 1967).

Difficulties in rearing host plants or aphids may impose limits on mass-rearing operations, and workers may be tempted to use short-cut methods. The use of rapid-growing alternative aphid host plants for the target aphids, or the use of a more easily reared alternative aphid as host for a parasite, may be possible. However, a parasite's ability to locate and utilise the target aphid in the target environment, may be affected by a history of rearing in an alternative aphid or in the target aphid on an alternative host plant. In attempts to control *A. kondoi* on *M. sativa* (1.G.), the aphid was often reared on *Lens culinaris*, the lentil plant, which takes only a few weeks to grow to full size compared with the 12 weeks necessary for the lucerne. Milne (1986a) suggests that the use of *L. culinaris* in the first attempt at least delayed the establishment of the parasite, *Aphidius ervi* Haliday, in lucerne crops in New South Wales, compared with subsequent operations using *M. sativa* as host plants.

Section 11.2.1 references, p. 194

Another way to get sufficient insects for large-scale releases of natural enemies, is to produce them over a long period and to store them progressively until numerous enough. It is difficult to stop development completely at an immature stage and store the material without deterioration, but it is usually possible to store adult insects, for a few days at least, with little adverse effect. Usually sources of water and sugar must be supplied. Cool ambient temperatures, approaching, but not below, the development threshold of the insect concerned, will extend the storage life of adult natural enemies or perhaps prevent development of eggs. Although little evidence of any reduction in the activity of natural enemies kept in storage has been reported, reared material should generally be released as quickly as possible to minimise the risk.

Left to themselves, natural enemies will spread slowly from release points, each species according to its adaptations. However, if caged adults are released, even relatively static insects may initially disperse explosively for quite long distances before settling down to their normal behaviour. Aphid parasites can also be dispersed over greater distances by the immature stages being carried in recently parasitised winged hosts. Fungal pathogens, too, are probably dispersed over long distances by the same process (R.J. Milner, personal communication, 1984), the evidence being that infected individuals are mostly winged females at the periphery of spreading epizootics and at the foci of new ones. Except where such long-distance dispersal is a regular event, the unassisted natural spread of the natural enemies away from the release points may be slow. Assistance may be provided accidentally, e.g. the transport of infested crops may help disperse the natural enemies. More often assistance is deliberate, and greater dispersal is another reason for multipoint release of the enemy (see p. 178) if available resources allow. If the target aphid is a major pest, the resources of the growers of the target crop may be used to help spread the natural enemies rapidly over large regions.

During the attempt to control *T. trifolii* f. *maculata* on *M. sativa* using *T. complanatus* in Australia (1.F.a), both methods of assisting dispersal were used. In the limited lucerne-growing areas of Queensland, multipoint release of mass-reared material was the method preferred (B. Franzman, personal communication, 1978). Over the more extensive distribution of the crop in South Australia, Victoria and New South Wales, local extension services and lucerne growers themselves helped disperse the parasite. Material mass-reared in base insectaries was inoculated onto a few large (5–20 ha) nursery plots of *M. sativa* (e.g. Berg and Ridland, 1979), using the methods described above. When such a crop was well-infested (e.g. 10% parasitisation in South Australia (Wilson et al., 1982)), the local extension services organised a field day at the site for growers in the region. After a talk on how to make best use of the material, the growers collected freshly cut lucerne on which the aphids were heavily parasitised and returned to their farms, thereby dispersing the parasites over thousands of square kilometres in a single day. Where a significant lucerne region was too far from a nursery site, the freshly cut lucerne was loaded onto road transports with minimum disturbance and with as little compression of the material as possible (Berg et al., 1979). The loads were then sent to the region, where again local distribution by the growers was organised.

The use of "artificial focus units" (see p. 177) will usually circumvent temporarily adverse conditions at the time of release, and even if adverse conditions persist there may still be time for a later release. However, in some situations the problem is not so easily dealt with. In Australia and elsewhere the effects of droughts on host plants, or other factors, may result in target aphids being extremely scarce for the whole season. If a mass-rearing and release programme has been scheduled for such a season, the result can be a disastrous waste. If a known alternative host aphid is available on plants unaffected by the drought, an attempt to establish the enemy on it can be made. However, the chance of success seems slight, probably because either genetic or phenetic adaptation to the target aphid usually prevents the other host being utilised effectively. In Australia, the release of several specific parasites of *A. pisum* (1.H.) probably failed because of the scarcity of the host in the year following its first appearance. Again in an attempt to use *Lysiphlebus testaceipes* (Cresson), originally derived from *Aphis nerii* Boyer de Fonscolombe to attack *Aphis craccivora* Koch (1.J.b.) on legumes, the actual releases coincided with a prolonged drought. There were no *A. craccivora* available on legume crops so an attempt was made to release on *Aphis gossypii* Glover infesting *Hibiscus* spp. bushes. Although *A. gossypii* was one of a range of preferred host aphids, no parasites were recovered in the following year from either *A. craccivora* or *A. gossypii*, but they appeared the next year attacking *A. nerii* on *Oleander*.

Step 6. Establishment

Although the natural enemy may continue to be observed around the release points, it is not until individuals of the F_1, or of a later generation, can be positively identified that initial establishment can be claimed. Systematic observation or sampling is usually instituted at and around the sites from the time of release until fresh specimens have been consistently collected over a period during which no further releases have been made, e.g. for *T. complanatus* (3.C.a) in New South Wales (Walters and Dominiak, 1984).

Initial establishment precludes any immediate need for further releases and marks the end of the attempt phase of the biological control project. It is possible to stop work on projects at this point, but most proceed to the verification of long-term establishment as the first step of the assessment phase. This involves restarting the systematic observation or sampling following the adverse season(s) subsequent to initial establishment. Adverse seasons for natural enemies are of two types: either those imposed by direct effects of the environment, e.g. cold, wet winters, or those during which the target aphid becomes scarce, e.g. during a hot summer or dry season. The first type is a test of the climatic adaptation of the natural enemy in terms of development and survival, the second type tests the enemy's mechanism for enduring or circumventing the shortage of host aphids.

Even if the enemy survives the first sequence of seasons it may die out subsequently in more severe seasons or during natural disasters such as a drought. In the attempt to biologically control *T. trifolii* f. *maculata* on *M. sativa* in Australia (1.F.a,b,c), three parasites were introduced and were all initially established. Their subsequent history varied, only *T. complanatus* eventually becoming numerous (Hughes et al., 1987), perhaps due to the difference in their response to the 4-year drought which ensued after their release. Not only natural disasters threaten the establishment of an enemy. Aerial spraying or other broad-acre applications of insecticide to control polyphagous pests attacking the whole range of crops in a region may wipe out the natural enemy population on and immediately around the target crop. The resilience of *T. trifolii* f. *maculata* in such a situation was demonstrated in Australia in the spring of 1980, when blanket spraying of lucerne in whole agricultural districts was undertaken to control an outbreak of *Heliothis* spp. After an absence of a few months, *T. trifolii* f. *maculata* re-established itself on our observation plot and elsewhere (R.D. Hughes, unpublished data).

Imported biological control agents are screened to prevent their own adapted natural enemies being released as well (see p. 176). Nevertheless, they will have to interact with local native competitors (predators and parasites), with

Section 11.2.1 references, p. 194

hyperparasites and of course with any other introduced natural enemies of the target aphid. If an attempt is being made to biologically control the target aphid it is unlikely that any native (or exotic) parasites or predators are already operating effectively enough to prevent the establishment of a newly introduced natural enemy. An intermittently effective pathogenic fungus disease may affect the target aphid during an attempt to introduce a parasite, and this may be a factor in the parasite's failure to establish.

Partially effective native predators of the target aphid may affect the long-term establishment of an introduced parasite by reducing the survival of its immature stages to dangerously low levels. Hymenopterous parasites are, in general, not the target of general predators, but they can be the target of native hyperparasites. These hymenopterous hyperparasites are usually relatively polyphagous, and native species will often attack the immature stages of an introduced parasite. Their effects may be relatively minor. In Australia, over 2 years, more than 2000 mummies of *T. complanatus* in *T. trifolii* f. *maculata* were collected from *M. sativa* in South Australia (3.C.b). Four species of native hyperparasite were reared, totalling only 8.7% of the emerging insects (Wilson and Swincer, 1984). A similar result was seen in a survey in Victoria (Berg et al., 1979). The low levels may have been due to previous lack of aphid hosts in the *M. sativa* crop. However, in countries where a wide range of aphid host species is available, hyperparasites have the potential to be a serious hazard to biological control programmes. The rapid build-up of hyperparasitisation of *Trioxys pallidus* (Haliday) when introduced into California to control *Chromaphis juglandicola* (Kaltenbach) (3.F.b) is well documented (Frazer and Van den Bosch, 1973; Van den Bosch et al., 1979).

Where more than one attempt to control the target aphid has been made, the most recently introduced enemy may have to compete with the previously introduced, partially effective, enemies. The interaction between them may be modified by the local environment, so that the new enemy becomes established only in a restricted area. A good example of this was seen with the successive introduction of two new parasites of *A. pisum* into North America (3.D). The pea aphid was already parasitised by a few native parasites, including the nearctic *Aphidius ervi pulcher* Baker. When *A. smithi* was introduced in 1958, it rapidly spread and during the next 10 years became the dominant parasite in the pea aphid complex over a wide area of North America, from Mexico to Ontario, Canada (Angalet and Fuester, 1977). *Aphidius ervi ervi* Haliday, the palearctic subspecies, was introduced from France over a longer period, from 1959 to 1963. By 1970 *A. ervi ervi* had displaced *A. smithi* as the dominant parasite in all the cooler, wetter regions of North America (Campbell and Mackauer, 1975; Angalet and Fuester, 1977).

After *T. complanatus* had been introduced into Australia to biologically control *T. trifolii* f. *maculata* (1.F.a), the pathogenic fungus *Z. radicans* was also introduced (1.F.d). There have been no reports of the spatial separation of these two natural enemies, and a laboratory study of their interaction suggests that neither one nor the other is likely to be favoured over prolonged periods of time (Milner et al., 1984).

Failure to establish an enemy is a regular event in attempts at biological control. Even enemies which appear to succeed initially may fail to persist in the longer term. However, the reverse also occurs. Enemies that do not seem to produce an F_1 generation, or apparently die out soon after initial establishment has been claimed, may suddenly reappear one or more seasons later. Milne (1986a) quotes the example of *A. ervi* introduced to control *A. kondoi*: the initial releases apparently failed, but the parasite was found to be present in small numbers when a new attempt was made 2 years afterwards. Such cases of dealyed establishment of an enemy can usually be explained as follows: the original material released was not adapted to the target aphid or to its environment, but a few survivors allowed the gradual adaptation of the species to the new conditions. The lack of any adaptation may have been inherent in the imported material or may have resulted from selection by adverse culturing procedures (see p. 180) before release.

Step 7. Assessment of impact

The long-term establishment of a natural enemy is no guarantee of immediate or eventual biological control. Even if the enemy has become widespread and is obviously present within infestations of the target aphid, the mortality seen may be compensated for by density-dependent changes in natality or by reductions in other sorts of mortality within the aphid population. Biological control may be said to occur only when the primary objective – a reduction in the pest problem caused by the aphid – is achieved.

Decline in pest problem. Where the target aphid causes major damage or nuisance, any decline in the pest problem may be very obvious. Then, for major effects at least, it may be possible to rely on the opinions of growers themselves or of the extension officers whom they consult. Reduction in the number of complaints can give a more quantitative indirect assessments of the effect. This method was used by Olkowski et al. (1982b) to assess the impact of *Trioxys curvicaudus* Mackauer on *Eucallipterus tiliae* (Linnaeus) (3.J.a), infesting street trees. Better still is a record of reduction in the specific use of alternative methods of control, e.g. fewer insecticide applications.

Where the decline in the pest problem is less obvious, i.e. where the control develops slowly or is only partly effective, or the need for control is subjective, it is better to monitor the target aphid infestation itself. Ideally it is advantageous to have aphid population data from before the biological control attempt for comparison with subsequent population levels, e.g. as for *T. trifolii* f. *maculata* in New South Wales (Walters and Holtkamp, 1984). Nevertheless, population monitoring from time of release of the enemies should show a decline to a significantly lower level, e.g. as for *T. trifolii* f. *maculata* at Canberra (Hughes et al., 1987). The effects of enemies are usually cumulative over a few seasons at least, as the chance of reinfestation by the target aphid is progressively reduced (see Fig. 11.2.1.1).

In this regard it is as well to be aware of other factors that may be causing a concurrent decline in the pest problem, as is discussed in section 11.5. In their claims for biological control of aphids on street trees in California (3.J. and K.a) Olkowski et al. (1982a,b) fail to mention that the aphids were deprived of attendant ants by the additional use of sticky banding (Olkowski et al., 1974). With a newly apparent problem the incentive to attempt its biological control may also apply to other control approaches. Changes in agricultural practice, e.g. by avoidance of sites with alternative host-plants, or by changing planting times, may also cause a decline in pest problems. The discovery that it is possible to select aphid-resistant cultivars of many crops has led to increasing use of that approach (see section 11.4). Lingering doubts as to the impact of *Aphidius salicis* Haliday on *Cavariella aegopodii* (Scopoli) on carrots (3.E.a) derive mainly from the concurrent general acceptance of aphid-resistant host plants. The events following the appearance of *T. trifolii* f. *maculata* on *M. sativa* (3.C) in New Mexico in 1954 and its devastating spread through North America provide an interesting case of different approaches being concurrent. The part played by each method in the subsequent control of the problem was a matter of some debate between their protagonists.

Section 11.2.1 references, p. 194

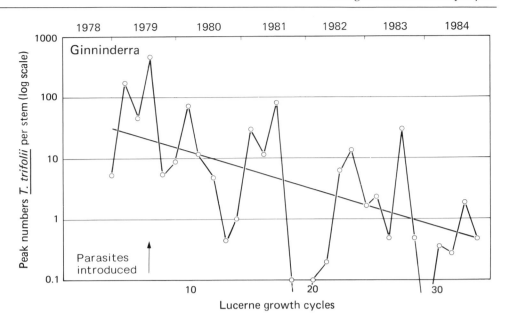

Fig. 11.2.1.1. The decline in *Therioaphis trifolii* f. *maculata* (Buckton) infestations (peak numbers per stem) of successive crops of lucerne, *Medicago sativa*, cv. Hunter River, at Ginninderra Farm near Canberra, Australia, over the 5 years after the introduction of the aphidiid *Trioxys complanatus* Quilis (after Hughes et al., 1987).

Verification of the role of the natural enemy. Even if a significant decline in the pest problem can be demonstrated, the role of the introduced natural enemy in that decline must be elucidated. Firstly, the enemy must have become sufficiently widespread within the range of target crop and target aphid. Unless effective means of dispersal are available, it is unlikely that any decline will be immediate or rapid. The spread of a natural enemy is usually relatively easy to document if the resources are available. Surveys showing a rapid natural spread of *T. complanatus* (3.C.b) to many kilometres from release sites in South Australia were made by Wilson et al. (1982). This was contrary to experience in California (3.C.a) where spreads of only a few kilometres seemed to be achieved in the first years (Hagen et al., 1958). The similar rapid spread of *A. ervi* attacking *Acyrthosiphon* spp. on lucerne in New South Wales (Milne, 1986b) suggests that conditions for the dispersal of parasites of lucerne aphids may be particularly favourable in Australia.

Besides the demonstration that the enemy is widespread throughout the area where the pest problem has declined, it is also necessary to show that the phenology of the enemy in relation to that of the target aphid could explain the decline. Out-of-phase cycles in aphid and enemy populations are unlikely to be able to account for any immediate decline.

Direct demonstration of the numerical impact of a natural enemy on a target aphid in the field is never easy and extremely difficult for predators that consume their prey entirely. Estimating their contribution involves long periods of direct observation (Frazer and Gilbert, 1976; see also section 9.2). With a parasitic hymenopteran, the aphids they attack remain available for observation and counting as mummified remains in which the parasite pupates. The difficulties in assessing the impact of such parasites on target aphids (i.e. the percentage parasitised before reproduction) results from the overlapping generations of the aphids, and the resultant problems of timing.

Because the parasites do not kill their hosts until long ater oviposition, the number of mummies found is not directly related to current aphid numbers, the to the number present at the time of attack. Percent parasitisation, derived from the ratio of mummy counts to the total of live aphids plus mummies, will be over- or under-estimated, unless allowance is made for the change in the live aphid population since the time of oviposition. Mummified aphid skins tend to persist after the new adult parasite has emerged, so care must be taken to count only occupied mummies. Any error in that procedure will over-estimate the rate of parasitisation. Even if only those mummies containing parasites are counted, relative changes in the development rate of parasite and aphid will cause further errors – over-estimating the percentage parasitisation with any relative slowing of parasite development. In the extreme case, diapause of immature parasites within the mummy will cause their numbers to accumulate within an aphid population and lead to a great over-estimation of parasitisation (George, 1957). With some aphid species there is a major cause of under-estimation of parasitisation: just before mummification, most parasitised aphids tend to wander from their last feeding sites, but in extreme cases, the aphids leave their host plants entirely to pupate in litter or soil (Behrendt, 1968; Dean et al., 1980; Lykouressis and Van Emden, 1983).

The difficulties in getting valid measures of the percentage of parasitisation from counts of aphids and mummies has led workers to seek alternative measures. The proportion of live aphids parasitised can be determined by dissection of a sample of aphids and recognition of any immature stage of the parasites within them. If parasite impact is taken to be the number dying before reproduction, there are still problems of timing. The younger aphid nymphs in the sample will not have been exposed to parasite attack for their whole period of immature development, so the proportion found to contain immature parasites will underestimate the parasite impact. It would seem best to dissect late-stage nymphs (i.e. fourth instar) and use the percentage parasitisation in these as an estimate of impact, e.g. as in the study of biological control of *C. juglandicola* by *T. pallidus* (3.F.b) (Van den Bosch et al., 1979). However, there is still a timing problem, because aphid nymphs attacked in the early instars often mummify before the end of the fourth instar (e.g. Roberts, 1979). These attacked aphids cannot be included with the live fourth-instar aphid sample, because the relative ages of the mummified aphids in a sample cannot be estimated. So even using fourth-instar dissection there is a potential for underestimation of the impact.

A better estimate of the numerical impact of a parasite may be obtained by collecting a sample of live aphids from the field and rearing them under carefully controlled conditions in the laboratory for a period which must be less than the duration of immature development under these conditions. The proportion of aphids that mummify during that period can then be used, with estimates of the mean period of exposure of the sampled aphids to parasite attack, to work out a *rate* of parasitisation (Hughes et al., 1982). That rate can then be extrapolated over the whole immature life of the aphid to estimate the impact of the parasite.

The problems of assessing the impact of fungal pathogens on aphids in the field will be very similar to those arising with the assessment of hymenopterous parasites. Estimates of rates of infection based on the ratio of counts of infected cadavers to those of living aphids, or based on the proportion of a sample of living aphids which subsequently die of disease when reared in the laboratory, will both have the same problems as the analogous estimates of parasitisation.

Attempts to verify directly the role of any sort of natural enemy, in the decline of target aphid populations in the field, by an experimental approach, are fraught with difficulties. Only the results of a concurrent valid *exclusion* experiment would be a direct verification. Several techniques for overcoming microclimatic differences through the use of exclusion cages have been des-

Section 11.2.1 references, p. 194

cribed (e.g. Way and Banks, 1968; Sparks et al., 1966). So the difficulties lie in excluding the natural enemy without excluding other native enemies and competitors which are natural components of the target aphid's environment. The disruption of immigration and emigration of the target aphid itself is another problem.

An alternative is to use an insecticidal check method (DeBach, 1964). In such methods, differential effects of contact insecticides on the target aphid and on the introduced natural enemy are exploited to, at least, markedly alter their relative numbers. However, the methods still have the problems associated with effects of the insecticide used on the survival of other interactive faunal components and perhaps on plant growth.

There may be a better way of demonstrating the role of a natural enemy in the decline of a target aphid population in the field. The nature of the attack by the enemy would, in some cases at least, be expected to change the composition of the target aphid population (Hodek et al., 1972), and evidence for this may be found in samples from the field. In the Australian attempt to control *T. trifolii* f. *maculata* using *T. complanatus* (1.F.a), it was reasoned that the high attack rates estimated by rearing aphid samples from the field should have been reflected in a substantial reduction of adults in the target aphid population. Re-examination of the age structures in the routine population samples of *T. trifolii* f. *maculata* taken before and after the establishment of *T. complanatus* showed this indeed to be the case (Hughes et al., 1987).

Even more reliance can be placed in the role of a natural enemy as a causal agent of any decline of a target aphid population, if the population parameters of the host and enemy can be combined with environmental information into a numerical model of their interaction to explain or (preferably) simulate the decline. A numerical comparison of the innate capacities for increase over a range of environmental conditions was used by Force and Messenger (1964b) to explain the relative success of *T. complanatus* as a parasite of *T. trifolii* f. *maculata* (3.C.a) as compared with *P. exsoletum* and *A. asychis* introduced at about the same time. Unfortunately, as was pointed out later by one of the authors (in Huffaker et al., 1976) such a comparison did not include the numerical effects of the interaction on the reproduction of the target aphid. To take this and other equally relevant factors into account, a simulation model of the population dynamics of the target aphid, incorporating the interaction with the natural enemy, may be necessary. One of the best examples of such a technique, using a well known target aphid, *A. pisum*, and its native predator in Canada, *Coccinella trifasciata* Mulsant, has been documented by Frazer and Gilbert (1976). Hughes et al. (1987) have constructed such a model for *T. trifolii* f. *maculata* and *T. complanatus* to show that their numerical interaction could explain the observed declines in pest numbers.

THE ECOLOGICAL ATTRIBUTES OF A SUCCESSFUL NATURAL ENEMY

There have been several reviews of case histories of the successful biological control of target insect pests (e.g. DeBach, 1964; Huffaker et al., 1976; Hokkanen, 1986) and there seems to be agreement about the characteristics of the natural enemies used in successful attempts. Most of them have had:
(a) A substantial degree of host specificity or preference for the target insect;
(b) Good phenological and behavioural adaptation to the target insect population;
(c) An ability to locate individuals in all parts of a target insect population and to utilise them effectively;

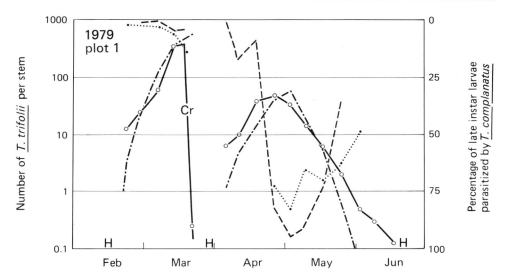

Fig. 11.2.1.2. The use of a simulation model based on the biological properties of parasite and host, to show that *Trioxys complanatus* Quilis can account for the patterns of decline in *Therioaphis trifolii* f. *maculata* (Buckton) observed in the field. The model was adjusted to simulate the aphid population growth and the percentage parasitisation in the environment of the first crop and then adjusted to simulate the patterns in the succeeding crop. H = harvest. Aphid numbers: (O—O) observed. (–·–·–) simulated. Percentage parasitisation: (·····) observed, (– – –) simulated.

(d) A capacity for increase which in the normal course of events allows them to overtake and suppress the target insect population;

(e) Some density-dependent mechanism involved in the interaction with the target insect population.

Nearly all the reviewers are convinced that it would have been difficult or impossible to predict the success of the *first* attempt to use a given natural enemy to control a particular target pest even if a great deal had been known about the population dynamics of both organisms. So far, experience seems to be the only satisfactory predictor! However, because of the truistic nature of most of the attributes given above, it should have been possible to predict failure of biological control in some instances, with only limited knowledge of both organisms, if some characteristics were the reverse of those cited above. (see also Fig. 11.2.1.2.)

The interdependence of the attributes listed above suggests that they can probably be accurately abridged. Huffaker et al. (1976) point out that the attributes are most likely to be possessed by an enemy capable of maintaining itself on low-density populations of the target insect alone. Clausen (1951) presented the empirical hypothesis that introduced natural enemies will either establish successful biological control around the point of release very quickly (in about three generations) or not at all.

AN ANALYSIS OF ATTEMPTS TO BIOLOGICALLY CONTROL TARGET APHID SPECIES

In Table 11.2.1.3 those attempts that have been adequately documented in the literature are presented, together with a synopsis of background information, practice used and the assessments in the form of subjective symbolic representation of levels, etc.

The total of attempts is surprisingly small (although up to fifty have been made to control the woolly apple aphid, *Eriosoma lanigerum* (Hausmann), in

TABLE 11.2.1.3

Synoptic summary of adequately-documented attempts to biologically control target aphids, in approximate chronological order (explanations of abbreviations on p. 191)

Target aphid and target environment	Selected natural enemy	Country of attempt	Climate code	Complexity of aphid life cycle	Stability of environment	Type of enemy	Specificity of enemy	Climatic match	Sophistication of practice	Establishment	Reduction of target aphid problem	Verification of role of enemy
(A) *Eriosoma lanigerum* (Hausmann) Woolly apple aphid	(a) *Aphelinus mali* Haldemann Aphelinid wasp	Wor[a]	WTM[b]	++[c]	+++	Para[d]	+++	++	+	+++	++	+
Pyrus malus, P. communis Apple, pear Orchard trees	(b) *A. mali*	Wor	CTM	++	+++	Para	+++	-	++	-+	+	+
	(c) *A. mali*	Wor	WTM	+++	+++	Para	+++	+++	++	+++	(+)	++
(B) *Adelges piceae* (Ratzeburg) Balsam woolly aphid	(a) *Leucopis* spp. Chamaemyiid fly	Can	CCL	++	+++	Pred	++	++	+	+++	-	
Abies spp. True fir trees Forest trees	(b) *Cremifannia nigrocellulata* Czerny	Can	CCL	++	+++	Pred	++	++	+++	+++	-	
	(c) *Aphidoletes thompsoni* Mohn	Can	CCL	++	+++	Pred	-	++	+++	++	-	
	(d) *Pullus impexus* (Mulsant) Coccinellid beetle	Can	CCL	++	+++	Pred	+	++	+++	+-	-	
	(e) *Aphidecta obliterata* (Linnaeus) Coccinellid beetle	Can	CCL	++	+++	Pred	++	+	+++	-+	-	
	(f) *Laricobius erichsonii* Rosenhauer Derodontid beetle	Can	CCL	++	+++	Pred	++	++	+++	+++	-	

Target aphid / control agent	Country	Climate			Type								
(C) Therioaphis trifolii f. maculata (Buckton) Spotted alfalfa aphid; *Medicago sativa*, Alfalfa, lucerne, Perennial crop													
(a) *Trioxys complanatus* Quilis — Aphidiid wasp	USA	MDN	+	++	Para	+++	+++	+++	+++	+++	+	+
(b) *T. complanatus*	Aus	MDN	+	++	Para	+++	+++	+++	+++	+++	++	—
(c) *Praon exsoletum* (Nees) — Aphidiid wasp	USA	MDN	+	++	Para	+++	+++	+++	+	+++	++	—
(d) *Aphelinus asychis* Walker — Aphelinid wasp	USA	MDN	+	++	Para	—	+	+++	+	+	++	—
(e) *Zoophthora radicans* Batko — Entomophthoran fungus	Aus	WTM	+	++	Path	+++	+++	+++	++	+++	++	—
(D) Acyrthosiphon pisum (Harris) Pea aphid													
(a) *Aphidius smithi* Sharma and Rao — Aphidiid wasp	USA	WTM	+	—	Para	+++	+	—	—	—	+	
(b) *A. smithi*	USA	MDN	+	++	Para	+++	++	++	+++	+++	—(+)	+
(c) *A. smithi*; *Pisum sativum, M. sativa*, Peas, alfalfa, Annual crop, perennial crop	USA	WTM	+	—	Para	+++	++	++ ?	IR[e]	+	—	
(d) *Aphidius ervi ervi* Haliday — Aphidiid wasp	USA	WTM	+	—	Para	+	+	+	+++	+++	—	
(e) *Aphidius eadyi* Starý — Aphidiid wasp	NZ	CTM	+	++	Para	+++	+++	+++	+++	+++	++	++
(E) Cavariella aegopodii (Scopoli) Carrot aphid													
(a) *Aphidius salicis* Haliday — Aphidiid wasp; *Daucus carota* etc., Carrot etc., Annual crops, weeds; *Salix* spp., Willows	Aus	WTM	+++	—+	Para	+++	++	—	+++	—	+++	
(F) Chromaphis juglandicola (Kaltenbach) Walnut aphid													
(a) *Trioxys pallidus* Haliday — Aphidiid wasp; *Juglans* spp., Walnut, Plantation tree	USA	MDN	++	+++	Para	+++	—	+++	+++	+++	—	—
(b) *Trioxys pallidus* Haliday — Aphidiid wasp	USA	MDN	++	+++	Para	+++	+++	+	+++	+++	+++	++

Section 11.2.1 references, p. 194

TABLE 11.2.1.3 (*Continued*)

Target aphid and target environment	Selected natural enemy	Country of attempt	Climate code	Complexity of aphid life cycle	Stability of environment	Type of enemy	Specificity of enemy	Climatic match	Sophistication of practice	Establishment	Reduction of target aphid problem	Verification of role of enemy
(G) *Myzus persicae* (Sulzer) *Macrosiphum euphorbiae* (Thomas) *Aulacorthum solani* (Kaltenbach) Potato aphids *Solanum tuberosum* Potato Annual crop	(a) *Coccinella septempunctata* Linnaeus Coccinellid beetle	USA	WTM	+	–	Pred	–	+	+++	IR[e]	–	
(H) *Acyrthosiphon kondoi* (Shinji) Blue-green aphid *Medicago sativa* Lucerne Perennial crop	(a) *Aphidius ervi* Haliday Aphidiid wasp	Aus	WTM	+	++	Para	+	+	+++	+++	++	
(I) *Schizaphis graminum* (Rondani) Greenbug *Sorghum* spp. Sorghum Annual crop	(a) *Aphelinus asychis* Walker Aphelinid wasp	USA	STL	+	–	Para	–	–	++	?	–	

(J) *Eucallipterus tiliae* (Linnaeus) Linden (lime) aphid *Tilia* spp. Linden, lime Street tree										
(a) *Trioxys curvicaudus* Mackauer — Aphidiid wasp	USA	MDN	++	+++	Para	+++	++	+	+++	++
(K) *Tinocallis platani* (Kaltenbach) Elm aphid *Ulmus* spp. Elm Street trees										
(a) *Trioxys tenuicaudus* Starý — Aphidiid wasp	USA	MDN	++	+++	Para	+++	++	+	+++	++

(A) Greathead and Pschorn-Walker (1976), Chacko (1967), Massee (1944), Bodenheimer (1947), Evenhuis (1958).

(B) Smith and Coppel (1957), McGugan and Coppel (1962), Clark et al. (1971), Clark and Brown (1958).

(C) Van den Bosch (1957), Van den Bosch et al. (1959a,b, 1964), Finney (1960), Finney et al. (1960), Schlinger and Hall (1959, 1961), Force and Messenger (1964a,b, 1965, 1968), Flanders (1959), Messenger (1968), Wilson et al. (1982), Woolcock (1979), Milner (1984), Kenneth and Olmert (1975), Milner et al. (1982).

(D) Angalet and Coles (1966), Hagen and Schlinger (1960), Van den Bosch et al. (1966, 1967), Starý (1966, 1971), Mackauer (1973, 1983), Campbell and Mackauer (1975), Halfhill and Featherstone (1973), Angalet and Fuester (1977), Cameron et al. (1979, 1981), Bates and Miln (1982).

(E) Stubbs (1966), Mackauer and Stary (1967), Hughes et al. (1964, 1965), Carver (this volume, section 11.2), E. Martyn, G. O'Loughlin (unpublished data).

(F) Schlinger et al. (1960), Sluss (1967), Van den Bosch et al. (1962, 1970, 1979), Messenger and Van den Bosch (1971), Frazer and Van den Bosch (1973).

(G) Shands et al. (1970, 1972), Shands and Simpson (1972).

(H) Gonzales et al. (1979), Milne (1982, 1986a,b).

(I) Jackson and Eikenbary (1971), Jackson et al. (1971), Starks et al. (1976).

(J) Olkowski et al. (1982b).

(K) Olkowski et al. (1982a).

[a] Aus = Australia, Can = Canada, NZ = New Zealand, USA = United States of America, Wor = many countries throughout the world.

[b] CCl = cold continental, CTM = cold temperate, WTM = warm temperate, MDN = mediterranean, STL = subtropical.

[c] See p. 192 for explanation of + and −.

[d] Para = parasitic Hymenoptera, Path = pathogenic fungus, Pred = predatory insect.

[e] IR = inundative release.

Section 11.2.1 references, p. 194

various countries throughout the world). The listed countries in which attempts were made are also few. Both deficiencies reflect the criteria of adequate
documentation used in this review. For inclusion, some information at least
must be available on the five main steps of the biological control project (Table
11.2.1.2), including positive or negative evidence regarding both establishment
and decline of the pest problem.

Analysing Table 11.2.1.3, the limited climatic zones represented reflect both
the few countries involved and the conditions for aphid occurrence, with their
avoidance of environmental extremes. The complexity of life cycles ranged
from anholocycly (+) to simple holocycly on one host (+ +) and complex
holocycly, involving a primary host or a significantly different microhabitat
(+ + +). The stability of environment ranged from unstable, short-term crops
(−), via primary hosts available for *part* of the year (+), to perennial crops
disrupted by harvesting etc. (+ +), and long-lived host plants, particularly
trees (+ + +). The specificity of the enemy ranged from polyphagy (−), to
polyphagy within restricted environments (+), to preferred in target environment (+ +) and more or less specific (+ + +). The range of climatic match is
self-explanatory. Sophistication of practice is, of course, very subjective and
symbols indicating it to be less than the satisfactory level (+ + +) are based
only on increasingly obvious shortcomings. Establishment ranged from rapid
and persistent (+ + +) to slow but eventual (+ +), to initial with subsequent
decline (+) to initial then failure (+ −), apparent failure but eventual (− +)
and total failure (−). In two cases the enemy was already present in the
country of attempt (?). Declines in the pest problem ranged from massive
(+ + +), proportionally down to some general reduction (+), and none (−).
Some decline under special conditions, e.g. crop manipulation or in some
seasons is denoted by (+). Verification of role of the enemy would be maximal
(+ + +), if laboratory, experimental, field and simulation studies carried out
all indicated a positive role. Laboratory and field studies only (+ +), or experimental studies alone (+), indicate lesser degrees of certainty of a positive role
in the observed decline. Any indication that the enemy did too little, or did not
have the capacity to cause a decline is denoted by (−). No symbol indicates that
no attempt at verification has yet been reported.

Substantial success seems to be claimable for: *A mali* against *E. lanigerum*
in some circumstances; *T. complanatus* and *Z. radicans* against *T. trifolii* f.
maculata; *Aphidius eadyi* Stary against *A. pisum*; *A. ervi* against *A. kondoi*; *T.
curvicaudus* against *E. tiliae*; *Trioxys tenuicaudus* Stary against *Tinocallis
platani* (Kaltenbach), and particularly *T. pallidus* against *C. juglandicola*. Too
many doubts remain unresolved in the case of *A. salicis* against *C. aegopodii* to
judge whether other direct factors are involved in the decline of the pest
problem.

CONCLUSIONS

The eight "successes" form a substantial proportion of the adequately
documented attempts, say about 30%. However, any relaxation of the criteria
for inclusion of attempts in Table 11.2.1.3 would bring in a much larger proportion of failures – the successes have been better documented!

The first conclusion from this review is that the attributes of successfully
natural enemies of aphid pests seem to be closely similar to those of the
successful enemies of all insects (Huffaker et al., 1976). Apart from cases where
the complex life cycle of a target aphid may result in some parts of the population being inaccessible to attack (see below), there seem to be no special
considerations for aphid pests as biological control targets.

With one possible exception, i.e. *A. ervi* attacking *A. kondoi* (3.H.a), the agents for which success has been claimed are those in which a strong preference or requirement makes them more or less specific to the target aphid. This must be due to necessary behavioural adaptations to the target aphid, to the biotype and to phenological adaptations to the climate of the target environment. There appear to be cases where climatic matching of enemy and target seemed less important, e.g. *A. mali* attacking *E. lanigerum* (3.A), and others where it seemed crucial, e.g. *T. pallidus* attacking *C. juglandicola* (3.F.a,b). More care in climatic matching may be needed where extreme conditions affecting the active stages of insects are a regular feature of the target environment. The unavailability of the underground stages of *E. lanigerum* to attack by *A. mali* (3.A.c) gives cause to consider whether the complex life cycles of some target aphids, or the host plant alternation in others, generally provide any immunity from attack for part of the population. The rate of reproduction of an aphid pest is usually very high, especially at the start of an infestation. A successful enemy must attack enough of the aphids at least to stabilise, but often also first to reduce, the target population. Other things being equal, without getting involved in the mathematical complexities, a successful enemy will consistently attack and kill a majority of the aphids of the target population. Nearly all the enemies claimed to be successful showed attack rates (parasitisation level) estimated at 50% or more. Remembering that even the dissection of fourth-instar aphids tends to underestimate parasitisation, the somewhat lower percentage estimated for *T. curvicaudus* attacking *E. tiliae* (3.J.a) by dissection of third instars (Olkowski et al., 1982b) probably represent attacks on the majority of aphids. Very high levels of attack, (above 90%), observed in the field when *T. complanatus* attacks *T. trifolii* f. *maculata* (1.F.a) (Lehane, 1982), indicate the potential of that parasite for control, in contrast to the moderate levels (around 30%), observed when *D. rapae* attacks *B. brassicae* (1.B.a) (Hughes, 1963). Only by greatly increasing the initial ratio of parasites to aphids and getting the timing right was it possible to simulate a control situation for *B. brassicae* using *D. rapae* (Gilbert and Hughes, 1971). Very high enemy/target aphid ratios may occur naturally if the enemy has a phenology or behaviour which cause it to be aggregated in regions of relatively low aphid density. Manipulation of the ratio is sometimes attempted, as with *Coccinella septempunctata* Linnaeus attacking potato aphids (3.G.a) (Shands and Simpson, 1972) and with *A. smithi* attacking *A. pisum* (3.D.c) (Halfhill and Featherstone, 1973). If a very high background mortality was a feature of the target aphid population dynamics, a moderate rate of attack by an introduced natural enemy could theoretically suppress the population. Of more practical importance are the situations where the rate of reproduction of the target aphid declines naturally, allowing the attacks of the natural enemy to overtake and suppress the target population. This is the explanation of most cases of suppression of aphids *after* peak populations have been reached. With *B. brassicae*, the parasite *D. rapae* was able to suppress populations only after a density-induced decline in the reproductive rate had occurred (Hughes, 1963). Such declines in the reproductive rate may also be induced by independent changes in the host-plant condition which, if seasonal, may regularly allow the natural enemy to suppress the target aphid. This may be the explanation for the apparent success of *A. ervi* attacking *A. kondoi*, an aphid which is greatly affected by the maturity of the host plant, *M. sativa* (3.H.a), compared with the failure of the same parasite attacking *A. pisum*, an aphid not so affected (3.D.d). Such an explanation has been proposed already for cases where *A. mali* is effective in controlling summer populations of *E. lanigerum* on apple trees (3.A.a) (Evenhuis, 1958). Analogous seasonal changes in the reproductive rate of aphids infesting trees may account for the undue preponderance of tree aphids among the claimed successes for biological control. To maintain biological control, a natural enemy needs to have more than just the potential to suppress the target aphid. Mechanisms to make enemies less effective at low target densities and more effective at high target densities are needed. These may occur as a result of functional and numerical responses of the enemy, but equivalent effects may be induced indirectly by other factors. For example, hyperparasites attacking relatively high natural enemy populations after suppression of target aphid numbers may reduce the enemy/aphid ratio to a point where the aphid can increase again. No study of a claimed successful attempt to biologically control a target aphid has yet documented the action of any such stabilising mechanisms in the field.

In summary, the best predictor of success for biological control of aphids is previous successful experience with the natural enemy in a similar environment. Otherwise, the best chances appear to be with a climatically adapted, host-specific natural enemy that is known to attack a high proportion of the target aphid in the field in its native or expanded range. The parasitic Hymenoptera generally seem to be the best type of natural enemy for aphids – probably because of the high level of specificity many of them exhibit.

Target aphids in which a reduction in reproductive rate is induced by factors other than crowding seem to be the most likely candidates for biological control. It certainly seems worth giving consideration to the appropriateness of attempting to biologically control any particular target aphid. While it seems impossible to predict success, prediction of failure of a particular enemy may be much more certain, and to embark on expensive mass-rearing and/or release programmes in such cases could lead to costly failures.

If the conditions for an attempt are otherwise auspicious, it is probably not worth persisting with a stock of a natural enemy that fails to establish in its first season. However, if an apparently good natural enemy fails, it could be worth seeking the same enemy from a more appropriate source, e.g. one with a better climatic match, or using an F_1 generation from field-collected material rather than material kept at a long time in insectary culture.

With all biological control attempts it is very important to record them carefully and fully. Even more important is to preserve specimens of the imported enemies *and* examples of the progeny released. Documenting any decline in the pest problem is also important, either by monitoring target aphid populations, or by getting indirect indications from growers or extension workers. Where an obvious decline occurs, the more that can be done to verify the role of the enemy the better. With such information it will be possible to update, from time to time, the evaluation of the biological control approach to aphid problems.

REFERENCES

Aeschlimann, J-P., 1981. Occurrence and natural enemies of *Therioaphis trifolii* Monell and *Acyrthosiphon pisum* Harris (Homoptera: Aphididae) on lucerne in the Mediterranean region. Oecologia Applicata, 2: 3–11.

Angalet, G.W. and Coles, L.W., 1966. The establishment of *Aphidius smithi* in the Eastern United States. Journal of Economic Entomology, 59: 769–770.

Angalet, G.W. and Fuester, R., 1977. The *Aphidius* parasites of the pea aphid *Acyrthosiphon pisum* in the eastern half of the United States. Annals of the Entomological Society of America, 70: 87–96.

Bates, L.H. and Miln, A.J., 1982. Parasites and predators of lucerne aphids at Flock House, Bulls. Proceedings 35th New Zealand Weed and Pest Control Conference, Palmerton North, 1982, pp. 123–126.

Behrendt, K., 1968. Das Abwandern parasitierter Aphiden von ihren Wirtspflanzen und eine Methode zu ihren Erfassung. Beiträge zur Entomologie, 18: 293–298.

Berg, G. and Ridland, P., 1979. Field rearing and distribution of a spotted alfalfa aphid parasite in Victoria. Australian Applied Entomological Research Conference, Queensland Agricultural College, Lawes, June 1979. Canberra, 1979, pp. (5)2–3.

Berg, G., Ridland, P. and Blackstock, J., 1979. Field rearing of *Trioxys complanatus* in Victoria. In: Lucerne Aphid Workshop, Tamworth, 1978. New South Wales Department of Agriculture, Sydney, pp. 75–76.

Bodenheimer, F.S., 1947. Studies of the physical ecology of the woolly aphid (*Eriosoma lanigerum*) and its parasite, *Aphelinus mali*, in Palestine. Palestine Agricultural Research Station Rehovot, Bulletin No. 43 20 pp.

Brieze-Stegeman, R., 1979. Parasite release program – Tasmania. Lucerne Aphid Workshop, Tamworth, 1978, New South Wales Department of Agriculture, Sydney, pp. 90–91.

Cameron, P.J., Thomas, W.P. and Hill, R.L., 1979. Introduction of lucerne aphid parasites and a preliminary evaluation of the natural enemies of *Acyrthosiphon* spp. (Hemiptera: Aphididae) in New Zealand. In: T.K. Crosby and R.P. Pottinger (Editors), Proceedings of the 2nd Australasian Conference on Grassland Invertebrate Ecology. Government Printer, Wellington, pp. 219–223.

Cameron, P.J., Walker, G.P. and Allan, D.J., 1981. Establishment and dispersal of the introduced parasite *Aphidius eadyi* (Hymenoptera: Aphidiidae) in the Northern Island of New Zealand, and its initial effect on the pea aphid. New Zealand Journal of Zoology, 8: 105–112.

Campbell, A. and Mackauer, M., 1975. Source climatic effects on the spread and abundance of two parasites of the pea aphid in British Columbia (Hymenoptera: Aphidiidae–Homoptera:Aphidiidae). Zeitscrhift für Angewandte Entomologie, 74: 47–55.

Carver, M., 1984. The potential host-ranges in Australia of some imported aphid parasites (Hym.: Ichneumonoidea: Aphidiidae). Entomophaga, 29: 351–359.

Carver, M. and Starý, P., A preliminary review of the Aphidiidae (Hymenoptera: Ichneumonoidea) of Australia and New Zealand. Journal of the Australian Entomological Society, 13: 235–400.

Carver, M. and Woolcock, L.T., 1986. The introduction into Australia of biological control agents of *Hyperomyzus lactucae* (L.) (Homoptera: Aphididae). Journal of the Australian Entomological Society, 25: 65–69.

Chacko, M.J., 1967. Establishment of *Aphelinus mali* (Haldeman) at Shillong, Assam. Current Science, 36: 298–299.

Clark, R.C. and Brown, N.R., 1958. Studies of predators of balsam woolly aphid, *Adelges piceae* (Ratz.) (Homptera; Adelgidae). V. *Laricobius erichsonii* Rosen. (Coleoptera, Derodontidae), an introduced predator in Eastern Canada. Canadian Entomologist, 90: 657–672.

Clark, R.C., Greenbank, D.O., Bryant, D.G. and Harris, J.W.E., 1971. 46. *Adelges piceae* (Ratz.), balsam woolly aphid (Homoptera; Adelgidae). In: Biological Control Programmes Against Insects and Weeds in Canada 1959–1968. Commonwealth Institute of Biological Control, Trinidad, Technical Communication No. 4, pp. 113–127.

Clausen, C.P., 1951. The time factor in biological control. Journal of Economic Entomology, 44: 109.

Cordingly, C.L., Swincer, D.E. and Walden, K., 1979. Blue-green aphid parasite release program. Lucerne Aphid Workshop, Tamworth, 1978. New South Wales Department of Agriculture, Sydney, pp. 85–86.

Dean, G., Dewar, A.M. and Powell, W., 1980. Integrated control of cereal aphids. International Organisation for Biological Control, West Palaearctic Region Section Bulletin 1980/III/4, pp. 30–47.

DeBach, P., 1964. Biological Control of Insect Pests and Weeds. Chapman and Hall, London, 844 pp.

Evenhuis, H.H., 1958. Ecological investigations of the woolly aphid. *Eriosoma lanigerum* (Hausm.) and its parasite *Aphelinus mali* (Hald.) in the Netherlands. Tijdschrift over Plantenziekten, 64: 1–103.

Finney, G.L., 1960. A ventilation unit for aerating insectary sleeve cages. Journal of Economic Entomology, 53: 959–960.

Finney, G.L., Puttler, B. and Dawson, L., 1960. Rearing of three spotted alfalfa aphid hymenopterous parasites for mass release. Journal of Economic Entomology, 53: 656–659.

Flanders, S.E., 1959. The employment of exotic entomophagous insects in pest control. Journal of Economic Entomology, 52: 71–75.

Force, D.C. and Messenger, P.S., 1964a. Duration of development, generation time and longevity of three hymenoptera parasites of *Therioaphis maculata*, reared at various constant temperatures. Annals of the Entomological Society of America, 57: 405–413.

Force, D.C. and Messenger, P.S., 1964b. Fecundity, reproductive rates and innate capacity for increase of three parasites of *Therioaphis maculata* (Buckton). Ecology, 45: 706–715.

Force, D.C. and Messenger, P.S., 1965. Laboratory studies on competition among three parasites of the spotted alfalfa aphid, *Therioaphis maculata* (Buckton). Ecology, 46: 835–839.

Force, D.C. and Messenger, P.S., 1968. The use of laboratory studies of three hymenopterous parasites to evaluate their field potential. Journal of Economic Entomology, 61: 1375–1378.

Frazer, B.D. and Gilbert, N., 1976. Coccinellids and aphids: A quantitative study of the impact of adult ladybirds (Coleoptera: Coccinellidae) preying on field populations of pea aphids (Homoptera: Aphididae). Journal of the Entomological Society of British Columbia, 73: 33–56.

Frazer, B.D. and Van den Bosch, R., 1973. Biological control of the walnut aphid in California: The interrelationship of the aphid and its parasite. Environmental Entomology, 2: 561–568.

George, K.S., 1957. Preliminary investigations on the biology and ecology of the parasites and predators of *Brevicoryne brassicae* (L.). Bulletin of Entomological Research, 48: 618–629.

Gilbert, N. and Hughes, R.D., 1971. A model of the an aphid population – three adventures. Journal of Animal Ecology, 40: 525–534.

Gonzalez, D., Miyazaki, M., White, W., Takada, H., Dickson, R.C. and Hall, J.C., 1979. Geographical distribution of *Acyrthosiphon kondoi* Shinji (Homoptera: Aphididae) and some of its parasites and hyperparasites in Japan. Kontyû, 47: 1–7.

Greathead, D.J. and Pschorn-Walcher, P., 1976. Apple Woolly Aphis – *Eriosoma lanigerum* (Hausm.) (Aphididae, Hem.). In: D.J. Greathead (Editor), A Review of Biological Control in Western and Southern Europe. Commonwealth Institute of Biological Control, Trinidad, Technical Communication No. 7, pp. 4–13.

Hagen, K.S. and Schlinger, E.I., 1960. Imported Indian parasite of the pea aphid established in California. California Agriculture, 14(9): 5–6.

Hagen, K.S., Holloway, F.E., Skinner, F.E. and Finney, G.L., 1958. Aphid parasites established: Natural enemies of spotted alfalfa aphid brought from Middle East expected to be established throughout the State in 1958. California Agriculture, 12(2): 3.

Halfhill, J.E. and Featherstone, P.E., 1973. Inundative releases of *Aphidius smithi* against *Acyrthosipyon pisum*. Environmental Entomology, 2: 469–472.

Hodek, I., Hagen, K.S. and Van Emden, H.F., 1972. Methods for studying the effectiveness of natural enemies. In: H.F. van Emden (Editor) Aphid Technology. Academic Press, London and New York, pp. 147–188.

Hokkanen, H.M.T., 1986. Success in classical biological control CRC Critical Reviews in Plant Sciences 3: 35–72.

Huffaker, C.B., Luck, R.F. and Messenger, P.S., 1976. Proceedings of the XVth International Congress of Entomology. Washington 1976, pp. 560–586.

Hughes, R.D., 1963. Population dynamics of the cabbage aphid. Journal of Animal Ecology, 32: 393–424.

Hughes, R.D., Casimir, M., O'Loughlin, G.T. and Martyn, E.J., 1964. A survey of aphids flying over eastern Australia in 1961. Australian Journal of Zoology, 12: 174–200.

Hughes, R.D., Carver, M., Casimir, M., O'Loughlin, G.T. and Martyn, E.J., 1965. A comparison of the numbers and distribution of aphid species flying over eastern Australia in two successive years. Australian Journal of Zoology, 13: 823–839.

Hughes, R.D., Morton, R. and Roberts, J.A., 1982. Assessment of parasitisation of aphids. Proceedings 3rd Australasian Conference on Grassland Invertebrate Ecology. Adelaide 1981. S.A. Government Printer, Adelaide, pp. 183–189.

Hughes, R.D., Woolcock, L.T., Roberts, J.A. and Hughes, M.A., 1987. The biological control of the spotted alfalfa aphid, *Therioaphis trifolii* (Monell) f. *maculata*, on lucerne crops in Australia, by the introduced parasitic hymenopteran, *Trioxys complanatus* (Quilis). Journal of Applied Ecology, 1987 (in press).

Jackson, H.B. and Eikenbary, R.D., 1971. Bionomics of *Aphelinus asychis* (Hymenoptera: Eulophidae) an introduced parasite of the sorghum greenbug. Annals of the Entomological Society of America, 64: 81–85.

Jackson, H.B., Rogers, C.E. and Eikenbary, R.D., 1971. Colonisation and release of *Aphelinus asychis*, an imported parasite of the greenbug. Journal of Entomology, 64: 1435–1438.

Jenkins, C.F.H., 1948. Biological control in Western Australia. Journal of the Royal Society of Western Australia, 32: 1–17.

Kain, W.M., Atkinson, D.S., Marsden, R.S., Oliver, M.J. and Holland, T.V., 1977. Blue-green lucerne damage in lucerne crops with southern North Island. In: Proceedings of the 30th New Zealand Weed and Pest Control Conference. New Zealand Weed and Pest Control Society, 1977, pp. 177–181.

Kenneth, R. and Olmert, I., 1975. Entomopathogenic fungi and their insect pests in Israel: Additions. Israel Journal of Entomology, 10: 105–109.

Kircher, H.W. and Lieberman, F.V., 1967. Toxicity of tobacco smoke to the spotted alfalfa aphid *Therioaphis maculata* (Buckton). Nature, London, 215: 97–98.

Lehane, L., 1982. Biological control of lucerne aphids. Rural Research (CSIRO, Australia), 114: 4–10.

Lodge, G.M., 1980. Effects of spotted alfalfa aphids and blue-green aphids on the dry matter production of some lucerne varieties. In: Pathways to Productivity. Proceedings of Australian Agronomy Conference, 1980, p. 261.

Lykouressis, D.P. and Van Emden, H.F., 1983. Movement away from the feeding site of the aphid *Sitobion avanae* (F.) (Hemipteral Aphididae) when parasitised by *Aphelinus abdominalis* (Dalman) (Hymenoptera: Aphenlinidae). Entomologia Hellenica, 1: 59–63.

Mackauer, M., 1973. Host selection and host suitability in *Aphidius smithi*. (Hymenoptera: Aphidiidae). In: A.D. Lowe (Editor), Perspectives in Aphid Biology. Entomological Society of New Zealand, Bulletin No. 2, pp. 20–29.

Mackauer, M., 1983. Quantitative assessment of *Aphidius smithi* (Hymenoptera: Aphidiidae): fecundity, intrinsic rate of increase, and functional response. Canadian Entomologist, 115: 399–415.

Mackauer, M. and Starý, P., 1967. Hym. Ichneumonoidea. World Aphidiidae. In: V.L. Delucchi and G. Remaudière: Index of Entomophagous Insects. Le Francois, Paris, 195 pp.

Massee, A.M., 1944. Further notes on the woolly aphid parasite (*Aphelinus mali* Hald.). East Malling Research Station Annual Report, 1943, pp. 65–67.

McGugan, B.M. and Coppel, H.C., 1962. Part II – Biological control of forest insects 1910–1958. In: A Review of Biological Control Attempts Against Insects and Weeds in Canada. Commonwealth Institute of Biological Control, Trinidad, Technical Communication No. 2, pp. 43–51.

Messenger, P.S., 1968. Bioclimatic studies of the aphid parasite *Praon exsoletum*. 1. Effects of temperature on the functional response of females to varying host densities. Canadian Entomologist, 100: 728–741.

Messenger, P.S. and Van den Bosch, R., 1971. Introduced biological control agents. In: C.B. Huffaker (Editor), Biological Control. Plenum Press, New York, pp. 83–84.

Milne, W.M., 1982. Imported parasites help control lucerne aphids. Agricultural Gazette, New South Wales, 93: 15–17.

Milne, W.M., 1986a. The release and establishment of *Aphidius ervi* Haliday (Hymenoptera: Ichneumonoidea) in lucerne aphids in eastern Australia. Journal of the Australian Entomological Society, 25: 123–130.

Milne, W.M., 1986b. The establishment and dispersal of *Aphidius ervi* in blue-green aphid populations of lucerne in New South Wales, Australia. Ind: Hodex, I. (Editor), Ecology of Aphidophaga 1986. Academia, Prague and Dr W. Junk, Dordrecht, pp. 459–464.

Milner, R.J., 1984. Pathogen importation for biological control – risks and benefits. In: A.J. Gibbs and H.R.C. Meischke (Editors), Pests and Parasites as Migrate – An Australian Perspective 1985. Australian Academy of Science, Canberra, p. 115–121.

Milner, R.J., Soper, R.S. and Lutton, G.G., 1982. Field release of an Israeli strain of the fungus *Zoophthora radicans* (Brefeld) Batko for biological control of *Therioaphis trifolii* (Monell) f. *maculata*. Journal of the Australian Entomological Society, 21: 113–119.

Milner, R.J., Lutton, G.G. and Bourne, J., 1984. A laboratory study of the interactions between aphids, fungal pathogens, and parasites. In: P. Bailey and D. Swincer (Editors), Proceedings of the Fourth Applied Entomology Research Conference, September 1984. South Australian Department of Agriculture, Adelaid, pp. 375–381.

Olkowski, W., Pinnock, C., Toney, W., Mosher, G., Neasbit, W., Van den Bosch, R. and Olkowski, H., 1974. An integrated insect control program for street trees. California Agriculture, 28(1): 3–4.

Olkowski, W., Olkowski, H., Van den Bosch, R., Hom, R., Zuparko, R. and Klitz, W., 1982a. The parasitoid *Trioxys tenuicaudatus* Starý (Hymenoptera: Aphidiidae) established on the elm aphid *Tinocallis platani* Kaletenbach (Homoptera: Aphididae) in Berkeley, California. Pan-pacific Entomologist, 58: 59–63.

Olkowski, W., Olkowski, H. and Van den Bosch, R., 1982b. Linden aphid parasite establishment. Environmental Entomology, 11: 1023–1025.

Roberts, J.A., 1979. Studies on the biology of *Trioxys complanatus*. In: Lucerne Aphid Workshop, Tamworth, 1978. New South Wales Department of Agriculture, Sydney, pp. 99–101.

Schlinger, E.I. and Hall, J.C., 1959. A synopsis of the biologies of three imported parasites of the spotted alfalfa aphid. Journal of Economic Entomology, 52: 154–157.

Schlinger, E.I. and Hall, J.C., 1961. The biology, behaviour and morphology of *Trioxys (Trioxys) utilis*, an internal parasite of the spotted alfalfa aphid, *Therioaphis maculata* (Hymenoptera, Braconidae, Aphidiinae). Annals of the Entomological Society of America, 54: 34–45.

Schlinger, E.I., Hagen, K.S. and Van den Bosch, R., 1960. Imported French parasite of walnut aphid established in California. California Agriculture 14(11): 3–4.

Shands, W.A. and Simpson, G.W., 1972. Insect predators for controlling aphids on potatoes. 7. A pilot test of spraying eggs of predators on potatoes in plots separated by bare fallow land. Journal of Economic Entomology, 65: 1383–1387.

Shands, W.A., Holmes, R.L. and Simpson, G.W., 1970. Improved laboratory production of eggs of *Coccinella septempunctata*. Journal of Economic Entomology, 63: 315–317.

Shands, W.A., Gordon, C.C. and Simpson, G.W., 1972. Insect predators for controlling aphids on potatoes. 6. Development of a spray technique for applying eggs in the field. Journal of Economic Entomology, 65: 1099–1103.

Sluss, R.R., 1967. Population dynamics of the walnut aphid, *Chromaphis juglandicola* (Kalt.) in northern California. Ecology, 48: 41–58.

Smith, B.C. and Coppel, H.C., 1957. Releases in North America and reviews of bionomics in Europe of insect predators of the balsam woolly aphid, *Adelges piceae* (Ratz.) (Homoptera, Adelgidae). Canadian Entomologist, 89: 410–420.

Snowball, G.J. and Lukins, R.G., 1979. Rearing and liberation of parasites of blue-green aphid (BGA). Lucerne Aphid Workship, Tamworth, 1978. New South Wales Department of Agriculture, Sydney pp. 87–89.

Sparks, A.N., Chiang, H.C., Burkhardt, C.C., Fairchild, M.L. and Weckman, G.T., 1966. Evaluation of the influence of predation on corn borer populations. Journal of Economic Entomology, 59: 104–107.

Starks, K.J., Burton, R.L., Teetes, G.L. and Wood, E.A., 1976. Release of parasitoids to control greenbugs on sorghum. ARS-S-91 Agricultural Research Service, United State Department of Agriculture, 12 pp.

Starý, P., 1966. Aphid Parasites of Czechoslovakia. Academia, Prague and Dr W. Junk, The Hague, 242 pp.

Starý, P., 1968. The creation of artificial foci of parasites – a new method of aphid parasite release. Acta Entomologica Bohemoslovaca, 65: 76–77.

Starý, P., 1971. Alternative hosts of *Aphidius smithi*, an introduced parasite of the pea aphid in Central Europe (Hom. Aphidiidae; Hym. Aphidiidae). Annals of the Entomological Society of France (N.S.), 8: 351–355.

Stubbs, L.L., 1966. Biological control of *Cavariella aegopodii* Scopoli, by an introduced parasite, *Aphidius salicis* Haliday. Proceedings Australian Plant Pathology Conference, Toowoomba, Queensland, 1966.

Van den Bosch, R., 1957. The spotted alfalfa aphid and its parasites in the Mediterranean region, Middle East and East Africa. Journal of Economic Entomology, 50: 352–356.

Van den Bosch, R., Schlinger, E.I., Dietrick, E.J., Hagen, K.S. and Hollaway, J.K., 1959a. The colonisation and establishment of imported parasites of the spotted alfalfa aphid in California. Journal of Economic Entomology, 52: 136–141.

Van den Bosch, R., Schlinger, E.I., Dietrick, E.J. and Hall, I.M., 1958b. The role of imported parasites in the biological control of the spotted alfalfa aphid in southern California in 1957. Journal of Economic Entomology, 52: 142–154.

Van den Bosch, R., Schlinger, E.I. and Hagen, K.S., 1962. Initial field observations in California on *Trioxys pallidus* (Haliday) a recently introduced parasite of the walnut aphid. Journal of Economic Entomology, 55: 857–862.

Van den Bosch, R., Schlinger, E.I., Dietrick, E.J., Hall, J.C. and Puttler, B., 1964. Studies on succession, distribution and phenology of imported parasites of *Therioaphis trifolii* (Monell) in southern California. Ecology, 45: 602–621.

Van den Bosch, R., Schlinger, E.I., Lagace, C.F. and Hall, J.C., 1966. Parasitisation of *Acyrthosiphon pisum* by *Aphidius smithi*, a density dependent process in nature (Homoptera: Aphididae) (Hymenoptera: Aphidiidae). Ecology, 47: 1049–1055.

Van den Bosch, R., Lagace, C.F. and Stern, V.F., 1967. The interrelationship of the aphid, *Acyrthosiphon pisum*, and its parasite, *Aphidius smithi*, in a stable environment. Ecology, 48: 993–1000.

Van den Bosch, R., Frazer, B.D., Davies, C.S., Messenger, P.S. and Hom, R., 1970. An effective walnut aphid parasite from Iran. California Agriculture, 24(11): 8–10.

Van den Bosch, R., Hom, R., Matteson, P., Frazer, B.D., Messenger, P.S. and Davis, C.S., 1979. Biological control of the walnut aphid in California: Impact of the parasite, *Trioxys pallidus*. Hilgardia, 47: 1–3.

Walters, P.J. and Dominiak, B.C., 1984. The establishment in New South Wales of *Trioxys complanatus* Quilis (Hymenoptera: Aphidiidae), an imported parasite of *Therioaphis trifolii* (Monell) f. *maculata*, spotted alfalfa aphid. General and Applied Entomology, 16: 65–67.

Walters, P.J. and Holtkamp, R.H., 1984. The monitoring of lucerne aphids in the major lucerne growing areas of New South Wales (1977–80). New South Wales, Department of Agriculture Sydney, Science Bulletin 90, pp. 1–62.

Way, M.J. and Banks, C.J., 1968. Population studies on the active stages of the black bean aphid, *Aphis fabae* Scop., on its winter host *Euonymus europaeus* L. Annals of Applied Biology, 62: 177–197.

Wilson, C.G. and Swincer, D.E., 1984. Hyperparasitism of *Therioaphis trifolii* f. *maculata* (Homoptera: Aphididae) in South Australia. Journal of the Australian Entomological Society, 23: 47–50.

Wilson, C.G., Swincer, D.E. and Walden, K.J., 1982. The introduction of *Trioxys complanatus* Quilis (Hymenoptera: Aphidiidae), an internal parasite of the spotted alfalfa aphid, into South Australia. Journal of the Australian Entomological Society, 21: 13–27.

Wilson, F., 1960. A review of the biological control of insects and weeds in Australia and Australian New Guinea. Commonwealth Institute of Biological Control, Ottawa, Technical communication No. 1, 102 pp.

Woolcock, L.T., 1979. Introduction and rearing of SAA (spotted alfalfa aphid) parasites. In: Lucerne Aphid Workshop, Tamworth, 1978. New South Wales Department of Agriculture, Sydney, pp. 72–74.

11.2.2 Biological Control in Greenhouses

P.M.J. RAMAKERS

INTRODUCTION

For most crops grown in protected cultivation one or more aphid species are potential pests. *Myzus persicae* (Sulzer) is the most common species, attacking a wide range of host plants, with the Cucurbitaceae family as a notable exception. Other important aphid species are: *Aphis gossypii* Glover on Cucurbitaceae, *Nasonovia ribis-nigri* (Mosley) on lettuce, *Aulacorthum solani* (Kaltenbach) and *Macrosiphum euphorbiae* (Thomas) on various – mainly solanaceous – crops, and a complex of species on chrysanthemum, including *M. persicae, Macrosiphoniella sanborni* (Gillette), *Brachycaudus helichrysi* (Kaltenbach) and *Aphis* spp. (Hussey et al., 1969; De Brouwer, 1976).

On annual greenhouse crops aphids reproduce by parthenogenesis only. The overlapping of different types of cultures (vegetable production, plant propagation, year-round ornamentals) and the short duration of interruptions between successive cultures in modern greenhouse industry allow polyphagous species like *M. persicae* to maintain a year-round reproduction, producing numerous generations per year. Populations originating from immigrating alatae will also complete more generations than in the field, because of the artificial prolongation of favourable conditions.

If greenhouse populations are allowed to complete several generations undisturbed, the occurrence of natural enemies is very likely, particularly – but not exclusively – in summertime, when ventilators are opened and outdoor populations are active. It is generally found that without the protection of insect-proof cages, aphid rearings in greenhouses are sooner or later destroyed by predators or hymenopterous parasitoids. Less often pathogenic fungi will decimate aphid populations.

The interest in natural enemies was stimulated by the successful introduction of biological control methods for the main greenhouse pests: the two-spotted spider mite, *Tetranychus urticae* (Koch), and the greenhouse whitefly, *Trialeurodes vaporariorum* (Westwood) (Hussey and Bravenboer, 1971). When the Dutch company Koppert B.V. started to introduce these methods commercially, for *T. urticae* in cucumbers in the late 1960s and for *T. vaporariorum* in tomatoes in the early 1970s, an immediate practical need arose for selective biological (or chemical) agents to control simultaneously occurring aphids. In addition, an English research group tried to introduce the biocontrol concept in ornamental crops, through the development of integrated control programmes, including biological control of aphids, for year-round chrysanthemum (Cross et al., 1983).

Although various natural enemies have been tested, some of which were promising candidates for controlling aphids, so far none have been widely adopted by greenhouse growers. Apart from the general problems involved in developing biocontrol methods (see sections 11.2 and 11.2.1), the following additional considerations are thought to be relevant for greenhouse conditions:

(a) In spite of resistance problems, chemical control of aphids is still relatively easy. Sufficient effective insecticides from different chemical groups are available, and greenhouse growers can apply techniques that are more convenient than spraying, such as thermal fogging, aerosols and smokes, and addition of systemic insecticides to hydroponic systems. No aphid instar is protected in any particular way. The phenomenon of resistance (see section 11.1.3), though well known in both *M. persicae* and *A. gossypii*, is obviously not severe or widespread enough to make chemical control impossible.

(b) The problem of chemical aphid control within integrated pest management programmes was relieved by the registration of pirimicarb in the early seventies. This carbamate is sufficiently selective in the field, though it does harm delicate adult wasps such as *Encarsia formosa* Gahan, the control agent for greenhouse whitefly (Delorme and Angot, 1983). The replacement of broad-spectrum chemicals by this compound gives natural aphid control a better opportunity, thus reducing the selection pressure by chemicals and the risk of resistance.

(c) Aphid populations in greenhouses are favoured by extreme reproductive capacities, due to a high rate of development (from birth to adulthood in little more than one week) and the absence of weather-determined mortality. An up to 6-fold increase in one week was found for populations of *M. persicae* on chrysanthemum (Wyatt, 1970a), and a 7-fold increase on eggplant (Rabasse et al., 1983). *A. gossypii* populations on cucumber grow even faster, increasing about 10-fold in one week (Wyatt, 1970a). Except for micro-organisms, few natural enemies will have similar reproductive potentials. Biocontrol exercises must therefore be based on repeated inundative releases (usually too expensive for practical use), or on a combined use with other limiting factors, such as crowding, plant resistance or chemical control. In theory, it might be possible to achieve a balance between the host and a natural enemy with a similar or lower reproductive capacity, but this is unlikely to be stable.

PARASITOIDS

Aphidius matricariae Haliday is the most widespread hymenopterous aphid parasitoid in greenhouses. It is found worldwide, mainly in temperate climates, but its occurrence in the New World is assumed to be the result of one or several accidental introductions (Schlinger and Mackauer, 1963). Tremblay (1974) reviewed the studies of the potency of this aphidiid as a control agent for greenhouse cultures.

A. matricariae was reared from as many as 40 different hosts, including all the species mentioned above, but is most effective against *M. persicae* (Schlinger and Mackauer, 1963). It develops faster than any other parasitoid mentioned in this section (Scopes and Biggerstaff, 1977), but still slower than its main host (12 and 9 days, respectively, at 21°C). This is compensated for by a higher oviposition frequency and total fecundity. One female wasp can produce up to 400 eggs, with an average of approximately 300 (Vevai, 1942; Shalaby and Rabasse, 1979; Giri et al., 1982). Assuming that half of this progeny are males, the number of females per female is thus 150, compared to 80 for *M. persicae*.

Moreover, the aphidiid is able to parasitize dozens of aphids per day, whereas the daily production of the aphid is "only" 5 or 6 nymphs.

Rabasse et al (1983) found populations of this parasitoid to increase even faster than its host on eggplant in spring. In contrast, Wyatt (1970b), studying the same parasitoid–aphid relation in winter, observed the opposite on aphid-susceptible chrysanthemum varieties, and therefore this author discouraged inoculative releases. Consequently, Scopes (1970) recommended preventive introduction of both parasitized and unparasitized aphids in similar numbers on young plants, in order to impair the aphid's reproductive capacity from the start.

Repeated attacks of parasitoids on an aphid colony cause a disturbance, resulting in a considerable mortality in addition to, and sometimes more important than, the parasitization itself (Tremblay, 1974). Therefore the final breakdown of an aphid population is often spectacular. Disturbance of colonies, as well as the fact that some aphids leave the plant before mummification (Hofsvang and Hågvar, 1978), explain why after such a breakdown the number of mummies left on the leaves is surprisingly low.

Several other hymenopterous parasitoids, usually Aphidiidae, appear to be potential control agents for *M. persicae*. Results are difficult to compare, since authors have studied different aspects and used different methods. In Norway, two native species *Aphidius ervi* Haliday and *Ephedrus cerasicola* Starý, and an accidentally introduced one, *Aphidius colemani* Viereck, were compared by the same authors. Similar to *A. matricariae* ('t Hart et al., 1978), *E. cerasicola* was shown to be able to discriminate between parasitized and unparasitized hosts (Hofsvang and Hågvar, 1986; this aspect was not included in the study of the other species). All three species were inferior to *A. matricariae* with regard to both rate of development and fecundity (Hofsvang and Hågvar, 1975a, b, c). Nevertheless, satisfactory and lasting control of *M. persicae* on sweet peppers, *Capsicum annuum*, was obtained by introducing both the aphid and *E. cerasicola* (Hofsvang and Hågvar, 1978). Other promising parasitoid species are *Diaeretiella rapae* (M'Intosh) (Tremblay, 1974), *Aphelinus asychis* Walker (Rabasse, 1980), *Aphidius gifuensis* Ashmead (Nakazawa, 1983) and *Praon volucre* (Haliday) (Plotnikov, 1981), all used to control *M. persicae*. Few attempts have been described to use parasitoids for the control of *A. gossypii* on Cucurbitaceae, an aphid that reproduces much faster than any of its parasitoids (Wyatt, 1970a). A research project on an aphelinid species is mentioned by Scopes (1970) but was apparently not continued. Probably predators are more suitable tools to destroy the very dense colonies characteristic of this aphid.

Obviously, all authors started from the observation that a certain parasitoid species destroyed aphid populations spontaneously on crops or in aphid rearings in greenhouses, thus avoiding the difficulty of predicting the success of an antagonist from another environment (see Table 11.2.1.2, no. 3). The interaction of aphid and parasitoid at rather high densities is usually the best-described part of the experiments or observations. It is not always clear, however, whether the pest was checked by the action of the parasitoids alone, or if self-limitation by crowding was required first (Wyatt, 1970b). Nor is it evident whether or not lasting control was really based on a permanent interaction between low numbers of the pest and parasitoid. Other possibilities are: total pest elimination, as might occur in a limited space; semi-permanent effects (possibly sufficient to protect short-growing cultures) caused by an overproduction of parasitoids during a previous aphid peak; immigration of parasitoids (or parasitized aphids) from the surrounding vegetation; or additional effects of other natural enemies. Wyatt (1970a) observed control to last for as long as 2 years in a chrysanthemum nursery; however, he does not give details. In contrast, Lyon (1976) found repeated or continuous introduction of parasitoids to be necessary to protect tomatoes and sweet peppers. More precisely described is the experiment of Hofsvang and Hågvar (1978) on sweet pepper; control was obtained over a total period of 5 months, consisting of a 2-month period of true interaction, followed by an aphid-free period and finally a re-infection which came too late to cause trouble.

The usefulness of aphid parasitoids is questionable if a crop is likely to bear different aphid species, as is often the case (Rabasse, 1980). For such cultures, the use of polyphagous predators seems to be a more feasible alternative than combinations of parasitoids (Lyon, 1976).

Most authors do not mention hyperparasitization as a problem. Tremblay (1974) warned of this phenomenon in general terms, but at the same time stated that *A. matricariae* was not attacked by hyperparasitoids in his greenhouses in Italy. Similarly, Lyon (1968) reported that *D. rapae*, though readily attacked by hyperparasitoids in outdoor populations in southern France, was little affected in greenhouses. However, in long-term experiments with *A. matricariae* in the Netherlands, we found hyperparasitism to be very severe, resulting in new aphid outbreaks at a moment when the pest was thought to be under control. Alloxystinae spp. attacked immobilized aphids shortly before mummification, but Ceraphronidae ovipositing in mummies were usually more numerous. A *Praon* species, spontaneously occurring in the same experiments, was equally attacked. Most probably these hyperparasitoids would affect any other aphidiid used. Only during the winter months were hyperparasitoids obviously absent in (heated) greenhouses, may be because of diapause (Rabasse and Dedryver, 1983). The apparent absence of diapause in greenhouse populations of some primary parasitoids allows them to be effective during at least this part of the growing season.

Aphid parasitoids might be unable to control their hosts completely but can contribute considerably to aphid control, particularly in programmes with a reduced input of chemicals. The protection provided by the mummy allows them to survive treatments with short-acting chemicals. Since aphids do not have such a protection, it is quite possible to carry out "chemical reparations" when the parasitoid–host ratio is too low. Pirimicarb is commonly used for this purpose.

PREDATORS

Predators from all the important families attacking aphid populations in the field are found in the greenhouse environment as well. Several authors have studied the use of (in order of increasing importance) Syrphidae (see section 9.2.3) (Fig.11.2.2.1), Coccinellidae (section 9.2.1), Chrysopidae (section 9.2.2) and Cecidomyiidae (section 9.2.4).

A general advantage of predators compared to parasitoids is their ability to control several aphid species simultaneously. Their effectiveness is usually limited by host plant qualities and climatic factors rather than by the prey species they have to deal with. For example, Scopes (1969) found that *Chrysopa carnea* Stephens feeds readily on both *M. persicae* and *A. gossypii* but cannot control the latter species on cucumber. In the experiments of Hämäläinen (1977), *Coccinella septempunctata* (Linnaeus) was superior to *Adalia bipunctata* (Linnaeus) in controlling *M. persicae* on chrysanthemum, but the opposite result was found with the same aphid on sweet pepper, whereas neither species was effective in controlling *M. euphorbiae* on roses. Both authors ascribe these differences to direct effects of the vegetation type on the predators.

A drawback of using predators is their need for a rather high prey density, requiring their introduction and thus mass-rearing in considerable numbers.

Fig. 11.2.2.1. Larva of the hoverfly, *Scaeva pyrastri* (Linnaeus) feeding on the aphid *Hyalopterus pruni* (Geoffroy) (photograph by A. van Frankenhuyzen, Wageningen).

Moreover, in aphid predators so far, though often penetrating greenhouses spontaneously, are obviously not well adapted to the environmental conditions in protected cultivations, since the adults oviposit poorly and tend to leave the greenhouse. Therefore, biocontrol must be based on repeated introductions of vast numbers of eggs or larvae.

In experimental plots, successful control was obtained in this way by several authors, both with coccinellids (see Quilici et al., 1986) and chrysopids (see Hassan, 1977). Mass-rearing, however, is the limiting factor for commercialization, though methods have been improved step by step (see Tulisalo, 1978). *C. carnea* is the most widely used species, since the adults can be fed on an artificial diet. The production of larvae is still expensive; they are usually fed with *Sitotroga* eggs and must be reared individually to prevent cannibalism.

Only the cecidomyiid predator *Aphidoletes aphidimyza* (Rondani) has been found to produce self-perpetuating populations in greenhouses. Therefore it is the most preferred predator at this moment, in spite of the fact that it must be mass-reared on its natural host and the predation capacity of an individual larva is low in comparison with that of other predators. It has been suggested that the absence of wind and the high humidity in greenhouses favour the fragile adults of this predator, which is usually inconspicuous outdoors. It often invades greenhouses unnoticed in summer, and destroyes aphid rearings quickly and thoroughly, thus drawing the attention of entomologists studying aphids or aphid antagonists. After such an event (Bombosch, 1958), it has been studied intensively by a number of entomologists of Göttingen University to elucidate fundamental aspects of its behaviour. Dissertations and publications of this group are listed by Markkula et al (1979).

El Titi (1974) demonstrated the potentials and limitations of this predator in control experiments with *M. persicae* on small greenhouse plots. He described two basically different introduction methods, namely the introduction of relatively high numbers of predators after the pest was sufficiently established, and the maintenance of an open rearing system in the crop that has to be protected;

in the latter case it is elegant but not necessary to use an aphid species that does not feed on the crop.

In the Netherlands, *Aphidoletes* was used experimentally for several years on sweet pepper, a crop with severe aphid problems nearly year-round. These trials confirmed El Titi's description of the interaction between pest and predator populations. Density oscillations, typical for the performance of a specialized predator preying on all pest instars in a simplified environment, occurred both in small ($100\,m^2$) and large (1 ha) greenhouses. In small plots, damping of oscillations was obtained by frequent re-introductions of aphids every time it was suspected that the predator had run out of food, thus preventing the predator population from falling under a certain minimum. However, for large-scale commercial use, where continuous and intensive pest monitoring is difficult, occasional interference through (selective) chemicals seems to be a more feasible solution.

Spontaneous occurrence of *Aphidoletes*, originating from outdoor populations, is usually too late (July) to be of practical significance. Artificial introductions in greenhouses are worth considering, since from March onwards conditions are suitable for the prevention of diapause. Attempts have been made to spread the use of *Aphidoletes* among commerical farms in Russia (Bondarenko and Moiseev, 1978) and Finland (Markkula et al., 1979), with limited results as far as Finland is concerned, whereas the information from Russia is difficult to verify. It seems that the Russian authors adopted the latter, more reliable method for cucumber, whereas the Finns use the former, more practicable method, mainly on tomato. It should be noticed that aphids on tomato develop relatively slowly. Furthermore, the numerous farms using this predator represent only a few hectares together, and not all of them are professionally managed. Commercial production of predators, started in the late 1970s by the Finnish company Kemira Oy but given up already, provided no more than 10^5 individuals per year, and the total annual production of *Aphidoletes* is probably less than 10^6.

Summarizing, we may conclude that at the moment *A. aphidimyza* is the most useful predator known for aphid control in greenhouses, but that certain limiting factors exist: (a) mass rearing can be done on natural hosts exclusively; (b) the predator maintains a permanent (for one season) population, but does not prevent new temporary outbreaks of aphids; (c) the predator needs long-day conditions to prevent diapause (Havelka, 1980). It may be possible to extend the operative period by selection of non-diapausing strains; progress along this line is reported from Canada (Gilkeson and Hill, 1986).

FUNGI

It is not unusual to find individual aphids infected by Entomophthorales in greenhouses, but usually without the severe epizootics that sometimes develop in field crops. However, many authors assume that greenhouse cultures have optimal conditions for the study of the epidemiology of such fungi. Few attempts have been made to use them as control agents; this is in part due to the technical problems inherent to handling this difficult material (see further section 9.3).

Some encouraging results in controlling greenhouse aphids with natural (Dedryver, 1979) or artificial (Latgé et al., 1983) inoculum were achieved. In order to create what are thought to be optimal conditions for a mycosis, water sprinklers were used frequently in such experiments (Dedryver, 1979; Rabasse et al., 1983). If such measures to increase humidity should appear to be inevitable, a possible conflict with the control of plant-pathogenic fungi must be faced.

The study of Entomophthorales as control agents is probably too early a stage to draw final conclusions about applicability. More developed is the study of another mycosis, caused by *Verticillium lecanii* (Zimmermann), a facultative pathogen of many arthropods and some fungi. Laboratory studies on the susceptibility of several aphid species, followed by successful control experiments on chrysanthemums (Hall and Burges, 1979), showed that this fungus is able to bring about persistent control; this led to the development of a commercial product called Vertalec, manufactured by the British company Tate & Lyle Ltd. (Gardner et al., 1984). This product was further improved by a substrate-containing formulation, allowing the fungus to produce additional inoculum in a saprobic manner after being sprayed (Hall, 1982), and by strain selection (Hall, 1984).

Within the concept of integrated control on chrysanthemum, *V. lecanii* was meant to be the successor of the hymenopterous parasitoid *A. matricariae*, the latter being too specific to control the total aphid complex. Surprisingly, this fungus, though basically a broad-spectrum agent, affects aphid species differently. While the main pest *M. persicae* is controlled well and suppressed persistently, control of the – fortunately less common – species *M. sanborni* is unsatisfactory (Hall and Burges, 1979; Gardner et al., 1984). Since the susceptibility of several aphids in laboratory tests was similar, the authors ascribe this to differences in site preference (microclimate) and mobility (infection risk) between the aphid species.

A more important impediment for the practical use of this pathogen is its dependence on considerable periods of high humidity (Ekbom, 1981). The best results were obtained in chrysanthemum crops, where the daily periods of high humidity were lengthened by sprinkling water and covering the plants with polythene black-out sheets (Hall and Burges, 1979) normally used for flower initiation. In other crops grown without such facilities, a persistent mycosis is not easily obtained, while the short-term effect of a spray is uncertain. This lack of reliability diminishes the attractiveness of Vertalec for commercial growers and is an objection for official registration as an effective pesticide.

According to data from the British advisory service ADAS, a small number of their chrysanthemum growers used Vertalec to control aphids within integrated control schemes, on a total area of 6.5 ha (Gould, 1984). Dutch chrysanthemum growers are more dependent on exports than their British colleagues and therefore have to comply with quarantine demands that are so strict that any kind of integrated control is out of the question. Since the company could not enlarge the market for this biological insecticide, the production was stopped recently. In various countries *V. lecanii* is used on a more or less experimental scale, but more often against whiteflies than against aphids, with varying results.

SUMMARY

Although greenhouse crops bear a relatively poor entomofauna, natural enemies of aphids can play an important role in the population dynamics of these pests, especially in long-lasting cultures, such as fruiting vegetables and some ornamentals, with persistent aphid problems.

Research on artificial manipulations of natural enemies have stimulated commercialization of the predator *A. aphidimyza* and the fungus *V. lecanii*, but so far both attempts failed. An experimental method for controlling aphids with a combination of natural enemies is outlined in Fig. 11.2.2.2, but it is too complicated for practical use.

Section 11.2.2 references p. 206

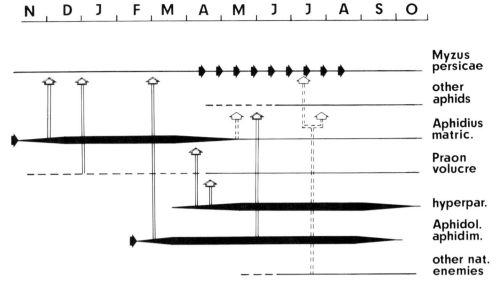

Fig. 11.2.2.2. Simplified outline of experimental aphid control on winter plantings of sweet pepper, *Capsicum annuum*, using natural enemies only: (– – – –) possible presence; (———) presence; (▬▬), abundance; black arrows, artificial introduction; open arrows, attack.

For the time being, spontaneously occurring natural enemies in control schemes with little or no broad-spectrum chemicals have more impact than biological control sensu stricto. At present, the most fruitful approach is to stimulate the use of selective control agents (Hassan et al., 1987) and to provide information on the appearance and function of aphid antagonists. In specific situations artificial introduction of mass-produced antagonists is worth considering.

ACKNOWLEDGEMENTS

The author wishes to express his thanks to Dr. A. El Titi (Stuttgart, Federal Republic of Germany), Dr. E.B. Hågvar (Ås-NLH, Norway), Dr. R.A. Hall (Littlehampton, Great Britain), and Dr. J.-M. Rabasse (Antibes, France) for criticizing and supplementing the manuscript.

REFERENCES

Bombosch, S., 1958. Die Ursache eines eigenartigen Blattlaussterbens. Zeitschrift für Pflanzenkrankheiten und Pflanzenschutz, 65: 694–695.
Bondarenko, N.V. and Moiseev, E.G., 1978. Predacious cecidomyiid on cucumbers in glasshouses. Zashchita Rastenil, 1978 (2): 30–31 (in Russian).
Cross, J.V., Wardlow, L.R., Hall, R., Saynor, M. and Bassett, P., 1983. Integrated control of chrysanthemum pests. Bulletin OILB/SROP, 1983/VI/3: 181–185.
De Brouwer, W.M.Th.J., 1976. Observations on aphids in the glasshouse district of the province of Zuid-Holland. Gewasbescherming, 7: 31–39 (in Dutch).
Dedryver, C.A., 1979. Déclenchement en serre d'une épizootie á *Entomophthora fresenii* sur *Aphis fabae* par introduction d'inoculum et regulation de l'humidité relative. Entomophaga, 24: 443–453.
Delorme, R. and Angot, A., 1983. Toxicités relatives de divers pesticides pour *Encarsia formosa* Gahan (Hym., Aphelinidae) et pour son hôte, *Trialeurodes vaporariorum* Westw. (Hom., Aleyrodidae). Agronomie, 3: 577–584.

Ekbom, B.S., 1981. Humidity requirements and storage of the entomopathogenic fungus *Verticillium lecanii* for use in greenhouses. Annales Entomologici Fennici, 47: 61–62.

El Titi, A., 1974. Auswirkung von der räuberischen Gallmücke *Aphidoletes aphidimyza* (Rond.) (Itonididae: Diptera) auf Blattlauspopulationen unter Glas. Zeitschrift für Angewandte Entomologie, 76: 406–417.

Gardner, W.A., Oetting, R.D. and Storey, G.K., 1984. Scheduling of *Verticillium lecanii* and benomyl applications to maintain aphid (Homoptera: Aphidae) control on chrysanthemums in greenhouses. Journal of Economic Entomology, 77: 514–518.

Gilkeson, L.A. and Hill, S.B., 1986. Genetic selection for and evaluation of nondiapause lines of predatory midge, *Aphidoletes aphidimyza* (Rondani) (Diptera). Canadian Entomologist, 118: 869–879.

Giri, M.K., Pass, B.C., Yeargan, K.V. and Parr, J.C., 1982. Behavior, net reproduction, longevity, and mummy-stage survival of *Aphidius matricariae* (Hym. Aphidiidae). Entomophaga, 27: 147–153.

Gould, H.J., 1984. Survey of biological control on tomatoes, cucumbers and chrysanthemums in England and Wales, 1983, In: J. Woets and J.C. van Lenteren (Editors), Sting; Newsletter on biological control in greenhouses. no. 7, pp. 5–7.

Hall, R.A., 1982. Control of whitefly, *Trialeurodes vaporariorum* and cotton aphid, *Aphis gossypii* in glasshouses by two isolates of the fungus, *Verticillium lecanii*. Annals of Applied Biology, 101: 1–11.

Hall, R.A., 1984. Epizootic potential for aphids of different isolates of the fungus, *Verticillium lecanii*. Entomophaga, 29: 311–321.

Hall, R.A. and Burges, H.D., 1979. Control of aphids in glasshouses with the fungus, *Verticillium lecanii*. Annals of Applied Biology, 93: 235–246.

Hämäläinen, M., 1977. Control of aphids on sweet peppers, chrysanthemums and roses in small greenhouses using the ladybeetles *Coccinella septempunctata* and *Adalia bipunctata* (Col., Coccinellidae). Annales Agriculturae Fenniae, 16: 117–131.

Hassan, S.A., 1977. Untersuchungen zur Verwendung des Prädators *Chrysopa carnea* Steph. (Neuroptera, Chrysopidae) zur Bekämpfung der Grünen Pfirsichblattlaus *Myzus persicae* (Sulzer) an Paprika im Gewächshaus. Zeitschrift für Angewandte Entomologie, 82: 234–239.

Hassan, S.A., Albert, R., Bigler, F., Blaisinger, P., Bogenschütz, H., Boller, E., Brun, J., Chiverton, P., Edwards, P., Englert, W.D., Huang, P., Inglesfield, C., Naton, E., Oomen, P.A., Overmeer, W.P.J., Rieckmann, W., Samsøe-Petersen, L., Stäubli, A., Tuset, J.J., Viggiani, G., Vanwetswinkel, G., 1987. Results of the third joint pesticide testing programme by the IOBC/WPRS-Working Group "Pesticides and Beneficial Organisms". Journal of Applied Entomology, 103: 92–107.

Havelka, J., 1980. Some aspects of the photoperiodism of the carniverous gall-midge *Aphidoletes aphidimyza* Rond. (Diptera, Cecidomyiidae). Entomologicheskoe Obozrenie, 59: 241–248 (in Russian).

Hofsvang, T. and Hågvar, E.B., 1975a. Developmental rate, longevity, fecundity, and oviposition period of *Ephedrus cerasicola* Starý (Hym., Aphidiidae) parasitizing *Myzus persicae* Sulz. (Hom., Aphididae) on paprika. Norwegian Journal of Entomology, 22: 15–22.

Hofsvang, T. and Hågvar, E.B., 1975b. Duration of development and longevity in *Aphidius ervi* and *Aphidius platensis* (Hym.: Aphidiidae), two parasites of *Myzus persicae* (Hom.: Aphididae). Entomophaga, 20: 11–22.

Hofsvang, T. and Hågvar, E.B., 1975c. Fecundity and oviposition period of *Aphidius platensis* Brèthes (Hym., Aphidiidae) parasitizing *Myzus persicae* Sulz. (Hom., Aphididae) on paprika. Norwegian Journal of Entomology, 22: 113–116.

Hofsvang, T. and Hågvar, E.B., 1978. Effect of parasitism by *Ephedrus cerasicola* Stary on *Myzus persicae* (Sulzer) in small glasshouses. Zeitschrift für Angewandte Entomologie, 85: 1–15.

Hofsvang, T. and Hågvar, E.B., 1986. Oviposition behaviour of *Ephedrus cerasicola* (Hym.: Aphidiidae) parasitizing different instars of its aphid host. Entomophaga, 31: 261–267.

Hussey, N.W. and Bravenboer, L., 1971. Control of pests in glasshouse culture by the introduction of natural enemies. In: C.B. Huffaker (Editor), Biological Control. Plenum Press, New York, pp. 195–216.

Hussey, N.W., Read, W.H. and Hesling, J.J., 1969. The Pests of Protected Cultivation. Section "Family Aphididae: Greenflies". Edward Arnold, London, pp. 106–121.

Latgé, J.P., Silvie, P., Papierok, B., Remaudière, G., Dedryver, C.A. and Rabasse, J.M., 1983. Advantages and disadvantages of *Conidiobolus obscurus* and of *Erynia neoaphidis* in the biological control of aphids. In: R. Cavalloro (Editor), Aphid Antagonists. A.A. Balkema, Rotterdam, pp. 20–32.

Lyon, J.P., 1968. Remarques préliminaires sur les possibilités d'utilisation pratique d'Hyménoptères parasites pour la lutte contre les pucerons en serre. Annales des Épiphyties, 19: 113–118.

Lyon, J.P., 1976. Les populations aphidiennes en serre et leur limitation par utilisation expérimentale de divers entomophages. Bulletin OILB/SROP, 1976/4: 64–76.

Markkula, M., Tiittanen, K., Hämäläinen, M. and Forsberg, A., 1979. The aphid midge *Aphidoletes aphidimyza* (Diptera, Cecidomyiidae) and its use in biological control of aphids. Annales Entomologici Fennici, 45: 89–98.

Nakazawa, K., 1983. Factors preventing the development of biological control in Japanese greenhouses. Bulletin OILB/SROP, 1983?VI/3: 32–35.

Plotnikov, V.F., 1981. Efficacité de l'utilisation d'insectes aphidiphages pour la protection des cultures maraîchéres sous verre. In: Lutte Biologique et Intégrée Contre les Pucerons. Colloque Franco-Soviétique, Rennes, September 1979, pp. 59–65.

Quilici, S., Iperti, G. and Rabasse, J.M., 1986. Essais de lutte biologique en serre d'aubergine à l'aide d'un prédateur aphidiphage: *Propylea quatordecimpunctata* L. (Coleoptera, Coccinellidae). Frustula Entomologica N.S., 7–8 (20–21): 9–25.

Rabasse, J.M., 1980. Implantation d'*Aphidius matricariae* dans les populations de *Myzus persicae* en culture d'aubergines sous serre. Bulletin OILB/SROP, 1980/III/3: 175–185.

Rabasse, J.M. and Dedryver, C.A., 1983. Overwintering of primary parasites and hyperparasites of cereal aphids in Western France. In: R. Cavalloro (Editor), Aphid Antagonists. A.A. Balkema, Rotterdam, pp. 57–64.

Rabasse, J.M., Lafont, J.P., Delpuech, I. and Silvie, P., 1983. Progress in aphid control in potected crops. Bulletin OILB/SROP, 1983/VI/3: 151–162.

Schlinger, E.I. and Mackauer, M.J.P., 1963. Identity, distribution, and hosts of *Aphidius matricariae* Haliday, an important parasite of the green peach aphid, *Myzus persicae* (Hymenoptera: Aphidiidae – Homoptera: Aphidoidea). Annals of the Entomological Society of America, 56; 648–653.

Scopes, N.E.A., 1969. The potential of *Chrysopa carnea* as a biological control agent of *Myzus persicae* on glasshouse chrysanthemums. Annals of Applied Biology, 64: 433–439.

Scopes, N.E.A., 1970. Control of *Myzus persicae* on year-round chrysanthemums by introducing aphids parasitized by *Aphidius matricariae* into boxes of rooted cuttings. Annals of Applied Biology, 66: 323–327.

Scopes, N.E.A. and Biggerstaff, S.B., 1977. The use of a temperature integrator to predict the developmental period of the parasite *Aphidius matricariae*. Journal of Applied Ecology, 14: 799–802.

Shalaby, F.F. and Rabasse, J.M., 1979. On the biology of *Aphidius matricariae* Hal. (Hymenoptera, Aphidiidae), parasite on *Myzus persicae* (Sulz.) (Homoptera, Aphididae). Annals of Agricultural Science, Moshtohor, 11: 75–97.

't Hart, J., Jonge, J. de, Collé, C., Dicke, M., Lenteren, J.C. van and Ramakers, P., 1978. Host selection, host discrimination and functional response of *Aphidius matricariae* Haliday (Hymenoptera: Braconidae), a parasite of the green peach aphid, *Myzus persicae* (Sulz.). Mededelingen van de Faculteit Landbouwwetenschappen, Rijksuniversiteit Gent, 43: 441–453.

Tremblay, E., 1974. Possibilities for utilization of *Aphidius matricariae* Hal. (Hymenoptera: Ichneumonoidea) against *Myzus persicae* (Sulz.) (Homoptera: Aphidoidea) in small glasshouses. Zeitschrift für Pflanzenkrankheiten und Pflanzenschutz, 81: 612–619.

Tulisalo, U., 1978. An improved rearing method for *Chrysopa carnea* Steph. Annales Agriculturae Fenniae, 17: 143–146.

Vevai, E.J., 1942. On the bionomics of *Aphidius matricariae* Hal., a braconid parasite of *Myzus persicae* Sulz. Parasitology, 34: 141–151.

Wyatt, I.J., 1970a. Progress towards integrated control under glass in Britain. Proceedings of the Conference on Integrated Control in Glasshouses, Naaldwijk, September 1970, pp. 49–62.

Wyatt, I.J., 1970b. The distribution of *Myzus persicae* (Sulz.) on year-round chrysanthemums. II. Winter season: the effect of parasitism by *Aphidius matricariae* Hal. Annals of Applied Biology, 65: 31–41.

11.3 Modifying Aphid Behaviour

R.W. GIBSON and A.D. RICE

INTRODUCTION

Biotypes of several aphid species, including the peach-potato aphid *Myzus persicae* (Sulzer) and the damson-hop aphid *Phorodon humuli* (Schrank), are resistant to aphicides, and this resistance is often against several types of aphicide (see section 11.1.1). Neither aphicides nor host plant resistance are generally effective at preventing spread of non-persistently transmitted plant viruses. Indeed, some aphicides may even promote the brief probing during which such viruses are transmitted and increase dispersal. Biological control of vectors is unlikely to hinder virus spread, as it is usually effective only after aphids have colonized crops. Consequently alternative methods of control have been sought, including the manipulation of behaviour.

It is possible to influence the host-seeking behaviour of alatae whilst they are airborne so that few alight on crops, and it may be possible to prevent probing by using repellents such as aphid alarm pheromone. It may even be possible to modify the probing so that viruses are not transmitted. Such potential means of control have the dual advantages of diminishing aphid numbers in crops and hindering the spread even of non-persistently transmitted viruses, as well as being environmentally safe. In certain circumstances, pesticides have increased virus spread, presumably by modifying aphid behaviour, an aspect that is often ignored but which could give important clues as to how the maximum benefit from their use could be obtained. In this section, successes to date and the future potential of modifying aphid behaviour to protect crops are examined.

PREVENTING FLYING APHIDS FROM COLONIZING PLANTS

Flying aphids may be attracted to or repelled from plants by light of particular wavelength or by volatile substances (Kring, 1972).

The effects of light on aphid behaviour

When alatae first fly from their host plant they are attracted by the blue–ultraviolet light from the sky. After a period of flight they stop flying upwards and are attracted to orange–yellow–green light reflected from leaves; this characteristic is exploited to catch them in yellow traps. The size, shape and contrast of a plant or trap against its background affects its attractiveness to alatae. Large plants attract more aphids than smaller ones (Dunn, 1969) and the number of aphids caught increases with yellow trap size, though small traps catch more per unit area (Costa and Lewis, 1967). Yellow traps and plants which contrast with their background attract more aphids than those camouflaged against a green background (Smith, 1976). Crops that are sparse are often colonized by relatively more aphids and infected more by viruses than crops which present a closed canopy to incoming alatae (Heathcote, 1969).

The principal ways in which the responses of alatae to light are exploited to diminish aphid settling involve either attracting them to a yellow trap, or repelling them from a crop with mulches or screens which reflect short-wave light (Loebenstein and Raccah, 1980; Harpaz, 1982). Sheets of yellow polyethylene covered with transparent glue have been used in Israel to trap aphids. The incidence of non-persistent viruses was decreased by an average of 60% in plots surrounded by a 0.7-m high sticky yellow sheet compared with untreated plots (Cohen and Marco, 1973); this treatment is now routinely used for the control of potato virus Y and cucumber mosaic virus in peppers in Israel (Harpaz, 1982). An obvious drawback to their use is that the sticky glue gradually becomes covered by trapped insects and dust and their attractiveness decreases. Yellow sheets treated with a fast-acting, persistent, contact insecticide such as a synthetic pyrethroid might also be effective, more convenient and allow the texture of the surface of the sheet to be adjusted to give maximum attractiveness. The colour could also be adjusted to allow for different wavelength responses of different vector species.

An alternative to attracting aphids away from a crop is to repel them. Aphids are not attracted to plants growing close to white or reflective surfaces, and aphid catches in yellow water traps placed on reflective sheets were decreased by up to 99% compared with conventionally placed traps (Loebenstein and Raccah, 1980). This can be exploited to protect crops. The reflective material is placed either on the soil surface as a mulch, over the crop or even sprayed onto the crop as a powder. Aluminium foil, often strengthened with a plastic backing, or grey or white polyethylene sheets are commonly used as soil mulches; white plastic netting can also be put over the crop. These treatments commonly diminish the incidence of non-persistent viruses by 70–90% (Table 11.3.1).

TABLE 11.3.1

Some examples of non-persistently transmitted viruses controlled by reflective material

Crop	Virus	Material	Control (%)	Reference
Squash (*Cucurbita* spp.)	Watermelon mosaic	Aluminium[a]	94	Wyman et al. (1979)
		White plastic[a]	77	
Gladiolus (*Gladiolus* spp.)	Cucumber mosaic	Aluminium foil[a]	69–79	Johnson et al. (1967)
		White plastic[a]	65	
		Aluminium powder[c]	54	
Lettuce (*Lactuca sativa*)		Aluminium foil[a]	94	Nawrocka et al. (1975)
Sweet pepper (*Capsicum annuum*)	Cucumber mosaic + Potato virus Y	White netting[b]	87–100	Cohen (1981)
Potato (*Solanum tuberosum*)	Potato virus Y	Clear film[b]	62	Gibson and Gunenc (1981)

[a] Mulches applied to the soil.
[b] Netting or film covering the crop.
[c] Powder sprayed on the plot.

Reflective sheets which repel aphids and yellow sheets which trap them are economic for the protection of high-unit-value horticultural crops such as peppers, tomatoes and cucurbits. They are used primarily to control spread of non-persistently transmitted viruses for which there are few alternative treatments. The choice of method may be influenced by the crop and its environment. For example, a reflective soil mulch can give added benefit where water conservation and unhindered harvesting and pesticide application are important, but may be unsuitable in a crop such as melons which grows horizontally and quickly covers the mulch.

Plant odours and aphid behaviour

The importance of volatile substances produced by plants in host selection by aphids is still largely unknown. However, aphid pests of agricultural crops are often polyphagous and are therefore unlikely to depend upon specific odours to attract them to each of their many host plants. Aphids that colonize perennial crops and wild plants are commonly monophagous or oligophagous, and for these species more specific cues than plant colour to seek out their hosts would seem advantageous. One of the few reports of flying aphids being attracted by a specific chemical involves the oligophagous species *Cavariella aegopodii* (Scopoli), a pest of carrot crops whose summer hosts also include wild Umbelliferae. Traps baited with carvone, a volatile component of several Umbelliferae, caught about 50 times more alate *C. aegopodii* then unbaited traps; linalool, another volatile component of many plants, diminished the attractiveness of carvone (Chapman et al., 1981). Pettersson (1970) demonstrated that *Rhopalosiphum padi* (Linnaeus) was attracted by the odour of its primary host *Prunus padus*, and the black bean aphid, *Aphis fabae* Scopoli, was attracted by the odour of bean plants (Alikhan, 1960). Attractant chemicals could improve the efficiency of yellow water traps and increase their selectivity, while repellent chemicals may have potential as aphid deterrents, as illustrated by the counteracting effect of linalool on carvone. Leaves of *Lantana* sp. diminish colonization of potato tubers by *M. persicae* (Raymundo and Alcazar, 1983), presumably because of a repellent odour released by the leaves. Other repellent odours may be responsible for some of the numerous "amateur" reports that plants grown in association with other species are colonized by fewer aphids than when grown alone.

Aphid alarm pheromone

Alarm pheromone is released by aphids attacked by predators (Nault et al., 1973). There would seem to be obvious advantage to an alate aphid in avoiding plants where aphid colonies are being attacked by predators, and few alates settled on plants treated with alarm pheromone (Wohlers, 1981; 1982). The main chemical component of the alarm pheromone of most aphid pest species is (*E*)-β-farnesene (Nault and Bowers, 1974). Some aphids react to as little as 20 pg, although the sensitivity differs with species (Montgomery and Nault, 1977) and morph; alates are particularly sensitive (Montgomery and Nault, 1978). (*E*)-β-farnesene can be prepared readily from commercially available nerolidol (Dawson et al., 1982b) but it is very volatile and persists only briefly in a crop. This disadvantage may be overcome by using slow release formulations.

Closely related derivatives of (*E*)-β-farnesene have been synthesized, and some elicit the alarm response (Bowers et al., 1977). Assuming that a practical means of ensuring the long-term presence of aphid alarm pheromone or an active derivative can be developed, it is still doubtful whether the hoped-for benefits would be achieved. Aphids that do alight may quickly accommodate to the presence of the pheromone and remain to colonize the crop (Wohlers, 1981) and strains of aphids insensitive to alarm pheromone are already known (Dawson et al., 1983; Müller, 1983). *M. persicae* always appears to probe soon after it alights and the presence of *(E)-β*-farnesene may not prevent virus transmission (Griffiths et al., 1982; Phelan and Miller, 1982). Furthermore, treating aphids with alarm pheromone did not diminish their subsequent ability to transmit a non-persistent virus (Yang and Zettler, 1975). These results were obtained in laboratory tests, but the spread of potato virus Y was also unaffected in field plots of potato treated with pheromone (Hille Ris Lambers and Schepers, 1978). However, because it is difficult to ensure a continuously adequate release of this volatile and unstable chemical, it is difficult to know whether failure to achieve control is inherent to the technique or a result of inadequate application. Although use of aphid alarm pheromone has not yet achieved a practical status, the need to apply only a calculated 300 mg/ha of *(E)-β*-farnesene to modify aphid behaviour (Briggs et al., 1983) represents such an attractive prospect that effort to test this approach fully in crops continues.

REPELLING APHIDS THAT HAVE ALIGHTED

Aphid repellents provide a non-toxic alternative to insecticides. They should preserve aphid predators and parasites, they are less hazardous to apply and less disruptive in the environment than insecticides. Furthermore, they may prove necessary replacements for insecticides, should widespread resistance develop.

The present-day aphicides rarely act fast enough to prevent aphids making the brief probe needed to acquire a non-persistently transmitted virus from an infected plant or to infect a healthy plant. By contrast, a chemical repellent might be detected quickly enough to act *before* an aphid probes. Such a chemical would be commercially valuable because its potential to control the spread of non-persistently transmitted viruses would command a market additional to that of conventional aphicides. However, a repellent acting *after* an aphid had probed might increase virus spread by increasing aphid movement among crop plants; this might also occur if a repellent is applied to crops with an established aphid population. Thus, to compete commercially with aphicides, a repellent would have to act *before* an aphid probes and be sufficiently persistent to allow its application before the main period of aphid immigration.

Aphids possess external chemoreceptors, making it as least theoretically possible to repel them before they probe. Antennal chemoreceptors (Dunn, 1975; see also section 2.6) react both to plant volatiles (Bromley and Anderson, 1982; Visser, 1983) and to aphid alarm pheromone (Nault et al., 1973). Accordingly, aphids characteristically hold their antennae forwards close to the leaf surface whilst "walking". The sensilla at the tip of the rostrum are also well-situated to act as chemoreceptors, especially during the rostral "dabbing" behaviour often exhibited before probing. However, opinion is divided as to whether these sensilla are mechanoreceptors (Tjallingii, 1978) or chemoreceptors (Tarn and Adams, 1982). There also appear to be chemoreceptors on the tarsi (Wensler, 1963). A greater knowledge of aphid chemoreception might allow electrophysiological assay of putative repellents.

TABLE 11.3.2

Control of the aphid *Myzus persicae* (Sulzer) by synthetic alarm pheromone and the contact insecticide permethrin (Griffiths and Pickett, 1980)

Treatment	Mean change in aphid numbers[a] after 24 h (%)
Water	+ 12 ± 3.3
Synthetic pheromone and water	− 21 ± 5.3
Permethrin	− 38 ± 5.7
Synthetic pheromone and permethrin	− 92 ± 2.9

[a] Starting population ca. 60 aphids

Aphid alarm pheromone

Alarm pheromone may prevent aphids from alighting and so prevent virus spread, as has been already discussed. Although repellency after alighting would not be so effective, since it seems likely that aphids would probe before flying off (Griffiths et al., 1982; Phelan and Miller, 1982), more of the alatae placed on plants flew or fell from (E)-β-farnesene-treated (46%) than from untreated plants (12%), and high-molecular-weight alarm pheromone derivatives, although failing to alarm settled aphids, did diminish colonization, nymph production and transmission of beet yellows virus and potato virus Y (Dawson et al., 1982a). It is not clear whether decreased virus transmission is associated with effects on aphid behaviour or whether the derivatives somehow interfere directly with virus transfer (Gibson et al., 1984).

(E)-β-farnesene may also be of practical value in improving the effectiveness of contact aphicides. Applied in combination with the aphicide, the volatile alarm pheromone penetrates to protected feeding sites and provokes aphids to move onto treated areas (Griffiths and Pickett, 1980) (Table 11.3.2). This technique is being tested in field trials and, if successful, will be the first demonstration of a practical use of aphid alarm pheromone.

It is also possible to inhibit the response of aphids to alarm pheromone by (-)-β-caryophyllene. Inhibition is detectable down to a ratio of 0.03 parts caryophyllene to 1 part (E)-β-farnesene (by weight) (Dawson et al., 1984). Such an inhibitor might minimize aphid dispersal from colonies attacked by predators, thus diminishing virus spread (Roitberg and Myers, 1978) and enhancing predator efficiency.

Other repellent chemicals

The so-called secondary plant substances are important in determining the host range of many insects – often acting as repellents, feeding deterrents or phagostimulants, some even acting as a deterrent to some insects and a phagostimulant to others (Kogan, 1977; Van Emden, 1978). Aphids are also affected by such substances, and chemicals such as phlorizin (Montgomery and Arn, 1974), and DIMBOA (2,4-dihydroxy-7-,methoxy-1.4-benzoxine-3-one) (Long et al., 1977) are feeding deterrents for various aphid species. Sinigrin is a feeding deterrent for some aphids and a phagostimulant for others (Nault and Styer, 1972). An extract from neem (*Azadirachta indica*) seeds also acts as a feeding deterrent: the most concentrated application tested more than halved the numbers of *M. persicae* on treated leaves (Griffiths et al., 1978). Feeding experiments using artificial media suggest that certain of these chemicals can be detected by aphids when they probe, but it is not clear whether probing can be deterred.

Section 11.3 references, p. 221

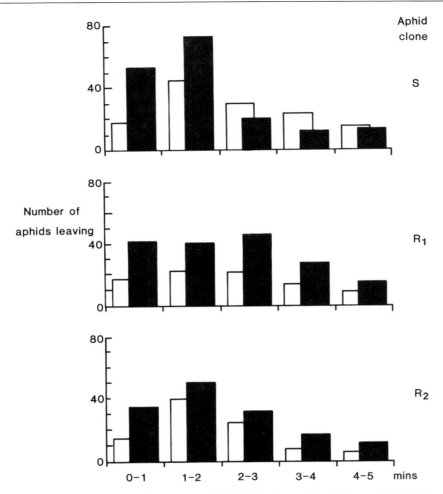

Fig. 11.3.1. Departure from untreated (white bars) or polygodial-treated (black bars) leaf discs of 200 insecticide-susceptible (S), moderately-resistant (R_1) or highly-resistant (R_2) *Myzus persicae* (Sulzer).

Polygodial

Apart from those discussed above, plants contain many chemicals which may also be repellent. Polygodial is a natural constituent of *Polygonum hydropiper*, the marsh pepper plant, from which it can be extracted, although it has also been synthesized. Apterous *M. persicae* are repelled when it is painted on glass, which shows it can be detected without ingestion (Rice et al., 1983a). Apterae quickly walked off polygodial-treated leaves and were repelled irrespective of their susceptibility to insecticides (Fig. 11.3.1). The non-persistently-transmitted potato virus Y (PVY) and the semi-persistently-transmitted beet yellows virus (BYV) were acquired less frequently from polygodial-treated leaves than from untreated leaves (Gibson et al., 1982a) (Table 11.3.3). This suggests that aphids were repelled by polygodial before a large number of them had probed or fed. In these experiments starved apterae were used, but Griffiths et al. (1982) reported that polygodial did not prevent alatae that had flown from making an initial probe into treated leaves.

Polygodial itself seems unlikely to be a valuable crop protectant because it has limited persistence, but its activity has demonstrated that the search for similar but more stable chemicals, which both repel aphids and hinder the transmission of non-persistently-transmitted viruses, is worth pursuing.

TABLE 11.3.3

Numbers of plants infected by *Myzus persicae* (Sulzer) after access to leaves with either potato virus Y (PVY) or beet yellows virus (BYV) and treated or untreated with polygodial at $1 \, \text{g} \, \text{l}^{-1}$

Aphid clone[a]	PVY test		BYV test	
	Treated leaves	Untreated leaves	Treated leaves	Untreated leaves
S	33/144 + + +	96/144	11/135 + + +	78/135
R_1	10/144 + + +	88/144	14/135 + +	62/135
R_2	16/144 + + +	95/144	10/135 + + +	39/135

+ +, + + + Treated means differ from corresponding untreated means at $P < 0.01$ and $P < 0.001$ respectively.
[a]S = susceptible, R_1 = moderately resistant, R_2 = highly resistant to insecticides (see section 11.1.1).

Carboxylic acids

The behavioural effects of carboxylic fatty acids have been investigated in considerable detail (Greenway et al., 1978; Sherwood et al., 1981). Monocarboxylic acids from carbon chain lengths C_8 to C_{13} were repellent to *M. persicae*, which showed most response to dodecanoic acid (C_{12}). In contrast, those with chain lengths greater than C_{16} enhanced settling. Interestingly, in a series of derivatives of aphid alarm pheromone with carboxylic acid side-chains of different lengths, one with a C_{11} side-chain was also the most repellent (Briggs et al., 1983). Dodecanoic acid was repellent when applied to leaf surfaces, and also when sandwiched between membranes enclosing artificial diet, so aphids may sense it during, rather than before, probing. Accordingly, neither undecanoic nor dodecanoic acid decreased the probability of an aphid making a test probe, although each shortened the duration of the probe from 148 to 37 and 39 s respectively (Phelan and Miller, 1982). This is ample time to acquire non-persistent viruses, and although acquisition of persistent and semi-persistent viruses was decreased, acquisition of potato virus Y was enhanced by treatment with dodecanoic acid, emphasizing the danger of using repellents that act after probing (Gibson et al., 1982a).

HOST PLANT RESISTANCE

Effect on behaviour

Non-preference is a well-recognized and important form of plant resistance to insects, and aphids often behave differently on host and non-host plants. Aphids rarely remain on a non-host plant and an example in which aphids remain until they die (Taylor, 1959) is sufficiently rare to merit recording. Commonly, modifications to aphid behaviour enhance other aspects of resistance. Even on a plant on which an aphid can survive but reproduces only slowly there may be an increased possibility that incoming alatae will leave without settling, as occurs with alate *A. fabae* on *Vicia faba* cv. Rastatter (Müller, 1958) and *Aphis gossypii* Glover on resistant muskmelon (Kennedy and Kishaba, 1977). Specific phagostimulants (Smith, 1957; Wensler, 1962) and repellents (Van Emden, 1978) are probably important in determining the host range of some aphid species, although the polyphagous aphids colonizing crop

Section 11.3 references, p. 221

plants may require only common nutrients as phagostimulants (Mittler, 1967; Akey and Beck, 1971; see also section 11.4, p. 229).

Some of the best examples of plants modifying aphid behaviour are those in which visual attractiveness is decreased. Brown cultivars of lettuce (Müller, 1964) and red cultivars of cabbage (Radcliffe and Chapman, 1965) are colonized less than green ones by incoming alatae. Virus spread may also be diminished, as occurs in forms of cucurbit in which the epidermal and palisade cells separate to provide a silvery, light-reflecting surface (Davis and Shifriss, 1983). The host plant can also influence the proportion of winged aphids produced (Harrewijn, 1978) with concomitant effects on aphid population behaviour, and more alatae were produced on cereals infected with barley yellow dwarf virus than in healthy plants (Gildow, 1983). There is even an experimental example of a wild potato plant producing sufficient aphid alarm pheromone to repel aphids (Gibson and Pickett, 1983), although its effect on aphids colonizing the plant naturally has yet to be investigated.

In many instances the way in which a plant resists aphid colonization is unknown, making it difficult to categorize or predict how aphid behaviour will be affected. However, the reproductive rate is generally less on resistant plants, and aphids may leave quickly so that few remain one day after artificial infestation. Thus, resistance of sugar beet to colonization by *A. fabae* and *M. persicae* involves both resistance to settling and multiplication. Settling and multiplication are sometimes linked, but resistance to settling of alatae is not always matched by resistance to colonization by apterae (Lowe and Russell, 1969). There are several examples of resistance to aphids being determined by a single gene, although polygenically-inherited resistance is also common. Resistance is usually effective against only a single species and not always against all biotypes of a species (Gibson and Plumb, 1977; see also section 11.4).

Host plant resistance and virus control

As with other means of aphid control, a main purpose of breeding resistant plants is to limit spread of viruses (Gibson and Plumb, 1977). This has often proved very successful, as for example in raspberry cultivars resistant to the raspberry aphid *Amphorophora idaei* Börner (Jones, 1979), especially if the virus is semi-persistently or persistently transmitted. These require – as we have seen in section 10.3 – long acquisition and inoculation feeds for efficient virus transmission. However, resistance and even immunity, except when based on a physical mechanism, generally provide little or no protection against non-persistently-transmitted plant viruses, as these are transmitted so quickly. Therefore, non-colonizing aphids are important vectors of such viruses in several crops and increased restlessness on non-host plants may actually cause spread.

There are, however, a few examples of resistance to aphids conferring protection against non-persistent viruses. Resistance to *M. persicae* in peach gives partial protection against inoculation of plum pox virus by *M. persicae*, although the plants remain susceptible to inoculation by *Brachycaudus helichrysi* (Kaltenbach) (Maison and Massonie, 1982) and resistance to *A. gossypii* in melon apparently protects completely against inoculation of cucumber mosaic virus (CMV) and several potyviruses by a French biotype of *A. gossypii*, although the plants were susceptible to inoculation of these viruses by other aphid species (Lecoq et al., 1979, 1980). The mechanism by which resistance to aphids in peach and melon is translated into resistance to virus inoculation has not been explained. Plant resistance can change probing behaviour (Tarn and Adams, 1982) possibly in such a way as to prevent inoculation. Even though this is only one of several explanations, the possibility of being able to modify aphid probing behaviour in such a way as to prevent inoculation of non-persistant viruses is so important that the mechanism(s) by which resistance to aphids provides virus resistance needs to be elucidated.

EFFECTS OF APHICIDES ON APHID BEHAVIOUR AND VIRUS SPREAD

Aphids are often exposed to sublethal doses of aphicides. Uneven spray penetration into a crop and misapplication can result in delivery of a sublethal amount of chemical, or, even when an ultimately lethal dose is delivered, slow intrinsic speed of action or insecticide resistance allows time for behaviour to be modified before the aphids die. As persistent, systemic or residual aphicides lose potency with time due to breakdown, redistribution and plant growth, their effectiveness may decline below that necessary to kill, but remain sufficient to affect behaviour. Behaviour-modifying effects may interact synergistically or antagonistically with toxic ones. For example, aphicide-induced restlessness may increase contact with chemical residues and enhance kill (Rice et al., 1983b) whereas aphid avoidance of treated foliage decreases it (Aliniazee, 1983).

Changes in behaviour are important because they affect both the control of direct aphid damage and, more especially, aphid-borne virus spread. The virus transmission process is behaviour-dependent and requires a sequence of events in which probing, movement and timing are critical. Small changes in this sequence can restrict or favour virus spread. In practice, behavioural effects which reduce spread are not easily separable from toxic ones, although Highwood (1979) observed a decrease in spread of viruses in pyrethroid-treated crops and suggested that a repellent/irritant effect on the aphid vectors could be important. On the other hand, reports of aphicides enhancing spread – particularly of non-persistently-transmitted viruses – are numerous, although evidence that behavioural changes are directly involved is often circumstantial (Broadbent, 1957; Shanks and Chapman, 1965). Ferro et al. (1980) noted apparent enhancement by aldicarb of maize dwarf mosaic virus infection in sweetcorn, and Gabriel et al. (1981) reported increased potato virus Y infection in dimethoate- and pirimicarb-treated plots. Foster et al. (1981) recorded increased spread of the persistently-transmitted potato leafroll virus in disulfoton- and thiofanox-treated crops infested with insecticide-resistant aphids. Although behavioural effects are strongly implicated, increased spread could have been caused by the build-up of large aphid populations in treated plots due to the killing of predators or to increases in aphid fecundity and longevity (Parry and Ford, 1971; Ritcey et al., 1982; Coombes, 1983; Neubauer et al., 1983).

Effects on aphid probing

The aim of virus control by aphicides has been to reduce the numbers of aphid vectors moving within crops. It has commonly been assumed by growers and insecticide manufacturers that virus control automatically follows from aphid control. As a result, much emphasis has been placed on aphicide toxicity and persistence. Furthermore, short-term, killing-power of aphicides is much more readily assessed than possible long-term effects on aphid behaviour and virus transmission. Some studies, however, have recognized the importance of behavioural effects and attempts to define them in the laboratory have illustrated some complex interactions. For example, in tests on aphid probing. Lehman et al. (1975) found that methyl parathion and dimethoate halved the

Section 11.3 references, p. 221

number of short "trial" probes made on various host plants by *M. persicae, A. fabae* and *Acyrthosiphon pisum* (Harris) and they suggested this decreased the probability of transmission of non-persistently-transmitted viruses. However, the number of probes made by *Macrosiphum euphorbiae* (Thomas) was sometimes increased, presumably increasing the chances of transmission. The duration of the first "feed", which might affect semi-persistent or persistent transmission, was increased in *A. fabae* and *A. pisum* but decreased in *M. euphorbiae* and *M. persicae*. Pirimiphos-methyl and pirimicarb prolonged initial probing and decreased feeding periods of *M. persicae* and *A. fabae*, but permethrin had the reverse effect (Matthieu and Verhoyen, 1980). The duration and frequency of probes by *A. pisum, M. persicae* and *Aphis craccivora* Koch was decreased due to the "deterrent" effect of the pyrethroids deltamethrin or permethrin, but transmission of the non-persistently-transmitted bean yellow mosaic virus was only slightly decreased, whereas that of the persistent potato leafroll virus was greatly decreased (Sassen, 1983). It is clear from these results that aphicides can affect probing in ways which influence virus transmission. Griffiths et al. (1982) suggest that these behavioural changes may be particularly subtle.

Effects on aphid movement

Whether changes induced in behaviour restrict or enhance virus spread depends not only on whether or not changes in probing favour transmission,

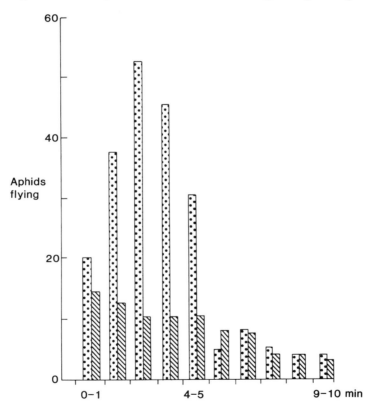

Fig. 11.3.2 Departure from untreated (striped bars) or deltamethrin-treated (10 mg a.i. per litre) (dotted bars) leaves of moderately-resistant (R₁) *Myzus persicae* (Sulzer).

but also on whether aphid movement is stimulated or deterred. In glasshouse tests, Shanks and Chapman (1965) showed that different insecticides, as well as having different effects on probing, differently affected movement and the transmission of potato virus Y by *M. persicae*. After initial probes, wingless aphids made longer feeding probes on parathion-treated than on untreated foliage, but the reverse was true with DDT. Similarly, winged aphids stayed longer on parathion- or phorate-treated than on untreated leaves, but for a shorter time on those treated with DDT. The irritant action of DDT did not, in these tests, affect the rate of transmission of PVY, although parathion applied to source plants greatly diminished PVY-dissemination by winged aphids; this is probably due, in part, to its arrestant action.

Although a pesticide may induce behavioural changes, its toxicity often limits the time available for virus transfer. Non-persistently-transmitted viruses can be transmitted in a few minutes, whereas several days may be required for persistently-transmitted viruses. The speed of knockdown or kill is therefore crucial and is dependent on the type of aphicide, the amount that reaches the aphid and the aphid's sensitivity to the insecticide. With insecticide-resistant *M. persicae* (Sawicki and Rice, 1978; Devonshire and Moores, 1982) an application rate of deltamethrin equivalent to that used in crops made pyrethroid-resistant *M. persicae* move within a few minutes of treatment (Rice et al., 1983a) (Fig. 11.3.2). Moderately and strongly resistant aphids walked or flew from treated leaves, and when those that dispersed from PVY-infected leaves were transferred to healthy, untreated seedlings the moderately resistant strain did not infect, whereas the strongly resistant one did. Lack of transmission may have been due to deterrent effects of delta-methrin on probing caused either directly (Sassen, 1983) or as a result of induced restlessness. However, although deltamethrin initially stimulates walking or flight this is rapidly followed by a period of paralysis and incapacitation (Gibson et al., 1982b), during which PVY-infectivity is probably lost.

Pyrethroids probably do not enhance virus spread because they modify aphid behaviour only after direct contact with the aphid. It is not clear whether responses provoked by this contact result from detection of the chemical by sense organs or from direct action of the chemical on the nerves, since either "true" repellency or hyperactivity could result in movement. Similarly, probing behaviour could be altered either through stimulation of chemoreceptors

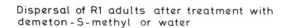

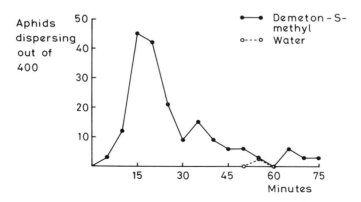

Fig. 11.3.3. Dispersal of moderately-resistant (R_1) adult apterae of *Myzus persicae* (Sulzer) after treatment with demeton-S-methyl (500 mg a.i. per litre) (●—●) or water (○—○).

Section 11.3 references, p. 221

and gustatory sense organs or by affecting the nerves. In practice, how behaviour is modified is probably not important, although direct action on the nervous system may be more "dose"-dependent than sensory detection and perhaps less likely to stimulate insecticide-resistant aphids.

Whatever the basis for behavioural changes induced by pyrethroids, the need for contact means that beneficial, toxicity-associated effects are maximized. In contrast, organophosphorus and carbamate aphicides can drastically change the behaviour of wingless *M. persicae* without contact, by inducing alarm pheromone release. In laboratory tests (Rice et al., 1983b), treatment of aphid-infested leaves with concentrations of demeton-S-methyl or pirimicarb equivalent to those used on crops caused secretion of the pheromone. The young and the more insecticide-susceptible individuals secreted more quickly than other aphids and soon died but "triggered" dispersal of nearby aphids (Fig. 11.3.3), often before these had suffered any ill-effects from the aphicide. Resistant aphids responded readily to the pheromone and were able to survive exposure to demeton-S-methyl-treated foliage long enough to move to adjacent plants and infect them with PVY or BYV. The extent to which they were able to do this depended, as it did with deltamethrin, on their level of resistance and also, in this case, on transmission characteristics of the virus. Both viruses were readily transmitted when only source leaves were treated, but when indicator seedlings were also sprayed the non-persistently-transmitted PVY was spread by moderately resistant aphids and the semi-persistently-transmitted BYV only by strongly resistant aphids.

It is difficult to judge the practical significance of the behaviour-modifying properties of aphicides demonstrated in the laboratory. Organophosphorus and carbamate aphicides decrease the numbers of aphids moving within crops rather than prevent acquisition or inoculation, and so it might be expected that pre- or sub-lethal movement-provoking properties such as induction of alarm pheromone release might, in some instances, increase rather than decrease virus spread. Although there is no direct field evidence of this, increased virus spread has been associated with aphicides of these groups (Gabriel et al., 1981).

Pyrethroids are used increasingly against aphid-borne viruses. Their rapid, residual "knockdown" action not only curtails movement over treated foliage but may also restrict virus transmission, even of non-persistently-transmitted viruses (Asjes, 1981; Gibson et al., 1982b; Rice et al., 1983a). Although movement is initially provoked, incapacitation ensues so rapidly that enhanced virus spread seems unlikely. This conclusion is supported by results from flight-chamber and field experiments (Rice et al., 1983a). Nevertheless, it must be emphasized that because *M. persicae* and several other aphid species are now resistant to organophosphorus, carbamate and pyrethroid aphicides, reduced kill and slower speed of action favours enhancement of virus spread. Whether and to what extent spread is enhanced may depend largely on the numbers of highly resistant aphid vectors present in the crop.

Although induced changes in aphid mobility are important in the control of virus spread, so also are changes in probing behaviour. The different roles of winged and wingless aphids should also be recognized. Behavioural responses induced in wingless aphids in the crop at spraying occur when conditions are most adverse. Winged aphids arriving later, when toxicity is declining, are more mobile and more sensitive, and so may be more susceptible to behavioural influence. It is perhaps significant that enhanced virus spread reported by Ferro et al. (1980) and Foster et al. (1981) was in crops treated with systemic aphicides at planting.

Greater understanding of the behavioural effects of pesticides would enable a more rational choice to be made of aphicides for particular virus–vector problems based not only on toxicity but also after consideration of pre- and sub-lethal behaviour-modifying properties. Furthermore, it may be possible, by careful choice of aphicide or aphicide mixtures, to manipulate aphid behaviour advantageously. For example, applying mixtures of systemic, alarm-pheromone-inducing organophosphorus or carbamate aphicides with contact pyrethroids could "flush out" aphids from refuges within the crop not reached by the pyrethroid onto spray-exposed foliage where it could act. This should maximize kill and minimize virus spread (Rice et al., 1983b). At present, however, we fail to make good use of the behavioural properties of aphicides. We are also failing adequately to recognize, understand and avoid those circumstances in which they frustrate, or even reverse, the desired aim of control.

CONCLUSIONS

The use of reflective materials to repel alates from crops is an established technique for controlling spread of aphid-borne viruses in high-value horticultural crops. Chemical repellents have already shown great potential value and warrant continued research, although they have not yet been developed to a stage at which any are used commercially. This may, at least in part, be due to our lack of basic knowledge of aphid chemoreception and of the behavioural responses of flying aphids. In the short term, considerable benefit may be derived from eliminating adverse usage of pesticides. These chemicals are already used extensively on our crops with little or no knowledge of their behavioural effect on aphids. When these effects have been investigated, as to a limited extent they have been with insecticides, behaviour has frequently been found to be changed in a manner likely to affect the spread of viruses.

The likelihood of other pesticides, such as herbicides or fungicides, affecting aphid behaviour largely remains uninvestigated, although these chemicals are used more extensively than insecticides. It is vital that this subject should receive more attention in the future, in order to ensure that appropriate pesticides are used for particular crop situations and in a manner providing maximum control not only of aphids but also of virus diseases.

REFERENCES

Akey, D.H. and Beck, S.D., 1971. Continuous rearing of the pea aphid, *Acyrthosiphon pisum*, on a holidic diet. Annals of the Entomological Society of America, 64: 353–356.

Alikhan, M.A., 1960. The experimental study of the chemotactic basis of host-specificity in a phytophagous insect, *Aphis fabae* Scop. (Aphididae, Homoptera). Annales Universitatis Mariae Curie - Sklowdowcks, Section C, 12: 117–159.

Aliniazee, M.T., 1983. Carbaryl resistance in the filbert aphid (Homoptera: Aphididae). Journal of Economic Entomology, 76: 1002–1004.

Asjes, C.J., 1981. Control of stylet-borne spread of aphids of tulip breaking virus in lilies and tulips, and hyacinth mosaic virus in hyacinths by pirimicarb and permethrin sprays versus mineral-oil sprays. Mededelingen van de Faculteit Landbouwwetenschappen Rijksuniversiteit Gent, 46: 1073–1077.

Bowers, W.S., Nishino, C., Montgomery, M.E. and Nault, L.R., 1977. Structure-activity relationships of analogs of the aphid alarm pheromone, (E)-β-farnesene. Journal of Insect Physiology, 23: 697–701.

Briggs, G.G., Dawson, G.W., Gibson, R.W., Griffiths, D.C., Pickett, J.A., Rice, A.D., Stribley, M.F. and Woodcock, C.M., 1983. Compounds derived from the aphid alarm pheromone of potential use in controlling colonisation and virus transmission by aphids. In: J. Miyamoto et al. (Editors), IUPAC Pesticide Chemistry. Pergamon Press, pp. 117–122.

Broadbent, L., 1957. Insecticidal control of the spread of plant viruses. Annual Review of Entomology, 2: 339–353.

Bromley, A.K. and Anderson, M., 1982. An electrophysiological study of olfaction in the aphid *Nasonovia ribis-nigri*. Entomologia Experimentalis et Applicata, 32: 101–110.

Chapman, R.F., Bernays, E.A. and Simpson, S.J., 1981. Attraction and repulsion of the aphid, *Cavariella aegopodii*, by plant odors. Journal of Chemical Ecology, 7: 881–888.

Cohen, S., 1981. Reducing the spread of aphid-transmitted viruses in peppers by coarse-net cover. Phytoparasitica, 9: 69–76.

Cohen, S. and Marco, S., 1973. Reducing the spread of aphid-transmitted viruses in peppers by trapping the aphids on sticky yellow polyethylene sheets. Phytopathology, 63: 1207–1209.

Coombes, D.S., 1983. Sublethal effects of demeton-S-methyl on the development and fecundity of *Myzus persicae* (Sulzer). In: Tests of Agrochemicals and Cultivars, 4: 38–39 (Annals of Applied Biology, 102, Supplement).

Costa, C.L. and Lewis, T., 1967. The relationship between the size of yellow water traps and catches of aphids. Entomologia Experimentalis et Applicata, 10: 485–487.

Davis, R.F. and Shifriss, O., 1983. Natural virus infection in silvery and nonsilvery lines of *Cucurbita pepo*. Plant Disease, 67: 379–380.

Dawson, G.W., Gibson, R.W., Griffiths, D.C., Pickett, J.A., Rice, A.D. and Woodcock, C.M., 1982a. Aphid alarm pheromone derivatives affecting settling and transmission of plant viruses. Journal of Chemical Ecology, 8: 1377–1388.

Dawson, G.W., Griffiths, D.C., Pickett, J.A., Smith, M.C. and Woodcock, C.M., 1982b. Improved preparation of (*E*)-β-farnesene and its activity with economically important aphids. Journal of Chemical Ecology, 8: 1111–1117.

Dawson, G.W., Griffiths, D.C., Pickett, J.A. and Woodcock, C.M., 1983. Decreased response to alarm pheromone by insecticide-resistant aphids. Naturwissenschaften, 70: 254–255.

Dawson, G.W., Griffiths, D.C., Pickett, J.A., Smith, M.C. and Woodcock, C.M., 1984. Natural inhibition of the aphid alarm pheromone. Entomologia Experimentalis et Applicata, 36: 197–199.

Devonshire, A.L. and Moores, G.D., 1982. A carboxylesterase with broad substrate specificity causes organophosphorus, carbamate and pyrethroid resistance in peach-potato aphids (*Myzus persicae*). Pesticide biochemistry and physiology, 18: 235–246.

Dunn, J.A., 1969. The colonisation by *Cavariella aegopodii* Scop. of carrot plants of different sizes. Annals of Applied Biology, 63: 318–324.

Dunn, J.A., 1978. Antennal sensilla of vegetable aphids. Entomologia Experimentalis et Applicata, 24: 348–349.

Ferro, D.N., Mackenzie, J.D. and Margolies, D.C., 1980. Effect of mineral oil and a systemic insecticide on field spread of aphid-borne maize dwarf mosaic virus in sweet corn. Journal of Economic Entomology, 73: 730–734.

Foster, G.N., McKinlay, R.G., Shaw, M.W., Aveyard, C., Gordon, S.C. and Woodford, J.A.T., 1981. The control of the aphids and leaf roll virus disease of potatoes by granular insecticides. Proceedings: Conference on Crop Protection in Northern Britain, Dundee, 1981, Scottish Crop Research Institute, pp. 91–96.

Gabriel, W., Szulc, J. and Wislocka, M., 1981. Influence de la distance des sources d'infection sur l'effet de traitements à l' aide d'insecticides systémiques sur la propagation des virus Y et M de la pomme de terre. Potato Research, 25: 1–11.

Gibson, R.W. and Gunenc, Y., 1981. Effects of covering potato crops with clear polyethylene film on spread of potato virus Y. Plant Pathology, 30: 233–235.

Gibson, R.W. and Pickett, J.A., 1983. Wild potato repels aphids by release of aphid alarm pheromone. Nature, 302: 608–609.

Gibson, R.W. and Plumb, R.T., 1977. Breeding plants for resistance to aphid infestation. In: K.F. Harris and K. Maramorosch (Editors), Aphids as Virus Vectors. Academic Press, New York London, pp. 473–500.

Gibson, R.W., Rice, A.D., Pickett, J.A., Smith, M.C. and Sawicki, R.M., 1982a. Effects of the repellents dodecanoic acid and polygodial on the acquisition of non-, semi- and persistent plant viruses by *Myzus persicae*. Annals of Applied Biology, 100: 55–59.

Gibson, R.W., Rice, A.D. and Sawicki, R.M., 1982b. Effects of the pyrethroid deltamethrin on the acquisition and inoculation of viruses by *Myzus persicae*. Annals of Applied Biology, 100: 49–54.

Gibson, R.W., Pickett, J.A., Dawson, G.W., Rice, A.D. and Stribley, M.F., 1984. Effects of aphid alarm pheromone derivatives and related compounds on non- and semi-persistent plant virus transmission by *Myzus persicae*. Annals of Applied Biology, 104: 203–209.

Gildow, F.E., 1983. Influence of barley yellow dwarf virus-infected oats and barley on morphology of aphid vectors. Phytopathology, 73: 1196–1199.

Greenway, A.R., Griffiths, D.C. and Lloyd, S.L., 1978. Response of *Myzus persicae* to components of aphid extracts and to carboxylic acids. Entomologia Experimentalis et Applicata, 24: 369–374.

Griffiths, D.C. and Pickett, J.A., 1980. A potential application of aphid alarm pheromones. Entomologia Experimentalis et Applicata, 27: 199–201.

Griffiths, D.C., Greenway, A.R. and Lloyd, S.L., 1978. The influence of repellent materials and aphid extracts on settling behaviour and the larviposition of *Myzus persicae* (Sulzer). (Hemiptera: Aphididae). Bulletin of Entomological Research, 68: 613–619.

Griffiths, D.C., Pickett, J.A. and Woodcock, C.M., 1982. Behaviour of alatae of *Myzus persicae* (Sulzer) (Hemiptera: Aphididae) on chemically treated surfaces after tethered flight. Bulletin of

Entomological Research, 72: 687–693.

Harpaz, I., 1982, Nonpesticidal control of vector-borne viruses. In: K.F. Harris and K. Maramorosch (Editors), Pathogens, Vectors and Plant Diseases: Approaches to Control. Academic Press. New York, London, pp. 1–21.

Harrewijn, P., 1978. The role of plant substances in polymorphism of the aphid *Myzus persicae*. Entomologia Experimentalis et Applicata, 24: 198–214.

Heathcote, G.D., 1969. Cultural factors affecting colonisation of sugar beet by different aphid species. Annals of Applied Biology, 63: 330–331.

Highwood, D.P., 1979. Some indirect benefits of the use of pyrethroid insecticides. In: Proceedings of the 1979 British Crop Protection Conference, Pests and Diseases Brighton, 1979, pp. 361–369.

Hille Ris Lambers, D. and Schepers, A., 1978. The effect of trans-β-farnesene, used as a repellent against landing aphid alatae in seed potato growing. Potato Research, 21: 23–26.

Johnson, G.V., Bing, A. and Smith, F.F., 1967. Reflective surfaces used to repel dispersing aphids and reduce spread of aphid-borne cucumber mosaic virus in Gladiolus plantings. Journal of Economic Entomology, 60: 16–19.

Jones, A.T., 1979. Further studies on the effect of resistance to *Amphorophora idaei* in raspberry (*Rubus idaeus*) on the spread of aphid-borne viruses. Annals of Applied Biology, 92: 119–123.

Kennedy, G.G. and Kishaba, A.N., 1977. Response of alate melon aphids to resistant and susceptible muskmelon lines. Journal of Economic Entomology, 70: 407–410.

Kogan, M., 1977. The role of chemical factors in insect/plant relationships. In: Proceedings of the XV International Congress of Entomology, Washington, 1976, pp. 211–227.

Kring, J.B., 1972. Flight behaviour of aphids. Annual Review of Entomology, 17: 461–492.

Lecoq, H., Cohen, S., Pitrat, M. and Labonne, G., 1979. Resistance to cucumber mosaic virus transmission by aphids in *Cucumis melo*. Phytopathology, 69: 1223–1225.

Lecoq, H., Labonne, G. and Pitrat, M., 1980. Specificity of resistance to virus transmission by aphids in *Cucumis melo*. Annales de Phytopathologie, 12: 139–144.

Lehman, W., Claus, S. and Karl, E., 1975. Das Verhalten von Aphiden unter dem Einfluss verschiedener Pflanzenarten und ihrer chemischen Behandlung bezogen auf den Anteil einstechender Tiere und die Kurzfristigen Probesaustiche. Archiv für Phytopathologie und Pflanzenschutz, 11: 273–282.

Loebenstein, G. and Raccah, B., 1980. Control of non-persistently transmitted aphid-borne viruses. Phytoparasitica, 8: 221–235.

Long, B.J., Dunn, G.M., Bowman, J.S. and Routley, D.G., 1977. Relationship of hydroxamic acid content in corn and resistance to the corn leaf aphid. Crop Science, 17: 55–58.

Lowe, H.J.B. and Russell, G.E., 1969. Inherited resistance of sugar beet to aphid colonisation. Annals of Applied Biology, 63: 337–344.

Maison, P. and Massonie, G., 1982. Premières observations sur la spécificité de la resistance du pêcher à la transmission aphidienne du virus de la Sharka. Agronomie, 2: 681–683.

Matthieu, J.L. and Verhoyen, M., 1980. Facteurs influençant les aphides dans leur comportement de sélection de l'hôte. I. Les pucerons et les plantes. Mededelingen van de Faculteit Landbouwwetenschappen, Rijksuniversiteit Gent, 45: 485–511.

Mittler, T.E., 1967. Effect of amino acid and sugar concentrations on the food uptake of the aphid *Myzus persicae*. Entomologia Experimentalis et Applicata, 10: 39–51.

Montgomery, M.E. and Arn, H., 1974. Feeding response of *Aphis pomi*, *Myzus persicae*, and *Amphorophora agathonica* to phlorizin. Journal of Insect Physiology, 20: 413–421.

Montgomery, M.E. and Nault, L.R., 1977. Comparative response of aphids to the alarm pheromone, (*E*)-β-farnesene. Entomologia Experimentalis et Applicata, 22: 236–242.

Montgomery, M.E. and Nault, L.R., 1978. Effects of age and wing polymorphism on the sensitivity of *Myzus persicae* to alarm pheromone. Annals of the Entomological Society of America, 71: 788–790.

Müller, F.P., 1983. Differential alarm pheromone responses between strains of the aphid *Acyrthosiphon pisum*. Entomologia Experimentalis et Applicata, 34: 347–348.

Müller, H.J., 1958. The behaviour of *Aphis fabae* in selecting its host plants, especially different varieties of *Vicia faba*. Entomologia Experimentalis et Applicata, 1: 66–72.

Müller, H.J., 1964. Über die Auflugdichte von Aphiden auf farbige Salatpflanzen. Entomologia Experimentalis et Applicata, 7: 85–104.

Nault, L.R. and Bowers, W.S., 1974. Multiple alarm pheromones in aphids. Entomologia Experimentalis et Applicata, 17: 455–457.

Nault, L.R. and Styer, W.E., 1972. Effects of sinigrin on host selection by aphids. Entomologia Experimentalis et Applicata, 15: 423–437.

Nault, L.R., Edwards, L.J. and Styer, W.E., 1973. Aphid alarm pheromones: secretion and reception. Environmental Entomology, 2: 101–105.

Nawrocka, B.Z., Eckenrode, C.J., Uyemoto, J.K. and Young, D.H., 1975. Reflective mulches and foliar sprays for suppression of aphid-borne virus in lettuce. Journal of Economic Entomology, 68: 694–698.

Neubauer, I., Raccah, B., Aharonson, N., Swirski, E. and Ishaaya, I., 1983. Systemic effect of aldicarb, dimethoate and ethiofencarb on mortality and population dynamics of the spirea aphid *Aphis citricola* Van der Goot. Crop Protection, 2: 211–218.

Parry, W.M. and Ford, J.B., 1971. The artificial feeding of phosphamidon to *Myzus persicae*. Entomologia Experimentalis et Applicata, 14: 389–398.

Pettersson, J., 1970. Studies on *Rhopalosiphum padi* (L.). (1) Laboratory studies of olfactometric responses to the winter host *Prunus padus* L. Landbrukshogskolans Annaler, 36: 381–399 (in Swedish, with English summary).

Phelan, P.L. and Miller, J.R., 1982. Post-landing behaviour of alate *Myzus persicae* as altered by (*E*)-β-farnesene and three carboxylic acids. Entomologia Experimentalis et Applicata, 32: 46–53.

Radcliffe, E.B. and Chapman, R.K., 1965. Seasonal shifts in the relative resistance to insect attack of eight commercial cabbage varieties. Annals of the Entomological Society of America, 58: 892–897.

Raymundo, S.A. and Alcazar, J., 1983. Some components of integrated pest management on potatoes. In: Proceedings of 10th International Congress of Plant Protection, Brighton, p. 1206.

Rice, A.D., Gibson, R.W. and Stribley, M.F., 1983a. Effects of deltamethrin on walking, flight and potato virus Y-transmission by pyrethroid-resistant *Myzus persicae*. Annals of Applied Biology, 102: 229–236.

Rice, A.D., Gibson, R.W. and Stribley, M.F., 1983b. Alarm pheromone secretion by insecticide-susceptible and -resistant *Myzus persicae* treated with demoton-S-methyl; aphid dispersal and transfer of plant viruses. Annals of Applied Biology, 103: 375–381.

Ritcey, G., McGraw, R. and McEwen, F.L., 1982. Insect control on potatoes in Ontario from 1973 to 1982. Proceedings of the Entomological Society of Ontario, 113: 1–6.

Roitberg, B.D. and Myers, J.H., 1978. Effect of adult Coccinellidae on the spread of a plant virus by an aphid. Journal of Applied Ecology, 15: 775–779.

Sassen, B., 1983. The effect of two pyrethroids on the feeding behaviour of three aphid species and on transmission of two different viruses. Journal of Plant Diseases and Protection, 90: 119–126.

Sawicki, R.M. and Rice, A.D., 1978. Response of susceptible and resistant peach-potato aphids *Myzus persicae* (Sulz.) to insecticides in leaf-dip bioassays. Pesticide Science, 9: 513–516.

Shanks, C.H. and Chapman, R.K., 1965. The effects of insecticides on the behaviour of the green peach aphid and its transmission of potato virus Y. Journal of Economic Entomology, 58: 79–83.

Sherwood, M.H., Greenway, A.R. and Griffiths, D.C., 1981. Responses of *Myzus persicae* (Sulzer) (Hemiptera: Aphididae) to plants treated with fatty acids. Bulletin of Entomological Research, 71: 133–136.

Smith, B.D., 1957. A study of the factors affecting the populations of aphids on *Sarothamnus scoparius* L. Doctoral Thesis, University of London.

Smith, J.G., 1976. Influence of crop background on aphids and other phytophagous insects on Brussels sprouts. Annals of Applied Biology, 83: 1–13.

Tarn, R.T. and Adams, J.B., 1982. Aphid probing and feeding, electronic monitoring, and plant breeding. In: K.F. Harris and K. Maramorosch (Editors), Pathogens, Vectors, and Plant Diseases: Approaches to Control. Academic Press, New York, London, pp. 221–246.

Taylor, L.R., 1959. Abortive feeding behaviour in a black bean aphid of the *Aphis fabae* group. Entomologia Experimentalis et Applicata, 2: 143–153.

Tjallingii, W.F., 1978. Mechanoreceptors of the aphid labium. Entomologia Experimentalis et Applicata, 24: 731–737.

Van Emden, H.F., 1978. Insects and secondary plant substances – an alternative viewpoint with special reference to aphids. In: H.J. Harborne (Editor), Biochemical Aspects of Plant and Animal Coevolution. Academic Press, New York, London, pp. 309–323.

Visser, J.H., 1983. Differential sensory perceptions of plant compounds by insects. In: P.A. Hedin (Editor), Plant Resistance to Insects. American Chemical Society Symposium, Series 288, pp. 215–230.

Wensler, R.J.D., 1962. Mode of host selection by an aphid. Nature, 195: 830–831.

Wensler, R.J.D., 1963. Sensory physiology of host selection in some aphids. Doctoral Thesis, Cambridge University, 74 pp.

Wohlers, P., 1981. Aphid avoidance of plants contaminated with alarm pheromone (*E*)-β-farnesene. Zeitschrift für Angewandte Entomologie, 92: 329–336.

Wohlers, P., 1982. Effect of alarm pheromone (*E*)-β-farnesene on aphid behaviour during flight and after landing on plants. Zeitschrift für Angewandte Entomologie, 93: 102–108.

Wyman, J.A., Toscano, N.C., Kido, K., Johnson, H. and Mayberry, K.S., 1979. Effects of mulching on the spread of aphid-transmitted watermelon mosaic virus to summer squash. Journal of Economic Entomology, 72: 139–143.

Yang, S.L. and Zettler, F.W., 1975. Effects of alarm pheromones on aphid probing behaviour and virus transmission efficiency. Plant Disease Reporter, 59: 902–905.

11.4 Host Plant Resistance

JACQUES L. AUCLAIR

INTRODUCTION

In his classical book, Painter (1951) defined resistance of plants to insect attack as the relative amount of heritable qualities possessed by the plant which influence the ultimate degree of damage done by the insect. This broad concept of resistance reflects the complexity of the phenomenon, and a plant is no longer a host to an insect when it is immune to its attacks. Plant resistance is heritable and controlled by one or more genes; it is relative and can be measured only by comparison with other genotypes; it is variable and can be modified by physical, chemical and biological factors. Painter proposed three General mechanisms to account for plant resistance to insect damage: (1) Non-preference (the terms non-acceptance and antixenosis were proposed by Van Marrewijk and De Ponti, 1975, and Kogan and Ortman, 1978, respectively), which is shown by plants that are unattractive or unsuitable for colonization or oviposition by an insect; (2) antibiosis, which adversely affects the insect life history, such as reduced growth, reproduction or survival, when the insect uses a resistant host plant for food; and (3) tolerance, which enables a host plant to grow and reproduce itself or to repair injury to a marked degree in spite of supporting a population approximately equal to that damaging a susceptible host. These three mechanisms, although somewhat empirical and vaguely delimited, have been used by a majority of workers in the field. Russell (1978) suggested a fourth type of resistance, pest avoidance, which is a tendency to escape infestation, e.g. because the host plant is not at a susceptible stage when pest populations are at their peak.

Basic plant characteristics that may impart resistance or susceptibility to insects can be morphological, such as variation in f oliage size, shape, colour, pubescence, hardness or thickness of tissue, and biochemical, which includes the nutritional composition of plant tissue, and especially the proportion of essential nutrients, and also includes allelochemic factors such as allomones (e.g. repellents, oviposition and feeding deterrents, toxicants) and kairomones (e.g. attractants, arrestants, excitants). With the advent of integrated pest management, the use of insect-resistant plants in combination with other control measures is possibly the most convenient and economical method of insect control. Its desirable features include specificity to one or several pests; cumulative effectiveness, compounded in successive insect generations; persistence for several years; harmony with the environment; ease of adoption into normal farm operations, usually at no extra cost; and compatibility with other tactics in pest management (Pathak, 1970; Kogan, 1982). The utility of resistant varieties and the millions of dollars saved by growers has been documented by Luginbill (1969).

Section 11.4 references, p. 254

Many reviews on insect resistance in crop plants have appeared, (e.g. Snelling, 1941; Painter, 1951, 1958a; Maxwell et al., 1972; Panda 1979; Maxwell and Jennings, 1980), as well as on plant resistance to aphids (e.g. Painter, 1958b; Gibson and Plumb, 1977). One of the first significant contributions to the control of aphids by plant resistance occurred in 1890, when the French wine industry was protected from severe damage and possible destruction from the grape phylloxera, *Viteus vitifoliae* (Fitch), by the successful grafting of European grape vines onto resistant rootstock from North America (Maxwell and Jennings, 1980). Since this early contribution, aphids have become the most frequent group of insects for which plant resistance has been reported. During the period 1937–1956, varieties of alfalfa, barley, corn and sorghum resistant to aphids were already being used by farmers in many parts of the U.S.A., some varieties originating after long-planned research (Painter, 1958a).

Damage to plants caused by aphids includes sap withdrawal, injection of toxic saliva, dissemination of plant virus diseases, and excretion of usually sticky honeydew which may gum up the plant and serve as a medium for the growth of sooty mold fungus. This topic is covered at length under Chapter 10 of this book. Environmental effects affecting aphid populations are dealt with in section 5.2, and environmental factors influencing the expression of plant resistance to insects, including some twelve species of aphids, was well reviewed in Maxwell and Jennings (1980). Because of space limitations in relation to the huge number of publications on plant resistance to aphids, references discussed in this review include mainly selected examples from more recent work on the subject.

BREEDING PLANTS RESISTANT TO APHIDS

Standard books on plant breeding include those by Hayes et al. (1955), Allard (1960), and Mayo (1980); the general principles and methods of breeding for pest and disease resistance are detailed in Russell (1978). Some contemporary aspects of plant biotechnology, such as induced mutations, plant tissue culture, and the production and application of haploid plants, are reviewed in Mantell and Smith (1983) and Vose and Blixt (1984). According to Borlaug (1983), it is doubtful that significant production benefits will soon be forthcoming from the use of modern techniques with higher plants, especially polyploid species, and conventional breeding will continue for a long time to be the main research approach for crop improvement. A major barrier to a genetically engineered plant is our limited understanding of the molecular bases of gene expression (Barton and Brill, 1983).

A more optimistic view is endorsed by Schneiderman (1984), who forecasted that crops engineered to be resistant to insects and other pests will be available by the 1990s.

Breeding plants for resistance to aphids does not differ fundamentally from breeding for any other plant character. Methods suitable in a particular case will depend largely upon the breeding system of the host plant concerned, for example whether it is mainly open or cross-pollinated (e.g. alfalfa, apple, many Brassicaceae, some clovers, cucumber, maize, rye) or self-pollinated (e.g. barley, cotton, lettuce, oats, pea, sorghum, tobacco, tomato, wheat) and on the sources of resistance that are available. Several important crops, including most types of fruit and potato, are normally propagated vegetatively. In the cross-pollinated types, any individual plant selected for its resistance cannot by itself usually form the basis of a variety, unless it can be propagated vegetatively on a large scale. This is because many cross-pollinated plants are self-incompatible and thus cannot be selfed. If they are self-compatible, inbreeding usually results in a marked depression in vigour and yielding ability (inbreeding depression). This contrasts with self-pollinated crops, where inbreeding does not usually substantially decrease vigour and yield. The choice of methods to develop new cultivars depends on the objectives of the breeding programme, the reproductive characteristics of the plant, and the inheritance of the character to be incorporated into the breeding lines. Some of the most commonly used methods include mass selection (used in alfalfa with aphids, e.g. by Dahms and Painter, 1940; Harvey et al., 1972), backcrossing, pedigree selection, single-seed descent, early-generation testing, and recurrent selection. A description summarizing these methods is given by Niles (1980a), and plant breeding techniques for several crop plants are described in Maxwell and Jennings (1980).

Germplasm banks of plant species are a huge source of plant material in a search for resistance to insects; it is estimated that about two million plant samples in approximately 1500 collections have been located throughout the world (Gallun et al., 1975; Maxwell and Jennings, 1980). Sources of resistance within a crop species are the easiest for breeders to utilize: sources within other species or even genera may need special methods to produce fertile hybrids. A fruitful area for finding resistant plants is often where the pest is indigenous or is at the centre of origin of the host species, and this may be more so if the insect is a recent introduction. Readers interested in the genetics of plant resistance to aphids are referred to reviews by Gibson and Plumb (1977) and Maxwell and Jennings (1980), in which the genetics of resistance in alfalfa, apple, barley, lettuce, muskmelon, oats, raspberry, rye, sorghum, sugar beet, sweet clover and wheat, against some thirteen species of aphids are examined. Usually dominant genes are involved in resistance.

SCREENING APPROACHES FOR DETECTING RESISTANCE TO APHIDS

It is well-known that aphid distribution on most plants is far from uniform, and varies with the aphid species and race, the host plant and its stages of growth and maturation, the age of leaves, etc. The task of the evaluator is especially difficult when several aphid species with rather similar appearance, but with different food preferences, infest the same plant, e.g. on apple or potato. The various problems encountered and the numerous techniques utilized for the evaluation of aphid populations on plants were described and reviewed by Heathcote (1972), Hughes (1972) and Vickerman and Wratten (1979), and should serve as useful guides for screening plant germplasm for resistance to aphids. More specific information on screening techniques for plant resistance are presented by Wood (1961a) and Johnson et al. (1976) for evaluating resistance of small grains and sorghum to greenbugs, *Schizaphis graminum* (Rondani) (Fig. 11.4.1), by Barnes et al. (1974) for alfalfa and *Acyrthosiphon pisum* (Harris) and *Therioaphis trifolii* f. *maculata* (Buckton), by Dewar (1977) for cereals and three aphid species, and by Mabbett and Nachapong (1979) for cotton and *Aphis gossypii* Glover. Johnson and Teetes (1980) adapted the technique of Wood (1961a) for sorghum and *S. graminum*.

Because it is often necessary to evaluate plants for resistance in the absence of natural infestations, mass-rearing of aphids must be carried out to supply insects for artificial infestation, and such a procedure has been widely used. Greenhouse tests should be conducted with uniform illumination, temperature, water and fertilizers, since these and other factors may alter overall and relative resistance (Schweissing and Wilde, 1979; Maxwell and Jennings, 1980). In screening tests, it is often important to control infestation rates so that

Section 11.4 references, p. 254

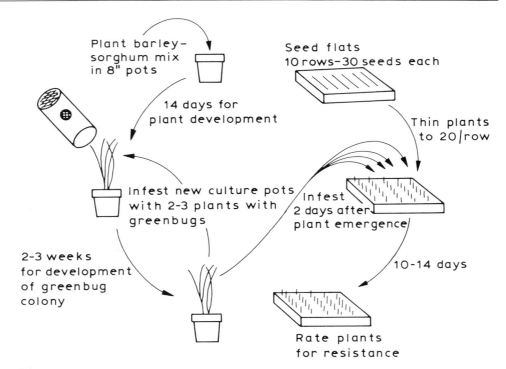

Fig. 11.4.1. Screening techniques for plant resistance to the greenbug, *Schizaphis graminum* (Rondani) (after Starks and Burton, 1977b).

aphid population pressure does not destroy or obscure potential germplasm sources. Results from greenhouse tests must usually be confirmed under field conditions; indoor trials alone may not provide adequate measures of resistance, since plant structure may be an important factor in the field as compared to isolated plants in protected conditions (Watson and Dixon, 1984). With some variants greenhouse screening procedures are similar for cotton against *A. gossypii*, alfalfa against *A. pisum* and *T. trifolii* f. *maculata*, but seedling screening procedures would be inadequate, for instance for corn against the corn leaf aphid, *Rhopalosiphum maidis* (Fitch). Field screening is preferred for *Acyrthosiphon kondoi* Shinji, as mass-rearing of the latter indoors over extended periods is more difficult, although the use of lentils for rearing this species is recommended (E. Horber, personal communication, 1984).

Figure 11.4.2 (Maxwell and Jennings, 1980) outlines various phases in developing plants for resistance to insects, most of which apply to aphids. Aphids can be reared on susceptible host plants, or on artificial media (section 8.13) but the latter is not recommended for mass-rearing of aphids through several generations, as it is still impractical, time-consuming and costly. Furthermore, aphid performance on such diets is often marginal, and this may affect their behaviour when transferred to host plants. However, rearing on artificial media may be recommended for more refined basic studies, and as a bioassay method to evaluate the effect of various plant extracts and allelochemics on aphid performance.

Homogeneous populations should be used to determine the genetics and nature of resistance in the plant, and individual progenies of clones or specific biotyes are recommended for more basic studies. Biotypes have been reported in at least ten species of aphids. Reared populations of aphids can be used along with heterogeneous populations to evaluate plant resistance, the latter being used to verify resistance under field conditions and corroborate those findings in the laboratory. When available (e.g. in *S. graminum* and *T. trifolii* f. *maculata*), different biotypes can be used to evaluate resistance in new plant introduc-

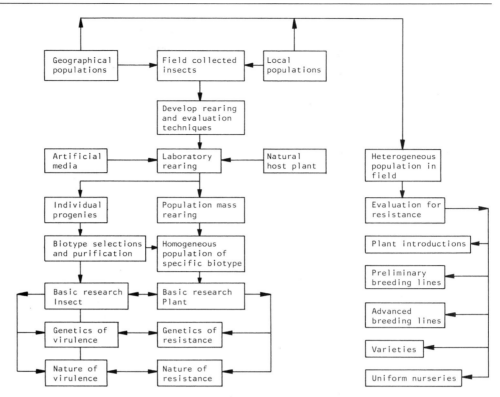

Fig. 11.4.2. Phases of research in the development of plants resistant to aphids (from Maxwell and Jennings, 1980).

tions, varieties, preliminary or advanced breeding lines, and uniform nurseries. The value of biotypes as opposed to natural field populations is that more critical genetic and physiological research can be conducted when the purity of the aphid population is established. Finally, it is important that the assay technique represents the insect–host relationships as they occur in the field, and especially the correlation of resistance in the seedling and more mature stages of the plants.

The various detailed measures of aphid performance on plants described by Adams and Van Emden (1972) include survival, population development, individual fecundity, reproductive rate, embryo number, as well as measures of individual development, such as length of instars, aphid size and weight, mean relative growth rate, potential increase rate (Hughes, 1972), and food utilization. Uptake of food can be determined with radioisotopes (see section 8.7), and aphid behaviour associated with probing, salivation and ingestion can be recorded electronically as discussed in section 8.8 and by Tarn and Adams (1982). Honeydew collection is another method for indirectly measuring food uptake by aphids (Auclair, 1959; Maxwell and Painter, 1959), or other plant-sucking insects (Paguia et al., 1980; Auclair et al., 1982) that feed on susceptible and resistant host plants. Microliter samples of phloem sap for biochemical analyses can be collected from the sectioned and exuding stylet bundle of aphids feeding on host plants, using a microscalpel (Kennedy and Mittler, 1953), microscissors (Van Soest, 1955), high-frequency radio waves (Unwin, 1978), or a laser beam (Barlow and Randolph, 1978) to section the stylet bundle in situ. Detailed studies on antibiosis usually require caging techniques for whole plants, stem cuttings, or leaves (e.g. whole leaf cages and leaf clip-on cages). Descriptions of the various techniques for rearing aphids and measuring their performance on host plants can also be found in papers cited under particular plant species in the next part of this section.

APHID RESISTANCE IN PARTICULAR HOST PLANTS

Leguminosae

Medicago spp.

Alfalfa or lucerne (*Medicago* spp.) occurs in all major agricultural areas of the world, particularly in the temperate zone, with a total of more than 33 million hectares under cultivation. Three aphid species are key pests of this plant (Nielson and Lehman, 1980): the spotted alfalfa aphid, *T. trifolii* f. *maculata*, the pea aphid, *A. pisum*, and the so-called blue alfalfa aphid (or blue-green lucerne aphid in Australian literature), *A. kondoi*. Key pests are highly noxious, perennially occurring species that dominate control practices. Occasional pests include the clover aphid, *Nearctaphis bakeri* (Cowen), and the cowpea aphid, *Aphis craccivora* Koch. Occasional pests usually cause economic damage only in localized areas or at certain times and exceed economic injury levels only sporadically.

T. trifolii f. *maculata* was first described from India in 1899; it occurs in Europe, North Africa, the Middle East, China, Australia, Mexico and the U.S.A. In this latter country, this aphid was introduced in central New Mexcio in 1954, by 1957 it was present from coast to coast in a broad area extending from Wisconsin to Mexico, and it finally appeared in Alberta, Canada, in 1979. According to Nielson and Lehman (1980), the development of aphid-resistant alfalfa cultivars through selection and breeding methods had a spectacular success in the U.S.A. during the past 25 years, and this can be attributed mainly to the introduction of *T. trifolii* f. *maculata* in that country in 1954. This major pest provoked extensive research (described in part by Maxwell et al., 1972), which led to the development and release of resistant alfalfas (Table 11.4.1), many of which have multiple resistance and high forage yields. Annual savings to farmers in the U.S.A. from the use of resistant alfalfa cultivars were estimated at a conservative 35 million dollars in 1963, with yield increases of 22–27 tonnes per hectare and a reduction in the use of insecticides (Schalk and Ratcliffe, 1977). By 1980, most of the alfalfa grown in western U.S.A. was resistant to *T. trifolii* f. *maculata* (Sailer, 1981).

After the discovery of this aphid in Australia in 1977, three resistant varieties of lucerne from the U.S.A., i.e. cvs. WL 451, DeKalb 167 and Condura 73, were reported by Hamilton et al. (1978) to be resistant also in Australia. Other varieties from the U.S.A. shown to hold their resistance in Australia include cvs. CUF 101, PS 572, UC Cargo, WL 508, WL 512, WL 514 and WL 600 (Lloyd et al., 1980b). Oram (1980) registered eight cultivars of *Medicago* spp. resistant or

TABLE 11.4.1

Some varieties of alfalfa, *Medicago sativa*, resistant to aphids[a]

Aphid	Resistant alfalfa
Acyrthosiphon kondoi Shinji	CUF 101, KS 80, WL-514, Paine, Fortin Pergamino
Acyrthosiphon pisum (Harris)	Apex, Arc, BIC-5-PA, BIC-5WH-PA, CUF 101, Dawson, KS76, KS 77, K78-10, KS145, KS 80, KS 167, Kanza, Mesilla, PA-1, Riley, Paine, Team, Washoe, WL-512
Therioaphis trifolii f. *maculata* (Buckton)	BIC-5-SAA, BIC-5WH-SAA, Bonanza, Caliverde 65, Cody, CUF-101, Dawson, KS 76, KS 77, K78-10, KS 145, KS 80, KS-167, Kanza, Lahontan, Mesa-Sirsa, Mesilla, Moapa[b], Riley, Sonora, UC-Cargo, Washoe, WL-504, Zia

[a] Mainly from Nielson and Lehman (1980) and some other sources.
[b] First cultivar developed specifically for resistance.

tolerant to *T. trifolii* f. *maculata* and/or *A. kondoi* in Australia. In the spring of 1977, some fifteen varieties of alfalfa newly imported from the U.S.A. and resistant to *T. trifolii* f. *maculata* were made available for large-scale commercial growing in New South Wales (Lodge et al., 1981). In their recommendations for lucerne planting in Queensland in 1983, Gramshaw et al. (1982) listed 31 winter-dormant or winter-active cultivars, 27, 16 and 4 of which were classified as resistant to *T. trifolii* f. *maculata, A. pisum* and *A. kondoi*, respectively. They strongly recommended the use of aphid-resistant cultivars especially for extensive pasture or grazing situations where chemical control of aphids is hazardous or uneconomical.

The pea aphid, *A. pisum*, is a cosmopolitan insect that can colonize many species of Leguminosae. It was introduced in North America about 1887 and soon became a common pest of alfalfas and peas. It has recently been observed in Victoria, Australia, by Ridland and Berg (1981). An exhaustive bibliography up to December 1976 (Harper et al., 1978) listed a total of 1756 publications on *A. pisum*, of which some 162 references were on plant resistance. To the above total must be added about 2000 mentions of *A. pisum* published in the USDA Cooperative Economic Insect Pest Report (1951–1975), and the Canadian Economic Insect Pest Review (1923–1966), illustrating its great economic importance. Investigations appraising pea aphid resistance to alfalfa indicated that Flemish and Turkestan types offered the best sources of resistance, and methods for evaluating resistance were reviewed in Barnes et al. (1974) and Nielson and Lehman (1980). Some alfalfa varieties (mostly from the U.S.A.) resistant to *A. pisum* are given in Table 11.4.1. The resistant cultivar Apex was susceptible when tested in France against two local pea aphid biotypes; but, on the other hand, cv. Lahontan, which is considered susceptible in the U.S.A. (Pedersen et al., 1975), was resistant to the French biotypes. Further studies (Bournoville, 1980) indicated that resistant cv. Team (Table 11.4.1) was susceptible when tested in France, but cv. Kanza retained its resistance to *A. pisum*. Using six wild-type glandular-haired *Medicago* species and two *M. sativa* cultivars, i.e. resistant cv. Team and susceptible cv. Vernal, Shade and Kitch (1983) noted significantly lower aphid performance on glandular-haired species than on cv. Vernal, and there was some relationship between hair exudate viscosity and larval mortality.

A. kondoi is native to the Far East (e.g. Mongolia, China, Japan), and was described from Manchuria in 1938 where it occurs on species of *Medicago, Melilotus* and *Trifolium*. In the New World, it was first observed in California in 1974 and, by 1977, was reported from nine other States. During that period, it also spread to Argentina, Australia and New Zealand (Nielson and Lehman, 1980). Like the pea aphid, *A. kondoi* is a cool-season species, important on alfalfa primarily from April through June in central U.S.A., whereas *T. trifolii* f. *maculata* is often the prevalent aphid from midsummer through fall. *A. kondoi* is more damaging to alfalfa than *A. pisum* (Stern et al., 1980), and some alfalfas resistant to it are given in Table 11.4.1. Out of 23 lucerne lines introduced into Australia from the U.S.A., Lloyd et al (1980a) reported considerable tolerance to *A. kondoi* by seedlings of all lines infested at the 2–3 trifoliate leaf stage. However, by using the suppression of both plant height and dry matter production as indices, six imported lines exhibited greater resistance than the others. Cultivars CUF 101 and Rere were also resistant in New Zealand to *A. kondoi* (Farrell and Stufkens, 1981).

Many alfalfas resistant to aphids (Table 11.4.1) carry multiple resistance, e.g. cvs. CUF 101 and KS 80 are resistant to the three aphid species, cvs. Dawson, Mesilla, Washoe and all KS's carry resistance to both *T. trifolii* f. *maculata* and *A. pisum*. Some alfalfas resistant to one species may be highly susceptible to another, e.g. cv. Team is resistant to *A. pisum* but highly suscept-

Section 11.4 references, p. 254

ible to four biotypes of *T. trifolii* f. *maculata* (Barnes et al., 1970). Alfalfas resistant to aphids in North America may keep their resistance when transferred to another continent or, exceptionally, may become susceptible, and the reverse can also occur. A strong association between resistance to *T. trifolii* f. *maculata* as seedlings and mature plants is usual (e.g. Lloyd et al., 1983), although mature plants are more tolerant than seedlings.

Alfalfa varieties resistant to aphids have been successfully bred without much knowledge of the underlying basic nature of resistance. Antibiosis is often the principal mechanism, although antixenosis and tolerance can occur. McMurtry and Stanford (1960) reported that *T. trifolii* f. *maculata* on resistant alfalfa plants soon assumed a feeding position, but became restless in 1–4 h and eventually died or left the plants. Aphids confined to highly resistant plants died at nearly the same rate as those deprived of food. Aphid stylets entered the phloem less frequently in resistant than in susceptible plants, suggesting that the aphids' mechanism for locating the phloem was interfered with in the resistant host. Maxwell and Painter (1959, 1962) and Kircher et al. (1970) investigated several compounds in alfalfa plants that elicited certain behavioral responses in *T. trifolii* f. *maculata*. Correlations with host resistance were usually poor, but Maxwell and Painter pointed out that auxins may be primarily involved in the tolerant component of resistance; they also reported a much higher honeydew excretion from aphids feeding on susceptible hosts. Vegetative and reproductive structures of several *Medicago* species, mostly annual, are covered with simple and with glandular trichomes. In free-choice tests, Ferguson et al. (1982) reported that five glandular-haired species (including one subspecies) and a resistant *M. sativa* control clone were less attractive to adult *T. trifolii* f. *maculata* than the susceptible *M. sativa* control. Adult aphids were not immobilized by exudate from glandular hairs, but these may have provided an olfactory or gustatory repellent. However, the authors concluded that factors other than glands may be important in resistance.

At least eight biotypes of *T. trifolii* f. *maculata* have been recognized in the U.S.A. (reviewed by Nielson and Lehman, 1980), and most of these were differentiated on the basis of their biological response to parent clones of the Moapa and Washoe cultivars, and in some cases on the basis of their resistance to organophosphate insecticides. Studies on the probing behaviour of four biotypes of *T. trifolii* f. *maculata* by Nielson and Don (1974) revealed no differences in probing, but some biotypes were unable to ingest sap when the stylets reached the phloem of resistant plants. Differences between biotypes in their ability to feed were attributed to the presence or absence of detoxifying mechanisms, and the described mechanism of resistance is similar to that of the phytoalexin concept of interaction between pathogens and plants. Although well characterized, biotypes of *T. trifolii* f. *maculata* have not become widespread or economically important (Sailer, 1981); nevertheless, efforts are being made to develop horizontal resistance in alfalfa against them (e.g. Nielson and Kuehl, 1982; Nielson and Olson, 1982) (see Fig. 11.4.3)

Little is known about the nature of resistance to *A. pisum* in alfalfa. The multiplication rate of aphids is lower and mortality is higher on resistant plants. Resistance is usually associated with a high concentration of saponin fractions in the host plant (e.g. Horber et al., 1974). However, Pedersen et al. (1975) reported that the average foliage saponin index of six pea-aphid-resistant alfalfa varieties did not differ significantly from that of five susceptible ones, suggesting that other factors beside saponins are involved in resistance. Biotypes of *A. pisum* in relation to alfalfa have been reported (Cartier et al., 1965; Frazer, 1972). A red form of this aphid has been known in Europe for many years, but it was recently observed in North America by Kugler and Ratcliffe (1983). These authors evaluated 22 alfalfas, 19 of which had been previously

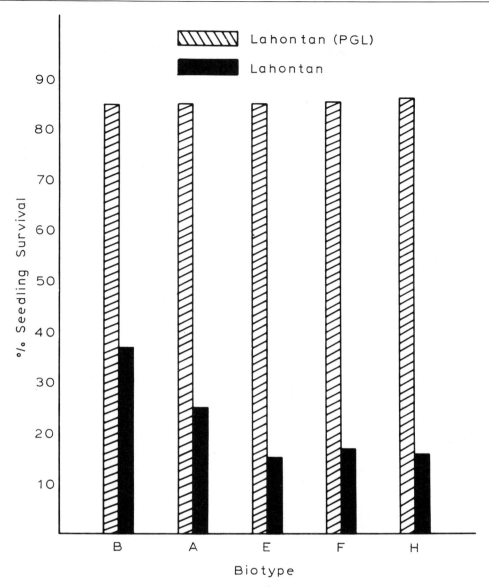

Fig. 11.4.3. Percent seedling survival of progeny of 'Lahontan' and 'PGL' (polygenic Lahontan) to five biotypes of the spotted alfalfa aphid, *Therioaphis trifolii* f. *maculata* (Buckton) (from Nielsen and Olson, 1982).

classified as resistant to the green form. Only the lines BAA-15 and PA-1 and the cultivar CUF 101 exhibited high resistance to the red form. Cultivars such as Arc, Kanza and Team, which are resistant to the green form, (Table 11.4.1) were susceptible to the red one.

Pisum sativum

A most important insect feeding on cultivated peas, *Pisum sativum*, is the pea aphid, *A. pisum*, already considered in this section as a pest of alfalfa. Early experiments on screening peas for aphid resistance (Maltais, 1937; Harrington et al., 1943; Cartier, 1963) had revealed the existence of resistant varieties such as cvs. Champion of England and Onward as compared to the susceptible cvs. Perfection or Daisy. From a cross between cvs. Champion of England and Lincoln, a semi-resistant hybrid (H-103) was developed in 1942 by Maltais (1950) and named Laurier. Exhaustive screening tests on 1250 pea accessions from

Section 11.4 references, p. 254

germplasm banks in the U.S.A. by Newman and Pimentel (1974) revealed that twenty peas were significantly more resistant than cv. Onward, and two peas originating from Poland were more susceptible than cv. Perfection, *A. pisum* on them being about 29% heavier than on cv. Perfection.

Searls (1935) concluded that peas with light green or yellowish foliage (e.g. cv. Onward) were more resistant to *A. pisum* than those with dark green foliage, such as cv. Perfection. However, from screening tests on thirteen pea varieties, Cartier (1963) observed that yellow-green plants were preferred over green ones, and that plant height had two opposite effects: a barrier effect, contributing to a higher initial infestation of tall plants by alate migrants, and an exposure effect that reduced ensuing populations from an adverse environment. Biochemical analyses carried out at the stages of growth and in the parts of the plants usually infested in the field showed that susceptible varieties contained more nitrogen and less sugar than resistant ones (Maltais and Auclair, 1957), and a higher concentration of free and total amino acids (Auclair et al., 1957), including the ten amino acids usually reported as essential in the nutrition of insects, and the amides asparagine and glutamine. *A. pisum* fed at a higher rate on susceptible plants, with a greater proportion of the material ingested being excreted and a lesser proportion assimilated on resistant varieties (Auclair, 1959). Calculations indicated that on a susceptible variety, aphids ingested a sap containing about 3.75% free amino acids, whereas on a resistant variety they ingested a sap containing about 2% amino acids (Auclair, 1976). Similar effects on aphid growth and reproduction were obtained when the aphids were given access to a resistant variety, to chemical diets with reduced amino acid level or when aphids were starved 10 hours daily on a susceptible variety (Auclair and Cartier, 1960; Srivastava and Auclair 1974; Auclair 1986).

Several biological races or biotypes of *A. pisum* have been described (e.g. Harrington, 1945; Cartier, 1963; Frazer, 1972; Auclair, 1978) as well as a red form in Europe (Markkula, 1963; Lowe and Taylor 1964). Green pea aphids were generally more successful on peas than the red ones. Two green biotypes from Québec (J and C) were shown by Simon et al. (1982) to differ in superoxide dismutase isozyme patterns, and in their amino acid requirements (Srivastava et al., 1985). Out of sixteen pea varieties tested by Markkula (1970) against three pea aphid biotypes originating from red clover in Finland, all proved resistant or semi-resistant to biotypes 1b (a red form) and 16 (green), including the usually susceptible cv. Perfection. However, biotype 1a (green) reproduced well on all varieties (76–108 descendants per aphid). Results of tests on 115 pea varieties and lines with biotype 1a showed that all of them were susceptible, including the usually resistant cvs. Onward and Witham Wonder. This prompted the author to conclude that it may be impossible to find a variety resistant to this biotype.

Vicia faba

Broad bean or faba bean, *Vicia faba*, also called field bean in Great Britain, is considered a susceptible host plant of *A. pisum* and is commonly used for aphid mass rearing by experimenters. In Finland, however, *V. faba* variety Hangdown was the least infested by *A. pisum* when grown in field plots with 28 pea varieties (Markkula and Roukka, 1971); but when caged on different host plants of similar age, the fecundity of ten aphid clones (four red, six green) was usually higher on broad bean (85–102 larvae per female) than on peas or red clover, although five clones reproduced almost as well on resistant peas as on broad bean (Markkula and Roukka, 1970). Frazer et al. (1976) reported no significant differences in the performance of *A. pisum* between twelve cultivars of *V. faba*, none of which showed any significant resistance to the aphid.

Aphids other than *A. pisum* can be pests of *V. faba*, including *A. craccivora* and *Aphis fabae* Scopoli, and, according to Bouchery (1977), the latter species can be more destructive to broad beans than *A. pisum*. Müller (1953) tested forty varieties of *V. faba* against *A. fabae* and studied the behaviour and performance of this aphid on the most resistant cv. Rastatt and the most susceptible cv. Schlanstedt (Müller, 1958); he concluded that both non-preference and antibiosis were involved as resistance mechanisms. Cultivar Rastatt was later crossed with local varieties of English horse beans. Beans resistant to *A. fabae* slightly out-yielded the controls in infested field trials, but the reverse occurred in uninfested trials (Bond and Lowe, 1975). The authors suggested that resistance might be improved by crossing the resistant lines with *Vicia narbonensis* if the incompatibility between these *Vicia* species can be overcome. In the meantime, greater use could be made of the cv. Rastatt level of resistance by transferring it to higher yielding varieties. In the Near East, Tahhan and Hariri (1981) screened 552 BPL lines of faba beans for resistance to *A. fabae*; the average numbers of aphids per plant after 3–4 weeks of artificial infestation in the field ranged from 9 for BPL 1076 to 6650 for BPL 678.

Phaseolus vulgaris

Bean, *Phaseolus vulgaris*, also called snap beans or dry beans in America, and French beans in Great Britain, has a specialized type of epidermal appendage, the hooked trichome, that imparts some resistance to *Myzus persicae* (Sulzer) (McKinney, 1938) and *A. craccivora* (Johnson, 1953). These trichomes entrap small, soft-bodied arthropods, usually by impalement through unsclerotized regions of the body wall. According to Pillemer and Tingey (1976), the levels of resistance increase with the density of these hooked trichomes, and the degree of erectness relative to the leaf blade also affects capture efficiency. A review by Levin (1973) on the role of trichomes in plant defence and a useful annotated bibiography on plant hairs and insect resistance by Webster (1975) include aphids on beans, cotton, lupins and Solanaceae.

Trifolium spp.

The cultivar Dollard of red clover, *Trifolium pratense*, was found to be more resistant to *A. pisum* than the cultivar Wegener, resistance being due to nonpreference and antibiosis (Wilcoxson and Peterson, 1960). When mechanically inoculated, both cultivars were equally susceptible to virus infection, but under field conditions a much lower incidence of virus occurred in cv. Dollard than in cv. Wegener, suggesting that breeding red clover for resistance to aphids may be a more successful approach than trying to breed for resistance to virus diseases.

El-Kandelgy and Wilcoxson (1964) found some plants in the red clover cultivar Lakeland to be resistant to *A. pisum*, but this cultivar and cv. Dollard were susceptible to the clover aphid, *N. bakeri*, whereas cv. Wegener was resistant. This shows an opposite reaction of these varieties to the two aphid species and suggests different resistance mechanisms. Out of ten varieties of native and foreign red clover tested against three pea aphid biotypes from Finland, all were highly resistant to biotype 1a, which never produced more than ten offsprings per aphid on these plants (in contrast to 115 peas susceptible to this biotype, as seen earlier); the red clovers were all susceptible to biotype 16, which produced over 80 offsprings per aphid (Markkula, 1970). In their review, Manglitz and Gorz (1972) concluded that there is a pressing need for the development of clovers resistant to aphids and other insects, since only two varieties (by chance) resistant to aphids were available to farmers, as compared to some twenty alfalfa varieties resistant, either singly or in combination, to five serious insect pests, including two species of aphids. In Australia, some lines of *Trifolium subterraneum* were found by Gillespie (1983) to be more resistant to *A. kondoi* than some commercial varieties of the same species.

In field and greenhouse tests, the variety Spanish and the breeding line N 13 of sweet clover, *Melilotus* spp., were the most resistant to the sweet clover aphid, *Therioaphis riehmi* (Börner), and resistance usually increased with the age of the plant (Manglitz and Gorz, 1961).

At the International Institute for Tropical Agriculture in Nigeria, varieties of cowpea, *Vigna unguiculata*, were identified as being resistant to *A. craccivora* (Singh, 1977) and, in Kenya, three cultivars of cowpea showed resistance to *A. gossypii* as a result of non-preference and antibiosis (Karel and Malinga, 1980).

To Wink et al (1982), species of *Lupinus* which accumulate quinolizidine alkaloids (so-called bitter lupins) are usually devoid of polyphagous aphids, whereas so-called sweet, alkaloid-free plants are often heavily infested by aphids and some other insects.

This antifeeding device of some host plants has apparently been overcome by the broom aphid, *Aphis cytisorum* Hartig, feeding on the broom plant, *Cytisus scoparius*, which contains quinolizidine alkaloids such as sparteine and its derivatives. The aphids would seem to select broom plants with about 50% less alkaloids than the unifested plants. The lower alkaloid content of infested plants may have resulted, however, from the aphid infestation. Aphids feeding on this species sequester up to 500 μg alkaloid per g fresh weight, indicating that these compounds were translocated via the phloem, which was confirmed by phloem content analyses. For *A. pisum* (= *Acyrthosiphon spartii* (Koch)) feeding on broom, sparteine was a feeding stimulant (Smith, 1966).

Gramineae

Sorghum spp.

Sorghum ranks fifth in acreage and production among the world's major cereal crops, and Miller (1980) reviewed the rather complex breeding of this plant. Aphids that attack sorghum (Teetes 1980) include a key pest, the greenbug, *S. graminum*, and occasional pests such as the corn leaf aphid, *R. maidis*, the yellow sugarcane aphid, *Sipha flava* (Forbes), and the sugarcane aphid, *Aphis sacchari* Zehntner (Young and Teetes, 1977). In greenhouse studies on some eight cereal crops against *S. flava*, corn was the least preferred and supported few and small nymphs, and all entries of common wheat and *Sorghum bicolor* were susceptible, as well as most of the barley tested; although the aphid takes one week longer than *S. graminum* to mature, it causes great plant necrosis to these three cereals (Starks and Mirkes, 1979). Hsieh (1982) has reported that PI257295 sorghum has a high level of resistance to *A. sacchari*, conditioned by a single dominant gene pair.

The greenbug, *S. graminum*, a cosmopolitan aphid observed in Europe and Africa, was first reported in Virginia in 1882, and has been a pest of small grains in the U.S.A. for a century. Greenbug biotypes (A) and (B) were already established as serious pests of wheat and barley when a new biotype (C) was recognized on sorghum in 1968 (Harvey and Hackerott, 1969). It became important in most areas of the U.S.A. where the crop is grown, especially the Great Plains. Walker et al. (1972) published a bibliography on *S. graminum* in which some 75 references on host plant resistance, mostly from North America, are cited. Biotype (C) has apparently largely replaced (A) and (B) and it infests sorghum as well as small grains (Young and Teetes, 1977). These authors mention that infestations are detectable by reddish spots on the leaves, which are caused by toxins injected into the plant; the leaves turn brown from the outer edges toward the center and finally die. The greenbug may be a pest of seedlings, though it usually reaches damaging proportions after heading of the plant. Breeding programmes to develop resistant sorghums were initiated in 1969 by the Federal (USDA) and several State Experiment Stations and private agencies; greenbug-resistant germplasms were found from sources such as IS 809, PI 264453 (*S. bicolor*), KS 30, KS 41 to KS 44 (*S. virgatum*) and SA 7536-1 (*S. nigricans*) (Wood et al., 1969; Hackerott et al., 1972; Weibel et al., 1972; Schuster and Starks, 1973; Johnson et al., 1976; Starks and Burton, 1977a, b; DePew and Witt, 1979) and resistant breeding material was released to commercial seed companies. Substantial yield increases in grain sorghum, coupled with a reduction in the use of insecticides, were reported by Schalk and Ratcliffe (1977) from the use of cultivars resistant to this aphid. Johnson and Teetes (1980) reviewed the breeding of sorghum for arthropod resistance, including *S. graminum*, and described several sources of resistance to this aphid in grain, forage and grassy types of sorghum. From laboratory and field experiments, it appears that moderate levels of nonpreference and antibiosis exist, but the primary resistance mechanism is tolerance, which is ecologically advantageous and increases the value of resistant sorghum as a component in pest management (Teetes, 1980). By 1980, at least 90% of the grain sorghum acreage in the Southern plains of the U.S.A. was sown to hybrids with greenbug resistance, derived mainly from SA 7536-1 and KS 30 (Starks et al., 1983). According to Peters and Starks (1981), plant resistance had largely replaced insecticides as the main input for greenbug management in the U.S.A., where the development of resistant hybrids of sorghum took about 5 years from the time the greenbug first became a pest. This belies the argument that plant resistance by nature must have projected payoffs of 10 to 20 years. They further stated that the economics of sorghum production make plant resistance the most desirable approach in all sorghum production areas; it is especially suitable for developing countries, where educational and financial barriers limit the use of other methods of control. In Brazil, Galli et al. (1981) reported that six sorghum genotypes (out of 47 assayed) were resistant to *S. graminum*, whereas five others were highly susceptible.

Cress and Chada (1971) showed that when reared on a chemically defined diet, biotype (A) had a lower rate of growth, development and reproduction than biotype (B), and Starks and Burton (1977b) described characteristics for separating the four major biotypes then known (A, B, C, D), the latter two including sorghum as host in the field. However, Porter et al. (1982) demonstrated that KS 30 and SA 7536-1 sorghums, among other resistant genotypes, were susceptible to *S. graminum* collected from a wheat field in Texas in 1979–1980, but that PI 220248 and PI 264453 sorghums represented useful sources of resistance to this new biotype, designated (E). These results were confirmed by Starks et al. (1983), who concluded that the occurrence of biotype (E) had affected many of the small grains and sorghum genotypes previously developed for resistance to *S. graminum*. The cytogenetics of greenbug biotypes have been studied and new biotypes are being discovered in the U.S.A. (Z.B. Mayo and K.J. Starks, personal communication, 1984).

Benzyl alcohol administered by uptake through the roots or by injection into stems of wheat, sorghum or barley plants that serve as hosts to *S. graminum* (Juneja et al., 1975), and *p*-hydroxybenzaldehyde, dhurrin, procyanidin and other compounds from sorghum (Dreyer et al., 1981) that were included in artificial diets, were reported to have detrimental and/or feeding-deterrent effects on the aphid. However, the presence of benzyl alcohol in barley was not confirmed (Dreyer et al., 1981), and whether the other compounds are actually present in the phloem and/or affect aphid feeding in situ has not been established. Using the electronic method for monitoring aphid probing, Campbell et al. (1982) reported that biotype (C) fed for significantly less time from the phloem of resistant cultivars (i.e. TAM 2567, IS 809) compared to susceptible cultivars. Using the same method, Montllor et al. (1983) monitored the probing and feeding behaviour of biotypes (C) and (E) on sorghum IS 809, on which biotype (E) grew and reproduced about twice as fast as (C). Biotype (E) established committed phloem ingestion (CPI, i.e. ingestion from the phloem lasting longer than 15 min) in a significantly shorter time than (C), and the total duration of phloem ingestion in 24 h was significantly longer for (E) than (C). However, when (C) was exposed to IS 809 for at least 24 h prior to monitoring on this cultivar, the time to reach the phloem and to establish CPI was shorter, and periods of feeding from phloem were longer. During stylet penetration of the intercellular pectin of host plants, biotype (E) secretes a pectin methylesterase whose activity is about twice as high as that of biotype (C), and, as a result, it can depolymerize a highly methylated pectin more efficiently than biotype (C), even though the extractable pectin of resistant IS 809 is twice as high as that of a susceptible sorghum line (Dreyer and Campbell, 1984).

The corn leaf aphid, *R. maidis*, which is distributed widely in the world, infests sorghum throughout the growing season, but opinions differ as to its importance in reducing yield (Teetes, 1980). An early report by McColloch (1921) identified sorghum resistance to the corn leaf aphid in Sudangrass. This was confirmed by Howitt and Painter (1956) when they tested 595 varieties of sorghum and selected a highly resistant plant from Piper Sudan. Four aphid biotypes were then described (Pathak and Painter, 1958), based on their differential responses to sorghum and barley; a fifth biotype and its effect on some cereal and small grains was reported by Wilde and Feese (1973). *R. maidis* had a markedly poorer growth on shorter (and younger) plants of sorghum than on taller ones, and this was associated with less time spent feeding and more movement on shorter plants (Fisk, 1978). This aphid reproduced less on seedlings of these varieties of sorghum than on older plants, with fewer probes terminating in the phloem of seedlings (Kimmins, 1982). New sources of resistance to *R. maidis* were reported from converted exotic lines, especially TAM 428, and Teetes (1980) concluded that by limiting populations of *R. maidis* to non-injurious levels, resistant sorghums would be useful in pest management programmes by providing stability and maintaining natural biological agents. Adults and nymphs of *R. maidis* were apparently not affected by high concentrations of phenolic acids in sorghum cultivars (Woodhead et al., 1980).

Zea mays

Corn or maize, *Zea mays*, is the world's third leading cereal after wheat and rice. Breeding methods and resistance to many insects are reviewed by Ortega et al. (1980). Corn genotypes are known to react differently to an infestation by the corn leaf aphid, *R. maidis* (Painter, 1951; Rhodes and Luckmann, 1967), and a correlation is reported between susceptibility to aphids and damage by the European corn borer, *Ostrinia nubilalis* (Hübner) (Huber and Stringfield, 1942). The Kansas Experiment Station developed a corn cultivar, KS 1859, which is resistant to this aphid, and was first distributed to farmers in 1950 (Painter, 1958a). Observations and studies of damage by *R. maidis* to corn and host resistance were reviewed by Long et al. (1977) and Beck et al. (1983). The former authors reported a highly significant correlation between aphid infestation and hydroxamate concentration in stem tissues of twelve inbred lines of corn, and *R. maidis* mortality increased with higher concentrations of one hydroxamate, 2, 4-dihydroxy-7-methoxy-1, 4-benzoxazin-3-one (DIMBOA), added to synthetic diets fed to the aphids. They concluded that DIMBOA was an active agent in the resistance of corn to *R. maidis* and that selection for a higher DIMBOA concentration in corn should increase its resistance to several important pests. Beck et al. (1983) confirmed the results and conclusions of Long et al. (1977) when tassels of eleven corn inbreds were analysed for DIMBOA. The fact that corn seedlings are generally resistant to aphids (Villanueva and Strong, 1964; Denéchère, 1979) could be correlated to DIMBOA distribution in relation to plant age and plant parts (see studies of Argandoña et al. (1983) under small grains).

Small grains

Small grains such as wheat, *Triticum* spp., oats, *Avena* spp., barley, *Hordeum* spp., and rye, *Secale cereale*, are infested by several aphid species. The greenbug, *S. graminum*, already discussed under sorghum, is a serious pest in many parts of the world and has been reported on 62 species of Gramineae (Dahms et al., 1955); damage to small grains by this aphid is usually severe. Electron photomicrographs showing feeding damage caused by *S. graminum* in wheat cells were published by Al-Mousawi et al. (1983), who also cited earlier works on aphid feeding mechanisms. Dahms et al. (1955) tested the reaction of several hundred varieties and hybrids of small grains to *S. graminum*. None of the wheat and rye varieties showed a high degree of resistance, but several, especially some *T. durum* types, were considerably more tolerant than varieties grown commercially, and a hexaploid selection from Dickinson No. 485, C.I. 3707, was highly resistant. None of the oat varieties showed high resistance but many of the barley varieties did, including cvs. Dicktoo, Kearney, and several originating from China, Korea and Japan. Dickinson Sel. 28 A wheats, a selection from C.I. 3707, was reported to be susceptible to a new biotype (Wood 1961b), later designated (B). In their excellent review, Gallun et al. (1975) wrote that varieties of wheat resistant to the greenbug have yet to be released. Six wheats, some with dense trichomes and already selected from screening tests, were found to be of doubtful value as sources of resistance against biotype (C) (Starks and Merkle, 1977). The classification of wheats and breeding strategies and methods are discussed in Maxwell and Jennings (1980).

Back in 1954, Arriaga in Argentina had observed resistance to *S. graminum* in rye and in *Triticum tauschii*, and he developed and registered a resistant rye cultivar, Insave F.A. (Arriaga, 1956). Wood et al. (1974) reported that cv. Gaucho and octoploid Triticale developed from a cross between the susceptible Chinese Spring wheat and cv. Insave F.A. were resistant to greenbug biotype (C). The source of resistance in cv. Insave F.A. was translocated (using X-rays) to finally give cv. Amigo wheat germplasm, released in 1977 (Sebesta and Wood, 1978). The complex and exciting story of this rare achievement by induced mutation to obtain practical insect resistance from an intergeneric cross is related by Starks and Sebesta (1978). Amigo wheat is resistant to biotypes (A), (B), and (C) of the greenbug, and seed of this wheat was requested by breeders from several countries, e.g. South and Central America, Europe, South Africa, Australia, and the U.S.A. More recently, Joppa produced an amphiploid of *T. turgidum* × *T. tauschii*, and a selection from this cross was registered as cv. Largo (Jopp and Williams, 1982): it is resistant to all known biotypes of *S. graminum*, including biotype (E) (Porter et al., 1982), and is being used in several breeding programmes in the U.S.A. (Hollenhorst and Joppa, 1983). Other sources of resistance were produced in synthetic hexaploid wheats derived from *T. tauschii* (Harvey et al., 1980), and two lines of hard red winter wheats, *T. aestivum*, resistant to biotypes (C) and (E) of *S. graminum*, were registered (Martin et al., 1982).

Over a nine-year period, Chada et al. (1961) screened some 5105 varieties and strains from the world collection of oats, and ten showed the greatest amount of resistance. With the appearance of biotype (C), some 4343 oat selections from the USDA world collection were tested from 1970 through 1977, and 31 were found to be resistant (Daniels, 1978); none of these had been reported by Chada et al. (1961) to be resistant to biotype (B). In his review, Gardenhire (1980) concluded that there are at present no commercial oat varieties with effective levels of resistance to *S. graminum*. With the discovery of biotype (E), 51 oat selections resistant to biotype (C) were re-evaluated (Chedester and Michels, 1982) and ten rated as moderately resistant to biotype (E), whereas 24 were susceptible.

Several barley varieties resistant to *S. graminum* are available in the U.S.A., for example cvs. Era, Kearney, Kerr, Nebar, Post and Will (e.g. Dahms et al. 1955; Chada et al., 1961; Gardenhire, 1980), and many barley lines were reported to be resistant to biotypes (E) and (C), especially cvs. Post and PI 426756 (Webster and Starks, 1984). Substantial yield increases in barley (e.g. 805 kg per ha) were reported from the use of a greenbug-resistant cultivar, accompanied by a significant reduction in insecticide applications (Schalk and Ratcliffe, 1977). Bioassays in which biotype (B) of *S. graminum* was reared on chemically defined diets containing phenolic and flavonoid compounds showed that many of these compounds (e.g. catechol, tannic acid, quercetin, chlorogenic acid, cis-caffeic acid) were detrimental to greenbug growth, fecundity and survival. Since several of these compounds are constituents of barley leaves, they may confer part of the resistance of some varieties to the aphid (Todd et al., 1971).

Out of 474 barley varieties from the Canadian Genetic Stock of Barley tested in the greenhouse against the bird-cherry–oat aphid, *Rhopalosiphum padi* (Linnaeus), 127 showed some resistance (antibiosis or tolerance), but only 43 of these demonstrated both antibiosis and tolerance when tested in the field (Hsu and Robinson, 1963). However, Robinson (1965) reported that all varieties found resistant from the two previous tests did not demonstrate antibiosis (as measured by fecundity) in a plant growth room and discussed various environmental factors and attributes that could influence significantly the reproduction of aphids, which may explain the contradictory results obtained. In the Netherlands, differences in resistance to aphids, mainly *Sitobion avenae* (Fabricius), observed in the field between hundreds of barley accessions were generally not confirmed by tests in the greenhouse or on single plants, with the exception of resistant barley accession CI-16145 (Van Marrewijk and Dieleman, 1980).

Four biotypes of *R. maidis*, differing in their damage to Spartan barley, were described in the U.S.A. by Pathak and Painter (1958). During 1963–1967, a world collection comprising 3011 varieties of barley screened for resistance to *R. maidis* showed that resistant varieties, including cvs. Balkan and Kreta, were mostly from the Mediterranean area, with some from Taiwan. Other studies, which evaluated 1300 barleys, showed again that the Mediterranean area included several resistant varieties along with two highly resistant ones, cvs. Omugi Shin from Japan and Guided Real from Spain (Trehan et al., 1970). Increases in barley yield following the use of cultivars resistant to *R. maidis* were reported in the U.S.A. (Schalk and Ratcliffe, 1977), as was a reduction in the use of insecticides.

Two important reviews on the pest status and ecology of cereal aphids, mainly in Europe, are those by Vickerman and Wratten (1979) and Carter et al. (1980). The latter authors present a synthesis of the information gathered from the literature and incorporate the results into a simulation model for forecasting cereal aphid outbreaks. According to Vickerman and Wratten (1979), aphids, although recorded as attacking crops since at least the 18th century, were not regarded as being of great economic importance until the 1950s, when they were found to be vectors of viruses, and later when severe outbreaks of the English grain aphid, *S. avenae*, were recorded during 1968–1978 in several countries of Europe. Although over 40 species of Aphididae in Europe are associated with Gramineae, only seven are commonly found on cereals and grasses. Vickerman and Wratten (1979) were critical of methods used for assessment of host-plant preferences and resistance towards aphids, possibly because of the variability in the material tested, and the sometimes insufficient replications and statistical analyses. Nevertheless, some elaborate studies carried out by Rautapää (1970), on the preference of alatae of *S. avenae* and *R. padi* for seedlings and 3–4-week-old plants of some 86–165 cereal varieties and 47–59 species of Gramineae, Cyperaceae and Juncaceae, showed that cereals were generally preferred to wild hosts, with plant age occasionally affecting preference. In Britain, Lowe reported significant differences in the resistance between winter wheat varieties after caging the aphids *S. avenae* and *Metopolophium dirhodum* (Walker) on the upper part of mature tillers. However, it was concluded that assessment methods better suited to the plant breeder were needed, and, in further work, Lowe (1980) devised the walkabout method, in which five immature aphids were released at the base of each plant during the tillering stage, and aphids counted when the oldest progeny of the introduced aphids were about half grown. With this superior method , further refined by Lowe, (1984), which assessed differences in resistance to *S. avenae* before they are confused with the effects due to plant age and ear development between varieties, Lowe (1982) reported at least nine wheat cultivars resistant in greenhouse tests to *S. avenae*, including cv. Amigo already resistant in the U.S.A. to *S. graminum*. He indicated that aphid infestations in the field, at least in some cases, would be reduced by these cultivars. The ancient wheat cv. Einkorn was the most resistant, in terms of nonpreference and antibiosis and at all growth stages examined, to *S. avenae* and *M. dirhodum*; some resistance was also shown by other ancient wheats (Sotherton and Van Emden, 1982).

Extracts of several Gramineae contain cyclic hydroxamic acids, and one of them, DIMBOA, is found in maize, wheat and rye. It occurs naturally in the glycosidic form, may be converted to the toxic aglycone by bacterial, fungal or insect injury, and has been implicated in host plant resistance to these organisms. A direct correlation was observed between hydroxamic acids content in Gramineae (wheat, rye and barley) and resistance to *M. dirhodum*, and DIMBOA fed to this aphid via a synthetic diet was toxic at concentrations lower than those found in resistant plants. Survival and fecundity of *S. graminum* also were reduced on such diets, DIMBOA being more deleterious than its glycoside, but *R. maidis* was less affected (Corcuera et al., 1982). The content of hydroxamic acid is always higher in younger wheat plants and in younger leaves, and tests combining electrically-monitored assays of *S. graminum* feeding on synthetic diets containing DIMBOA (Argandoña et al., 1983) showed that the deleterious effects of DIMBOA on aphids were due to feeding deterrency and toxicity. The authors proposed that cyclic hydroxamic acids are important factors of resistance in wheat and rye to aphid attack. Earlier in this section a similar conclusion reached by Long et al. (1977) is mentioned concerning the concentration of DIMBOA in several inbred lines of corn in relation to resistance to *R. maidis*. However, it is apparently not known whether these compounds are present in phloem and mesophyll tissue, or whether they are ingested by aphids. A chemical analysis of honeydew should clarify this situation.

Triticale (generic name × *Triticosecale* Wittmack) is a man-made cereal obtained from a cross between wheat and rye (see review by Hulse and Spurgeon, 1974). The ability of cereal aphids to survive and reproduce on 20 triticale strains showed that *S. avenae* reproduced equally well on seedling and mature stages of all strains; *S. graminum* reproduced more on seedlings than on mature plants, whereas *R. padi* failed to reproduce on mature plants. *R. maidis* was poorly adapted to both growth stages; thus the first three aphid species have

Section 11.4 references, p. 254

Fig. 11.4.4. (A) Plants of a susceptible lettuce variety (foreground) have been badly damaged or killed by the root aphid, *Pemphigus bursarius* (Linnaeus), whereas plants of the resistant cv. Avoncrisp (background) are unaffected.
(B) Susceptible lettuce plants (right) show heavy infestation with *P. bursarius*, whereas very few aphids are found on the roots of the resistant cv. Avoncrisp. (Both photographs used with permission of the Institute of Horticultural Research, Wellesbourne, Warwick, United Kingdom.)

the potential for causing significant losses especially in triticale seedlings (Kieckhefer and Thysell, 1981). Six lines of triticale were selected as possible new genetic sources of resistance to greenbug biotype (E) for triticale and wheat (Webster and Inayatullah, 1984).

According to Ratcliffe and Murray (1983), individual plants from 36 cultivars of Kentucky bluegrass, *Poa pratensis* L., showed moderate to high levels of antibiosis and/or tolerance to a Maryland form of *S. graminum*. The latter developed well on susceptible Kentucky bluegrass and barley, while greenbugs from Ohio did not develop well on barley. Four among 23 cultivars of *P. pratensis*, and Canada bluegrass, *Poa compressa* (L.), showed a high level of resistance to both biotypes (C) and (E) of *S. graminum*, and the very high level of resistance of Canada bluegrass could be transferred to selected cultivars of Kentucky bluegrass by interspecific hybridization (Kindler et al., 1983). Differences in host preferences and reproduction of four cereal aphids in relation to certain species of range grasses were reported by Stoner and Kieckhefer (1979) and Kieckhefer (1983).

Oryza sativa. The ecology and control of rice aphids were reviewed by Tanaka (1961) and Yano et al. (1983), and one test on 36 rice varieties did not reveal any resistance to rice root aphids.

Compositae

Chrysanthemum spp.

Tests on six cultivars of this host plant for resistance against *M. persicae* confined on leaves showed poorest aphid reproduction on cultivar Princess Anne and highest on cv. Tuneful. Aphids reproduced significantly better on the under as compared to the upper surface of leaves (Markkula et al., 1969). Wyatt (1965) had shown in free infestations in a greenhouse that cv. Tuneful supported six times as many *M. persicae* as cv. Yellow Princess Anne, but aphids preferred the upper surface of leaves of the latter cultivar. Furthermore, *M. persicae* reared on the resistant cv. Princess Anne was more susceptible to three insecticides than when reared on the susceptible cv. Tuneful (Selander et al., 1972), and this, although less pronounced, was also observed on two other aphid species.

Lactuca spp.

Lettuce, *Lactuca* spp., sown in England between mid-April and the end of May was most severely infested by the lettuce root aphid, *Pemphigus bursarius* (Linnaeus), while lettuce sown in July escaped attack (Dunn, 1959). Several varieties of lettuce were also resistant through antibiosis, and more commercially acceptable cultivars such as Avoncrisp and Avondefiance, developed for resistance to downy mildew, were highly resistant (Dunn and Kempton, 1974); their resistance, derived from a California lettuce (Imperial 45634-M), gave better aphid control than insecticides (Fig. 11.4.4). In crossing experiments of these two varieties with susceptible ones, Dunn (1974) showed that resistance is controlled both by cytoplasmic and genetic factors. A series of papers by Eenink and Dieleman from 1977 to 1982, on resistance of *Lactuca* spp. to *M. persicae* and *Nasonovia ribis-nigri* (Mosley), are summarized by these authors Eenink and Dieleman, 1982). They reported that *M. persicae* grew better on older plants (30–40 days) of two cultivars of *L. sativa*, and the use of plants of this age gave a better discriminative selection. Partial resistance was found in several cultivars, and, through genetic recombination, the resistance level could be enhanced. Accessions of the wild species *L. virosa* appeared to be completely resistant to *N. ribis-nigri*, and, through a number of interspecific crosses, resistance was transferred to the cultivated lettuce, *L. sativa*.

Section 11.4 references, p. 254

Helianthus spp.

Twelve species of sunflower, *Helianthus* spp., were found to be significantly more resistant to *Illinoia masoni* (Knowlton), than six other species or the cultivar Hybrid 896, used as control. As this aphid is a potential threat to cultivated sunflower, *H. annuus*, in the U.S.A., Europe and Argentina, several of the resistant *Helianthus* species that are cross-compatible may be good sources of germplasm for development of aphid-resistant cultivars (Rogers and Thompson, 1979).

Cruciferae

Brassica spp.

On *Brassica* crops, e.g. Brussels sprouts, cabbage, kale, rape and turnip, the cabbage aphid, *Brevicoryne brassicae* (Linnaeus), is a serious pest: it forms large dense colonies and can be an important virus vector. It is often uneconomical to control aphids by insecticides on some *Brassica* crops, hence the use of resistant cultivars has been exploited, especially in Brussels sprouts, rape and turnip. Seasonal shifts in the resistance of commerical varieties of cabbage, *B. oleracea* var. *capitata*, to *B. brassicae* were reported by Radcliffe and Chapman (1966). Two red varieties proved less susceptible to host selection by alatae during the early season, but susceptibility increased as the season progressed. Dwarf Essex rape, *B. napus*, was very susceptible, marrow-stem kale, *B. oleracea* var. *acephala*, fairly susceptible, and turnips *B. rapa*, resistant to *B. brassicae* (Lamb, 1960). Furthermore, a marrow-stem kale selection of the non-waxy type was not colonized by *B. brassicae*, whereas normal waxy plants had large colonies (Thompson, 1963). Of these, turnips have light green leaves with less wax, whereas the others have generally waxy blue-green leaves. Brussels sprouts, *B. oleracea* var. *gemmifera*, with waxy leaves also generally supported larger populations of the cabbage aphid than those with glossy or non-waxy leaves, but *M. persicae* preferred to settle on glossy leaves (Way and Murdie, 1965). It would appear therefore that resistance to *B. brassica* may be determined in part by the non-waxy nature of *Brassica* plants.

M. persicae responded more to changes in nitrogen and potassium levels in Brussels sprouts grown in nutrient solution than did *B. brassicae* (Van Emden, 1966). An increase in nitrogen or a decrease in potassium resulted in an increase of soluble N in leaves, and fecundity of *M. persicae* was correlated with these changes. Dodd and Van Emden (1979) selected for further study Winter Harvest and Early Half Tall as susceptible and resistant cultivars, respectively, to *B. brassicae*. In younger plants, the concentration of mustard oil glycosides was over twice as high in leaves of cv. Winter Harvest than in those of the resistant cultivar, and the total concentration of all amino acids (especially the "favourable amino acids") in leaf samples correlated with aphid reproduction. Mustard oil glycosides are apparently prerequisites to colonization of certain Cruciferae by *B. brassicae* (e.g. Nault and Styer, 1972; Van Emden, 1978). Nonpreference, antibiosis and tolerance were identified in some European varieties and inbreds of Brussels sprouts vis-á-vis *B. brassicae*, and the leaves of susceptible varieties were prone to incurling (Dunn and Kempton, 1971). These authors selected individual plants and their progenies for resistance to the cabbage aphid; biotypes of this aphid from different localities of England differed in their ability to colonize various sprout clones, and the authors were not very optimistic about the long-term prospects of breeding for aphid resistance (Dunn and Kempton, 1972). However, Brussels sprouts varieties resistant or tolerant to *B. brassicae* have been successfuly grown in Europe (Russell, 1978); combination of resistance mechanisms should control biotypes of *B. brassicae* for a long time, and resistant varieties as part of integrated control measures against aphids and other pests are always desirable.

Aphids, principally *B. brassicae*, are serious pests on cruciferous crops in parts of New Zealand, and some varieties of swedes, *B. napus* var. *napo brassicae*, such as cv. Calder are resistant to *B. brassicae* (Palmer, 1953). Cross-breeding of resistant cv. Calder with aphid-susceptible rape stocks, followed by selection, produced a high-yielding strain called Aphid Resistant Rape, resistant to *B. brassicae* but not to *M. persicae* or to *Lipaphis erysimi* (Kaltenbach) (Palmer, 1960). By 1966, it occupied more than 90% of the rape acreage in New Zealand. Subsequently, a higher yielding aphid-resistant variety, Rangi, also more palatable to sheep, became widely grown in that country. However, a new biotype of the cabbage aphid (NZII), capable of attacking formerly resistant lines of rape, including the above two varieties, was reported by Lammerink (1968). Dunn and Kempton (1969) confirmed that the variety Aphid Resistant Rape was resistant in England, due to nonpreference and antibiosis. However, plants lost their resistance to *B. brassicae* on flowering, but, with maturity, became slightly resistant to *M. persicae*. In Germany, no significant differences in development and reproduction of *B. brassicae* were observed when aphids were maintained in the greenhouse on eleven varieties of rape (Hinz and Daebeler, 1981). Jarvis (1982) found all accessions of *B. napus* and *B. carinata* to be highly susceptible to *B. brassicae*, and plants most susceptible to aphid feeding were also the most attractive to alate aphids. *Crambe (Brassica) juncea* Bieb. was highly resistant and may not be a host for this aphid.

The turnip cultivar Shogoin, although possessing some undesirable horticultural qualities, was resistant to the turnip aphid, *L. erysimi* (Barnes and Cuthbert, 1975). Recent breeding efforts that included cv. Shogoin have resulted in the release of two cultivars more acceptable horticulturally and resistant to *L. erysimi*, i.e. cvs. Roots and Charlestowne (Kennedy, 1978). Field tests showed three times more aphids on a susceptible variety than on cv. Shogoin, although the latter appears to be susceptible to *M. persicae* and *B. brassicae* (Kennedy and Abou-Ghadir, 1979).

In Brazil, some Portuguese varieties of kale showed non-preference and antibiosis when exposed to *B. brassicae*, which reproduced more abundantly on young than on older leaves (Lara et al., 1979). At the Asian Vegetable Research and Development Center in Taiwan, close to 700 *Brassica* accessions were screened over a period of five years for resistance to several insect pests, and four accessions belonging to *B. juncea* had moderate resistance to aphids, presumably *M. persicae* and *L. erysimi* (= *Rhopalosiphum pseudobrassicae* (Davis)); interspecific crosses with heading chinese cabbage, *B. campestris* ssp. *pekinensis* can be made to transfer resistance. The leaves of these accessions are thicker and smoother than those of susceptible ones, and their surfaces have higher concentrations of long carbon chain waxes, which seems to be associated with resistance to aphids (Talekar, 1980). We have seen that *M. persicae* avoided waxy leaves of Brussels sprouts (Way and Murdie, 1965).

Cucurbitaceae

Cucumis melo, C. anguria

Muskmelon or cantaloupe, *Cucumis melo*, breeding line LJ90234 was resistant to the melon (or cotton) aphid, *A. gossypii*, exhibiting nonpreference, tolerance and antibiosis (Bohn et al., 1972). Kishaba et al. (1976) observed a lower reproduction rate and higher mortality of *A. gossypii* on resistant plants, with less alatae being produced, and they studied genetic aspects of resistance to this aphid. Other field and laboratory tests revealed resistance in plant introductions of several *Cucumis* spp., as well as of *C. melo* and gherkin, *C.*

anguria, against *A. gossypii* (MacCarter and Habeck, 1974); introductions of *C. anguria* and its close relatives from Ethiopia and Arabia were generally more resistant than those from southern Africa. Electronic recordings revealed that on resistant line 91213 of *C. melo* a significantly higher percentage of the aphid probes contacted the phloem sieve cells, but a smaller proportion of these contacts resulted in ingestion than on susceptible line Top Mark; furthermore, ingestion periods from the sieve cells were 2–3 times longer on susceptible plants (Kennedy et al., 1978). The resistance factor(s) appeared to operate by interfering with ingestion by the aphid from the sieve cells. In France, resistant cultivars in the Charentais type of *C. melo* were developed against virus transmission by aphids. A line of *C. melo* (P.I. 161375) was resistant to transmission of three viruses by *A. gossypii*, although other aphid species remained efficient vectors in some cases (Lecoq et al., 1980). Less virus infections vectored by *A. gossypii* on aphid-resistant lines of *C. melo* were also reported by Kishaba et al. (1983).

Citrullus spp.

Out of 54 world accessions of watermelon, *Citrullus* sp., two (P.I. 296335 and P.I. 299563) were resistant to *A. gossypii* in field and laboratory tests, aphid populations remaining small on such plants; seven commercial varieties from the U.S.A. showed intermediate resistance (MacCarter and Habeck, 1973).

Rosaceae

Malus spp.

At least three species of aphids are considered pests of apple, *Malus* spp., in North America: the apple aphid, *Aphis pomi* de Geer, the rosy apple aphid, *Dysaphis plantaginea* (Passerini) (= *Pomaphis mali* (Ferrari), and the woolly apple aphid, *Eriosoma lanigerum* (Hausmann) (Madsen and Morgan, 1970). The first two species infest the aerial parts of the plant, whereas *E. lanigerum* also infests the roots. According to Croft and Bode (1983), host plant resistance has not been used as a major tactic in the control of *A. pomi* and *D. plantaginea* in North America. Although some apple varieties in England have showed a high level of resistance to *A. pomi*, *D. plantaginea* and *Dysaphis devecta* (Walker) (e.g. Alston and Briggs, 1977), more work has been carried out on *E. lanigerum*.

Painter (1951) reviewed the literature on resistance in apple against *E. lanigerum*, and mentioned that the varieties Northern Spy and Winter Majetin had been resistant for some 100 years. Another world review by Knight et al. (1962) listed 18 varieties immune or highly resistant to this aphid; 19 additional varieties were highly resistant in most localities where they were evaluated, but were susceptible in a few localities; cvs. Winter Majetin and Northern Spy maintained outstanding resistance. This latter variety was used extensively as a parent in breeding projects to obtain apple rootstocks resistant to *E. lanigerum*, and many of these were propagated commercially and identified as Malling (M) and Malling-Merton (MM) series, from the East Malling Research Station in the United Kingdom (e.g. Preston, 1955). However, propagation beds of MM rootstocks in Australia, of MM, Northern Spy and Merton rootstocks in South Africa and of M and MM rootstocks in North Carolina (Rock and Zeiger, 1974) became infested with *E. lanigerum*, providing evidence for the existence of resistance-breaking biotypes of this aphid in different parts of the world. At least two biotypes of *E. lanigerum* have been recorded in Australia (Sen Cupta and Miles, 1975) and in the U.S.A. (Young et al., 1982). Since many cultivars with resistance derived from cv. Northern Spy were susceptible to the North Carolina biotype, the latter authors stated that cv. Robusta 5 was being used more heavily as a source of resistance to *E. lanigerum* in the New York apple rootstock breeding programme than the Northern Spy derivatives. Despite the occurrence of biotypes, the effectiveness of control of *E. lanigerum* by plant resistance is still valid (Russell, 1978), and this aphid was not considered a major pest of apple in several regions of North America. Cummins et al. (1981) in New York State and Taylor (1981) in New Zealand carried out extensive screening tests of species, hybrids and clones of *Malus* and reported several sources of resistance to *E. lanigerum*.

Sen Gupta and Miles (1975) reported that between apple varieties, an increasing alpha-amino nitrogen content tended to be associated with increasing susceptibility to *E. lanigerum*, and there was a clear-cut inverse correlation between the susceptibility of plant tissues and the ratio of phenolics to alpha-amino nitrogen in the tissues. The authors postulated that a higher content of toxic phenolic compounds could conceivably be tolerated provided the tissues were relatively rich in nutrients.

Prunus spp., *Pyrus* spp.

In Europe, at least two aphids, *M. persicae* and *Hyalopterus amygdali* (Blanchard) are pests of peach trees, *Prunus* spp., and these aphids, as well as *Myzus varians* Davidson, are vectors of the sharka (plum pox) virus disease. Several *Prunus persica* varieties from various countries and with inferior fruit quality were highly resistant to *M. persicae* and *M. varians*, as well as myrobolan, *P. cerasifera*, and one hybrid *P. persica* × *P. davidiana* (Massonié et al., 1982); two varieties proved to be resistant also to the virus (Maison et al., 1982). Several pear seedlings, *Pyrus* spp., were resistant to root infestation by the woolly pear aphid, *Eriosoma pyricola* Baker and Davidson (Westwood and Westigard, 1969).

Rubus spp.

Studies on resistance in raspberry, *Rubus* spp., have, in Europe, mainly been done on resistance against *Amphorophora rubi* subsp. *idaei* (Börner) and, in North America, on that against *Amphorophora agathonica* Hottes (These two species were sometimes misidentified as *Aphis rubi* Kaltenbach); *Aphis rubicola* Oestlund can also be a pest and a virus vector. The misidentifications of the aphid species concerned led to a confused situation until Blackman et al. (1977) published their important cytological and biometric clarifications; *A. rubi* is now recognized to infest blackberry rather than raspberry. Some European publications that appeared before this clarification, and citing *A. rubi* on raspberry, were in fact dealing with *A. rubi* subsp. *idaei*. Conversely, studies reported in America before the paper of Converse et al. (1971), and citing *A. rubi* on raspberry, may have been dealing with *A. agathonica*. This must be kept in mind in the references to aphids on raspberry mentioned in the following discussion.

After a 3-year period of testing some 27 varieties of red raspberries, *Rubus strigosus*, in the U.S.A., Huber and Schwartze (1938) found cvs. Lloyd George, Indian Summer, Pyne Imperial and Pyne Royal to be practically immune to *A. rubi*; however, this aphid reproduced abundantly on nineteen other varieties. Hill (1957), in Great Britain, tested some sixteen varieties of *Rubus*, comprising four species, and reported cvs. Lloyd George and Malling Promise to be susceptible to *A. rubi*; when samples of these varieties were tested in western Canada, they were immune or highly resistant to the local *A. rubi* (Hill, 1956). From what is now known, two aphid species were involved, and cv. Lloyd George was highly resistant in North America to *A. agathonica* but susceptible in Europe to *A. rubi* subsp. *idaei*. Rautapää (1967) reviewed the conflicting literature from both continents, and reported that raspberry varieties Rikala and Malling Promise were resistant to *A. rubi* subsp. *idaei* and that cvs. Chief

Section 11.4 references, p. 254

and Malling Landmark were resistant to *A. rubi*. Four strains of *A. rubi* were recognized in England (Briggs, 1965) (although two aphid species were probably involved) and the detailed relationships between the genes for resistance in various *Rubus* spp. and the strains of aphids were reviewed by Maxwell et al. (1972) and Russell (1978); major resistance genes commonly occurred in wild raspberries and should be very useful in breeding for resistance to aphids.

Resistant varieties usually have a much lower incidence of aphid-transmitted virus diseases than susceptible ones (Jones, 1979), showing that resistance to aphid vectors can be instrumental in controlling virus diseases in the field. A combination of virus and aphid resistance should be even more effective and provide an insurance against the development of resistance-breaking biotypes. Recently, some 64 seedlings of the native North American red raspberry, *Rubus idaeus strigosus*, screened from a total of 1041, were selected for resistance to colonization by *a. agathonica* (Daubeny and Starý, 1982), and the authors suggested that this subspecies be studied further as a resistance source and utilized more widely in breeding programmes.

Nonpreference and antibiosis types of resistance to *A. agathonica* were reported in several cultivars, and, in resistant cv. Canby, antibiosis may be related to a low concentration of sugars and nitrogenous compounds in the phloem sap of leaves (Kennedy and Schaefers, 1975). Antibiosis to *A. rubicola* was also observed in field evaluations of red raspberry cultivars and selections, cvs. NY 632, Willamette, Latham and Canby being least susceptible (Brodel and Schaefers, 1980).

Fragaria spp.

Some thirteen aphid species have been reported to colonize strawberries, *Fragaria*, in North America, and major pests include four *Chaetosiphon* species that feed on the leaves and petioles, and the strawberry root aphid, *Aphis forbesi* Weed (Schaefers, 1981). Clones Del Norte and Yaquina of *F. chiloensis* were highly resistant to the strawberry aphid,*Chaetosiphon fragaefolii* (Cockerell), and the resistance was heritable when clone Del Norte was crossed with commercial-type strawberries. Further screening of 557 clones of wild *F. chiloensis* from the Pacific Coast of the U.S.A. to *Chaetosiphon* spp. showed that 29 clones had fewer aphids than clone Del Norte, although the differences were not statistically significant (Crock et al., 1982). Benton, a commercial cultivar, consistently supported fewer aphids than the others, and it would be a better breeding source for resistance than clone Del Norte. Further breeding experiments are being carried out to combine strawberry aphid resistance in cultivars with commercially acceptable fruit quality (Shanks and Sjulin, 1983). Electrical measurements of feeding of *C. fragaefolii* showed no difference in feeding behaviour, length of feeding periods or time to first ingestion between resistant and susceptible strawberry plants (Shanks and Chase, 1976).

Solanaceae

Solanum spp.

The cultivated potato, *Solanum tuberosum*, ranks fourth in world agricultural production, and it has been known for a long time that aphids carry several important potato viruses, especially the green peach aphid, *M. persicae* and the potato aphid, *Macrosiphum euphorbiae* (Thomas). Some twelve other potato-infesting aphid species are mentioned in the comprehensive review by Radcliffe (1982), and Painter (1951) reviewed early observations and screening of potatoes for resistance to aphids. Detailed tests by Adams (1946) conducted over six years in New Brunswick, Canada, on some 80 varieties and species of *Solanum*, showed *Solanum polyadenium* to be almost immune to foliage infesta-

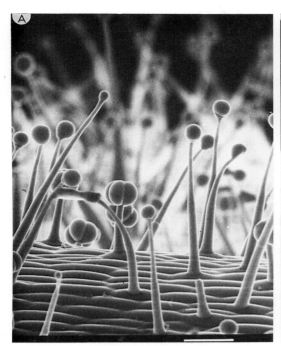

Fig. 11.4.5. (A) Scanning electron micrograph of the two types of glandular hairs on the leaf of *Solanum berthaultii* (scale bar = 95 μm).
(B) Scanning electron micrograph of an aphid entangled in the secretion of the glandular hairs on a leaf of *S. berthaultii* (scale bar = 290 μm). (Both photographs by Dr. R.W. Gibson, Harpenden, United Kingdom.)

tion by *M. persicae*, and various species of *Solanum* were more resistant than *S. tuberosum*, the variety Katahdin in this latter species being very susceptible. Between cultivars, differences in aphid infestations may be largely dependent upon temporal physiological circumstances and growth habit variations, but much greater differences are reported from wild potato species. In a series of papers published from 1966 to 1979 by Radcliffe and coworkers in Minnesota, U.S.A. (literature reviewed by Radcliffe et al., 1981), 1600 tuber-bearing *Solanum* species were screened for resistance to aphids. Some 134 and 223 entries were rated highly resistant or resistant to *M. persicae* and/or *M. euphorbiae*, respectively, and resistance appeared to be a highly heritable characteristic, stable to environment and seasons (Radcliffe, 1982). Some resistance to *M. persicae* was observed recently in commercially grown cultivars of potato (Jane et al., 1982).

Back in 1968, Gentile and Stoner had reported that *M. euphorbiae* became entangled in the exudate from glandular hairs of *Solanum pennellii*, a non-tuberiferous species, and resistance to aphids and some other small arthropods was shown to be due to glandular hairs or trichomes in *Solanum berthaultii*, *S. polyadenium*, *Solanum tarijense* and other species (Gibson and Plumb, 1977; Tingey et al., 1981). Two types of these glandular hairs on leaves and stems release a gummy exudate which stuck naturally-occurring aphids (Fig. 11.4.5). In greenhouse tests, the percentage of stuck aphids was positively correlated with the abundance of these hairs, especially on leaves, and type (A) and (B) trichomes seem to interact to enhance significantly the expression of resistance to aphids and leafhoppers (Tingey and Laubengayer, 1981). *S. berthaultii* is readily hybridized with the cultivated potato, and protection against aphids can be enhanced from such crossings (Mehlenbacher et al., 1983; Raman and Palacios, 1983). Furthermore, the presence of substantial quantities of the aphid alarm pheromone, (*E*)-β-farnesene, in air around *S. berthaultii* foliage and in ethanol washings from the foliage was identified by Gibson and Pickett (1983), who concluded that this compound is probably released from the glandular hairs (probably type B) on the foliage, acts as an allomone, and could protect against virus spread by alate aphids in particular (Gunenc and Gibson, 1980). Levels of glycoalkaloids in foliage and tubers of thirteen accessions of *S. berthaultii* and *S. berthaultii* × *S. tarijense* did not correlate with aphid resistance (Tingey and Sinden, 1982), but correlation with resistance to the potato leafhopper, *Empoasca fabae* (Harris) was reported in some *Solanum* species (reviewed by Tingey 1981).

Lycopersicon spp.

Tomato, *Lycopersicon* spp., is grown around the world, and it is a host plant for aphids, such as *M. persicae* and *M. euphorbiae*, that attack the seedlings, deform the fruit and may be vectors of viruses or mycoplasmas (Lange and Bronson, 1981). Secretion from glandular hairs on tomato foliage trapped aphids, e.g. *A. craccivora* and *M. persicae*, that eventually died, apparently from desiccation and exhaustion. However, attempts to maintain colonies on tomato plants from which the glandular cells had been removed were only partly successful, presumably due to some physiological incompatibility of the host plant (Johnson, 1956). Out of eight commercial varieties of tomato and 54 breeding lines, one variety, Campbell 16, and two breeding lines were resistant in the field to *M. euphorbiae* (Stoner et al., 1968), and five accessions of *Lycopersicon peruvianum* were resistant to this aphid. Quiros et al. (1977) reported that pubescence was not a resistance factor in normal genotypes of *Lycopersicon esculentum*; however, pink forms of *M. euphorbiae* in the field avoided both an excessively hairy mutant as well as the wild tomato *Lycopersicon hirsutum*, but under laboratory conditions the insects managed to feed on these plants. Comparative analyses of foliage of cultivars of *L. esculentum* and *Lycopersicon pimpinellifolium* revealed higher levels of sucrose, alanine and tyrosine, but lower levels of quinic acid, and a trend toward higher total amino acid concentrations in the susceptible plants. Furthermore, o-phosphoethanol amine was detected only in susceptible plants, and it was suggested that these differences in nutritional composition may play an important role in resistance to aphids (Quiros et al., 1977). A compound with insecticidal properties, 2-tridecanone, was isolated from the foliage of the wild tomato, *L. hirsutum* f. *glabratum*, where it is about 72 times more abundant than in *L. esculentum* (Williams et al., 1980). It is highly toxic to some insects, including *A. gossypii*. This compound is also found, although in much lower concentrations, in trichome exudate of *L. hirsutum* f. *glabratum* as well as in defensive secretions of a termite and a caterpillar. Since the above two species of tomato can be readily crossed, it should be possible to develop cultivars with increased quantities of this insecticide.

Nicotiana spp.

Tobacco introductions and wild *Nicotiana* species screened for resistance to *M. persicae* demonstrated that *Nicotiana gossei, N. repanda* and *N. trigonophylla* were highly resistant, but resistance was reduced after crossing *N. gossei* with *Nicotiana tabacum* (Thurston, 1961); the glucose content of leaves did not always correlate with resistance. Other screening tests of some 45 *Nicotiana* species and twelve related strains showed the range of *M. persicae* numbers to vary from 3500 individuals per plant on *N. knightiana* and *N. sylvestris* to only 30 per plant on *N. fragrans*. The highly resistant species were related phylogenetically, suggesting that the mechanism of resistance may have a common origin (Burk and Stewart, 1969). Although there were large differences in populations of apterous *M. persicae* between plants, there was no relationship between nicotine levels in leaves and populations of alate or apterous *M. persicae* in three cultivars and twelve haploid-derived experimental lines of burley tobacco, *N. tabacum* (Thurston et al., 1977). Of some 800 entries of *N. tabacum* evaluated in the field in South Carolina, 33 were resistant to *M. persicae*, and nonpreference and antibiosis were apparently involved, but the basic causes of resistance were unknown (Johnson, 1980).

Resistance in *N. gossei* to *M. persicae* and other aphid species apparently resulted from the production of a toxic material exuded from the leaf hairs, and poisoning symptoms on aphids closely resembled those observed from nicotine (Thurston and Webster, 1962). Although *N. tabacum* and *N. rustica*, two species heavily infested with aphids, contain fairly large concentrations of nicotine in their leaves, it was postulated that aphids were not affected, since they avoid the nicotine-containing xylem: they feed mainly on phloem and no nicotine was detected in their honeydew (Guthrie et al., 1962).

Other studies had shown the presence of nicotine, anabasine and probably nornicotine in the trichome secretions of various *Nicotiana* species, and it was postulated that different concentrations of these alkaloids in the plant secretions were the cause of resistance or susceptibility to aphids (Thurston et al., 1966). Johnson and Severson (1982) showed that leaf surfaces of *N. tabacum* entries low in duvatrienediols were resistant to *M. persicae*, and they had trichomes with little or no exudate. Washes of the leaf surface of some resistant tobaccos included high sucrose esters and duvatriene-ols. They suggested that more than one type of leaf surface chemistry is associated with resistance to aphids.

Miscellaneous host plants

Vaccinium spp.
Four cultivars of highbush blueberry, *Vaccinium* spp., supported much lower numbers of the aphid *Illinoia pepperi* (Mac Gillavray), the vector of blueberry shoestring virus, than twelve other cultivars (Hancock et al., 1982).

Daucus carota
Field and insectary tests with representative varieties of carrots, *Daucus carota*, from England, together with three Australian varieties, showed only slight differences in susceptibility to the aphid *Cavariella aegopodii* (Scopoli), and cv. Berlikum seemed to be the least susceptible (Dunn, 1970). Although the Australian variety Osborne Park had some resistance to *C. aegopodii* in New Zealand (Lamb, 1953), it proved of average susceptibility in England.

Gossypium spp.
A review by Niles (1980b) indicates that results on the relation of plant hairiness in cotton, *Gossypium* spp., with resistance to *A. gossypii* are contradictory. Laboratory tests found gossypol to be highly toxic to aphids.

Vitis spp.
Insects on grape, *Vitis* spp., were reviewed by Bournier (1977) who indicated that the grape phylloxera, *Viteus vitifoliae*, was formerly the most damaging pest of vine. This world-wide species produces, on *Vitis vinifera*, enormous tuberosities on the roots, thereby causing the death of the rootstock. On American vines (e.g. *V. rotundifolia*), the insect causes only very slight damage with the production of shallow wounds, and it swarms before winter. The origin of this resistance was studied by several workers (listed in Bournier, 1977). Control of *V. vitifoliae* is obtained through the grafting of cultivars of *V. vinifera* onto American *Vitis* stocks. Attempts have been made to hybridize

Section 11.4 references, p. 254

them to avoid grafting, but these hybrids can give an unpleasant taste to the wine, and their planting in France has been prohibited (Bournier, 1977). Damage to grape by *V. vitifoliae* in North America can vary with cultivars (e.g. Jubb, 1976); some hybrids (*V. vinifera* × *V. rotundifolia*) were resistant (Firoozabady and Olmo, 1982); at least two strains of the aphid exist in Ontario, Canada (Stevenson, 1970).

Humulus lupulus

Studies over several years at the Hop Research Department of Wye College in England and elsewhere in Europe on the resistance of cultivars of hop, *Humulus lupulus*, to the damson-hop aphid, *Phorodon humili* (Schrank), were reviewed by Campbell (1983).

Beta spp.

Differences in the resistance of sugar beet varieties (*Beta* spp.) to *M. persicae* and *A. fabae* were observed in 1960 by Russell (1978); varietal differences in aphid preference and tolerance were noted, but antibiosis was apparently the most important mechanism of resistance. Large differences existed between the colonizing ability and overall vigour of various aphid clones of *M. persicae*, but resistant varieties generally supported smaller aphid populations in field and greenhouse tests (Lowe, 1974); however, the expression of resistance can be affected by environmental factors, including certain nutrients to the host plant. Populations of *M. persicae* were reduced 60% by using the resistant *Beta vulgaris*, variety US H9B, which also reduced infection by the yellows viruses in Northern California (Hills et al., 1969). As measured with an electronic measuring system (Tarn and Adams, 1982), ingestion from sieve tubes by *M. persicae* was higher on two susceptible hosts (Chinese cabbage and a sugar beet hybrid) than on a resistant hybrid. Aphids on the resistant plants tended to leave the plants, or were restless when restricted to them. On Chinese cabbage, aphids spent much of their time ingesting from sieve-element cells (Haniotakis and Lange, 1974). Although only small quantitative differences in resistance to aphids were reported, they apparently can contribute significantly in the field to the control of aphids and viruses transmitted by them (Lowe, 1975), especially when plant resistance is combined with other methods of control. Two leaf spot and spider mite resistant varieties of sugar beet, GW 674 and GW 359, inhibited the development of the sugar beet root aphid, *Pemphigus betae* Doane (Harper, 1964).

Picea, Larix

Two reviews on the resistance of forest trees to insects and other pests (Hanover, 1980; Heybroek et al., 1982) include studies on the eastern spruce gall aphid, *Chermes abietis* (Linnaeus), a serious pest of at least two ornamental and Christmas tree species, the white spruce, *Picea glauca*, and the Norway spruce, *Picea abies*, in the northeastern U.S.A.; gall-free trees growing among heavily infested ones have been observed, and selection and production of resistant strains are in progress. From several years of study in Romania, clones of *Larix decidua* and *Larix leptolepis*, with varying polygenic resistance to *Adelges laricis* Vallot, were selected out for further improvement, possibly through hybridization.

Carya illinoensis

The pecan tree, *Carya illinoensis*, indigenous to the southwestern U.S.A., is commonly attacked by the pecan phylloxera, *Phylloxera devastatrix* Pergande, which produces conspicuous galls on petioles, leaflets and occasionally nutlets. Responses of some pecan cultivars have varied from resistant to susceptible depending on location, suggesting the existence of biotypes of this phylloxera. In their review, Carpenter et al. (1979) listed some seventeen selected pecan genotypes resistant or tolerant to this insect, three resistant to the pecan leaf phylloxera, *Phylloxera notabilis* Pergande, and at least two tolerant to the black pecan aphid, *Aphis fumipennella* Fitch (= *Melanocallis caryaefoliae* (Davis)); some of the tolerant genotypes have been included in a varietal breeding programme. Tolerance of some pecan genotypes to damage by infestations from yellow pecan aphids, *Monellia* spp. and *Monelliopsis* spp., was highly variable during the growing season, and no resistance factors were identified. Finally, some 32 pecan clones (cultivars or selections) were resistant to the southern pecan leaf phylloxera, *Phylloxera russellae* Stoetzel (Calcote, 1983).

Tilia spp.

Resistance of some species of lime trees, *Tilia* spp., to the lime aphid, *Eucallipterus tiliae* (Linnaeus), was associated with either leaf pubescence and/or the presence of small glandular prominences lying on the leaf cuticle (Carter, 1982).

CONCLUSION

There have been numerous and significant achievements in the control of aphids by host plant resistance, especially during the past two decades, and there remains tremendous potential in breeding aphid-resistant cultivars that could be included in integrated pest management programmes. Hundreds of resistant cultivars or selections, from close to fifty crop plants against as many aphid species, were reported in this review, which does not claim to be exhaustive. It has also been shown that in many cases, a cultivar resistant to one species is not necessarily resistant to the other, neither is it necessarily resistant to certain biotypes of the former aphid species. Furthermore, there is potential for restricting the spread of damaging viruses by breeding aphid-resistant cultivars, although seldom has virus control been a major aim of breeding programmes for vector control. It would undoubtedly be better to utilize a type of aphid resistance that would also restrict virus spread when the aphid is both a pest and a vector (see also section 11.5).

There has been some experience that the growing of near-immune varieties, especially with antixenosis and antibiosis, may impose a great selection pressure in favour of resistance-breaking variants of the insects concerned. Biotypes or races have occurred most frequently among aphids, mainly because of their particular reproductive biology. It appears therefore more sensible to concentrate on the development of partially resistant varieties with tolerance, which reduce the effects of parasitism to an acceptable level rather than attempt to breed almost immune varieties. Semi-resistant varieties give protection and allow colonization by small numbers of aphids that provide a livelihood to parasites and predators. On the other hand, it is particularly important that plant breeders do not develop varieties that are very susceptible to any potential pest. New varieties with an adequate level of resistance to almost any pest or disease could be developed in all the major agricultural crops, and this often within a few years, as long as sufficient resources, in terms of skilled manpower, research facilities and funds, are provided for such investigations. The use of resistant crop varieties should be of special economic benefit in developing countries, where individual acreages are usually small and farmers often not well acquainted with the scientific application of pest control methods.

A knowledge of the genetics and basic causes of resistance is not a prerequis-

Section 11.4 references, p. 254

ite in the production of resistant cultivars. Thus resistance of alfalfa to the spotted alfalfa aphid was highly successfully exploited by selection from populations with a few resistant plants, even though the genetics and the chemical bases of resistance were not known. Only few investigators have thought it important to study the basic causes of resistance, and entomologists and plant breeders have often ignored the subject, mainly because of its complexity and the lack of specialists with appropriate techniques to cope with the multivarious interrelationships between the plant, the insect, and the environment. However, sophisticated techniques and other developments in the fields of plant physiology and biochemistry, insect nutritional physiology and behaviour, chemical ecology, and related disciplines, have recently re-activated investigations on the more fundamental aspects of host-plant resistance to insects, and several examples related to aphids were presented in this section. Furthermore, health and environmental concerns require an understanding of the chemistry and possible toxicity to humans and animals of resistant factors in plants, and it will become a powerful stimulant in inspiring research into the basic nature of resistance. This research requires a team approach and a willingness by investigators to share with others in research accomplishments, as exemplified by progress described herewith from groups of scientists working together in the development of plants resistant to aphids.

ACKNOWLEDGEMENTS

Financial assistance from the Natural Sciences and Engineering Research Council of Canada to the author for the preparation of this review is gratefully acknowledged. Special thanks are due to my colleague, Dr. P.N. Srivastava, and to Professors Ernst Horber and Don C. Peters, from the Departments of Entomology, Kansas State University and Oklahoma State University, respectively, for critically reading the manuscript and making valuable suggestions.

REFERENCES

Adams, J.B., 1946. Aphid resistance in potatoes. American Potato Journal, 23: 1–22.

Adams, J.B. and Van Emden, H.F., 1972. The biological properties of aphids and their host plant relationships. In: H.F. van Emden (Editor), Aphid Technology. Academic Press, London, pp. 47–104.

Allard, R.W., 1960. Principles of Plant Breeding. Wiley, New York, 485 pp.

Al-Mousawi, A.H., Richardson, P.E. and Burton, R.L., 1983. Ultrastructural studies of greenbug (Hemiptera: Aphididae) feeding damage to susceptible and resistant wheat cultivars. Annals of the Entomological Society of America, 76: 964–971.

Alston, F.H. and Briggs, J.B., 1977. Resistance genes in apple and biotypes of *Dysaphis devecta*. Annals of Applied Biology, 87: 75–81.

Argandoña, V.H., Corcuera, L.J., Niemeyer, H.M. and Campbell, B.C., 1983. Toxicity and feeding deterrency of hydroxamic acids from Gramineae in synthetic diets against the greenbug, *Schizaphis graminum*. Entomologia Experimentalis et Applicata, 34: 134–138.

Arriaga, H.O., 1956. A rye synthetical hybrid "Insave F.A." resistant to the greenbug toxemia. Universidad Nacional de la Plata, Revista de la Facultad de Agronomia (Tercera Epoca), 32 (2a): 191–209 (in Spanish, with English summary).

Auclair, J.L., 1959. Feeding and excretion by the pea aphid, *Acyrthosiphon pisum* (Harr.) (Homoptera: Aphididae), reared on different varieties of peas. Entomologia Experimentalis et Applicata, 2: 279–286.

Auclair, J.L., 1976. Feeding and nutrition of the pea aphid, *Acyrthosiphon pisum* (Harris) with special reference to amino acids. Symposia Biologica Hungarica, 16: 29–34.

Auclair, J.L., 1978. Biotypes of the pea aphid, *Acyrthosiphon pisum*, in relation to host plants and chemically defined diets. Entomologia Experimentalis et Applicata, 24: 212–216.

Auclair, J.L., 1986. La nutrition des aphides, et en particulier d'*Acyrthosiphon pisum* (Harris), en relation avec leurs plantes-hôtes. Revue d'Entomologie du Québec, 31: 16–22.

Auclair, J.L. and Cartier, J.J., 1960. Effets comparés de jeûnes intermittents et de périodes équiv-
alentes de subsistance sur des variétés résistantes ou sensibles de pois, *Pisum sativum* L., sur
la croissance, la reproduction et l'excrétion du puceron du pois, *Acyrthosiphon pisum* (Harr.)
(Homoptères: Aphididés). Entomologia Experimentalis et Applicata, 3: 315–326.

Auclair, J.L., Maltais, J.B. and Cartier, J.J., 1957. Factors in resistance of peas to the pea aphid,
Acyrthosiphon pisum (Harr.) (Homoptera: Aphididae) II. Amino acids. Canadian Entomologist,
89: 457–464.

Auclair, J.L., Baldos, E. and Heinrichs, E.A., 1982. Biochemical evidence for the feeding sites of the
leafhopper *Nephotettix virescens* within susceptible and resistant rice plants. Insect Science and
its Application, 3: 29–34.

Barlow, C.A. and Randolph, P.A., 1978. Quality and quantity of plant sap available to the pea
aphid. Annals of the Entomological Society of America, 71: 46–48.

Barnes, D.K., Hanson, C.H., Ratcliffe, R.H., Busbice, T.H., Schillinger, J.A., Buss, G.R., Campbell,
W.V., Hemken, R.W. and Blickenstaff, C.C., 1970. The development and performance of Team
alfalfa: a multiple pest resistant alfalfa with moderate resistance to the alfalfa weevil. Crops
Research, Agricultural Research Service, Beltsville, 34–115, 41 pp.

Barnes, D.K., Frosheiser, F.I., Sorensen, E.L., Nielson, M.W., Lehman, W.F., Leath, K.T., Ratcliffe,
R.H. and Buker, R.J., 1974. Standard tests to characterize pest resistance in alfalfa varieties.
United States Department of Agriculture, Agricultural Research Service, NC-19, 23 pp.

Barnes, W.C. and Cuthbert, Jr., F.P., 1975. Breeding turnips for resistance to the turnip aphid.
HortScience, 10: 59–60.

Barton, K.A. and Brill, W.J., 1983. Prospects in plant genetic engineering. Science, 219: 671–676.

Beck, D.L., Dunn, G.M., Routley, D.G. and Bowman, J.S., 1983. Biochemical basis of resistance in
corn to the corn leaf aphid. Crop Science, 23: 995–998.

Blackman, R.L., Eastop, V.F. and Hills, M., 1977. Morphological and cytological separation of
Amphorophora Buckton (Homoptera: Aphididae) feeding on European raspberry and blackberry (*Rubus* spp.). Bulletin of entomological Research, 67: 285–296.

Bohn, G.W., Kishaba, A.N. and Toba, H.H., 1972. Mechanisms of resistance to melon aphid in a
muskmelon line. HortScience, 7: 281–282.

Bond, D.A. and Lowe, H.J.B., 1975. Tests for resistance to *Aphis fabae* in field beans (*Vicia faba*).
Annals of Applied Biology, 81: 21–32.

Borlaug, N.E., 1983. Contributions of conventional plant breeding to food production. Science, 219:
689–693.

Bouchery, Y., 1977. Les pucerons *Aphis fabae* Scop. et *Acyrthosiphon pisum* (Harris) (Homoptère:
Aphididae) déprédateurs de la féverole de printemps (*Vicia faba* L.) dans le nord-est de la
France: influence sur le rendement des cultures. Mécanisme de la déprédation. Annales de
Zoologie et Ecologie Animale, 9: 99–110.

Bournier, A., 1977. Grape insects. Annual Review of Entomology, 22: 355–376.

Bournoville, R., 1980. Varietal characteristics under French conditions of some lucerne cultivars
selected for resistance to two pest insects. European Plant Protection Organization, Bulletin
10: 317–322.

Briggs, J.B., 1965. The distribution, abundance and genetic relationships of four strains of the
rubus aphid (*Amphorophora rubi* Kalt.) in relation to raspberry breeding. Journal of Horticultural Science, 40: 109–117.

Brodel, C.F. and Schaefers, G.A., 1980. Evidence for antibiosis in red raspberry to *Aphis rubicola*.
Journal of Economic Entomology, 73: 647–650.

Burk, L.G. and Stewart, P.A., 1969. Resistance of *Nicotiana* species to the green peach aphid.
Journal of Economic Entomology, 62: 1115–1117.

Calcote, V.R., 1983. Southern pecan leaf phylloxera (Homoptera: Phylloxeridae): clonal resistance
and technique for evaluation. Environmental Entomology, 12: 916–918.

Campbell, B.C., McLean, D.L., Kinsey, M.G., Jones, K.C. and Dreyer, D.L., 1982. Probing behaviour
of the greenbug (*Schizaphis graminum*, biotype C) on resistant and susceptible varieties of
sorghum. Entomologia Experimentalis et Applicata, 31: 140–146.

Campbell, C.A.M., 1983. Antibiosis in hop (*Humulus lupulus*) to the damson-hop aphid, *Phorodon
humuli*. Entomologia Experimentalis et Applicata, 33: 57–62.

Carpenter, T.L., Neel, W.W. and Hedin, P.A., 1979. A review of host plant resistance of pecan, *Carya
illinoensis*, to Insecta and Acarina. Bulletin of the Entomological Society of America, 25:
251–257.

Carter, C.I., 1982. Susceptibility of *Tilia* species to the aphid *Eucallipterus tiliae*. In: J.H. Visser and
A.K. Minks (Editors), Insect–Plant Relationships. Centre for Agricultural Publishing and
Documentation, Wageningen, pp. 421–423.

Carter, N., McLean, I.F.G., Watt, A.D. and Dixon, A.F.G., 1980. Cereal aphids: a case study and
review. Applied Biology, 5: 271–348.

Cartier, J.J., 1963. Varietal resistance of peas to pea aphid biotypes under field and greenhouse
conditions. Journal of Economic Entomology, 56: 205–213.

Cartier, J.J., Isaak, A., Painter, R.H. and Sorensen, E.L., 1965. Biotypes of pea aphid *Acyrthosiphon pisum* (Harris) in relation to alfalfa clones. Canadian Entomologist, 97: 754–970.

Chada, H.L., Atkins, I.M., Gardenhire, J.H. and Weibel, D.E., 1961. Greenbug-resistance studies with small grains. United States Department of Agriculture, Publication B-982, 18 pp.

Chedester, L.D. and Michels, Jr., G.J., 1982. An evaluation of greenbug resistance in oats. Southwestern Entomologist, 7: 166–169.

Converse, R.H., Daubeny, H.A., Stace-Smith, R., Russell, L.M., Koch, E.J. and Wiggans, S.C., 1971. Search for biological races in *Amphorophora agathonica* Hottes on red raspberry. Canadian Journal of Plant Science, 51: 81–85.

Corcuera, L.J., Argandoña, J.H., Pena, G.F., Pérez, F.J. and Niemeyer, H.M., 1982. Effect of a benzoxazinone from wheat on aphids. In: J.H. Visser and A.K. Minks (Editors), Insect–Plant Relationships. Centre for Agricultural Publishing and Documentation, Wageningen, pp. 33–39.

Cress, D.C. and Chada, H.L., 1971. Development of a synthetic diet for the greenbug, *Schizaphis graminum*. 3. Response of greenbug biotypes A and B to the same diet medium. Annals of the Entomological Society of America, 64: 1245–1247.

Crock, J.E., Shanks, Jr., C.H. and Barritt, B.H., 1982. Resistance in *Fragaria chiloensis* and *F.* × *ananassa* to the aphids *Chaetosiphon fragaefolii* and *C. thomasi*. Hortscience, 17: 959–960.

Croft, B.A. and Bode, W.M., 1983. Tactics for deciduous fruit IPM. In: B.A. Croft and S.C. Hoyt (Editors), Integrated Management of Insect Pests of Pome and Stone Fruits. Wiley, New York, pp. 219–270.

Cummins, J.N., Forsline, P.L. and Mackenzie, J.D., 1981. Woolly apple aphid colonization on *Malus* cultivars. Journal of the American Society of Horticultural Science, 106: 26–30.

Dahms, R.G. and Painter, R.H., 1940. Rate of reproduction of the pea aphid on different alfalfa plants. Journal of Economic Entomology, 33: 482–485.

Dahms, R.G., Johnston, T.H., Schlehuber, A.M. and Wood, Jr., E.A., 1955. Reaction of small-grain varieties and hybrids to greenbug attack. Oklahoma Agricultural Experiment Station, Stillwater, Technical Bulletin No. T-55, 61 pp.

Daniels, N.E., 1978. Greenbug resistance in oats. Southwestern Entomologist, 3: 210–214.

Daubeny, H.A. and Starý, P., 1982. Identification of resistance to *Amphorophora agathonica* in the native North American red raspberry. Journal of the American Society of Horticultural Science, 107: 593–597.

Denéchère, M., 1979. Premiers résultats concernant une étude comparative de la résistance de plantules de deux variétés de maïs vis-á-vis de *Rhopalosiphum padi* L. (Homoptère – Aphididae). Annales de l'Amélioration des Plantes, 29: 545–556.

DePew, L.J. and Witt, M.D., 1979. Evaluation of greenbug-resistant sorghum hybrids. Journal of Economic Entomology, 72: 177–179.

Dewar, A.M., 1977. Assessment of methods for testing varietal resistance to aphids in cereals. Annals of Applied Biology, 87: 183–190.

Dodd, G.D. and Van Emden, H.F., 1979. Shifts in host plant resistance to the cabbage aphid, *Brevicoryne brassicae* exhibited by brussels-sprout plants. Annals of Applied Biology, 91: 251–262.

Dreyer, D.L. and Campbell, B., 1984. Association of the degree of methylation of intercellular pectin with plant resistance to aphids and with induction of aphid biotypes. Experientia, 40: 224–226.

Dreyer, D.L., Reese, J.C. and Jones, K.C., 1981. Aphid feeding deterrents in sorghum: bioassay, isolation and characterization. Journal of Chemical Ecology, 7: 273–284.

Dunn, J.A., 1959. The biology of lettuce root aphid. Annals of Applied Biology, 47: 475–491.

Dunn, J.A., 1970. The susceptibility of varieties of carrot to attack by the aphid, *Cavariella aegopodii* (Scop.). Annals of Applied Biology, 66: 301–312.

Dunn, J.A., 1974. Study on inheritance of resistance to root aphid, *Pemphigus bursarius*, in lettuce. Annals of Applied Biology, 76: 9–18.

Dunn, J.A. and Kempton, D.P.H., 1969. Resistance of rape (*Brassica napus*) to attack by the cabbage aphid (*Brevicoryne brassicae* L.). Annals of Applied Biology, 64: 203–212.

Dunn, J.A. and Kempton, D.P.H., 1971. Differences in susceptibility to attack by *Brevicoryne brassicae* (L.) on brussels sprouts. Annals of Applied Biology, 68: 121–134.

Dunn, J.A. and Kempton, D.P.H., 1972. Resistance to attack by *Brevicoryne brassicae* among plants of brussels sprouts. Annals of Applied Biology, 72: 1–11.

Dunn, J.A. and Kempton, D.P.H., 1974. Lettuce root aphid control by means of plant resistance. Plant Pathology, 23: 76–80.

Eenink, A.H. and Dieleman, F.L., 1982. Resistance of *Lactuca* accessions to leaf aphids: components of resistance and exploitation of wild *Lactuca* species as sources of resistance. In: J.H. Visser and A.K. Minks (Editors), Insect–Plant Relationships. Centre for Agricultural Publishing and Documentation, Wageningen, pp. 349–355.

El-Kandelgy, S.M. and Wilcoxson, R.D., 1964. Insect transmission of red clover vein-mosaic virus and resistance of clover to aphids. Journal of the Minnesota Academy of Science, 32: 33–36.

Farrell, J.A. and Stufkens, M.W., 1981. Field evaluation of lucerne cultivars for resistance to blue-green lucerne aphid and pea aphid (*Acyrthosiphon* spp.) in New Zealand, New Zealand Journal of Agricultural Research, 24; 217–220.

Ferguson, S., Sorensen, E.L. and Horber, E.K., 1982. Resistance to the spotted alfalfa aphid (Homoptera: Aphididae) in glandular-haired *Medicago* species. Environmental Entomology, 11: 1229–1232.

Firoozabady, E. and Olmo, H.P., 1982. Resistance to grape phylloxera in *Vitis vinifera* × *V. rotundifolia* grape hybrids.Vitis, 21: 1–4.

Fisk, J., 1978. Resistance of *Sorghum bicolor* to *Rhopalosiphum maidis* and *Peregrinus maidis* as affected by differences in the growth stage of the host. Entomologia Experimentalis et Applicata, 23: 227–236.

Frazer, B.D., 1972. Population dynamics and recognition of biotypes in the pea aphid (Homoptera: Aphididae). Canadian Entomologist, 104: 1729–1733.

Frazer, B.D., Raworth, D. and Gossard, T., 1976. Faba bean: low resistance to pea aphids, *Acyrthosiphon pisum* (Homoptera: Aphididae), in eleven cultivars. Canadian Journal of Plant Science, 56: 451–453.

Galli, A.J.B., Lara, F.M. and Barbosa, J.C., 1981. Resistance of genotypes of sorghum *Schizaphis graminum* (Rondani, 1852) (Homoptera, Aphididae). Anais da Sociedade Entomologica do Brasil, 10: 61–71 (in Portuguese, with English summary).

Gallun, R.L., Starks, K.J. and Guthrie, W.D., 1975. Plant resistance to insects attacking cereals. Annual Review of Entomology, 20: 337–357.

Gardenhire, J.H., 1980. Breeding for greenbug *Schizaphis graminum* (Rondani) resistance in wheat and other small grains. In: M.K. Harris (Editor), Biology and Breeding for Resistance to Arthropods and Pathogens in Agricultural Plants. Texas Agricultural Experiment Station MP-1451, pp. 237–244.

Gentile, A.G. and Stoner, A.K., 1968. Resistance in *Lycopersicon* and *Solanum* species to the potato aphid. Journal of Economic Entomology, 61: 1152–1154.

Gibson, R.W. and Pickett, J.A., 1983. Wild potato repels aphids by release of aphid alarm pheromone. Nature, 302: 608–609.

Gibson, R.W. and Plumb, R.T., 1977. Breeding plants for resistance to aphid infestation. In: K.F. Harris and K. Maramorosch (Editors), Aphids as Virus Vectors. Academic Press, New York, pp. 473–500.

Gillespie, D.J., 1983. Developing clovers for disease and insect resistance. Journal of Agriculture of Western Australia, 1: 14–15.

Gramshaw, D., Lowe, K.F. and Lloyd, D.L., 1982. Lucerne cultivars characteristics and planting recommendations for Queensland in 1983. Queensland Agricultural Journal, 108: 254–255.

Gunenc, Y. and Gibson, R.W., 1980. Effects of glandular foliar hairs on the spread of potato virus Y. Potato Research, 23: 345–351.

Guthrie, F.E., Campbell, W.V. and Baron, R.L., 1962. Feeding sites of the green peach aphid with respect to its adaptation to tobacco. Annals of the Entomological Society of America, 55: 42–46.

Hackerott, H.L., Harvey, T.L. and Ross, W.M., 1972. Registration of KS 41, KS 42, KS 43 and KS 44 greenbug-resistant grain sorghum germplasm. Crop Science, 12: 720.

Hamilton, B.A., Greenup, L.R. and Lodge, G.M., 1978. Seedling mortality of lucerne varieties in field plots subjected to spotted alfalfa aphid. Journal of the Australian Institute of Agricultural Science, March 1978: 54–56.

Hancock, J.F., Schulte, N.L., Siefker, J.H., Pritts, M.P. and Roueche, J.M., 1982. Screening highbush blueberry cultivars for resistance to the aphid *Illinoia pepperi*. HortScience, 17: 362–363.

Haniotakis, G.E. and Lange, W.H., 1974. Beet yellows virus resistance in sugar beets: mechanism of resistance. Journal of Economic Entomology, 67: 25–28.

Hanover, J.W., 1980. Breeding forest trees resistant to insects. In: F.G. Maxwell and P.R. Jennings (Editors), Breeding Plants Resistant to Insects. Wiley, New York, pp. 487–511.

Harper, A.M., 1964. Varietal resistance of sugar beets to the sugar-beet root aphid, *Pemphigus betae* Doane (Homoptera: Aphididae). Canadian Entomologist, 96: 520–522.

Harper, A.M., Miska, J.P., Manglitz, G.R., Irwin, B.J. and Armbrust, E.J., 1978. The Literature of Arthropods Associated with Alfalfa III. A Bibliography of the Pea Aphid, *Acyrthosiphon pisum* (Harris) (Homoptera: Aphididae). University of Illinois, Agricultural Experiment Station, Urbana-Champaign, Special Publication 50, 89 pp.

Harrington, C.D., 1945. Biological races of the pea aphid. Journal of Economic Entomology, 38: 12–22.

Harrington, C.D., Searls, E.M., Brink, R.A. and Eisenhart, C., 1943. Measurement of the resistance of peas to aphids. Journal of Agricultural Research, 67: 369–387.

Harvey, T.L. and Hackerott, H.L., 1969. Recognition of a greenbug biotype injurious to sorghum. Journal of Economic Entomology, 62: 776–779.

Harvey, T.L., Hackerott, H.L. and Sorensen, E.L., 1972. Pea aphid resistant alfalfa selected in the field. Journal of Economic Entomology, 65: 1661–1663.

Harvey, T.L., Martin, T.J. and Livers, R.W., 1980. Resistance to biotype C greenbug in synthetic hexaploid wheats derived from *Triticum tauschii*. Journal of Economic Entomology, 73: 387–389.

Hayes, H.K., Immer, F.R. and Smith, D.C., 1955. Methods of Plant Breeding. McGraw-Hill, New York, 2nd Edition, 551 pp.

Heathcote, G.D., 1972. Evaluating aphid populations on plants. In: H.F. Van Emden (Editor), Aphid Technology. Academic Press, London, pp. 105–145.

Heybroek, H.M., Stephan, B.R. and Von Weissenberg, K. (Editors), 1982. Resistance to diseases and pests in forest trees. Proceedings of the Third International Workshop on Genetics of Host–Parasite Interactions in Forestry, Wageningen, 14–21 September 1980. Centre for Agricultural Publishing and Documentation, Wageningen, 503 pp.

Hill, A.R., 1956. Observations on the North American form of *Amphorophora rubi* Kalt. (Homoptera, Aphididae). Canadian Entomologist, 88: 89–91.

Hill, A.R., 1957. Observations on the reproductive behaviour of *Amphorophora rubi* (Kalt.), with special reference to the phenomenon of insect resistance in raspberries. Bulletin of Entomological Research, 48: 467–476.

Hills, F.J., Lange, W.H. and Kishiyama, J., 1969. Varietal resistance to yellows, vector control, and planting date as factors in the suppression of yellows and mosaic of sugar beet. Phytopathology, 59: 1728–1731.

Hinz, B. and Daebeler, F., 1981. Untersuchungen zur Anfälligkeit verschiedener Rapssorten gegenüber der Mehligen Kohlblattlaus, *Brevicoryne brassicae* (L.), und zur Schadwirkung der Blattlaus an Winterraps. Archiv für Phytopathologie und Pflanzenschutz, 17: 115–125.

Hollenhorst, M.M. and Joppa, L.R., 1983. Chromosomal location of genes for resistance to greenbug in 'Largo' and 'Amigo' wheats. Crop Science, 23: 91–93.

Horber, E., Leath, K.T., Berrang, B., Marcarian, V. and Hanson, C.H., 1974. Biological activities of saponin components from DuPuits and Lahontan alfalfa. Entomologia Experimentalis et Applicata, 17: 410–424.

Howitt, A.J. and Painter, R.H., 1956. Field and greenhouse studies regarding the sources and nature of resistance of sorghum, *Sorghum vulgare* Pers., to the corn leaf aphid, *Rhopalosiphum maidis* (Fitch). Kansas Agricultural Experiment Station, Manhattan, Technical Bulletin 82, 38 pp.

Hsieh, J.H., 1982. Genetic studies on resistance in sorghum. Journal of the Agricultural Association of China (Taiwan), 117: 6–14.

Hsu, S.J. and Robinson, A.G., 1963. Further studies on resistance of barley varieties to the aphid, *Rhopalosiphum padi* (L.). Canadian Journal of Plant Science, 43: 343–348.

Huber, G.A. and Schwartze, C.D., 1938. Resistance in the red raspberry to the mosaic vector *Amphorophora rubi* Kalt. Journal of Agricultural Research, 57: 623–633.

Huber, L.L. and Stringfield, G.H., 1942. Aphid infestation of strains of corn as an index of their susceptibility to corn borer attack. Journal of Agricultural Research, 64: 283–291.

Hughes, R.D., 1972. Population dynamics. In: H.F. van Emden (Editor), Aphid Technology. Academic Press, London, pp. 275–293.

Hulse, J.H. and Spurgeon, D., 1974. Triticale. Scientific American, 231: 72–81.

Jane, E., Bintcliffe, E.J.B. and Wratten, S.D., 1982. Antibiotic resistance in potato cultivars to the aphid *Myzus persicae*. Annals of Applied Biology, 100: 383–391.

Jarvis, J.L., 1982. Susceptibility of some Brassicaceae plant introductions to the cabbage aphid (Homoptera: Aphididae). Journal of the Kansas Entomological Society, 55: 283–289.

Johnson, A.W., 1980. Resistance in tobacco to the green peach aphid. Journal of Economic Entomology, 73: 707–709.

Johnson, A.W. and Severson, R.F., 1982. Physical and chemical leaf surface characteristics of aphid resistant and susceptible tobacco. Tobacco Science, 26: 98–102.

Johnson, B., 1953. The injurious effects of the hooked epidermal hairs of French beans (*Phaseolus vulgaris* L.) on *Aphis craccivora* Koch. Bulletin of Entomological Research, 44: 779–788.

Johnson, B., 1956. The influence on aphids of the glandular hairs on tomato plants. Plant Pathology, 5: 130–132.

Johnson, J.W. and Teetes, G.L., 1980. Breeding for arthropod resistance in sorghum. In: M.K. Harris (Editor), Biology and Breeding for Resistance to Arthropods and Pathogens in Agricultural Plants. Texas Agricultural Experiment Station, MP-1451, pp. 168–180.

Johnson, J.W., Teetes, G.L. and Schaefer, C.A., 1976. Greenhouse and field techniques for evaluating resistance of sorghum cultivars to the greenbug. The Southwestern Entomologist, 1: 150–154.

Jones, A.T., 1979. Further studies on the effect of resistance to *Amphorophora idaei* in raspberry (*Rubus idaeus*) on the spread of aphid borne viruses. Annals of Applied Biology, 92: 119–123.

Joppa, L.R. and Williams, N.D., 1982. Registration of 'Largo', a greenbug resistant hexaploid wheat. Crop Science, 22: 901–902.

Jubb, Jr., G.L., 1976. Grape phylloxera: incidence of foliage damage to wine grapes in Pennsylvania. Journal of Economic Entomology, 69: 763–766.

Juneja, P.S., Pearcy, S.C., Gholson, R.K., Burton, R.L. and Starks, K.J., 1975. Chemical basis for greenbug resistance in small grains. II. Identification of the major neutral metabolite of benzyl alcohol in barley. Plant Physiology, 56: 385–389.

Karel, A.K. and Malinga, Y., 1980. Leafhopper and aphid resistance in cowpea varieties. Tropical Grain Legume Bulletin, 20: 10–11.

Kennedy, G.G., 1978. Recent advances in insect resistance of vegetables and fruit crops in North America: 1966–1977. Bulletin of the Entomological Society of America, 24: 375–384.

Kennedy, G.G. and Abou-Ghadir, M.F., 1979. Bionomics of the turnip aphid on two turnip cultivars. Journal of Economic Entomology, 72: 754–757.

Kennedy, G.G. and Schaefers, G.A., 1975. Role of nutrition in the immunity of red raspberry to *Amphorophora agathonica* Hottes. Environmental Entomology, 4: 115–119.

Kennedy, G.G., McLean, D.L. and Kinsey, M.G., 1978. Probing behavior of *Aphis gossypii* on resistant and susceptible muskmelon. Journal of Economic Entomology, 71: 13–16.

Kennedy, J.S. and Mittler, T.E., 1953. A method of obtaining phloem sap via the mouth-parts of aphids. Nature, 171: 528.

Kieckhefer, R.W., 1983. Host preferences and reproduction of four cereal aphids (Hemiptera: Aphididae) on certain wheatgrasses, *Agropyron* spp. Environmental Entomology, 12: 442–445.

Kieckhefer, R.W. and Thysell, J.R., 1981. Host preferences and reproduction of four cereal aphids on 20 Triticale cultivars. Crop Science, 21: 322–324.

Kimmins, F., 1982. The probing behaviour of *Rhopalosiphum maidis*. In: J.H. Visser and A.K. Minks (Editors), Insect–Plant Relationships. Centre for Agricultural Publishing and Documentation, Wageningen, pp. 411–412.

Kindler, S.D., Staples, R., Spomer, S.M. and Adeniji, O., 1983. Resistance of bluegrass cultivars to biotypes C and E greenbug (Homoptera: Aphididae). Journal of Economic Entomology, 76: 1103–1105.

Kircher, H.W., Misiorowski, R.L. and Lieberman, F.V., 1970. Resistance of alfalfa to the spotted alfalfa aphid. Journal of Economic Entomology, 63: 964–969.

Kishaba, A.N., Bohn, G.W. and Toba, H.H., 1976. Genetic aspects of antibiosis to *Aphis gossypii* in *Cucumis melo* from India. Journal of the American Society of Horticultural Science, 101: 557–561.

Kishaba, A.N., Coudriet, D.L., McCreight, J.D. and Castle, S., 1983. Melon aphid resistant *Cucumis melo*: resistance to virus infection when vectored by *Aphis gossypii*. Annual Plant Resistance to Insects Newsletter, 9: 7.

Knight, R.L., Briggs, J.B., Massee, A.M. and Tydeman, H.M., 1962. The inheritance of resistance to woolly aphid, *Eriosoma lanigerum* (Hsmnn.), in the apple. Journal of Horticultural Science, 37: 207–218.

Kogan, M., 1982. Plant resistance in pest management. In: R.L. Metcalf and W.H. Luckmann (Editors), Introduction to Insect Pest Management. Wiley, New York, pp. 93–134.

Kogan, M. and Ortman, E.F., 1978. Antixenosis – a new term proposed to define Painter's "Non-preference" modality of resistance. Bulletin of the Entomological Society of America, 24: 175–176.

Kugler, J.L. and Ratcliffe, R.H., 1983. Resistance in alfalfa to a red form of the pea aphid (Homoptera: Aphididae). Journal of Economic Entomology, 76: 74–76.

Lamb, K.P., 1953. Observations on yield and varietal susceptibility of some carrot varieties to insect attack in the field. New Zealand Journal of Science and Technology, (A) 34: 531–537.

Lamb, K.P., 1960. Field trial of eight varieties of *Brassica* field crops in the Auckland district. I. Susceptibility to aphids and virus diseases. New Zealand Journal of Agricultural Research, 3: 320–331.

Lammerink, J., 1968. A new biotype of cabbage aphid (*Brevicoryne brassicae* (L.)) on aphid resistant rape (*Brassica napus* L.). New Zealand Journal of Agricultural Research, 11: 341–344.

Lange, W.H. and Bronson, L., 1981. Insect pests of tomatoes. Annual Review of Entomology, 26: 345–371.

Lara, F.M., Coelho, A. and Mayor, Jr., J., 1979. Varietal resistance in kale to *Brevicoryne brassicae* (Linnaeus, 1758). II. Antibiosis. Anais da Sociedade Entomologica do Brasil, 8: 217–223 (in Portuguese, with English summary).

Lecoq, H., Labonne, G. and Pitrat, M., 1980. Specificity of resistance to virus transmission by aphids in *Cucumis melo* L. Annales de Phytopathologie, 12: 139–144.

Levin, D.A., 1973. The role of trichomes in plant defense. Quarterly Review of Biology, 48: 3–15.

Lloyd, D.L., Turner, J.W. and Hilder, T.B., 1980a. Effects of aphids on seedling growth of lucerne lines. I. Blue-green aphid in field conditions (*Acyrthosiphon kondoi* Shinji). Australian Journal for Experimental Agriculture and Animal Husbandry, 20: 72–76.

Lloyd, D.L., Turner, J.W. and Hilder, T.B., 1980b. Effects of aphids on seedling growth of lucerne lines. II. Spotted alfalfa aphid (*Therioaphis trifolii* f. *maculata* (Monell)) in field conditions. Australian Journal for Experimental Agriculture and Animal Husbandry, 20: 452–456.

Lloyd, D.L., Franzmann, B.A. and Hilder, T.B., 1983. Resistance of lucerne lines at different stages of plant growth to spotted alfalfa aphid and blue-green aphid. Australian Journal of Experimental Agriculture and Animal Husbandry, 23: 288–293.

Lodge, G.M., Brownlee, H. and Cregan, P.D., 1981. Experience proves worth of new lucerne varieties. Agricultural Gazette of New South Wales, 92: 13–15.

Long, B.J., Dunn, G.M., Bowman, J.S. and Routley, D.G., 1977. Relationship of hydroxamic acid content in corn and resistance to the corn leaf aphid. Crop Science, 17: 55–58.

Lowe, H.J.B., 1974. Intraspecific variation of *Myzus persicae* on sugar beet (*Beta vulgaris*). Annals of Applied Biology, 70: 15–26.

Lowe, H.J.B., 1975. Crop resistance to pests as a component of integrated control systems. Proceedings of the Eighth British Insecticide and Fungicide Conference, 1: 87–92.

Lowe, H.J.B., 1980. Resistance to aphids in immature wheat and barley. Annals of Applied Biology, 95: 129–135.

Lowe, H.J.B., 1982. Some observations on susceptiblity and resistance of winter wheat to the aphid *Sitobion avenae* (F.) in Britain. Crop Protection, 1: 431–440.

Lowe, H.J.B., 1984. Development and practice of a glasshouse screening technique for resistance of wheat to the aphid *Sitobion avenae*. Annals of Applied Biology, 104: 297–305.

Lowe, H.J.B. and Taylor, L.R., 1964. Population parameters, wing production and behaviour in red and green *Acyrthosiphon pisum* (Harris) (Homoptera: Aphididae). Entomologia Experimentalis et Applicata, 7: 287–295.

Luginbill, Jr., P., 1969. Developing resistant plants – the ideal method of controlling insects. United States Department of Agriculture, Agricultural Research Service, Production Research Report 3, 14 pp.

Mabbett, T. and Nachapong, M., 1979. The development of assessment methods and scouting techniques for *Aphis gossypii* Glover on cotton in Thailand. Thai Journal of Agricultural Science, 12: 317–322.

MacCarter, L.E. and Habeck, D.H., 1973. The melon aphid: screening *Citrullus* varieties and introductions for resistance. Journal of Economic Entomology, 66: 1111–1113.

MacCarter, L.E. and Habeck, D.H., 1974. Melon aphid resistance in *Cucumis* spp. Florida Entomologist, 57: 195–204.

Madsen, H.F. and Morgan, C.V.G., 1970. Pome fruit pests and their control. Annual Review of Entomology, 15: 295–320.

Maison, P., Kerlan, C. and Massonié, G., 1983. Sélection de semis de *Prunus persica* (L.) Batsch résistants à la transmission du virus de la Sharka par les virginopares aptères de *Myzus persicae* Sulzer. Comptes Rendus des Séances de l'Académie d'Agriculture de France, 69: 337–346.

Maltais, J.B., 1937. Resistance of some varieties of peas to the pea aphid, *Illinoia pisi* Kalt. Sixty-seventh Annual Report of the Entomological Society of Ontario, 1936: 40–45.

Maltais, J.B., 1950. New developments in breeding of peas for resistance to the pea aphid. Eightieth Annual Report of the Entomological Society of Ontario, 1949: 29–30.

Maltais, J.B. and Auclair, J.L., 1957. Factors in resistance of peas to the pea aphid, *Acyrthosiphon pisum* (Harr.) (Homoptera: Aphididae). I. The sugar–nitrogen ratio. Canadian Entomologist, 89: 365–370.

Manglitz, G.R. and Gorz, H.J., 1961. Resistance of sweetclover to the sweetclover aphid. Journal of Economic Entomology, 54: 1156–1160.

Manglitz, G.R. and Gorz, H.J., 1972. A review of insect resistance in the clovers *Trifolium* spp. Bulletin of the Entomological Society of America, 18: 176–178.

Mantell, S.H. and Smith, H. (Editors), 1983. Plant Biotechnology. Society for Experimental Biology, Seminar Series 18, Cambridge University Press, 334 pp.

Markkula, M., 1963. Studies on the pea aphid, *Acyrthosiphon pisum* Harris (Hom., Aphididae), with special reference to the differences in the biology of the green and red forms. Annales Agriculturae Fenniae, 2, Supplement 1, 30 pp.

Markkula, M., 1970. Resistance of pea and red clover to the pea aphid, *Acyrthosiphon pisum* Harris (Hom., Aphididae). European Plant Protection Organization, Publication Series A, 54: 81–86.

Markkula, M. and Roukka, K., 1970. Resistance of plants to the pea aphid *Acyrthosiphon pisum* Harris (Hom., Aphididae) I. Fecundity of the biotypes on different host plants. Annales Agriculturae Fenniae, 9: 127–132.

Markkula, M. and Roukka, K., 1971. Resistance of plants to the pea aphid, *Acyrthosiphon pisum* Harris (Homoptera: Aphididae) III. Fecundity on different pea varieties. Annales Agriculturae Fenniae, 10: 33–37.

Markkula, M., Roukka, K. and Tüttanen, K., 1969. Reproduction of *Myzus persicae* (Sulz.) and *Tetranychus telarius* (L.) on different chrysanthemum cultivars. Annales Agriculturae Fenniae, 8: 175–183.

Martin, T.J., Harvey T.L. and Hatchett, J.H., 1982. Registration of greenbug and Hessian fly resistant wheat germplasm. Crop Science, 22: 1089.

Massonié, G., Maison, P., Monet, R. and Grasselly, C., 1982. Résistance au puceron vert du pêcher, *Myzus persicae* Sulzer (Homoptera: Aphididae), chez *Prunus persica* (L.) Batsch et d'autres espèces de *Prunus*. Agronomie, 2: 63–70.

Maxwell, F.G. and Jennings, P.R., 1980. Breeding Plants Resistant to Insects. Wiley, New York, 683 pp.

Maxwell, F.G. and Painter, R.H., 1959. Factors affecting rate of honeydew deposition by *Therio-*

aphis maculata (Buck.) and *Toxoptera graminum* (Rond.) Journal of Economic Entomology, 52: 368–373.

Maxwell, F.G. and Painter, R.H., 1962. Auxin content of extracts of certain tolerant and susceptible host plants of *Toxoptera graminum, Macrosiphum pisi*, and *Therioaphis maculata* and relation to host plant resistance. Journal of Economic Entomology, 55: 46–56.

Maxwell, F.G., Jenkins, J.N. and Parrott, W.L., 1972. Resistance of plants to insects. Advances in Agronomy, 24: 187–265.

Mayo, O., 1980. The Theory of Plant Breeding. Oxford University Press, New York. 293 pp.

McColloch, J.W., 1921. The corn leaf aphis (*Aphis maidis* Fitch) in Kansas. Journal of Economic Entomology, 14: 89–94.

McKinney, K.B., 1938. Physical characteristics on the foliage of beans and tomatoes that tend to control some small insect pests. Journal of Economic Entomology, 31: 630–631.

McMurtry, J.A. and Stanford, E.H., 1960. Observations of feeding habits of the spotted alfalfa aphid on resistant and susceptible alfalfa plants. Journal of Economic Entomology, 53: 714–717.

Mehlenbacher, S.A., Plaisted, R.L. and Tingey, W.M., 1983. Inheritance of glandular trichomes in crosses with *Solanum berthaultii*. American Potato Journal, 60: 699–708.

Miller, F.R., 1980. The breeding of sorghum. In: M.K. Harris (Editor), Biology and Breeding for Resistance to Arthropods and Pathogens in Agricultural Plants. Texas Agricultural Experiment Station MP-1451, pp. 128–136.

Montllor, C.B., Campbell, B.C. and Mittler, T.E., 1983. Natural and induced differences in probing behaviour of two biotypes of the greenbug, *Schizaphis graminum*, in relation to resistance in sorghums. Entomologia Experimentalis et Applicata, 34: 99–106.

Müller, H.J., 1953. Über die Ursachen der unterschiedlichen Resistenz von *Vicia faba* L. gegenüber der Bohnenblattlaus *Doralis fabae* Scop. IV. Das Zustandekommen des unterschiedlichen Initialbefalls. Der Zuchter, 23: 176–189.

Müller, H.J., 1958. Über die Ursachen der unterschiedlichen Resistenz von *Vicia faba* L. gegenüber der Bohnenblattlaus, *Aphis (Doralis) fabae* Scop. V. Antibiotische Wirkungen auf die Vermehrungskraft. Entomologia Experimentalis et Applicata, 1: 181–191.

Nault, L.R. and Styer, W.E., 1972. Effects of sinigrin on host selection by aphids. Entomologia Experimentalis et Applicata, 15: 423–437.

Newman, W. and Pimentel, D., 1974. Garden peas resistant to the pea aphid. Journal of Economic Entomology, 67: 365–367.

Nielson, M.W. and Don, H., 1974. Probing behaviour of biotypes of the spotted alfalfa aphid on resistant and susceptible alfalfa clones. Entomologia Experimentalis et Applicata, 17: 477–486.

Nielson, M.W. and Kuehl, R.O., 1982. Screening efficacy of spotted alfalfa aphid biotypes and genic systems for resistance in alfalfa. Environmental Entomology, 11: 989–996.

Nielson, M.W. and Lehman, W.F., 1980. Breeding approaches in alfalfa. In: F.G. Maxwell and R.P. Jennings (Editors), Breeding Plants Resistant to Insects. Wiley, New York, pp. 277–311.

Nielson, M.W. and Olson, D.L., 1982. Horizontal resistance in 'Lahontan' alfalfa aphid (Homoptera: Aphididae). Environmental Entomology, 11: 928–930.

Niles, G.A., 1980a. The basics of plant breeding. In: M.K. Harris (Editor), Biology and Breeding for Resistance to Arthropods and Pathogens in Agricultural Plants. Texas Agricultural Experiment Station, College Station, MP-1451, pp. 9–22.

Niles, G.A., 1980b. Breeding cotton for resistance to insect pests. In: F.G. Maxwell and P.R. Jennings (Editors), Breeding Plants Resistant to Insects. Wiley, New York, pp. 337–369.

Oram, R.N., 1980. Register of Australian herbage plant cultivars. B. Legumes, 8. Lucerne. a. *Medicago sativa* L. (lucerne). Journal of the Australian Institute of Agricultural Science, 46: 200–204; 254–262.

Ortega, A., Vasal, S.K., Mihm, J. and Hershey, C., 1980. Breeding for insect resistance in maize. In: F.G. Maxwell and P.R. Jennings (Editors), Breeding Plants Resistant to Insects. Wiley, New York, pp. 371–419.

Paguia, P., Pathak, M.D. and Heinrichs, E.A., 1980. Honeydew excretion measurement techniques for determining differential feeding activity of biotypes of *Nilaparvata lugens* on rice varieties. Journal of Economic Entomology, 73: 35–40.

Painter, R.H., 1951. Insect Resistance in Crop Plants. The University Press of Kansas, Lawrence, KA, 520 pp.

Painter, R.H., 1958a. Resistance of plants to insects. Annual Review of Entomology, 3: 267–290.

Painter, R.H., 1958b. The study of resistance to aphids in crop plants. Proceedings of the Tenth International Congress of Entomology, Montreal, 1956, 3: 451–458.

Palmer, T.P., 1953. Resistance of swedes to aphids. I. Resistant varieties. New Zealand Journal of Science and Technology, 34 (A): 553–555.

Palmer, T.P., 1960. Aphis resistant rape. New Zealand Journal of Agriculture, 101: 375–376.

Panda, N., 1979. Principles of Host-Plant Resistance to Insect Pests. Allanheld/Universe, Montclair and New York, 386 pp.

Pathak, M.D., 1970. Genetics of plants in pest management. In: R.L. Rabb and F.E. Guthrie

(Editors), Concepts of Pest Management. Proceedings of a conference at North Carolina State University, Raleigh, NC, pp. 138–157.

Pathak, M.D. and Painter, R.H., 1958. Differential amounts of material taken up by four biotypes of corn leaf aphids from susceptible and resistant sorghums. Annals of the Entomological Society of America, 51: 250–254.

Pedersen, M.W., Sorensen, E.L. and Anderson, M.J., 1975. A comparison of pea aphid-resistant and susceptible alfalfas for field performance, saponin concentration, digestibility and insect resistance. Crop Science, 15: 254–256.

Peters, D.C. and Starks, K.J., 1981. Pest management systems for sorghum insects. In: D. Pimentel (Editor), CRC Handbook of Pest Management in Agriculture, III. CRC Press. Boca Raton, FL, pp. 549–562.

Pillemer, E.A. and Tingey, W.M., 1976. Hooked trichomes: a physical plant barrier to a major agricultural pest. Science, 193: 482–484.

Porter, K.B., Peterson, G.L. and Vise, O., 1982. A new greenbug biotype. Crop Science, 22: 847–850.

Preston, A.P., 1955. Apple rootstock studies: Malling-Merton rootstocks. Journal of Horticultural Science, 370: 207–218.

Quiros, C.F., Stevens, M.A., Rick, C.M. and Kok-Yokomi, M.K., 1977. Resistance in tomato to the pink form of the potato aphid (*Macrosiphum euphorbiae* Thomas): the role of anatomy, epidermal hairs, and foliage composition. Journal of the American Society for Horticultural Science, 102: 166–171.

Radcliffe, E.B., 1982. Insect pests of potato. Annual Review of Entomology, 27: 173–204.

Radcliffe, E.B. and Chapman, R.K., 1966. Plant resistance to insect attack in commercial cabbage varieties. Journal of Economic Entomology, 59: 116–120.

Radcliffe, E.B., Lauer, F.I., Lee, M.H. and Robinson, D.P., 1981. Evaluation of the United States potato collection for resistance to green peach aphid and potato aphid. Minnesota Agricultural Experiment Station, Technical Bulletin 331, 41 pp.

Raman, K.V. and Palacios, M., 1983. Potential of physical resistance mechanisms for the control of major insect virus vectors of potato. In: W.J. Hooker (Editor), Research for the Potato in Year 2000. International Potato Center, Lima, p. 156.

Ratcliffe, R.H. and Murray, J.J., 1983. Selection for greenbug (Homoptera, Aphididae) resistance in Kentucky bluegrass cultivars. Journal of Economic Entomology, 76: 1221–1224.

Rautapää, J., 1967. Studies of the host plant relationships of *Aphis idaei* v.d. Goot and *Amphorophora rubi* (Kalt.) (Hom., Aphididae). Annales Agriculturae Fenniae, 6: 174–190.

Rautapää, J., 1970. Preference of cereal aphids for various cereal varieties and species of Gramineae, Juncaceae and Cyperaceae. Annales Agriculturae Fenniae, 9: 267–277.

Rhodes, A.M. and Luckmann, W.H., 1967. Survival and reproduction of the corn leaf aphid on twelve maize genotypes. Journal of Economic Entomology, 67: 527–530.

Ridland, P.M. and Berg, G.N., 1981. Seedling resistance to pea aphid of lucerne, annual medic and clover species in Victoria. Australian Journal of Experimental Agriculture and Animal Husbandry, 21: 506–511.

Robinson, A.G., 1965. Variability of resistance of barley varieties to the aphid *Rhopalosiphum padi* (L.) in different environments. Proceedings of the XIIth International Congress of Entomology, London, 1964, p. 533.

Rock, G.C. and Zeiger, D.C., 1974. Woolly apple aphid infests Malling and Malling-Merton rootstocks in propagation beds in North Carolina. Journal of Economic Entomology, 67: 137–138.

Rogers, C.E. and Thompson, T.E., 1979. *Helianthus* resistance to *Masonaphis masoni*. Southwestern Entomologist, 4: 321–324.

Russell, G.E., 1978. Plant Breeding for Pest and Disease Resistance. Butterworths, London, 485 pp.

Sailer, R.I., 1981. Extent of biological and cultural control of insect pests of crops. In: D. Pimentel (Editor), CRC Handbook of Pest Management in Agriculture, II. CRC Press, Boca Raton, FL, pp. 57–67.

Schaefers, G.A., 1981. Pest management systems for strawberry insects. In: D. Pimentel (Editor), CRC Handbook of Pest Management in Agriculture, III. CRC Press, Boca Raton, FL, pp. 377–393.

Schalk, J.M. and Ratcliffe, R.H., 1977. Evaluation of the United States Department of Agriculture program on alternative methods of insect control: host plant resistance to insects. Food and Agriculture Organization, Plant Protection Bulletin, 25: 9–14.

Schneiderman, H.A., 1984. What entomology has in store for biotechnology. Bulletin of the Entomological Society of America, 30: 55–61.

Schuster, D.J. and Starks, K.J., 1973. Greenbugs: components of host-plant resistance in sorghum. Journal of Economic Entomology, 66: 1131–1134.

Schweissing, F.C. and Wilde, G., 1979. Temperature and plant nutrient effects on resistance of seedling sorghum to the greenbug. Journal of Economic Entomology, 72: 20–23.

Searls, E.M., 1935. The relation of foliage color to aphid resistance in some varieties of canning peas. Journal of Agricultural Research, 51: 613–619.

Sebesta, E.E. and Wood, Jr., E.A., 1978. Transfer of greenbug resistance from rye to wheat with X-rays. Agronomy Abstracts, 70: 61–62.

Selander, J., Markkula, M. and Tiittanen, K., 1972. Resistance of the aphids *Myzus persicae* (Sulz.), *Aulacorthum solani* (Kalt.) and *Aphis gossypii* Glov. to insecticides, and the influence of the host plant on this resistance. Annales Agriculturae Fenniae, 11: 141–145.

Sen Gupta, G.C. and Miles, P.W., 1975. Studies on the susceptibility of varieties of apple to the feeding of two strains of woolly aphis (Homoptera) in relation to the chemical content of the tissues of the host. Australian Journal of Agricultural Research, 26: 157–168.

Shade, R.E. and Kitch, L.W., 1983. Pea aphid (Homoptera: Aphididae) biology on glandular-haired *Medicago* species. Environmental Entomology, 12: 237–240.

Shanks, Jr., C.H. and Chase, D., 1976. Electrical measurement of feeding by the strawberry aphid on susceptible and resistant strawberries and nonhost plants. Annals of the Entomological Society of America, 69: 784–786.

Shanks, C.H. and Sjulin, T.M., 1983. Strawberry aphid on strawberries. Annual Plant Resistance to Insects Newsletter, 9: 44.

Simon, J.P., Parent, M.A. and Auclair, J.L., 1982. Isozyme analysis of biotypes and field populations of the pea aphid, *Acyrthosiphon pisum*. Entomologia Experimentalis et Applicata, 32: 186–192.

Singh, S.R., 1977. Cowpea cultivars resistant to insect pests in world germplasm collection. Tropical Grain Legume Bulletin, 1977, 9: 1–7.

Smith, B.D., 1966. Effect of the plant alkaloid sparteine on the distribution of the aphid *Acyrthosiphon spartii* (Koch.). Nature, 212: 213–214.

Snelling, R.O., 1941. Resistance of plants to insect attack. Botanical Research, 7: 543–586.

Sotherton, N.W. and Van Emden, H.F., 1982. Laboratory assessments of resistance to the aphids *Sitobion avenae* and *Metopolophium dirhodum* in three *Triticum* species and two modern wheat cultivars. Annals of Applied Biology, 101: 99–107.

Srivastava, P.N. and Auclair, J.L., 1974. Effect of amino acid concentration on diet uptake and performance by the pea aphid, *Acyrthosiphon pisum* (Homoptera: Aphididae). Canadian Entomologist, 106: 149–156.

Srivastava, P.N., Gao, Y., Levesque, J. and Auclair, J.L., 1985. Differences in amino acid requirements between two biotypes of the pea aphid, *Acyrthosiphon pisum*. Canadian Journal of Zoology, 63: 603–606.

Starks, K.J. and Burton, R.L., 1977a. Greenbugs: a comparison of mobility on resistant and susceptible varieties of four small grains. Environmental Entomology, 6: 331–332.

Starks, K.J. and Burton, R.L., 1977b. Greenbugs: determining biotypes, culturing, and screening for plant resistance with notes on rearing parasitoids. United States Department of Agriculture, Technical Bulletin 1556, 12 pp.

Starks, K.J. and Merkle, O.G., 1977. Low level resistance in wheat to greenbug. Journal of Economic Entomology, 70: 305–306.

Starks, K.J. and Mirkes, K.A., 1979. Yellow sugarcane aphid: plant resistance in cereal crops. Journal of Economic Entomology, 72: 486–488.

Starks, K.J. and Sebesta, E.E., 1978. Cereal crop resistance to insects in the United States: An example of induced mutation, In: Plant Breeding for Resistance to Insect Pests: Considerations about the use of Induced Mutations. International Atomic Energy Agency, Vienna, pp. 49–62.

Starks, K.J., Burton, R.L. and Merkle, O.G., 1983. Greenbugs (Homoptera: Aphididae) plant resistance in small grains and sorghum to biotype E. Journal of Economic Entomology, 76: 877–880.

Stern, V.M., Sharma, R. and Summers, C., 1980. Alfalfa damage from *Acyrthosiphon kondoi* and economic threshold studies in southern California. Journal of Economic Entomology, 73: 145–148.

Stevenson, A.B., 1970. Strains of the grape phylloxera in Ontario with different effects on the foliage of certain grape cultivars. Journal of Economic Entomology, 63: 135–138.

Stoner, W.N. and Kieckhefer, R.W., 1979. Survival and reproduction of four cereal aphids on certain range grasses. Environmental Entomology, 8: 694–695.

Stoner, A.K., Webb, R.E. and Gentile, A.G., 1968. Reaction of tomato varieties and breeding lines to aphids. HortScience, 3: 77.

Tahhan, O. and Hariri, G., 1981. Screening of aphid resistance in faba bean lines. FABIS Newsletter, 3: 57.

Talekar, N.S., 1980. Search for host plant resistance to major insect pests in Chinese cabbage. Proceedings of the Symposium on the Production and Insect Control of Cruciferous Vegetables in Taiwan, April 1980, Plant Protection Center, Taiwan, pp. 164–173.

Tanaka, T., 1961. The rice root aphids, their ecology and control. College of Agriculture, Utsunomiya University, Japan, Special Bulletin 10, 83 pp. (in Japanese, with English abstract).

Tarn, T.R. and Adams, J.B., 1982. Aphid probing and feeding, electronic monitoring, and plant breeding. In: K.F. Harris, and K. Maramorosch (Editors), Pathogens, Vectors and Plant Diseases: Approaches to control. Academic Press, New York, pp. 221–246.

Taylor, J.B., 1981. The selection of Aotea apple rootstocks for resistance to woolly aphis and to root canker, a decline and replant disease caused by basidiomycete fungi. New Zealand Journal of Agricultural Research, 24; 373–377.

Teetes, G.L., 1980. Breeding sorghums resistant to insects. In: F.G. Maxwell and P.R. Jennings (Editors), Breeding Plants Resistant to Insects. Wiley, New York, pp. 457–485.

Thompson, K.F., 1963. Resistance to the cabbage aphid (*Brevicoryne brassicae*) in Brassica plants. Nature, 198: 209.

Thurston, R., 1961. Resistance in *Nicotiana* to the green peach aphid and some other tobacco insect pests. Journal of Economic Entomology, 54: 946–949.

Thurston, R. and Webster, J.A., 1962. Toxicity of *Nicotiana gossei* Domin to *Myzus persicae* (Sulzer). Entomologia Experimentalis et Applicata, 5: 233–238.

Thurston, R., Smith, W.R. and Cooper, B.P., 1966. Alkaloid secretion by trichomes of *Nicotiana* species and resistance to aphids. Entomologia Experimentalis et Applicata, 9: 428–432.

Thurston, R., Kasperbauer, M.J. and Jones, G.A., 1977. Green peach aphid and tobacco flea beetle populations on tobacco cultivars and haploid-derived lines with various alkaloid levels. Tobacco Science, 21: 22–24.

Tingey, M., 1981. Potential for plant resistance in management of arthropod pests. In: J.H. LaShomb and R. Casagrande (Editors), Advances in Potato Pest Management. Hutchinson Ross, Stroudsberg, PA, pp. 268–288.

Tingey, W.M. and Laubengayer, J.E., 1981. Defense against the green peach aphid and potato leafhopper by glandular trichomes of *Solanum berthaultii*. Journal of Economic Entomology, 74: 721–725.

Tingey, W.M. and Sinden, S.L., 1982. Glandular pubescence, glycoalkaloid composition and resistance to the green peach aphid, potato leafhopper and potato fleabeetle in *Solanum berthaultii*. American Potato Journal, 59: 95–106.

Tingey, W.M., Mehlenbacher, S.A. and Laubengayer, J.E., 1981. Occurrence of glandular trichomes in wild *Solanum* species. American Potato Journal, 58: 81–83.

Todd, G.W., Getahun, A. and Cress, D.C., 1971. Resistance in barley to the greenbug, *Schizaphis graminum*. 1. Toxicity of phenolic and flavonoid compounds and related substances. Annals of the Entomological Society of America, 64: 718–722.

Trehan, K.B., Bhatnagar, V.K., Sharma, R.C. and Chandala, R.P., 1970. Note on genetic stocks of barley (*Hordeum vulgare* L.) resistant to the aphid *Rhopalosiphum maidis* (Fitch). Indian Journal of Agricultural Science, 40: 756–758.

Unwin, D.M., 1978. A versatile high frequency radio microcautery. Physiological Entomology, 3: 71–73.

Van Emden, H.F., 1966. Studies on the relations of insect and host plant. III. A comparison of the reproduction of *Brevicoryne brassicae* and *Myzus persicae* (Hemiptera: Aphididae) on Brussels sprout plants supplied with different rates of nitrogen and potassium. Entomologia Experimentalis et Applicata, 9: 444–460.

Van Emden, H.F., 1978. Insects and secondary plant substances – an alternative viewpoint with special reference to aphids. In: J.B. Harborne (Editor), Biochemical Aspects of Plant and Animal Coevolution. Academic Press, London, pp. 309–323.

Van Marrewijk, G.A.M. and De Ponti, O.M.B., 1975. Possibilities and limitations of breeding for pest resistance. Mededelingen van de Fakulteit Landbouwwetenschappen, Rijkuniversiteit Gent, 40: 229–247.

Van Marrewijk, G.A.M. and Dieleman, F.L., 1980. Resistance to aphids in barley and wheat. In: A.K. Minks and P. Gruys (Editors), Integrated Control of Insect Pests in the Netherlands. Centre for Agricultural Publishing and Documentation, Wageningen, pp. 165–167.

Van Soest, W., 1955. An instrument for the cutting of the stylets of feeding aphids. Tijdschrift over Plantenziekten, 61: 60–61 (in Dutch).

Vickerman, G.P. and Wratten, S.D., 1979. The biology and pest status of cereal aphids (Hemiptera: Aphididae) in Europe: a review. Bulletin of Entomological Research, 69: 1–32.

Villanueva, J.R. and Strong, F.E., 1964. Laboratory studies on the biology of *Rhopalosiphum padi* (Homoptera: Aphididae). Annals of the Entomological Society of America, 57: 609–613.

Vose, P.B. and Blixt, S.G., 1984. Crop Breeding. A contemporary basis. Pergamon Press, Oxford, 443 pp.

Walker, A.L., Bottrell, D.G. and Cate, Jr., J.R., 1972. Bibliography on the greenbug, *Schizaphis graminum* (Rondani). Bulletin of the Entomological Society of America, 18: 161–173.

Watson, S.J. and Dixon, A.F.G., 1984. Ear structure and the resistance of cereals to aphids. Crop Protection, 3: 67–76.

Way, M.J. and Murdie, G., 1965. An example of varietal variations in resistance of Brussels sprouts. Annals of Applied Biology, 56: 326–328.

Webster, J.A., 1975. Association of plant hairs and insect resistance. An annotated bibliography. United States Department of Agriculture, Agricultural Research Service, Miscellaneous Publication 1297, 18 pp.

Webster, J.A. and Inayatullah, C., 1984. Greenbug (Homoptera: Aphididae) resistance in Triticale. Environmental Entomology, 13: 444–447.

Webster, J.A. and Starks, K.J., 1984. Sources of resistance in barley to two biotypes of the greenbug *Schizaphis graminum* (Rondani), Homoptera: Aphididae. Protection Ecology, 6: 51–55.

Weibel, D.E., Starks, K.J., Wood, Jr., E.A. and Morrison, R.D., 1972. Sorghum cultivars and progenies rated for resistance to greenbugs. Crop Science, 12: 334–336.

Westwood, M.N. and Westigard, P.H., 1969. Degree of resistance among pear species to the woolly pear aphid, *Eriosoma pyricola*. Journal of the American Society of Horticultural Science, 94: 91–93.

Wilcoxson, R.D. and Peterson, A.G., 1960. Resistance of Dollard red clover to the pea aphid, *Macrosiphum pisi*. Journal of Economic Entomology, 53: 863–865.

Wilde, G. and Feese, H., 1973. A new corn leaf aphid biotype and its effect on some cereal and small grains. Journal of Economic Entomology, 66: 570–571.

Williams, W.G., Kennedy, G.G., Yamamoto, R.T., Thacker, J.D. and Bordner, J., 1980. 2-Tridecanone: a naturally occurring insecticide from the wild tomato *Lycopersicon hirsutum* f. *glabratum*. Science, 207: 888–889.

Wink, M., Hartmann, T., Witte, L. and Rheinheimer, J., 1982. Interrelationship between quinolizidine alkaloid producing legumes and infesting insects: exploitation of the alkaloid-containing phloem sap of *Cytisus scoparius* by the broom aphid, *Aphis cytisorum*. Zeitschrift für Natürforschung, Series 37 C: 1081–1086.

Wood, Jr., E.A., 1961a. Description and results of a new greenhouse technique for evaluating tolerance of small grains to the greenbug. Journal of Economic Entomology, 54: 303–305.

Wood, Jr., E.A., 1961b. Biological studies of a new greenbug biotype. Journal of Economic Entomology, 54: 1171–1173.

Wood, Jr., E.A., Chada, H.L., Weibel, D.E. and Davies, F.F., 1969. A sorghum variety highly tolerant to the greenbug, *Schizaphis graminum* (Rond.). Oklahoma Agricultural Experiment Station, Stillwater, Progress Report P-618, 7 pp.

Wood, Jr., E.A., Sebesta, E.E. and Starks, K.J., 1974. Resistance of 'Gaucho' Triticale to *Schizaphis graminum*. Environmental Entomology, 3: 720–721.

Woodhead, S., Padgham, D.E. and Bernays, E.A., 1980. Insect feeding on different sorghum cultivars in relation to cyanide and phenolic acid content. Annals of Applied Biology, 95: 151–157.

Wyatt, I.J., 1965. Th distribution of *Myzus persicae* (Sulz.) on year-round chrysanthemums. I. Summer season. Annals of Applied Biology, 56: 439–459.

Yano, K., Miyake, T. and Eastop, V.F., 1983. The biology and economic importance of rice aphids (Hemiptera: Aphididae): a review. Bulletin of Entomological Research, 73: 539–566.

Young, E., Rock, G.C., Zeiger, D.C. and Cummins, J.N., 1982. Infestation of some *Malus* cultivars by the North Carolina woolly apple aphid biotype. HortScience, 17: 787–788.

Young, W.R. and Teetes, G.L., 1977. Sorghum entomology. Annual Review of Entomology, 22: 193–218.

11.5 Integrated Control

11.5.1 Integrated Aphid Management: General Aspects

P. HARREWIJN and A.K. MINKS

INTRODUCTION

The success of aphids in the struggle for life is, to a great extent, due to parthenogenesis – with its telescoping of generations – and clonal polymorphism. The aphids' high reproductive rate is a great advantage when favourable agricultural habitats, hardly exploited by other insects, are temporarily colonized. Many aphid species have a restricted host plant range. However, this does not mean that they can only thrive on domesticated cultivars of wild plant species: most aphids have evolved on the natural vegetation and are present on herbs in the vicinity of the crop that is to be protected. It would seem to be very difficult to trace and eradicate all the aphids that can easily migrate into the crop – through the production of winged morphs – and transmit various non-persistent viruses long before any insecticide has killed them. As chemical control is not the ultimate answer to all aphid problems (see earlier sections in this Chapter), it is sensible to integrate all available techniques to prevent migration, to disturb alighting in the crop (see also section 11.3) and to reduce aphid numbers.

Aphids do have their weak points. As a consequence of their small size they can hardly control the direction of their flight, and with their limited number of ommatidia they can recognize a plant only at close range. Because of the simultaneous growth of more than one generation in a viviparous individual, there is a need for an almost continuous supply of energy. Aphids cannot survive for long without feeding, and particularly aphids with a broad host-plant range are sensitive to the nutritional quality of their hosts, because, among other constraints, they have to overcome the defense mechanisms of different species and detoxify various classes of allomones (see also section 4.2). Mackauer and Way (1976), in their discussion of alternative methods to control aphids, especially refer to "the unique sensitivity of *Myzus persicae* (Sulzer) to the physiological condition of its host plants". They state that manipulation of the physiology of the host could be invaluable in a programme of integrated control for this aphid.

Although nutritional demands can differ between aphid species, there are also similarities, with the result that manipulation of the host plant can have the same effect on several species. Leckstein and Llewellyn (1973) found that both for *M. persicae* and *Aphis fabae* Scopoli, methionine, histidine, the group asparagine/aspartic acid and alanine and cysteine are important for optimal development. The same adapted fertilization scheme reduces population development of both *M. persicae* and *Macrosiphum euphorbiae* (Thomas) (Harrewijn, 1983), and, as will be discussed later, the oligophagous aphid *Brevicoryne brassicae* (Linnaeus) also reacts to changes in the physiology of its hosts (see also section 4.3). In the case of the latter aphid, however, "integrated control" should be merely regarded as a way to obtain optimal protection with a minimum of insecticides, as is recommended by Davis (1985) for aphids and other pests in alfalfa.

The successful exploitation of the environment by aphids should be regarded on the species level: what is most efficient for survival of the species is not always optimal for the individual. Numerous migrants of host-alternating species will never find their primary host and perish, and there are limits to the adaptational possibilities of the individual with regard to host acceptation and the utilization of food.

HOW TO REDUCE APHID NUMBERS

Integrated control consists of an array of techniques and measures to disturb or interrupt the aphid's life cycles in different phases, i.e. (a) when the aphids overwinter on the primary host; (b) during their migration to secondary hosts and host plant finding; (c) when they are probing and feeding on secondary host plants of economic importance; or (d) during larviposition and reproduction on secondary hosts (see Fig. 11.5.1.1).

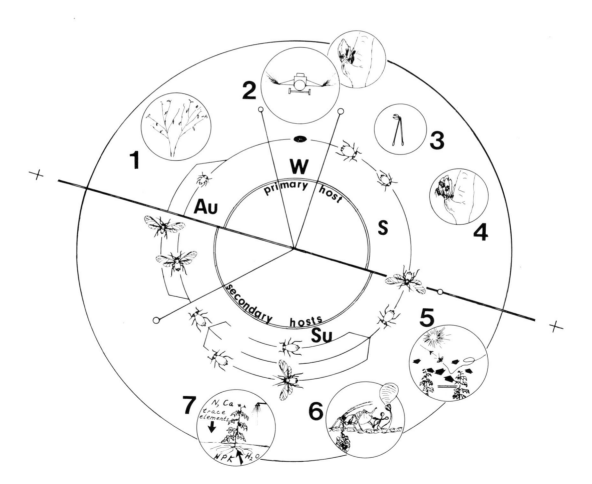

Fig. 11.5.1.1.. Integrated control of aphids at various stages of their life cycle. The control measures (numbered) are further discussed in the text. Au, autumn; W, winter; S, spring; Su, summer.

Control measures in the overwintering stage

The life cycles of many host-alternating aphids show an important agricultural feature: each species has only one or a few primary hosts, on which the oviparae have to complete their development (see section 4.1). The following control methods can be used (bold numbers refer to the corresponding numbers in Fig. 11.5.1.1): – Early defoliation of the primary host will strongly reduce oviposition (**1**). This can be done with a chemical defoliant that should be applied in such a way that natural leaf fall is accelerated by about two weeks. – Pruning of the primary host in spring (**3**) may also contribute to a reduction in the number of eggs that hatch. Tamaki (1981) found that low population levels of *M. persicae* in the autumn resulted in a relatively higher oviposition rate near the tips of peach tree branches. As a consequence, the percentage of eggs pruned out is inversely related to egg density, which makes pruning particularly effective in years with a low oviposition rate. – Egg mortality can be enhanced by spraying mineral oil (**2**). The fundatrices that hatch from the surviving eggs are soon attacked by predators (see section 9.2), such as Coccinellidae. – Sparing the habitat of predators in the vicinity of primary hosts (**4**) can substantially contribute to the suppression of aphid populations in spring.

Later in spring, the aphids that survive develop winged morphs, which migrate to weeds and cultivated plants. Some of these weeds belong to the same family as the cultivated plants and can be a source of various agriculturally important viruses. Chenopodiaceae, for instance, are present almost all over the world. The perennial Convolvulaceae are found in temperate and warmer zones, and the dangerous and omnipresent *M. persicae* accepts them as hosts. Tamaki et al. (1980) have shown that fundatrices of this species can develop and reproduce on these plants. Therefore it is advised to remove all pruned branches from peach orchards and burn them, or to replace the undergrowing herbs with permanent grass.

Different aphid species do not necessarily hatch in the same period on the same primary host. Whereas, on sycamore, *Periphyllus testudinaceus* (Fernie), can hatch as early as late January, *Drepanosiphum acerinum* (Walker) may not do so before early May. One should take this into consideration when introducing predators to reduce spring aphid populations on the primary host (see further section 9.2).

Control measures during migration and host-plant finding

Aphids that survive on the primary host will eventually produce winged migrants that invade crops or disperse to other secondary hosts of economic importance. Oligophagous and even monophagous species can form colonies on weeds scattered all over the landscape, where they are not easy to control. Here they produce winged virginoparae that alight on the crop.

Smith et al (1984) reported that a number of grass weeds found around or between cereal fields can be alternative hosts for *Rhopalosiphum padi* (Linnaeus) and *Sitobion avenae* (Fabricius). These grasses differ in their susceptibility to settling and reproduction, and the risk of virus spread to subsequent cereal crops can be affected by the choice of herbicides.

Many aphid species show a general orientation towards wavelengths exceeding 500 nm, which allows them to discriminate plants and soil from the sky (Van Emden et al., 1969). Burrows et al. (1983) found evidence that *M. euphorbiae* has one receptor sensitive to energy in the 590–595 nm range, and at least one in the 425–460 nm range. Enough is known about aphid colour perception to state that interspecific variations in the appreciation of UV, "blue" and

Section 11.5.1 references, p. 271

"yellow" colours determine alighting behaviour. *R. padi* has two spectral peaks of positive phototaxis, one in the green–yellow and one in the UV region. This aphid reacts strongly to UV (Rautapää, 1980). The distimulus ratio of yellow to blue may be high in *M. euphorbiae* and *M. persicae*, whereas some cereal aphids may be attracted to green plants by a different mixture of signals from yellow, blue, and UV receptors.

Aphids probably have poor vision, as they have so few ommatidia, but they react positively to the silhouette of a plant. Some aphid species are attracted by plant odours. Contrary to the opinion that these are not important for host selection, there is evidence that odour-conditioned anemotaxis exists in a number of aphid species (Visser and Taanman, 1987). However, different aphid species can react in very different ways, which means that the control measures discussed in section 11.3 should first be tried out individually for each species. *R. padi*, for instance, reacts positively to large, green-coloured surfaces and less so to individual plants (Åhman et al., 1985).

Besides control of weeds that act as a bridge between the primary host and the crop, precautionary measures can be thought of to prevent alighting in the crop: host plant masking (**5** in Fig. 11.5.1.1), the application of behaviour-modifying agents, placing nets over the crop to interrupt the landing phase, using UV-reflecting aluminium foil, spraying the crop with whitewash, surrounding a small field with a fence made of yellow plastic covered with insect glue, or intercropping with plants of different colour and attraction. Many of these techniques are discussed in section 11.3; none of them will give 100% protection, but they can be combined and do not negatively interact with each other.

Control during probing and feeding

Aphids can be deterred from probing and feeding on plants by a number of measures: use of resistant cultivars, adapted fertilization schemes – especially with respect to nitrogen and potassium – , a continuous water supply, and a proper timing of the planting data.

Host plant resistance to probing and feeding of aphids can be based upon three different mechanisms and their combinations (**6**): – The plant produces an allelochemic, a substance that can act as a deterrent or is toxic to the aphids (see section 4.2). – The phloem sap is insufficiently nutritious to sustain normal development of aphids (see section 11.4). – The amount of phloem sap available to the aphids is lower; this can be caused by a reduced flow of sap or by a blockage of the phloem cells as a reaction to stylet insertion by the aphids.

Host plant resistance is discussed in section 11.4. In many cases the causative factors are not known. Both the production of allelochemics and the nutritional quality of the host plant can be strongly affected by mineral nutrition (Harrewijn, 1977). Improvement of the nutrition of the host can render a suboptimal plant more suitable for aphids, but can at the same time enhance the production of allelochemics. The balance between these reactions determines whether the level of resistance is enhanced or reduced. An example of this is found in the resistance of a few lettuce selections against *Nasonovia ribis-nigri* (Mosley). High levels of nitrogen fertilizer did not alter the proportion of aphids which rejected the plants (rejection is probably caused by a phloem-bound allelochemic), but improved host quality for those aphids which accepted a plant regardless of a deterrence factor (Harrewijn and Dieleman, 1984).

When schemes for integrated control in a particular crop are developed, it is important to test whether mineral nutrition of the host affects the level of resistance. Resistance to probing and feeding can be measured ideally with the Electrical Penetration Graph technique discussed in section 8.8. This is of particular importance when the plants should have vector resistance. When the aphids transmit non-persistent viruses they should not probe at all, and when they transmit persistent viruses they should preferably not reach the phloem (continuous voltage drop), but if they do, a (partial) vector resistance can still be present, depending upon the duration of continuous sap uptake. This can be established when measuring the duration of an interrupted E pattern (Harrewijn, 1986).

Temporary water stress causes protein breakdown and increases the transport of free amino acids in the phloem. This can improve gustatory stimuli, especially for polyphagous aphids (see section 4.2), reducing the resistance of the host. Host plant resistance can vary strongly with the age of the plant. A planting date may be chosen in such a way that an optimal level of resistance has been achieved when the aphids start to immigrate into the crop.

Control during larviposition and reproduction

Growth and reproduction of aphids that have accepted a plant are influenced by the physiology of the host (see section 4.3; **7** in Fig. 11.5.1.1). In schemes for integrated control the quality of the host can be manipulated through water supply and mineral nutrition, particularly potassium and nitrogen fertilization. One should realize that adjusted fertilization schemes may at the same time alter the resistance of the plants to diseases (Anonymus, 1976). The effects of mineral nutrition on host plant resistance are best studied with methods that eliminate uncontrollable factors. A valuable technique is the use of soilless culture, but care should be taken to transfer the results to a field situation (Harrewijn, 1977). The effect of potassium on aphids is discussed by Perrenoud (1977). Out of 115 studies on the effect of an extra supply of potassium, 69 showed reduced development of the aphids, 21 gave no effect, and 25 showed a stimulating effect. The decrease in development or damage ranged from 4 to 87%. A positive effect of potassium on the development of *M. persicae* was predominantly found for tobacco; this may be of interest with respect to the production of nicotine, as this aphid can perform well on tobacco providing the level of nicotine is not too high. Tolerance of allelochemics can be tested with artificial diets (see section 8.13), and aphids could be used to indicate the level of allelochemics in, for instance, medicinal plants.

REFERENCES

Anonymus, 1976. Fertilizer Use and Plant Health. Proceedings of the 12th Colloquium of the International Potash Institute, Izmir, Turkey, 330 pp.

Åhman, I., Weibull, J. and Petterson, J., 1985. The role of plant size and plant density for host finding in *Rhopalosiphum padi* (L.) (Hem: Aphididae). Swedish Journal of Agricultural Research, 15: 19–24.

Burrows, P.M., Barnett, O.W. and Zimmerman, M.T., 1983. Color attraction and perception in *Macrosiphum euphorbiae*. Canadian Journal of Zoology, 61: 202–210.

Davis, D.W., 1985. Integrated pest management: new ways to manage aphids and other alfalfa pests. Utah Science, Spring 1985, pp. 24–28.

Harrewijn, P., 1977. Use of soilless culture in the establishment of host plant–insect relationships. In: Proceedings 4th International Congress on Soilless Culture, Las Palmas, 1976, pp. 347–358.

Harrewijn, P., 1983. The effect of cultural measures on behaviour and population development of potato aphids and transmission of viruses. Mededelingen Faculteit Landbouwwetenschappen, Rijksuniversiteit Gent, 48/3: 791–799.

Harrewijn, P., 1986. Vector resistance of potato aphids with respect to potato leafroll virus. In: Potato Research of Tomorrow. Proceedings of an International Seminar, Wageningen, pp. 85–96.

Harrewijn, P. and Dieleman, F.L., 1984. The importance of mineral nutrition of the host plant in resistance breeding to aphids. In: Proceedings 6th International Congress on Soilless Culture, Lunteren, The Netherlands, pp. 235–244.

Leckstein, P. and Llewellyn, M., 1973. Effect of dietary amino acids on the size and alary polymorphism of *Aphis fabae*. Journal of Insect Physiology, 19: 973–980.

Mackauer, M. and Way, M.J., 1976. *Myzus persicae* Sulz., an aphid of world importance. In: V.L. Delucchi (Editor), Studies in Biological Control. Cambridge University Press, London, pp. 51–119.

Perrenoud, S. 1977. Potassium and Plant Health. International Potash Institute, Research Topics, 3: 55–62, 175–179.

Rautapää, J. 1980. Light reaction of cereal aphids (Homoptera, Aphididae). Annales Entomologici Fennici, 46: 1–12.

Smith, S.D., Kendall, D.A. and Wright, M.A., 1984. Weed grasses as hosts of cereal aphids and effects of herbicides on aphid survival. In: Proceedings British Crop Protection Conference – Pests and Diseases, Brighton, 1984, pp. 19–24.

Tamaki, G., 1981. Exploiting the ecological interaction of the green peach aphid on peach trees. USDA Technical Bulletin, 1640: 1–6.

Tamaki, G., Fox, L. and Chauvin, R.L., 1980. Green peach aphid: orchard weeds are host to fundatrix. Environmental Entomology, 9: 62–66.

Van Emden, H.F., Eastop, V.F., Hughes, R.D. and Way, M.J., 1969. The ecology of *Myzus persicae*. Annual Review of Entomology, 14: 197–270.

Visser, J.H. and Taanman, J.W., 1987. Odour-conditioned anemotaxis of apterous virginoparae of *Cryptomyzus korschelti* in response to host plants. Physiological Entomology, 12: 473–479.

11.5.2 Integrated Control of Cereal Aphids

G.W. ANKERSMIT

INTRODUCTION

In the past 20 years an increase in the incidence of cereal pests and diseases has become apparent. Aphids were not considered of any importance until 1968 (Fletcher and Bardner, 1969). After that year a vast amount of literature on cereal aphids has accumulated, excellently reviewed by Vickerman and Wratten (1979) and Carter et al. (1980).

The increase in pest incidence coincided with an intensification in cereal growing. High sowing densities, split nitrogen fertilization (often partly applied during flowering), and the increased use of herbicides, plant growth regulators (e.g. CCC), fungicides and insecticides have resulted in wheat yields of 8–10 metric tonnes ha^{-1}. This is close to the theoretical maximal production capacity for wheat.

A question is whether the increase in cereal aphids coincided with these agricultural changes or whether a causal relationship exists. Carter et al. (1982) demonstrated, from (year-to-year) records of trap catches at Rothamsted, that the numbers of cereal aphids caught after 1968 were higher than before. In addition, it was found that the damage per aphid increases with higher yields (Mantel et al., 1982). So the economic threshold in wheat fields producing only 6 t ha^{-1} has been set at fifteen aphids per tiller, but when the production rises to 8 t ha^{-1} it should drop to about four aphids per tiller. Both this general increase in numbers and a greater sensitivity to damage in high-producing crops can have contributed to the present status of cereal aphids.

As has been described in section 11.1, in Western Europe *Sitobion avenae* (Fabricius) and *Metopolophium dirhodum* (Walker) are major pests of wheat, while *Rhopalosiphum padi* (Linnaeus) is only occasionally a pest on this crop. The latter species is more important in Scandinavia. In France, however, outbreaks of *R. padi* have been reported in maize (Moreau, 1983), following pyrethroid treatments against the European corn borer, *Ostrinia nubilalis* (Hübner).

M. dirhodum and *R. padi* are dioecious species with respectively *Rosa* spp. and *Prunus padus* as winter hosts. *S. avenae* is monoecious but can migrate from cereals to other Gramineae and vice versa. Depending on the climate, holocyclic and anholocyclic forms are known in all these species. In the maritime climate of Western France, *S. avenae* and *M. dirhodum* are mostly anholocyclic. The differences in annual cycles result in large differences in the epidemiology of these aphid species. The dioecious species hibernate outside the cereal fields on relatively few plants and often in large numbers. *S. avenae*, when holocyclic, hibernates in very low densities over a large area (Ankersmit and Carter, 1981).

Section 11.5.2 references, p. 277

Cereal aphids can be considered as belonging to the *r*-end of the *k–r* continuum (see Southwood, 1977). They rapidly exploit the temporal habitat of cereal fields and then emigrate to new hosts. They invade the cereal crops in autumn and spring. The first immigration does not result in a large epidemic during winter: aphid numbers remain rather low during that period and tend to decrease. It is difficult to assess the number of aphids because of the low population density, which is often reduced by natural enemies, such as ground beetles (Sunderland and Chambers, 1983). Too little is known about what actually happens to the aphid population during winter and spring. They are then around the "endemic ridge" in the model of Southwood and Comins (1976). The major outbreak occurs in June, when – often massive – immigration takes place, multiplication of *S. avenae* on the ears is fast and the apterous offspring of alates reproduce at a high rate. During this period the aphid population usually escapes from the natural enemies.

AGRICULTURAL PRACTICE AND CONTROL

The relative significance of the two immigration periods will determine much of our control strategy. In regions with a maritime climate such as the United Kingdom and western France, immigrations in the autumn seem to be of greater importance, than in more continental areas (Dedryver and Gelle, 1982). In continental regions, therefore, agricultural measures to restrict immigration have little effect in the autumn.

The following measures in cereal growing affect the development of aphid populations and can be used in a scheme for integrated control.

Change of sowing date

Early sowing, e.g. in the last decade of September in the Netherlands, results in a yield increase of approximately 300 kg per ha (Darwinkel, 1982) but enables aphids to immigrate early in the autumn, thus enhancing the spread of barley yellow dwarf virus (BYDV). Early-sown fields may have a larger hibernating aphid population than late-sown fields do. In the United Kingdom, Chambers and Sunderland (1983) found more parasite mummies in fields where more aphids had overwintered, but there was no evidence of their effectiveness and only a few diseased aphids were observed. However, Dedryver and Gelle (1982) found parasites present in anholocyclic *R. padi* in Brittany throughout the winter. The parasites could stop the growth of the colonies in the spring.

Increase of sowing density

High sowing densities, ranging from 135 to 200, or even to 250 kg per ha, give only a small increase in yield (Darwinkel, 1982), but the denser covering of the soil may benefit ground beetles. Nijveldt (1980) observed more predators in weedy plots. It should be established whether this is due to a better soil coverage or to the presence of alternative food sources on the weeds. Ground beetles (e.g. carabids, staphylinids) and spiders are good predators on aphids (see also section 9.2.6), and a lower peak in the aphid population in July is correlated with their presence (De Clercq and Pietraszko, 1983; Chambers et al., 1983).

Stimulation of parasite development by undersowing

Powell (1983) found more parasites from the *Aphidius uzbekistanicus*-group in fields undersown with rye-grass and infested with *Metopolophium festucae*

(Theobald) than in control fields without grass. Also, the numbers of *S. avenae* per tiller at the peak of aphid density were reduced. It can be compared to the undersowing of potatoes with grass, as mentioned by McKinlay (1985). However, economic aspects should be further investigated.

The above-mentioned methods are aimed at increasing the natural enemy complex that attacks the autumn immigrants and their offspring by creating a reservoir of natural enemies that may be capable of delaying the upsurge of the pest population in the following spring and early summer, which can substantially reduce the peak of the aphid population.

Manipulation of nitrogen fertilization

Nitrogen fertilization, often applied as split dressing, increases yields significantly because the leaves, in particular the flagleaf, remain green during the period of seed filling after flowering. Nitrogen retards the wing formation of cereal aphids and therefore postpones the moment of emigration (Vereijken, 1979); thus there are more apterous aphids, which have more offspring than alates. It is difficult to find the right compromise between higher yields and more aphids.

Breeding for host-plant resistance

Host-plant resistance to aphids has been amply discussed in section 11.4. Some cultivars show a certain degree of resistance to *S. avenae* and *M. dirhodum*. A major problem is the assessment of resistance. Greenhouse tests are of limited value only (Van Marrewijk and Dieleman, 1980), while field tests depend heavily upon the occurrence of the pest. A further problem is that at least two aphid species are involved, living on different parts of the plant and varying in significance each year.

Use of fungicides

Prior to flowering, control of cereal diseases is often necessary. Although most fungicides have little direct effect on aphids, they may affect pathogenic fungi, such as the Entomophthorales. In most years these fungi are found in the cereal crop. Usually three species are present: *Erynia neoaphidis* Remaudière and Hennebert, *Conidiobolus obscurus* (Hall and Dunn) Remaudière and Keller and *Entomophthora planchoniana* Cornu. In Brittany, Dedryver (1983) noticed epizootics by *E. neoaphidis* in only 2 out of 6 years. Wilding (1982) found kill of *E. neoaphidis* by benomyl, but other Entomophthorales were not adversely affected. Tridemorph, captafol, mancozeb and maneb had few direct effects. Sagenmüller (1977) observed only little effect of leaf treatments with tridemorph and chloraniformithane, although sometimes the population growth of *S. avenae* was higher. However, fungicides tend to prolong the longevity of the leaves and thus they also lengthen the period of aphid development. Most outbreaks of fungal diseases among cereal aphids occur after flowering, and therefore the pathogenic fungi tend to escape from direct fungicidal action. In maritime climates with an early aphid development this may be only partly true. The effects of fungicides on fungal diseases of cereal aphids should be studied in relation to the epidemiology of the aphids and the disease (see also section 9.3).

Use of insecticides

Chemical control of cereal aphids is usually done at the time of flowering. Farmers often mix the aphicide with the fungicide that has to be applied at the

Section 11.5.2 references, p. 277

same time. This treatment may seriously affect natural enemies, as it is done at the start of the rapid increase of the aphid population. The main natural enemies are coccinellids, syrphids and aphidiids. From these predators and parasites the coccinellids and aphidiids will suffer most. They are already present in the field at low aphid densities. The syrphids usually arrive somewhat later and are highly mobile. Selective insecticides are not harmful to hymenopterous parasites and coccinellids, and therefore the use of pirimicarb is recommended.

Use of insect pathogens for control

Much work has been done on the use of entomogenic fungi (Latgé et al., 1983; Latteur and Godefroi, 1983; section 9.3), but so far trials with preparations of Entomophthorales have generally been negative, mainly, it seems, because there is still insufficient knowledge of the abiotic factors that influence the development of the mycosis and because the pathogenicity of the preparations is low. For applications in the field the farmers' practice of using mixtures of fungicides and an aphicide may be an additional problem.

EPIPRE: A SPECIAL CASE OF SUPERVISED CONTROL OF CEREAL APHIDS

In 1977 a project for the optimalization of pesticide use in cereals was started in the Netherlands. This project, called EPIPRE (Epidemic Prevention), is an advisory system for each individual field (Zadoks, 1981). It takes into account (1) the disease and pest situation, based upon observations by individual farmers, (2) the stage of development of the crop, and (3) the yield expectation. From these data a "tailor-made" advice for pest and disease control can be derived. The EPIPRE system is a first step to integrated pest management. In a later phase the presence of natural enemies can also be taken into consideration and their development stimulated.

The system was evaluated by Rossing et al. (1985). They concluded that the system is educational for the farmers and – because of the reduced pesticide use – also has environmental advantages. No adverse effects on yield were found. In several, but not all, years labour costs in the EPIPRE system were higher because of the extra time spent in making the necessary observations. Aphid counts are done by counting the number of ears with aphid colonies. Rabbinge and Mantel (1981) found a correlation between the percentage of infected ears and the mean number of *S. avenae* per ear, and this allows the economic threshold to be established quite easily.

A problem in comparing the results of the farmers using EPIPRE and those using a conventional control system is that the latter group quickly adapts to new developments by introducing certain elements of the EPIPRE programme.

In a forthcoming integrated farming system farmers will not only be advised on disease and pest control, but also on weed control, fertilization and choice of crops. Programmes should become available on a central computer, and farmers will be able to consult them with the aid of their home computer. When such developments are introduced, winter wheat is a very suitable crop to start with. When wrong decisions have been taken, the financial risks are relatively small with this crop, because at present only the yield quantity is of major importance, and quality factors are largely neglected.

In such systems, pest population growth models like that for aphids developed by Rabbinge et al. (1979) and Carter et al. (1982) can be incorporated (see also section 8.2). Sometimes the effects of natural enemies can also be included, as well as the effects of cultural measures on pests and their enemies.

In the German Democratic Republic a similar agro-ecosystem model for winter wheat, AGROSIM-WHEAT, has been developed. It consists of component models for crop development, yield, pests and diseases. These models can be used independently to analyse and forecast environment-dependent specific growth, development and yield of winter wheat, and the dynamics of pest populations. PESTSIM-MAC is such a model for *S. avenae* (Rossberg et al., 1986).

According to Entwistle (1987), the EPIPRE system tends to overestimate the number of fields needing spraying; he developed a model for *S. avenae* based upon multiple regression equations.

The systems and models described above allow agricultural extension services to disseminate information much faster than is now possible. In the long term, real integrated control may be achieved, with farmers being fully informed about the consequences of their decisions and actions.

REFERENCES

Ankersmit, G.W. and Carter, N., 1981. Comparison of the epidemiology of *Metopolophium dirhodum* and *Sitobion avenae* on winter wheat. Netherlands Journal of Plant Pathology, 87: 71–81.

Carter, N., McLean, I.F.G., Watt, A.D. and Dixon, A.F.G., 1980. Cereal aphids – a case study and review. Applied Biology, 5: 271–348.

Carter, N., Dixon, A.F.G. and Rabbinge, R., 1982. Cereal Aphid Populations Biology Simulation and Prediction. Centre for Agricultural Publishing and Documentation, Wageningen, 91 pp.

Chambers, R.J. and Sunderland, K.D., 1983. The abundance and the effectiveness of natural enemies of cereal aphids on two farms in Southern England. In: R. Cavalloro (Editor), Aphid Antagonists. Balkema, Rotterdam, pp. 83–87.

Chambers, R.J., Sunderland, K.D., Wyatt, I.J. and Vickerman, G.P., 1983. The effects of predator exclusion and caging on cereal aphids in winter wheat. Journal of Applied Ecology, 20: 209–224.

Darwinkel, A., 1982. Intensifying of the winter wheat culture. Bedrijfsontwikkeling, 13: 400–402 (in Dutch).

De Clercq, R. and Pietraszko, R., 1983. Epigeal arthropods in relation to predation of cereal aphids. In: R. Cavalloro (Editor), Aphid Antagonists. Balkema, Rotterdam, pp. 88–92.

Dedryver, C.A., 1983. Field pathogenesis of three species of Entomophthorales of cereal aphids in Western France. In: R. Cavalloro (Editor), Aphid Antagonists. Balkema, Rotterdam, pp. 11–19.

Dedryver, C.A. and Gelle, A., 1982. Biologie des pucerons des céréales dans l'Ouest de la France IV. Etude de l'hivernation anholocyclique de *Rhopalosiphum padi* L., *Metopolophium dirhodum* Wlk, et *Sitobion avenae* F. sur repousses de céréales dans trois stations de Bretagne et du Bassin Pariesien. Acta Oecologia, Oecologia Applicata, 3: 321–342.

Entwistle, J.C., 1987. Forecasting cereal aphid outbreaks in England. Ph.D Thesis, University of East Anglia.

Fletcher, K.E. and Bardner, R., 1969. Cereal aphids on wheat. Report for Rothamsted Experimental Station, 1968, pp. 200–201.

Latgé, J.P., Sylvie, P., Papierok, B., Remaudière, C.A., Dedryver, C.A. and Rabasse, J.M., 1983. Advantages and disadvantages of *Conidiobolus obscurus* and of *Erynia neoaphidis* in the biological control of aphids. In: R. Cavalloro (Editor), Aphid Antagonists. Balkema, Rotterdam, pp. 20–32.

Latteur, G. and Godefroid, J., 1983. Trial of field treatments against cereal aphids with mycelium of *Erynia neoaphidis* (Entomophtorales) produced in vitro. In: R. Cavalloro (Editor), Aphid Antagonists, Balkema, Rotterdam, pp. 2–10.

Mantel, W.P., Rabbinge, R. and Sinke, J., 1982. Effects of aphids on the yield of winter wheat. Gewasbescherming, 13: 115–124 (in Dutch).

McKinlay, R.G., 1985. Effect of undersowing potatoes with grass on potato aphid numbers. Annals of Applied Biology, 106: 23–29.

Moreau, J.P.H., 1983. Aphid populations in maize crops. Damage induced, Natural control by antagonists. In: R. Cavalloro (Editor), Aphid Antagonists. Balkema, Rotterdam, pp. 93–99.

Nijveldt, W., 1980. Significance of predators on the development of insect pests in crops. In: A.K. Minks and P. Gruys (Editors), Integrated Control of Insect Pests in the Netherlands. Centre for Agricultural Publishing and Documentation, Wageningen, pp. 79–81.

Powell, W., 1983. The role of parasitoids on limiting cereal aphid populations. In: R. Cavalloro (Editor), Aphid Antagonists. Balkema, Rotterdam, pp. 50–56.

Rabbinge, R. and Mantel, W.P., 1981. Monitoring for cereal aphids in winter wheat. Netherlands Journal of Plant Pathology, 87: 25–29.

Rabbinge, R., Ankersmit, G.W. and Pak, G.A., 1979. Epidemiology of *Sitobion avenae* in winter wheat. Netherlands Journal of Plant Pathology, 85: 197–220.

Rossberg, D., Holz, F., Freier, B. and Wenzel, V., 1986. PESTSIM-MAC, a model for simulation of *Macrosiphum avenae* (Fabr.) populations. Tages Berichte Akademie der Landwirtschaftlichten Wissenschaften, Berlin, 242: 87–100.

Rossing, W.A.H., Schans, J. and Zadoks, J.C., 1985. The Epipre project. Landbouwkundig Tijdschrift, 97: 29–33 (in Dutch).

Sagenmüller, A., 1977. Untersuchungen über den Einfluss systemischer Fungicide auf der Getreideblattläuse, *Metopolophium dirhodum* Wlk. und *Sitobion avenae* F. Thesis, University of Giessen, F.R.G., 196 pp.

Southwood, T.R.E., 1977. The relevance of population dynamic theory to pest status. In: J.M. Cherrett and G.R. Sagar (Editors), Origin of Pest, Parasite, Disease and Weed Problems. Blackwell Scientific Publications, Oxford, pp. 35–44.

Southwood, T.R.E. and Comins, N.H., 1976. A synoptic population model. Journal of Animal Ecology, 45: 949–965.

Sunderland, K.D. and Chambers, R.J., 1983. Invertebrate polyphagous predators as pest control agents: some criteria and methods. In: R. Cavalloro (Editor), Aphid Antagonists. Balkema, Rotterdam, pp. 100–108.

Van Marrewijk, G.A.M. and Dieleman, F.L., 1980. Resistance to aphids in barley and wheat. In: A.K. Minks and P. Gruys (Editors), Integrated Control of Insect Pests in the Netherlands. Centre for Agricultural Publishing and Documentation, Wageningen, pp. 165–167.

Vereijken, P.H., 1979. Feeding and multiplication of three cereal aphid species and their effects on yield of winter wheat. Thesis, Agricultural University, Wageningen, 58 pp.

Vickerman, G.P. and Wratten, S.D., 1979. The biology and pest status of cereal aphids (Homoptera: Aphididae) in Europe: a review. Bulletin of Entomological Research, 69: 1–32.

Wilding, N., 1982. The effect of fungicides on field populations of *Aphis fabae* and the infection of the aphids by Entomophthoraceae. Annals of Applied Biology, 100: 221–228.

Zadoks, J.C., 1981. Epipre: a disease and pest management system for winter wheat developed in the Netherlands. Eppo Bulletin, 11: 365–369.

11.5.3 Integrated Control of Potato Aphids

P. HARREWIJN

INTRODUCTION

Aphids can cause two kinds of damage to the potato crop: direct damage to ware potatoes and indirect damage by the transmission of persistent and non-persistent viruses to seed potatoes (see also sections 10.1 and 10.3). As direct damage is correlated with population density, it is highly desirable that the field populations of potato aphids can be predicted (see section 8.2).

Some cultivars are particularly sensitive to the top-roll disorder caused by several aphid species, e.g. *Macrosiphum euphorbiae* (Thomas), probably because phloem transport is impaired and photosynthesis partly inhibited, resulting in an accumulation of carbohydrates in the leaves (Veen, 1985). Such cultivars have relatively low damage thresholds for these aphids.

If a good model of the development of aphid populations is available, control measures can be taken during the growth season, instead of precautionary measures beforehand. Whalon and Smilowitz (1979) developed a temperature-dependent model for field populations of *Myzus persicae* (Sulzer) in Pennsylvania, where weather conditions are more stable than they are in western Europe, for example. However, even short-term predictions are useful: within three days an aphid population can double its size, and in these few days an economic damage threshold could be crossed. In 1986, Entwistle and Dixon published a method that allows short-term forecasting of peak population densities of *Sitobion avenae* (Fabricius) on wheat on the basis of two field samplings. For potato the prediction accuracy should be estimated at various growth stages.

Population development of the cosmopolitan aphid *M. persicae* is determined to a great extent by the amount of nitrogen fertilizer given to the potato crop (Jansson and Smilowitz, 1986). Fertilizer levels can be adjusted in such a way that population development of *M. persicae* and *M. euphorbiae* is partly inhibited (Schepers et al., 1981). Table 11.5.3.1. shows how the population of *M. euphorbiae* on cv. Bintje was reduced when less nitrogen and more potassium were provided. Although work with artificial diets has revealed that the nutritional demands of *M. persicae* and *M. euphorbiae* are not identical, available amides like asparagine are necessary for reproduction, which is inhibited by "negative" amino acids like proline. The proportion of these two substances in the phloem sap is influenced by nitrogen and potassium fertilization.

Mineral oil, when used in weekly doses of 15 l per ha to reduce transmission of non-persistent viruses in seed potatoes, often stimulates growth and development of *M. euphorbiae*, especially in a young crop (Schepers et al., 1981). Its effect on *M. persicae* is much less pronounced. For ware potatoes this does not apply, because mineral oil is not used in this crop.

Section 11.5.3 references, p. 284

TABLE 11.5.3.1

Numbers of *Macrosiphum euphorbiae* (Thomas) per 100 leaves in cultivar Bintje in 1980

Date	Treatment[a]	Apterous adults	Larvae	Prealatae
3 July	Conventional methods	6	10	–
	Mineral oil	13	84	–
	Adapted fertilization	3	19	–
8 July	Conventional methods	14	69	3
	Mineral oil	34	255	6
	Adapted fertilization	6	42	–
16 July	Conventional methods	12	58	13
	Mineral oil	7	136	6
	Adapted fertilization	11	26	4
23 July	Conventional methods	19	44	29
	Mineral oil	6	45	17
	Adapted fertilization	4	19	4

[a] Mineral oil: weekly doses of 15 l per ha to reduce transmission of non-persistent viruses.
Adapted fertilization: less nitrogen and more potassium than usual; split nitrogen dressing.

WARE POTATOES

A scheme for integrated control of aphids in ware potatoes could be based upon the following measures:

Fertilization. Of the total amount of nitrogen to be applied to the crop (based on soil analysis), two-thirds should be given 3–4 weeks before the planting date, the remainder should be given about 4 weeks after the emergence of the crop, but in any case *before* the onset of the generative phase. In dry periods, sprinkler irrigation should prevent water stress and nitrogen deficits. Potassium (up to 1000 kg K_2O per ha) enhances plant resistance and at the same time reduces sensitivity to scab (Harrewijn, 1983).

Choice of cultivar. Resistant or partly resistant cultivars should be planted In ware potatoes the direct damage caused by aphids is more important than the indirect damage caused by virus transmission, and therefore vector resistance is not essential as long as there is some degree of aphid resistance. Hybrid populations have been developed from crossings of *Solanum berthaultii* and *S. tuberosum* ssp. *tuberosum* that bear glandular trichomes; this is a promising step towards the development of potato plants with increased aphid and even vector resistance (Xia and Tingey, 1980; see also section 11.4) Obrycki et al. (1983) have observed that both predators and parasitoids are active on potato clones with glandular hairs. Small predators may be partly hindered by a high density of trichomes, but in the breeding process some loss of glandular pubescence should occur.

   If it is a matter of choice, it may be wise to grow short-season cultivars. In contrast to their earlier expectations, Jansson et al. (1987) found populations of *M. persicae* to increase more rapidly on the longer-season cultivars Buckskin and Katahdin than on cultivars with a short maturation time.

Undersowing. The habitat for natural enemies in the crop itself can be improved by undersowing with grass. Perennial ryegrass, in particular, reduced the numbers of *M. euphorbiae* and *M. persicae* (Mckinlay, 1985).

Aphicides. On many cultivars the aphid density should not exceed 50 aphids per compound leaf. Therefore the action threshold for aphicide application is at approximately 25 aphids per compound leaf, when almost all leaves will be sustaining aphids. An aphicide that is not harmful to natural enemies is the non-systemic pirimicarb (see section 11.1.1). To stimulate sap utake and enhance the effect of future systemic but selective aphicides abrasives powders can be applied (Llewellyn and Eivaz, 1979).

Summary. Integrated pest management for the control of asphids in ware potato crops should consist of the following complementary measures:
(a) introduction of potato cultivars with glandular hairs;
(b) introduction of potato cultivars with other forms of aphid resistance;
(c) stimulation of the activity of natural enemies by habitat improvement, sprays of artificial honeydew etc.;
(d) adapted fertilization schemes, based upon soil analysis;
(e) irrigation;
(f) development of warning systems similar to that for cereals, EPIPRE (see section 1.5.2), tolerating 25 aphids per leaf;
(g) application of selective aphicides.

SEED POTATOES

Many of the measures mentioned above can also be used to reduce aphid populations in seed potato fields. However, to prevent the dispersal of viruses, it is more important to reduce the activity of the aphids than to reduce their numbers. Moreover, local treatments cannot prevent the transmission of viruses by immigrating aphids. Therefore control should consist of measures to control aphids in the crop itself and measures to prevent immigration of winged individuals that may be carrying a persistent or non-persistent virus.

Host-plant resistance. A diagnostic procedure to establish vector resistance of host plants has been developed by Harrewijn (1986). Especially when enhanced resistance to persistent viruses is required, the introduction of vector resistance can lead to resistant cultivars, which would otherwise have been only partly resistant to viruses.

In breeding for plants with vector resistance to non-persistent viruses, the best way is to prevent probing by the vector aphid, for example by selecting for the presence of glandular hairs especially those that release (*E*)-β-farnesene. Dawson et al. (1986) have shown that plant substances can act as synergists of this alarm pheromone. It should be noted that the presence of such substances, e.g. isothiocyanates, may be essential to obtain an optimal effect on alighting behaviour. The new Dutch cultivar Santé has a high level of resistance to non-persistent viruses, but not to persistent ones.

Cultivation measures. Winged residential aphids, developing on the crop, can disperse persistent and non-persistent viruses. Mineral oil, which is used to reduce transmission of non-persistent viruses, impairs wing production in *M. persicae*, as does adapted fertilization. In large fields of seed potatoes, where immigration from sites that accomodate viruses is less important than in small fields, a considerable reduction in virus infection can be obtained through fertilization at high potassium levels and minimal nitrogen levels, as this reduces the number of aphids (Harrewijn, 1983). However, the nitrogen should not be applied as a split dressing, as is recommended for ware potatoes, because this could lead to a high nutritiousness of the plants at the wrong time, leading to high population densities and the production of alatae.

Temporary water stress should be avoided, or cultivars should be selected for their degree of drought tolerance (Beekman and Bouma, 1986).

Section 11.5.3 references, p. 284

In seed potato growing one should plan for large, uninterrupted fields. The development of winged morphs of a few residential aphids can be postponed by the use of micro-elements, such as Li^+-ions (Harrewijn, 1983). In contrast to the situation in small fields, where the aphid pressure from the environment is relatively high, in large fields it is worthwhile to reduce aphid numbers (and the occurrence of winged morphs) by stimulating natural enemies. This predatory potential will be used in a proposed integrated pest management system for potatoes in the state of Washington (Tamaki, 1982).

Aphids that immigrate into the seed potato crop could have made a previous landing in an adjacent crop, where colonizing aphids (residents) adsorb both persistent and non-persistent viruses and where non-residents, such as cereal aphids, take up non-persistent viruses. The best way to prevent virus transmission is to disturb settling behaviour, as discussed above (see also section 11.3). Other possibilities are: early planting, in order to increase mature plant resistance and to have a closed crop (silhouette reduction) as soon as possible, and host plant masking with substances that reflect UV-light, such as aluminium mulch or whitewash. Cultivars with some degree of aphid resistance (not vector resistance) may be of little value, as frequent probing may be stimulated, which is unfavourable with regard to the dispersal of non-persistent viruses.

It is important that vector pressure be reduced by increasing the distance between the seed potato crop and any possible virus source, for example ware potatoes. Experiments with isotopes (see section 8.7) have shown that winged *M. persicae* and *M. euphorbiae*, originating in a potato field, can spread viruses over short distances (Harrewijn et al., 1981; unpublished results). When the distance between ware potatoes and the seed potato crop is less than 50 m, particular care should be taken to prevent the development of winged morphs in the ware potatoes.

When aphid populations are still small, a split nitrogen application could be given with the help of irrigation equipment: two-thirds about 4 weeks before planting and one third after the onset of tuber formation. Besides, such equipment is useful to prevent scab. Cultivars should have some degree of aphid resistance, and when winged morphs are produced a selective aphicide that is not harmful to natural enemies should be applied (see section 11.1).

A considerable distance between ware potato and seed potato fields could also reduce the pressure from non-resident vectors. However, too little is known about the flight behaviour of aphids that transmit non-persistent viruses, e.g. *Phorodon humuli* (Schrank) and *Cavariella theobaldi* (Gilette and Bragg), to allow a proper assessment of the usefulness of measures aimed at preventing aphid immigration, such as plant masking. The same holds true for the aphids' efficiency as virus vectors, as no more than some scattered laboratory data are available (see sections 8.6 and 10.3).

The presence of any virus sources in the vicinity of the seed potato crop is to be avoided, so beware of volunteer plants, as they can be a serious source of infection (Van Hoof, 1979).

Fig. 11.5.3.1 shows the various methods used in integrated control of virus transmission by aphids. For ware potatoes in the vicinity of seed potatoes, control measures should be taken in view of the benefits to the latter. For example, the above-mentioned split nitrogen dressing causes the plants' optimal suitability for growth and development of the aphid to occur later in the season, thereby reducing the number of winged morphs and thus reducing the transmission of persistent and non-persistent viruses in the crop itself and in adjacent seed potato fields. This measure has no effect, however, on the transmission of non-persistent viruses by alighting non-residents.

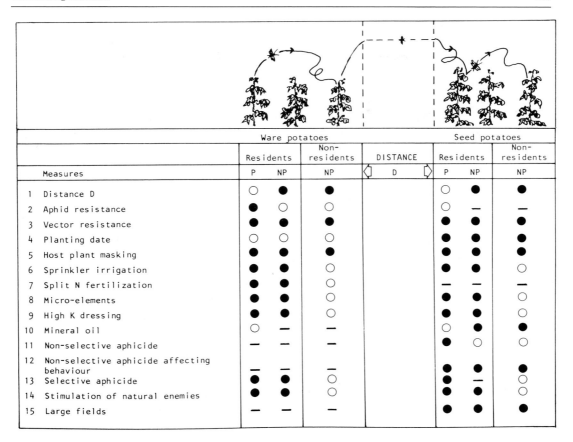

Measures	Ware potatoes Residents P	Ware potatoes Residents NP	Ware potatoes Non-residents NP	DISTANCE D	Seed potatoes Residents P	Seed potatoes Residents NP	Seed potatoes Non-residents NP
1 Distance D	○	●	●		○	●	●
2 Aphid resistance	●	○	○		○	—	—
3 Vector resistance	●	●	●		●	●	●
4 Planting date	○	○	○		●	●	●
5 Host plant masking	●	●	●		●	●	●
6 Sprinkler irrigation	●	●	○		●	●	○
7 Split N fertilization	●	●	○		—	—	—
8 Micro-elements	●	●	○		●	●	○
9 High K dressing	●	●	○		●	●	○
10 Mineral oil	○	—	—		○	●	●
11 Non-selective aphicide	—	—	—		●	○	○
12 Non-selective aphicide affecting behaviour	—	—	—		●	●	●
13 Selective aphicide	●	●	○		●	—	○
14 Stimulation of natural enemies	●	●	○		●	●	○
15 Large fields	—	—	—		●	●	●

Fig. 11.5.3.1. Integrated control of virus transmission by resident and non-resident aphids in seed potatoes and ware potatoes. P, persistent virus; NP, non-persistent virus.

The growth season of seed potatoes is shorter than that of ware potatoes, and therefore it is inadvisable to provide nitrogen in more than a single dressing, as this would improve the quality of the host rather than reduce it when the plants have not yet developed some degree of mature plant resistance.

Another example: a non-selective aphicide (11 in Fig. 11.5.3.1) kills both aphids and natural enemies. As a result, ware potatoes have to be treated more than once to keep them free of aphids. However, even one single treatment can give a considerable protection against transmission of persistent viruses, as later in the season the plants develop their own mature resistance. A single treatment gives no protection against the transmission of non-persistent viruses and should not be used in ware potatoes close to seed potatoes, as in those fields aphid populations should be continuously suppressed. A selective aphicide (13 in Fig. 11.5.3.1) has the advantage that natural enemies can keep the vector pressure at a low level.

CONCLUDING REMARKS

Aphids are not the only problem in the potato crop. Many diseases and pests are a threat to potato growing: history has proved that is is unrealistic to depend upon chemicals alone for control. We should therefore aim at integrated pest management, or even better: integrated production of ware and seed potatoes. An example of such an approach is given by Hollingsworth et al. (1986) for Massachusetts (U.S.A.).

Section 11.5.3 references, p. 284

REFERENCES

Beekman, A.G.B. and Bouma, W.F., 1986. A possible screening technique for drought tolerance in potato. In: Potato Research of Tomorrow: Proceedings of an International Seminar, Wageningen, pp. 55–67.

Dawson, G.W., Griffiths, D.C., Pickett, J.A., Wadhams, L.J. and Woodcock, C.M., 1986. Plant compounds that synergise activity of the aphid alarm pheromone. Proceedings British Crop Protection Conference, Brighton, 1985, Vol. 2, pp. 829–834.

Entwistle, J.C. and Dixon, A.F.G., 1986. Short-term forecasting of peak population density of the grain aphid (*Sitobion avenae*) on wheat. Annals of Applied Biology, 109: 215–222.

Harrewijn, P., 1983. The effect of cultural measures on behaviour and population development of potato aphids and transmission of viruses. Mededelingen Faculteit Landbouwwetenschappen, Rijksuniversiteit Gent, 48/3: 791–799.

Harrewijn, P., 1986. Vector resistance of potato aphids with respect to potato leafroll virus. In: Potato Research of Tomorrow: Proceedings of an International Seminar, Wageningen, pp. 85–96.

Harrewijn, P., Van Hoof, H.A. and Noordink, J.P.W., 1981. Flight behaviour of the aphid *Myzus persicae* during its maiden flight. Netherlands Journal of Plant Pathology, 87: 111–117.

Hollingsworth, C.S., Ferro, D.N. and Coli, W.M. (Editors), 1986. Potato Production in the Northeast. A Guide to Integrated Pest Management. Cooperative Extension Service, University of Massachusetts, Amherst, 93 pp.

Jansson, R.K. and Smilowitz, Z., 1986. Influence of nitrogen on population parameters of potato insects: abundance, population growth, and within-plant distribution of the green peach aphid, *Myzus persicae* (Homoptera: Aphididae). Environmental Entomology, 15: 49–56.

Jansson, R.K., Elliott, G.C., Smilowitz, Z. and Cole, R.H., 1987. Influence of cultivar maturity time and foliar nitrogen on population growth of *Myzus persicae* on potato. Entomologia Experimentalis et Applicata, 43: 297–300.

Llewellyn, M. and Eivaz, J., 1979. Abrasive dusts as a mechanism for aphid control. Entomologia Experimentalis et Applicata, 26: 219–222.

McKinlay, R.G., 1985. Effect of undersowing potatoes with grass on potato aphid numbers. Annals of Applied Biology, 106: 23–29.

Obrycki, J.J., Tauber, M.J. and Tingey, W.M., 1983. Predator and parasitoid interaction with aphid-resistant potatoes to reduce aphid densities: a two-year field study. Journal of Economic Entomology, 76: 456–462.

Schepers, A., Harrewijn, P., Van Rheenen, B. and Bus, C.B., 1981. Effect of mineral oil and fertilization on population development of two potato aphids. Abstract Papers, 8th Triennial EAPR Conference, München, pp. 44–45.

Tamaki, G., 1982. Biological control of potato pests. In: J.H. Lashhomb and R. Casagrande (Editors), Advances in Potato Pest Management. Hutchinson Ross, Strondsburg, 288 pp.

Van Hoof, H.A., 1979. Infection pressure of potato virus Y^N to and from potato fields. Netherlands Journal of Plant Pathology, 85: 31–37.

Veen, B.W., 1985. Photosynthesis and assimilate transport in potato with top-roll disorder caused by the aphid *Macrosiphum euphorbiae*. Annals of Applied Biology, 107: 319–323.

Whalon, M.E. and Smilowitz, Z., 1979. Temperature-dependent model for predicting field populations of green peach aphid, *Myzus persicae* (Homoptera: Aphididae). Canadian Entomologist, 111: 1025–1032.

Xia, Ji-Kang, and Tingey, W. 1980. Green peach aphid (Homoptera: Aphididae): developmental and reproductive biology on a *Solanum tuberosum* × *S. bertaultii* hybrid potato. Journal of Economic Entomology, 79: 71–75.

11.5.4 Integrated Control of Aphids in Field-Grown Vegetables

J. THEUNISSEN

INTRODUCTION

Aphid problems in field-grown vegetables are restricted to a few vegetables and aphid species, of which some cause damage almost every season, whilst others seldom occur in sufficiently large numbers to be injurious. Important species are: *Brevicoryne brassicae* (Linnaeus) in cruciferous crops; *Myzus persicae* (Sulzer), *Nasonovia ribis-nigri* (Mosley) and sometimes *Pemphigus bursarius* (Linnaeus), *Macrosiphum euphorbiae* (Thomas) and *Aulacorthum solani* (Kaltenbach) in leafy vegetables such as lettuce and endive; *Aphis fabae* Scopoli in bean and spinach; *M. persicae* in beetroot; *Acyrthosiphon pisum* (Harris) in pea; and sometimes *Cavariella aegopodii* (Scopoli) and *Semiaphis dauci* (Fabricius) in carrot.

BREVICORYNE BRASSICAE

The cabbage aphid, *B. brassicae*, is one of the most damaging and consistently present pests on cabbage crops in the temperate zone; it is part of the regular pest complex of this group of vegetables. Depending on the weather conditions, planted crops become infested by alate individuals that multiply and form small colonies which rapidly expand if not disturbed by adverse weather conditions, natural enemies or human interference. The first arrivals seem to be distributed at random. As flight continues more plants are infested, the newly established colonies perish or expand and a gradual infestation of the whole crop is established. This process has been followed through weekly observations of cabbage aphid infestation in fields of Brussels sprouts (see Fig. 11.5.4.1).

The population dynamics and distribution patterns on *M. persicae* and *B. brassicae* in broccoli have been studied in California (Trumble, 1982a, b, c). At low population densities aphids showed an aggregated distribution pattern, which became random at increasing population densities. Based on aphid counts a sampling method was developed which was not yet suitable for practical implementation (Trumble, 1982a). For another crop, strawberry, dispersion parameters of aphids were already determined by Trumble et al. (1983).

Tolerance levels

In the Netherlands, research has been done on the amount of *B. brassicae* infestation that can be tolerated in various cabbage crops, and tolerance levels have been defined and tested (Theunissen and Den Ouden, 1985, 1987). These

Section 11.5.4 references, p. 288

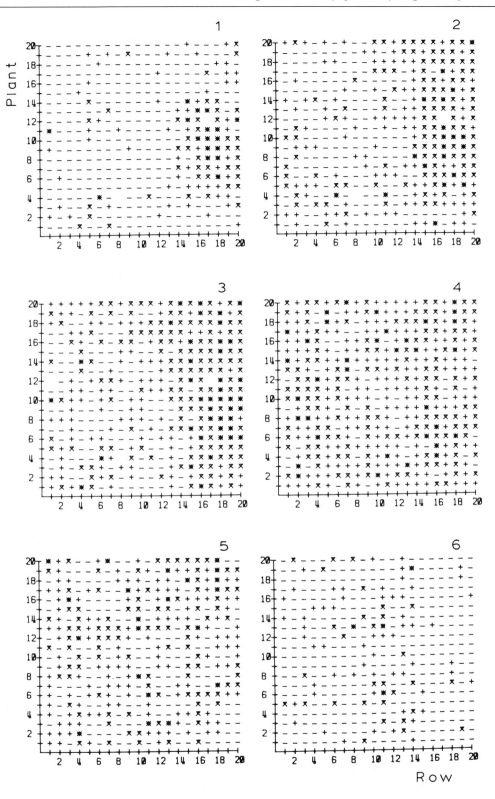

Fig. 11.5.4.1. Infestation of a field of Brussels sprouts (20 rows of 20 plants each) by *Brevicoryne brassicae* (Linnaeus). The degree of aphid infestation on each individual plant was assessed once a week and the results were mapped as shown above (1–6). Symbols: −, no infestation; +, light infestation; ×, moderate infestation; *, heavy infestation; "blank", no observation made (Theunissen, 1983).

TABLE 11.5.4.1

Tolerance levels (maximum percentage of infested plants) for *Brevicoryne brassicae* (Linnaeus) on three cabbage crops

Crop	Weeks after transplanting										
	2	4	6	8	10	12	14	16	18	20	etc.
Brussels sprouts	10	10	10	10	10	4	4	0	0	0	
White cabbage	5	5	5	5	5	2	0	0	0	.	
Red cabbage	20	20	20	5	5	20	20	5	5	0	

tolerance levels are a tool in the implementation of integrated pest management (IPM) in commercial cabbage growing. They have been defined according to the growth stage of the crop and the type of cabbage involved. For the sake of simplicity these tolerance levels are expressed as the percentage of plants found infested; the sampled plant is infested or not, regardless of the number of aphids or the rate of parasitization. For the same reason the growth stage of the crop is expressed as the number of weeks after transplanting, a thing every grower knows. Examples of these tolerance levels are given in Table 11.5.4.1.

The effects of taking tolerance levels into account when applying aphicides have been investigated over a number of seasons and locations and have been compared with results obtained with conventional crop protection measures. If an aphicide is used only when this is strictly necessary, as indicated by the tolerance level, a considerable reduction in the number of applications can be obtained, although this will vary over the years: in a warm and dry season more treatments will be needed then in a cool and wet one. The economic benefits are obvious, however, and farmers are quite willing to adopt this system of integrated pest management.

The system calls for a systematic sampling method, in which the farmers must go through their fields in order to assess the amount of infestation. This has the additional advantage that farmers become less worried about the presence of insects in their fields. When they become aware of the levels of infestation that can or cannot be tolerated in the crop, they gain a feeling of being in control and knowing when to take appropriate measures. The presence of a few insects will not worry them. Growers using the system sometimes even tolerate high densities of *B. brassicae* on individual plants, which thus act as reservoirs for parasites, because they perceive that usually no infestation of neighbouring plants occurs and they realize that chemical control would be uneconomical.

Other elements in IPM

Host plant. *B. brassicae*, although less sensitive to nutritional factors of the host plant than the polyphagous aphid *M. persicae* (Van Emden and Bashford, 1969), is affected by the physiological condition of rape-seed plants. According to Weissmann and Baran (1969), plant parts with a sugar/total nitrogen ratio lower than 1.4 are particularly suitable for this aphid. After the onset of the ripening phase the plants become unsuitable.

B. brassicae, which usually feeds on young parts, reacts positively to the amides asparagine and glutamine in cabbage leaves, although it responds primarily to sinigrin and secondarily to the nutritional quality of its hosts. The aphid develops faster at low (one fourth of normal) and high levels of potassium fertilization of cabbage (El-Tigani, 1962).

Section 11.5.4 references, p. 288

Temporary water stress stimulates development of *B. brassicae*, particularly on young leaves (Wearing, 1972) and should be avoided, whereas stormy weather may reduce population densities, as larvae that fall from the leaves cannot easily climb back on to the plants (Dadd and Van Emden, 1979).

In taller crops, such as Brussels sprouts, sampling takes a relatively long time. Therefore, sequential sampling schemes have been developed for Brussels sprouts (Theunissen, 1988). Wilson et al. (1983) published a presence–absence sampling method for *B. brassicae* and *M. persicae* in Brussels sprouts. They also related the numbers of aphids per plant to the percentage of aphid-infested plants in the sample.

Parasites. Parasitization of apterous *B. brassicae* in cabbage crops can approach 100%, especially later on in the growing season, when parasite populations have caught up with their hosts. But from a practical point of view the parasites are not sufficiently numerous and/or active to contribute significantly to the suppression of growing cabbage aphid populations at the time of rapid expansion after each flight. Generally populations exceed their tolerance level at least once a season, and in commercial cabbage growing chemical control is then inevitable. The current IPM policy is to advise the use of a selective compound which keeps the population curve of the aphids below the tolerance limit. Eradication of the cabbage aphid is out of the question, and a certain supply of hosts to parasites is necessary.

Intercropping. In intercropping experiments with Brussels sprouts, *B. brassicae* was shown to react strongly to the background colour of the crop (Smith, 1976). Aphid populations developed very differently on unsprayed crops that were either intercropped with *Spergula arvensis* or not intercropped (Theunissen and Den Ouden, 1980). At high aphid densities the differences in population size were the largest, which means that at the lower levels that are to be expected in commercial fields the intercropping effect would not be large enough to obtain a sufficient degree of cabbage aphid control. Although populations could be reduced, the remaining number of aphids is too high by commercial standards.

REFERENCES

Dadd, G.D. and Van Emden, H.F., 1979. Shifts in host plant resistance to the cabbage aphid (*Brevicoryne brassicae* L.) exhibited by Brussels sprout plants. Annals of Applied Biology, 91: 251–262.

El-Tigani, M.E., 1962. Der Einfluss der Mineraldüngung der Pflanzen auf Entwicklung und Vermehrung von Blattläusen. Wissenschaftliche Zeitschrift Universität Rostock, 11: 307–324.

Smith, J.G., 1976. Influence of crop background on aphids and other phytophagous insects on Brussels sprouts. Annals of Applied Biology, 83: 1–13.

Theunissen, J., 1983. Development of sampling methods of pests in cabbage crops. In: CEC Programme on Integrated and Biological Control, Progress Report 1979/1981, pp. 184–201.

Theunissen, J., 1988. Sequential sampling of insect pests in Brussels sprouts. FOBC/WPRS Bulletin, 1988/XI/1: 111–117.

Theunissen, J. and Den Ouden, H., 1980. Effects of intercropping with *Spergula arvensis* on pests of Brussels sprouts. Entomologia Experimentalis et Applicata, 27: 260–268.

Theunissen, J. and Den Ouden, H., 1985. Tolerance levels for supervised control of insect pests in Brussels sprouts and white cabbage. Zeitschrift für Angewandte Entomologie, 100: 84–87.

Theunissen, J. and Den Ouden, H., 1987. Tolerance levels and sequential sampling tables for supervised control in cabbage crops. Mitteilungen der Schweizerischen Entomologischen Gesellschaft, 60: 243–248.

Trumble, J.T., 1982a. Temporal occurrence, sampling, and within-field distribution of aphids on broccoli in coastal California. Journal of Economic Entomology, 75: 378–382.

Trumble, J.T., 1982b. Within-plant distribution and sampling of aphids (Homoptera: Aphididae) on broccoli in southern California. Journal of Economic Entomology, 75: 587–592.

Trumble, J.T., 1982c. Aphid (Homoptera: Aphididae) population dynamics on broccoli in an interior valley of California. Journal of Economic Entomology, 75: 841–847.

Trumble, J.T., Oatman, E.R. and Voth, V., 1983. Temporal variation in the spatial dispersion patterns of aphids (Homoptera: Aphididae) infesting strawberries. Environmental Entomology, 12: 595–598.

Van Emden, H.F. and Bashford, M.A., 1969. A comparison of the reproduction of *Brevicoryne brassicae* and *Myzus persicae* in relation to soluble nitrogen concentration and leaf age (leaf position) in the brussels sprout plant. Entomologia Experimentalis et Applicata, 12: 351–364.

Wearing, C.H., 1972. Responses of *Myzus persicae* and *Brevicoryne brassicae* to leaf age and water stress in Brussels sprouts grown in pots. Entomologia Experimentalis et Applicata, 15: 61–80.

Weissmann, L. and Baran, M., 1969. Einfluss von Veränderungen des physiologisches Zustandes des Sommerrapses auf die Populationsdichte der Kohlblattlaus (*Brevicoryne brassicae* L.). Biologia Bratislava, 24; 375–382.

Wilson, L.T., Pickel, C., Mount, R.C. and Zalom, F.G., 1983. Presence–absence sampling for cabbage aphid and green peach aphid (Homoptera: Aphididae) on Brussels sprouts. Journal of Economic Entomology, 76: 476–479.

11.5.5 From Integrated Control to Integrated Farming

P. VEREIJKEN

MOTIVATION

A generally accepted definition of integrated control is that presented by the FAO in 1967: "A pest management or integrated control system, in the context of the associated environment and the population dynamics of the pest species, utilizes all suitable techniques and methods in as compatible manner as possible and maintains pest population levels below those causing economic injury". Norton and Holling (1976) recommended to include the aim " to develop alternative, ecologically desirable tactics for use in suppressing major pests", because they wanted to emphasize the implicit need to reduce the use of broad-spectrum biocides and to decrease their adverse effects.

In a recent bulletin of IOBC/WPRS, Vereijken et al. (1986) concluded that "in Western Europe, with the exception of some glasshouse crops and in a restricted number of orchards and vineyards, integrated control as envisaged originally has not been put into practice to any significant extent. This also holds for other parts of the world such as North America, where considerable investment in research has led to relatively little adoption in practice. Nevertheless, it must strongly be emphasized that many of the techniques explored as parts of integrated control programmes, such as economic injury thresholds, insect pest monitoring with sex pheromones, use of selective pesticides and others, have already been incorporated into present-day pest control practices so far primarily based on chemicals and have led to a much more rational and efficient use of them".

But the question remains why so disappointingly few of the non-chemical control methods and techniques developed during the last decennia have been adopted by farmers. There are several reasons: (1) the non-chemical techniques are too complicated and time-consuming; (2) they are less reliable and/or effective; (3) they are more expensive than pesticides.

Frequently it has been noted that all these objections may be overcome by coordinated governmental actions, particularly by training, extension and subsidizing research and development of integrated control and by imposing levies or legal restrictions to chemical control.

However, even if this should happen, it still is doubtful whether practice eventually would change to "true" integrated control, with chemicals being used only as a last resort. Current cropping systems are aimed almost exclusively at obtaining maximal profits by attaining maximum physical yields. This requires high-yielding varieties and ample plant nutrition, which make crops attractive and vulnerable to most pests and diseases. As a consequence the potential yield cannot be realized without the protection of a so-called "chemical umbrella".

Section 11.5.5 references, p. 295

Therefore, specialists in crop protection have, up till now, mainly restricted themselves to rationalizing and optimizing chemical control. Biological control is only recommended if chemicals fail or get too expensive. Meanwhile world-wide intensification of agriculture is still proceeding, increasing pollution of the environment and destabilization of agro-ecosystems. This all seems even more absurd when taking into consideration the growing surplusses of agricultural products, the decreasing income and employment in rural areas and the growing concern of consumers about the quality of their food.

There is a growing awareness that these problems cannot be solved one by one on an ad-hoc basis, but that a comprehensive approach is needed. On an IOBC/WPRS meeting in 1976 (Anon., 1977), it was concluded that it is irrational to study crop protection in isolation from the other major farming system components such as fertilization, cultivation or crop rotation. Eventually, in 1981, an IOBC/WPRS Study Group was set up to examine the possibilities to develop programmes on "management of arable farming systems for integrated crop protection". This resulted in a report (Vereijken et al., 1986), in which a general design for experimental farming systems was presented, based on preliminary experiences in the Federal Republic of Germany (El Titi, 1986) and the Netherlands (Vereijken, 1986). Subsequently, an IOBC/WPRS Working Group for integrated arable farming has been founded with participation of researchers from nine European countries.

INTEGRATED FARM MANAGEMENT

In the Netherlands an experimental farm for the development and comparison of alternative agricultural systems was started near the village of Nagele in 1979. Three farming systems, namely (1) biodynamic, (2) integrated and (3) conventional, are being studied. These systems are managed independently of each other by one farmer and four other workers on a commercial basis.

The biodynamic farm (22 ha) is mixed and aims to be self-supporting in terms of fertilizers and fodder. Pesticides are not allowed. On the conventional and integrated farms (17 ha each) field crops only are grown, in the same 4-year rotation: potato $- \frac{1}{2}$ pea, $\frac{1}{4}$ onion, $\frac{1}{4}$ carrot $-$ sugar beet $-$winter wheat. The conventional farm serves as a reference. Its major objective is a maximum financial return. The integrated farm should at least produce a similar financial output, but this must be achieved with a minimum input of fertilizers, pesticides and machinery to avoid pollution of the environment and to save non-renewable resources. Therefore it can be characterized as an intermediate system.

Each year a detailed management scheme is made for every crop, based on the experience of previous years and knowledge available from the extension services and research institutes. Because chemical control is one of the most criticized aspects of conventional farming, integrated control of pests, diseases and weeds is given priority in the integrated farming system. The major elements of the integrated control strategies against pests and diseases are presented in Table 11.5.5.1. More specifically, aphids play a major role in wheat and seed potatoes only. In Table 11.5.5.2 the principal control measures are listed. Table 11.5.5.3 shows that unexpectedly large reductions in the applied amount of pesticides can be obtained if integrated control is really given priority. Table 11.5.5.4 shows that integrated farming may lead to yield reductions, which, however, can be compensated for by cost reductions. Finally it can be concluded from Table 11.5.5.5 that integrated arable farming with integrated crop protection as a major element does not necessarily lead to a lower income for the farmer, in contrast to what is generally assumed.

TABLE 11.5.5.1

Integrated control techniques for pests and diseases, in order of preference

Biological

Crop rotation (principally aimed at soil-borne pests and diseases,
    particularly nematodes)
Resistant cultivars
Nitrogen fertilization (some 40 kg per ha less than the economic optimum
    to enhance crop resistance against aphids and moulds: directly through
    reduction of the nutritional value of the crop, indirectly through a less favourable microcli-
    mate)
Stimulation of antagonists, parasites and predators (by green and organic
    manure, appropriate – e.g. minimum – soil tillage and selective chemical control)
Miscellaneous cases (for example application of sterile male technique
    against the onion fly)

Chemical

Monitoring and use of control thresholds
Choice of the pesticide based on its efficiency, selectivity, toxicity to
    man and animal, persistence, mobility and price[a]
Seed dressing and row application instead of broadcast treatment

[a]With respect to toxicity, persistence and mobility of the chemicals, a "black" list is carefully heeded.

TABLE 11.5.5.2

Integrated control measures against aphids in wheat and seed potato

Control measure	Wheat	Seed potato
Resistant cultivars	–	maximum against major viruses[a]
Less nitrogen	40 kg per ha[b]	
Monitoring in the field	weekly from the onset of flowering, 100 culms per field	before every selection round, 20 plants per field
Threshold of control	70% of culms infested	one *Myzus persicae* per plant[c]
Selective aphicide	pirimicarb	pirimicarb
Monitoring with suction trap, regional warning system	–	compulsory ultimate date for foliage destruction[d]

[a]Instead of cv. Bintje (conventional farm), cv. Santé is grown in the integrated system. This cultivar is little susceptible to potato blight and resistant to potato cyst nematodes and the major non-persistent viruses (A, X, Y^C, Y^O, Y^N).
[b]A gift of 40 kg per ha less than the optimum nitrogen dressing, based on the N-mineral method, results in slower growth rates of mildew, rust, *Septoria* and aphids. Reduced multiplication of the aphids allows parasites and predators more time to build up, thus increasing their significance as biological control agents.
[c]Cv. Santé is rather susceptible to the persistent leaf-roll virus mainly transmitted by *Myzus persicae* (Sulzer).
[d]If this were not compulsory, the regional warning system would not have been used because cv. Santé is sufficiently resistant against the major viruses. So such a drastic preventive measure results in an unnecessary substantial reduction of yield.

PERSPECTIVES AND RECOMMENDATIONS

   The Dutch are, and, have been, foremost in intensifying agriculture and therefore they were among the first to have to deal with the severe side-effects of such a policy. It is not surprising that the Netherlands leads the way in the development of alternative farming systems, with the intention of integrating various objectives in the field of economy, employment, environment, nature,

Section 11.5.5 references, p. 295

TABLE 11.5.5.3

Chemical control in conventional and integrated farming systems; 1982–1985

	Total number of treatments per field		Active ingredients (kg/ha)	
	Conventional	Integrated	Conventional	Integrated
Herbicides	2.1	1.0	4.15	1.57
Insecticides	4.4	2.0	5.50	2.21
Fungicides	1.3	0.7	0.35	0.18
Subtotal	7.8	3.7	10.00	3.96
Nematicide[a]	0.2	0.0	28.71	0.00
Total	8.0	3.7	38.71	3.96

[a]On the conventional farm, the soil has been fumigated against the potato cyst eelworm after each potato crop; on the integrated farm a resistant variety is grown.

TABLE 11.5.5.4

Average physical and financial yields of the three main marketable crops, potato (P), sugar beet (S) and winter wheat (W), in the three farming systems; 1982–1985

	Conventional			Integrated			Biodynamic		
	P	S	W	P	S	W	P	S	W
Physical yields[a]	52.1	9.39	7.9	36.6	9.53	7.0	23.6	8.26	5.3
Total returns[b]	10.45	7.10	4.22	10.94	7.10	3.80	11.02	6.06	4.94
Allocated costs[c]	5.51	1.82	1.28	4.49	1.61	1.08	2.43	1.11	0.62
Gross gain[d]	4.94	5.28	2.94	6.45	5.49	2.72	8.59	4.95	4.32

[a]Tonnes per ha (yield of sugar beet in tonnes of sugar per ha).
[b]Price per tonne x tonnes per ha (price of biodynamic wheat is 0.85 Dutch guilders per kg, that of biodynamic potato is 0.50 Dutch guilders per kg, respectively 70% and 100% that of wheat and potato grown on the conventional farm).
[c]Costs of pesticides, fertilizers, hired labour, sowing seeds, insurance, interest.
[d]Total returns minus allocated costs.

TABLE 11.5.5

Overall farm economic results of the three farming systems, in thousands of Dutch guilders, 1982–1985

	Conventional	Integrated	Biodynamic
Farm size (ha)	17	17	22
Labour (man/year)	0.6	0.6	1.8
Total returns	118.1	111.1	168.3
Total costs	132.6	121.2	265.4
including own labour[a]	(33.7)	(35.5)	(97.9)
Net surplus	− 14.5	− 10.1	− 97.1

[a]The mean cost of labour per hour is considered to be 27 Dutch guilders, but the real rewards per spent hour in the three systems were only 15, 19 and 0 Dutch guilders, respectively. The fact that farmers cannot get their labour properly paid for is quite normal, unfortunately.

landscape, quality of food-stuffs and well-being of man and animal. In this respect, it should be mentioned that in 1986 a second experimental farm for the development of integrated arable farming was started in the North-east of the country and that a third farm in the South-east is in preparation.

Currently, several other European countries appear to be planning similar long-term experiments. This means that integrated control is finally allowed to exchange its marginal role for a major one, thus getting a fair chance to realize its potential. This new development offers new opportunities for entomologists and aphidologists to put into practice their integrated control methods and techniques. Of course, they have to keep up with the other disciplines, to ensure that the proposed measures against single pests fit harmoniously into the total cropping system and do not interfere with other interests or cause new problems. Only in long-term projects like those in Lautenbach (Federal Republic of Germany) and Nagele (The Netherlands) can such delicate inter-disciplinary activities be brought to a successful end.

If eventually a more appropriate farming system has been elaborated, a new problem will arise. How must farmers handle sophisticated integrated systems? I am convinced that only with help of computerized guidance systems can the knowledge gap between research and practice be bridged. These systems should enable the farmers to act on the necessarily higher level of knowledge by providing detailed cropping programmes, based on a systematic survey of all possible decisions, with a logic key to choose the right decision in all imaginable situations, depending on specific farm, crop, soil and weather conditions.

In detail, these guidance systems must contain the following elements:
– Warning procedures; depending on crop characteristics, growth stage and previous cropping measures, attention is called to certain decisions that are to be taken.
– Decision rules and advice models; for control of pests, weeds and diseases but also for fertilization, timing of operations etc.
– General information; for example on disease observations, use of pesticides, properties of cultivars, susceptibility of diseases, pests and weeds for biocides etc.
– Registration; input/output, feedback to system.
Currently, guidance systems for the main crops in the area around Nagele are being developed. They will be tested and optimized with the help of a farmers' study-club before being made available to the whole region, probably by means of view-data.

REFERENCES

Anonymus, 1977. An approach towards integrated agricultural production through integrated plant protection. IOBC/WPRS Bulletin, 1977/4: 1–163.

El-Titi, A., 1986. Design and preliminary results of the Lautenbach project/West Germany. IOBC/WPRS Bulletin, 1986/IX/2, Appendix B: 16–22.

FAO, 1967. Report of the First Session of the FAO Panel of Experts on Integrated Pest Control, Rome, 18–22 September 1967.

Norton, G.A. and Holling, C.S. (Editors), 1976. Proceedings of the Conference on Pest Management, 25–29 October 1976. International Institute for Applied Systems Analysis, no. 2361, Laxenbourg, Austria.

Vereijken, P.H., 1986. From conventional to integrated agriculture. Netherlands Journal of Agricultural Science, 34: 387–393.

Vereijken, P.H., Edwards, C.A., El Titi, A., Fougeroux, A. and Way, M.J., 1986. Report of the Study Group "Management of Farming Systems for Integrated Control". IOBC/WPRS Bulletin, 1986/IX/2, 34 pp.

Epilogue

> ... "we must realize that in aphids
> speciation also takes place on plants
> that originally were *secondary* host plants" ...
>
> Dick HILLE RIS LAMBERS (1979)

As we already stated in the Preface of this book, knowledge of aphid biology is growing so rapidly that we found it difficult to choose the right time to close the editorial preparation of each of the three volumes. It was tempting to continue with an addendum, but that would not have been realistic and some concluding remarks must suffice.

In editing this book we aimed at serving the community of aphidologists, but we could do this only from our own view of nature. As part of the editing process we read each contribution several times. After such a wealth of information had been absorbed (and partly released again) by our brains, we were surprised how most scientists attempted to find the truth by setting up experiments based on existing knowledge. Inevitably metaphysics and myth play an important role in this process: scientific thinking often relies upon facts that as yet exist only in our minds and are still hidden in nature. Truth is often improbable, which may explain why the significance of facts is not absorbed by a pre-conditioned mind and often is neglected.

There are examples close at hand. Already in 1815, Kyber published a paper showing that on secondary hosts winged virginoparae predominantly produce apterae. This phenomenon is obviously programmed in the aphids and is not induced by the environment. Acquaintance with Kyber's paper might have made redundant much of the later work on the occurrence of winged and wingless morphs. It was not until 1951 that Bonnemaison, quoting Kyber, demonstrated that interaction of individuals is the major factor.

We hope that other interesting work does not need 136 years to become recognized. Odour perception by aphids in relation to host plant finding may be more important than we originally thought (Visser and Taanman, 1987) and the same holds for substances present on the surface of plants that affect settling behaviour. We give here another example: warburganal – a plant substance closely related to polygodial, which occurs naturally in *Polygodium hydropiper* – might interfere with probing which could improve resistance to both aphids and vectors. Improvement of host plant resistance operating at the onset of the attack phase appears to be essential since the few selective aphicides available, such as pirimicarb, hardly reduce probing, and relatively safe insecticides as benzoyl-phenyl ureas have little effect on adult insects and may, furthermore, disturb chitin deposition in natural enemies. Here, an increased understanding of the metabolism of mevalonic acid could be a promising development (see vol. 2A, section 2.3). Feed-back systems in the production of terpenoids and farnesenes may well interfere with endocrine organs and thus have a profound effect upon the aphid's life-cycle (Gut et al., 1988).

Augmentation of knowledge evokes new questions. What is the advantage to *Cavariella aegopodii* (Scopoli) in being attracted by carvone in caraway, while its winter host belongs to the genus *Salix* and not to the Umbelliferae? Why do aphids prefer, in summer, to go to secondary hosts that belong to a family quite different from that of their winter hosts? Why did these flexible aphids not exploit plants belonging to their "own" family of primary host plants, and why do they need their original host for oviposition?

As stated in the Preface, much attention has been given to host plant resistance. Although research has been increasingly focussed on the mechanisms of resistance, the *level* of resistance can be explained only partly by the elucidated factors. This is not surprising, since resistance to aphids can be located anywhere in the plant and, physiologically speaking, can be caused by factors playing a role at the onset of probing through to factors active after prolonged feeding. This inevitably argues for the use of a combination of sophisticated research techniques, as has been discussed in sections 8.4, 8.7, 8.8 and 8.13. However, the interpretation of data, in particular in off-plant experiments, can easily lead to false conclusions. Berenbaum (1986), in a book on insect–plant interactions, states that dose–response experiments should be based on concentrations recorded in plants when the insects are actively feeding. As Paracelsus, the founder of physiological chemistry (iatrochemistry), already knew: any effect found with dosages of a different order of magnitude than those existing in nature is of limited ecological significance.

REFERENCES

Berenbaum, M., 1986. Postingestive effects of phytochemicals on insects: On Paracelsus and plant products. In: J.R. Miller and T.A. Miller (Editors), Insect–Plant Interactions. Springer, Berlin, pp. 121–153.

Bonnemaison, L., 1951. Contribution à l'étude des facteurs provoquant l'apparition des formes ailées et sexuées chez les Aphidinae. Thèse de doctorat, Université de Paris.

Gut, J., Van Oosten, A.M., Harrewijn, P. and Van Rheenen, B., 1988. Additional function of alarm pheromones in developmental processes of aphids. Agriculture, Ecosystems and Environment, 21: 125–128.

Hille Ris Lambers, D., 1979. Aphids as botanists? Symbolae Botanicae Upsalienses, 22: 114–119.

Kyber, J.F., 1815. Einige Erfahrungen und Bemerkungen über Blattlaüse. Germar's Magazin für Entomologie, 1: 1–39.

Paracelsus (syn. von Hohenheim, T.B.P.A.), 1527. De modo pharmacandi. Capita Selecta, Universität Basel.

Visser, J.H. and Taanman, J.W., 1987. Odour-conditioned anemotaxis of apterous aphids (*Cryptomyzus korschelti*) in response to host plants. Physiological Entomology, 12: 473–479.

P. Harrewijn   A.K. Minks

General Index

(Asterisks refer to the indices to the aphids, the aphid parasites/pathogens, and the aphid predators)

Aacanthocnema casuarinae 156
Abies 35
Abies alba 5, 36
Abies amabilis 36
Abies balsamea 35, 149
Abies fraseri 36
Abies grandis 36, 53
Acacia 158
Accumulation 5
Acephate 93, 97, 100, 128
Acer pseudoplatanus 5, 25, 53, 57
Acetylcholinesterase 129
Acquisition efficiency 74
Action threshold 51, 95, 280
Actual population density 51
Actual yield 49, 51
Acyrthosiphon 155, 158, 184
Adelges 5, 31, 32
Adelgid 9, 31, 32, 34, 36, 168
Adelgid galls 12, 13, 41
African violet 25
Aglycone 241
Agricultural practice 183
Agrobacter tumefasciens 13
Agromyzid fly 143
Air permeability 12, 14
Alanine 250
Alarm pheromone 135
Aldicarb 92, 99, 102–105, 217
Alfalfa, *see Medicago sativa*
Alfalfa mosaic 69
Alfalfa mosaic virus 106, 109
Allelochemic factors 225
Allelochemics 270
Alloxystinae 202
Aluminium foil, 210, 270
Amino acid 7, 10, 15, 39, 234, 244, 250
Ammonia 39
Amrasca 93
Amylase 28
Anabasine 251
Antagonistic reflex 2
Anthriscus yellows virus 71, 73
Antibiosis 225, 232, 248
Antirrhinum 23
Ants 96, 104
Aphid abundance 58
Aphid alarm pheromone 249
Aphid biotype 228, 232, 235–240, 245
Aphid-borne viruses 66, 70
Aphid damage components 59
Aphid development rate 200
Aphid flight monitoring 78
Aphid infestation 56

Aphid parasites 173
Aphid performance 229
Aphid population density 57, 59
Aphid predators 173
Aphid repellents 212
Aphid resistance 280, 283
Aphidiidae 201
Aphidius 96
Aphidoletes 204
Aphis 158, 199
Apholate 110
Apion 98
Apium graveolens 112
Apple, *see Malus*
Apple aphid, *see Aphis pomi**
Apple-grass aphid, *see Rhopalosiphum insertum**
Apple twigs 174
Arachis hypogaea 109
Argentine ant 147, 174
Arrestant 2
Artificial focus units 177, 180
Arundo donax 155
Asparagine 234, 287
Asplenium nidis 25
Aster 110
Attainable yield 49
Attendant ants 174, 183
Auxins 8, 42, 232
Avena 239, 240
Avena byzantina 75
Avena sativa 93, 95, 107
Azadirachta indica 213
Azinphos methyl 101, 102

Balsam fir 35, 168
Balsam woolly aphid, *see Adelges piceae**
Banana 80
Banana bunchy top virus 159
Barley, *see Hordeum vulgare*
Barley yellow dwarf 82
Barley yellow dwarf virus 75, 78, 106, 159, 216
Barley yellow dwarf complex 74
Barrier crop 81
Bean, broad, faba, field, *see Vicia faba*
Bean, dry, French, snap, *see Phaseolus vulgaris* 235
Bean, English horse 235
Bean common mosaic virus 109
Bean yellow mosaic virus 109, 218
Beet mild yellowing virus 107
Beet mosaic virus 107, 109
Beet western yellows complex 74
Beet western yellows virus 107, 108, 152, 159

Beet western yellows 80
Beet yellows virus 72, 81, 107, 109, 213–215
Benzyl alcohol 237
Berlin blue 108
Beta vulgaris 23, 72, 80, 96, 107, 108, 216, 252
Biological control project 172
Biotopes 174
Bird cherry, *see Prunus serotina*
Bird-cherry-oat aphid, *see Rhopalosiphum padi**
Bird's-nest fern 25
Black bean aphid, *see Aphis fabae**
Black pecan aphid, *see Aphis fumipennella**
Black walnut, *see Juglans nigra*
Blackberry, *see Rubus*
Blue alfalfa aphid, *see Acyrthosiphon kondoi**
Blue disease 111
Blueberry shoestring virus 251
Bottlegourd 112
Brachycaudus 152
Brassica 81, 99
Brassica campestris 245
Brassica carinata 245
Brassica crops 99, 110
Brassica juncea 245
Brassica napus 56, 100, 244, 245
Brassica oleracea 100, 244
Brassica rapa 25, 31, 244, 245
Broad bean wilt 68, 69
Broadbean 75
Broccoli necrotic yellows virus 110
Bromelia 150
Bromophos 101
Broom aphid, *see Aphis cytisorum**
Broom plant, *see Cytisus scoparius*
Bruchus 98
Brussels sprouts, *see Brassica oleracea*
BVY 220

Cabbage 8, 31, 99, 216
Cabbage aphid, *see Brevicoryne brassicae**
Cabbage, Chinese, *see Brassica campestris*
Cabbage, *see Brassica oleracea* 244
Cactoblastis cactorum 143
Caffeic acid 38
Cannibalism 203
Cantaloupe mosaic virus 112
Capsicum annuum 93, 106, 201, 202, 204, 205, 206
Carabids 157
Carbamate insecticides 90
Carbofuran 99, 101, 104
Carbohydrate accumulation 58
Carbohydrate withdrawal 59
Carboxylesterase E4 126, 127, 131
Carlaviruses 68
Carnation latent virus 68
Carotene 25
Carrot, *see Daucus carota*
Carrot aphid, *see Cavariella aegopodii**
Carrot motley dwarf 82
Carrot motley dwarf virus 81, 147, 153
Carrot mottle virus 111
Carthamus tinctorius 101
Carvone 211

Carya illinoensis 40, 252, 253
Caryaceae 40
Caryophyllene, (-)-*β*- 213
Casuarina stricta 156
Catalase 15
Cauliflower 110
Cauliflower mosaic virus 66, 71, 73, 81, 110
Cecidogenesis 5–7
Cecidogenic substances 8
Cecidomyiid 149
Cecidomyiidae 202
Celery 112, 153
Celery mosaic virus, 80, 112
Ceraphronidae 202
Cereal aphid, *see Sitobion avenae**
Cereals 7, 26, 75, 93, 106, 150, 216
Certification programme 79
Chalcid 36
Chamaemyiidae 149
Chemical contamination 179
Chemical control 291, 292, 294
Chemoreceptors 212
Chemosterilants 110
Cherry 54
Chlorinated hydrocarbons 89
Chlorogenic acid 38
Chlorosis 7, 23, 26, 28
Chlorotic spots 26
Chondrilla juncea 160
Chrysanthemum 69, 101, 110, 149, 199, 200, 202, 205, 243
Chrysopidae 110, 149, 174, 202
Cichorium endivia 100, 110
Circulative virus 65
Citol oil 108
Citrus 104, 148, 155
Citrus tristeza virus 72
Classical biological control 167
Closteroviruses 72
Clover aphid, *see Nearctaphis bakeri**
Coccinellids 101, 110, 143, 147, 149, 151, 152, 156, 160, 174, 202
Colour perception 269
Colour preferences 82
Comoviruses 69
Compositae 29
Composite gall 13
Compression wood 35
Conidiospores 174
Coniferae 104
Conifers 24, 32
Cooley spruce aphid, *see Adelges cooleyi**
Corn leaf aphid, *see Rhopalosiphum maidis**
Cotton, *see Gossypium*
    Cotton aphid, *see Aphis gossypii**
Cotton cushion scale 143
Cottonwoods 29
Courgette 112
Cowpea, *see Vigna unguiculata*
Cowpea aphid, *see Aphis craccivora**
Crambe juncea 245
Cress 8
Crinkle virus 111
Crop growth 52

Crop loss 49
Crop resistance 81
Crop rotation 96, 293
Cross-pollinated plants 226
Cross-resistance 128, 129
Crowding effect 152
Crown gall 13
Cruciferous crops 23, 26
Cruciferous weeds 80
Cryptochaetum iceryae 143
Cryptolaemus montrouzieri 151
Cucumber, *see Cucumis sativus*
Cucumber mosaic 80
Cucumber mosaic virus 69, 106, 109–112, 210,
    216
Cucumis 245
Cucumis anguria 112, 244/245
Cucumis melo 215, 216, 245,
Cucumis sativus 25, 54, 112, 200, 202, 204
Cucumoviruses 69
Cucurbitaceae 199, 201
Currants 54
Cymbidium 23–25, 27
Cynareae 150
Cyperaceae 241
Cypermethrin 104, 111
Cyrtomium falcatum 25
Cytisus scoparius 236
Cytokinins 39

2,4-D 9
Dahlia mosaic virus 73
Damage thresholds 279
Damson-hop aphid, *see Phorodon humuli**
Daucus carota 111, 153, 211, 251
DDT 89, 219
Defensive reaction 7, 8, 15, 38, 39, 41
Defoliation 34, 269
Deltamethrin 104, 111, 130, 218–220
Demeton-S-methyl 100, 132, 219, 220
Derodontid beetle 149
Dhurrin 237
Diazinon 101
Dichlorophenoxyacetic acid, 2,4- 9
DIMBOA 213, 238, 239, 241
Dimethoate 96, 99, 101, 102, 104, 106, 110,
    125, 217
Dimeton-methyl 104
Diphenol 11, 15
Diplazon laetatorius 157
Direct damage 59, 168
Disulfoton 104, 217
DNA-ase 11
Dodecanoic acid 215
DOPA 11
Douglas fir 34
Dry matter 58
Dry matter production 52
Duvatrienediols 251
Dynamic simulation models 60

Early lifting 81
Economic damage 50, 57
Economic injury level 50, 57
Economic loss 49
Economic threshold 50, 57

Economic yield 49
(*E*)-*β*-farnesene 135, 211, 213, 249, 281
Egg plant, *see Solanum melongena*
Egg traps 155
Eitrullus 246
Electrophoresis 126
Electrophoretic techniques 131
Empoasca fabae 91, 250
Encarsia formosa 200
Endive, *see Cichorium endivia*
Endosulfan 101
English grain aphid, *see Sitobion*
    *avenae**
English horse beans 235
Entomogenous fungi 157
Entomophthorales 204, 205
Entomophthorous fungus 144, 154
Enzyme hydrolysis 127
Enzyme-linked immunoassay 132
EPIPRE 276, 277
Esterase activity 128, 132
Esterases 39, 125, 126
Ethiofencarb 104–106
Ethylene 9, 12, 36, 42
Eucalyptus 158
Eucalyptus trees 12, 14
Euonymus europaeus 98
European corn borer, *see Ostrinia*
    nubilalis
Exclusion experiment 185

False petioles 33
Farnesene, (*E*)-*β*- 134, 211, 213, 250,
    281
Feeding deterrents 213, 237
Fenitrothion 106
Fennel 153
Fenthion 101
Fenvalerate 93, 102, 111
Ferns 24
Fertilization 279
Field beans, *see Vicia faba*
Field peas, *see Pisum sativum*
Filbert aphid, *see Myzocallis coryli**
Fir 12
Flavonoid compounds 240
Flea beetles 110
Flight activity 78
Fluorouracil, 5- 110
Foeniculum vulgare 152
Food uptake 229
Formothion 99
Foxglove aphid, *see Aulacorthum*
    *circumflexus**
Fragaria 102
Fragaria chiloensis 248
Fruit trees 101, 110
Fungicides 275
Fungus 149

Gall formation 5
Gallicolae 37
Galls 5
Gene amplification 126
Genetic recombination 130

Geranium 25
Germplasm banks 227
Gherkin, *see Cucumis anguria*
Giallume disease 75
Gladiolus 111
Glandular hairs on potato 281
Glucose 250
Glutamine 234, 287
Glycosides 38
Gnorimoschema gallaesolidaginis 15
Gossypium 103, 111, 251
Gout disease 35
Gramineae 75
Granular systemic insecticides 103
Grape, *see Vitis*
Grape phylloxera, *see Viteus vitifoliae**
Grape vine 9
Grasses 150, 243
Green peach aphid, *see Myzus persicae**
Greenbug, *see Schizaphis graminum**
Greenhouse whitefly, *see Trialeurodes
      vaporariorum*
Ground beetles 274
Ground nuts 81, 109
Groundnut rosette virus 81
Groundsel 79

Hamamelis virginiana 13, 40
Harvesting date alternation 81
Hazel aphid, *see Myzocallis coryli**
Helianthus 244
Helianthus annuus 244
Heliothis 181
Helper factor 66
Helper role 73
Heptenophos 100, 101
Heracleum latent virus 72
Highbush blueberry, *see Vaccinium*
Histidine 10
Holly fern 25
Honeydew 54, 59
Hooked trichomes 235
Hop, *see Humulus lupulus*
Hordeum 239, 240, 243
Hordeum vulgare 53, 55, 56, 93, 94, 96, 107,
      236, 238
Host alternation 32
Host plant masking 270, 282, 283
Hoverfly 201
Humulus lupulus 103, 252
Hyacinth mosaic virus 111
Hyacinthus 111
Hybrid gall 13
Hydrolysis 39
Hymenoptera 101
Hypericum perforatum 160
Hyperomyzus 80
Hyperparasites 182
Hyperparasitization 202
Hypersensitive reactions 27, 28, 33, 36

IAA 9, 11, 12, 15, 39, 42
IAA-oxidase 11, 12, 15, 42
IAA-synergist 11
Icerya purchasi 143
Immigration periods 274

Immunoassay 134
Impatiens parviflora 150
Indirect damage 54, 59, 168
Indole-3-acetic acid (IAA) 8
Indoles 12
Infectivity retention periods 73
Infestation yield estimates 55
Inoculativity 71
Insect contamination 178
Insecticidal check method 186
Integrated farming 292
Intercropping 270
Inundative releases 149
Iridomyrmex humilis 174
Iris 111
Iris mild mosaic virus 111
Iris mild yellow mosaic virus 111
Isoperoxidases 39

Japanese beetle 40
Johnson grass 107
Juglans nigra 40
Juncaceae 241

Kale, *see Brassica oleracea*
Knockdown action 220

Lachnid 148
Lachnidae 24
Lactuca 242, 243
Lactuca sativa 29, 31, 80, 100, 110,
      147, 199, 216, 243
Lactuca virosa 243
Lagenaria siceraria 112
Lantana 211
Larix decidua 252
Larix leptolepis 252
Larval chambers 33
Latent period 71, 74, 76
Leaf galls 9, 11, 37–39
Leaf rolling 27
Leafhoppers 249
Leguminosae 231
Lens culinaris 180
Lentil 180
Lepidium sativa 8
Lettuce, *see Lactuca sativa*
Lettuce mosaic virus 79, 110
Lettuce necrotic yellows virus 80,
      110
Lettuce necrotic yellows 77, 147,
      152
Lettuce root aphid, *see Pemphigus
      bursarius**
Leucimodes 93
Leucopis spp. 176
Light reflective material 210
Light use efficiency 60
Lignin 12
Lilium 25, 111
Lime aphid, *see Eucallipterus tiliae**
Lime tree 24
Linalool 211
Lindane 96
Liriomyza 98
Lixus 98

Li⁺ -ions 282
Loblolly pine, *see Pinus taeda*
Lombardy poplar 13, 29, 30
Lonicera tatarica 150
Lucerne (alfalfa), *see Medicago sativa*
Lupins, bitter, *see Lupinus*
Lupinus 150, 236
Luteoviruses 74
Lycium 24
Lycium halimifolii 23, 27
Lycopersicon esculentum 25, 27, 93, 106, 204, 250
Lycopersicon hirsutum 250
Lycopersicon pruvianum 250
Lycopersicon pimpinellifolium 250
Lygaeid bug 12
Lysine 10

Macrosiphum 158
Maize, *see Zea mays*
Maize chlorotic dwarf virus 107
Maize dwarf mosaic virus 107, 217
Malathion 80, 101, 106
Malus 6, 14, 27, 28, 54, 101, 144, 246, 247
Mangold clamps 80
Marsh pepper, *see Polygonum hydropiper*
Mass rearing 148, 179, 203
Mature plant resistance 283
Mealy plum aphid, *see Hyalopterus pruni**
Mechanical stimulation 3
Mechanical wounding 3
Medicago 151
Medicago sativa 4, 5, 25, 26, 54, 55, 57, 97, 109, 144, 146, 167, 168, 173, 175, 177–181, 182, 184, 230
Megastigmus specularis 36
Melilotus 231, 236
Melon 112
Melon aphid, *see Aphis gossypii**
Melon, *see Cucumis melo*
Menazon 110
Mephosfolan 104
Metepa 110
Methamidofos 106
Methidathion 101
Methomyl 101, 128
Methyl parathion 99, 217
Metopolophium 155
Mild clover mosaic virus 73
Mild yellow edge virus 111
Mineral nutrition 270, 271
Mineral oil 82, 106, 107, 109–112, 269, 279, 280, 282, 283
Mixed plantings 81
Monellia 253
Monelliopsis 253
Monocrotophos 101
Monophenols 15
Monterey pine 168
Mosaic viruses 109
Mottle virus 111
Multilateral control 154
Multiple resistance 231
Mummified aphids 185
Muskmelon, *see Cucumis melo*
Mustard 100

Mustard oil glycosides 244
Myrobolan 247
Myzus spp. 24, 158

Natural enemies 274, 276, 282
Natural enemies, delayed establishment 183
Natural enemies, dispersal 180
Natural enemies, mass rearing 178
Natural enemies, natural impact 184, 185
Natural enemies, spread 184
Natural enemies, storage 180
Necrosis 5, 11
Neem, *see Azadirachta indica*
Nepo viruses 69
Nerolidol 211
Nettle aphids 155
Neuroptera 101
Nicotiana fragrans 250
Nicotiana gossei 250, 251
Nicotiana knightiana 250
Nicotiana repanda 250
Nicotiana rustica 251
Nicotiana spp. 92, 250
Nicotiana sylvestris 250
Nicotiana tabacum 4, 27, 69, 92, 106, 250, 251
Nicotiana trigonophylla 250
Nicotine 80, 89, 179, 251
Nitrogen 234, 244
Nitrogen budget 58
Nitrogen fertilization 275, 279, 293
Nitrogen levels 82
Nitrogen withdrawal 59
Nitrogenous 248
Non-persistent viruses 210
Nonpreference 215, 225
Non-residents 282
Nornicotine 251
Norway spruce, *see Picea abies*
Nucleic acid synthesis 39
Nutrient drain 7, 24, 28, 89
Nutrient sink 6, 7
Nutritive tissue I, II 32

Oats, *see Avena*
Oleander 155
Ononis 155
o-Phosphoethanol amine 250
Opuntia 143
Orchids 23, 24
Organoid gall 14, 38
Organophosphorous insecticides 90, 127
Ornamental crops 53
Oryza sativa 95, 243
Ostrinia nubilalis 238, 273
Oxidation–reduction reactions 39
Oxidative deamination 39
Oxidative phosphorylation 39
Oxidative reactions 39
Oxydemeton-methyl 92, 96, 101

Parasites 274
Parasitisation rate 185
Parathion 96, 125, 126, 134, 219
Parenchyma-feeder 2, 8
Parsley 153
Parsley virus 5, 67

Parsnip yellow fleck virus 71, 73
Pathogenic fungi 174
Pauesia 148
Pea, *see Pisum sativum*
Pea aphid, *see Acyrthosiphon pisum**
Pea enation mosaic virus 74–76, 109
Pea seed-borne mosaic 71
Pea seed-borne mosaic virus 72, 109
Pea top yellowing virus 109
Peach, *see Prunus persica*
Peanut stunt virus 69
Pear, *see Pyrus*
Pecan 40
Pecan galls 40
Pecan leaf phylloxera, *see Phylloxera notabilis**; *Phylloxera russellae**
Pecan phylloxera, *see Phylloxera devastatrix**
Pecan, *see Carya illinoensis*
Pectin 238
Pelargonium 8
Pemphigids 9, 13, 14, 29, 41, 42
Pepper 81
Permethrin 93, 102, 107, 111, 213, 218
Pest avoidance 225
Petiole galls 29, 42
pH 15
Phagostimulants 2, 213
Phaseolus vulgaris 235
Phenolases 11
Phenolic acids 238
Phenolic compounds 11, 38, 39, 41, 240
Phenolics 12, 29, 36, 247
Phenol–phenolase system 15, 27, 29
Phenylalanine 11
Phloem-feeder 2, 8
Phlorizin 213
Phorate 102, 103, 219
Phorate granules 98
Phosalone 101, 106
Phosmet 101
Phosphamidon 101, 104
Phosphatase 15
Phosphate 15
Photosynthesis 59
Photosynthesis, reduction 7
Phragmites 155
Phragmites communis 152
p-Hydroxybenzaldehyde 237
Phyllotetra 110
Phylloxera galls 39
Physiological damage 53
Physiological effect (feeding) 3
Physiological gall 5, 32
Phytoalexins 7
Phytoallactin 5, 28
Phytomyza 99
Phytotoxicosis 7
Picea 32, 34, 53
Picea abies 13, 26, 27, 33, 34, 54, 252
Picea alba 8
Picea orientalis 31
Picea rubens 13, 34
Picea sitchensis 26, 54
Pineapple galls 31, 32

Pineus 32
Pinus 148, 150
Pinus strobus 34
Pinus sylvestris 55
Pinus taeda 24
Piperonyl butoxide 125
Pirimicarb 93, 95–99, 100–102, 104, 106, 123, 132, 200, 217, 218, 276
Pirimicarb resistance 129
Pirimiphos-methyl 218
Pistacia 150
Pistacia terebinthus 14
Pisum sativum 55, 75, 99, 109, 146
Pisum spp. 99
Pittosporum tobira 155
Plant defensive reactions 7, 8
Plant growth hormones 8
Plant growth reduction 7
Plant growth substances 11
Plant odours 270
Plant production 57
Plant rhabdoviruses 76
Plant tissues; redifferentiation 7
Plantago 80, 150
Planting date alternation 81
Plastème nourricier 6, 28, 37, 39, 41
Plum 27, 101
Plum pox 247
Plum pox virus 79, 216
Poa compressa 243
Poa pratensis 243
Polyacrylamide gels 128
Polygodial 215
Polygonum hydropiper 214
Polyphenol oxidase 11, 39
Polyphenols 12
Polyploidization 42
Polyploidizing agents 10
Pomoidea 150
Poplar mosaic virus 68
Poplars 13, 29, 30, 41, 42, 54, 100
Populus deltoides 29
Populus nigra 100
Populus nigra var. *italica* 13, 30
Potassium 244
Potassium fertilization 279, 287
Potassium phosphate 10
Potato aphid, *see Macrosiphum euphorbiae**
Potato aucuba mosaic 68
Potato leafhopper, *see Empoasca fabae*
Potato leafroll virus 74–76, 105, 217, 218
Potato, *see Solanum tuberosum*
Potato virus 105
Potato virus S 68
Potato virus X 67
Potato virus Y (PVY) 67, 68, 78, 81, 106, 210, 212–215, 217, 219, 220
Potexviruses 67, 68
Potyviruses 67
Pouch galls 12, 27, 36, 40
Praon 202
Predators 269, 274
Predatory mites 101

Prickly pear cactus 143
Primitive yield 49
Procyanidin 237
Propagative virus 65
Propoxur 101, 104
Protease 10
Protective cropping 81
Protein 5, 15
Protein synthesis 39, 42
Pruning 269
Prunus 14, 27, 150, 152, 155
Prunus cerasifera 247
Prunus davidiana 247
Prunus padus 211
Prunus persica 23, 24, 27, 101, 152, 155, 216, 247
Prunus serotina 82
Pseudogalls 5, 12, 14, 23, 27, 40
Pseudotsuga menzeii 34
Psyllids 12
Psylliodes 110
Pteridium aquilinum 150
PVY, *see* Potato virus Y
Pyralid moth 143
Pyrethroids 90, 93, 170, 128, 129, 220
Pyrus 27, 101, 247

Quarantine 175
Quarantine programme 79
Quercetin 38
Quinic acids 38, 250
Quinolizidine alkaloids 236
Quinone 11, 15
Quinone reductase 39

Radicolae 37
Radish 53
Rape, *see* Brassica napus
Raphanus sativus 53
Raspberries, *see* Rubus
Raspberry aphid, *see* Amphorophora idaei*
Raspberry, *see* Rubus idaeus
Red beaked long bean 24
Red clover, *see* Trifolium pratense
Red spruce 13, 34
Redifferentiation 7
Redox potential 15
Reduction 7
Reed 155
Regurgitation 65
Reproductive rate 151
Resins 14
Respiration 12
Respiratory quotient 7
Rhopalosiphum 78, 159
Rice, *see* Oryza sativa
Rice root aphids 243
RNA 15
RNA synthesis 42
RNA-ase 11
RNA-transcriptase 76/77
Robinia mosaic 69
Robinia pseudoacacia 150
Rodolia cardinalis 143
Roguing 80
Root galls 38, 39

Rosa 24, 102, 150, 202
Rose aphid, *see* Macrosiphum rosae*
Rose-grain aphid, *see* Metopolophium dirhodum*
Rosette disease 109
Rosy apple aphid, *see* Dysaphis plantaginea*
[86]Rubidium 3
Rubus 102
Rubus idaeus 216, 247, 248
Rubus idaeus strigosus 248
Rubus strigosus 247
Rumex acetosella 150
Rumex crispus 4
Russian wheat aphid, *see* Diuraphis noxia*
Rusty plum aphid 150
Rutaceae 104
Rye, *see* Secale cereale

Safflower 101
Saliva 9, 10
Salivary phenolase 41
Salivary proteases 42
Salivary toxin 5, 26, 27
Salix 152, 153
Saponin 232
Scots pine 55, 168
Secale cereale 239, 241
Secondary defensive substances 7
Seed potatoes 80, 292, 293
Seed-potato crop 81
Self-pollinated plants 226
Senecio 110
Senecio vulgaris 79
Sesamex 125
Sharka 247
Simulation model 186, 187
Sinigrin 213, 287
Sitka spruce 26, 27, 54
Sitobion 155, 159
Sitona 98, 99
Sitotroga eggs 203
Skeleton weed 160
Soil application of insecticides 91
Solanum berthaultii 249, 280
Solanum melongena 93, 200
Solanum pennellii 249
Solanum polyadenium 248
Solanum tarijense 249, 250
Solanum tuberosum 14, 23–25, 27, 80, 81, 91, 104, 105, 211, 216, 248, 249, 292, 293
Solidago 15
Sonchus oleraceus 80, 147, 152, 157
Sorbus 150
Sorghum 94, 95, 147, 236, 238
Sorghum bicolor 236, 237
Sorghum halepense 107
Sorghum nigricans 237
Sorghum virgatum 237
Sowing date 56
Sowthistle 80, 152
Sowthistle aphid, *see* Hyperomyzus lactucae*
Sowthistle yellow vein rhabdovirus 76
Soybean dwarf virus 153
Soybean yellow dwarf group 74
Sparteine 236

Spergula arvensis 288
Spiders 157, 274
Spinacea oleracea 109
Spinach 109
Spiral galls 29, 30
Spot yield estimates 55
Spotted alfalfa aphid, *see Theriaophis trifolii* f. *maculata**
Spruce 31, 32, 34
Spruce gall aphid, *see Adelges abietis**
St John's wort 160
Staphylinids 157
Starch 40
Starch synthesis 39
Sticky yellow sheet 210
Strawberries, *see Fragaria*
Strawberry aphid, *see Chaetosiphon fragaefolii**
Strawberry crinkle virus 76, 77
Strawberry root aphid, *see Aphis forbesi**
Strawberry vein-banding virus 73
Stylet sheath 3
Stylet withdrawal 4
Stylet-borne virus 65
Subterranean clover red leaf 153
Sugar beet root aphid, *see Pemphigus betae**
Sugar beet, *see Beta vulgaris*
Sugarcane aphid, *see Aphis sacchari**
Sugars 234, 248, 250
Sulfone 103
Sulfoxide 103
Sunflower, *see Helianthus*
Swedes, *see Brassica napus*
Sweet clover, *see Melilotus*
Sweet clover aphid, *see Therioaphis riehmi**
Sweet peppers, *see Capsicum annuum*
Sweetcorn, *see Zea mays*
Sycamore aphid, *see Drepanosiphum platanoidis**
Sycamore tree 5
Symbionts 125
Syrphidae 101, 149, 174, 202
Systemic insecticides 105
Systemic toxin 5

Tagetes 110
Tannins 7, 41
Tetranychus urticae 199
Thelaxidae 40
Theoretical loss 49
Theoretical yield 49
Thiofanox 217
Thiometon 97, 101, 110
Thiotepa 110
Thrips 99
Thrips tabaci 92, 106
Thuja plicata 36
Tobacco, *see Nicotiana tabacum*
Tobacco smoke 179
Tobacco vein mottling virus 106
Tolerance 225, 237
Tolerance levels 285, 287
Tomato, *see Lycopersicon esculentum*
Tomato aspermy virus 69, 110
Tomato mosaic virus 106

Toproll 92
Toxic effects 25
Toxic reactions 4
Translocation 130
Trialeurodes vaporariorum 199, 200
Trichlorphon 110
Trifolium 150, 151, 231
Trifolium pratense 235
Trifolium subterraneum 236
Trioxys 174
Triterpenes 7, 14
Triticale 241, 243
Triticum 54, 93–95, 236, 239, 241, 242, 292, 293
Triticum aestivum 239
Triticum durum 239
Triticum tauschii 239
Triticum turgidum 239
Tryptophan 9, 10, 11
Tsuga heterophylla 36
Tulipa 111
Turnip, *see Brassica rapa*
Turnip aphid, *see Lipaphis erysimi**
Turnip mosaic virus 110
Turnip yellow mosaic virus 110
Two-spotted spiter mite, *see Tetranychus urticae*
Tyrosine 250

Ulmus 147
Umbelliferae 211
Undecanoic acid 215
Urtica 155
Urticaceae 103

Vaccinium 251
Vamidothion 101, 102
Vector pressure 282
Vector resistance 271, 280, 281, 283
Vein banding 26
Vicia faba 3, 54, 55, 57, 97–99, 109, 148, 211, 215, 234, 235
Vicia narbonensis 235
Vigna sesquipedalis 24
Vigna unguiculata 236
Vigna unguiculata subsp. *sesquipedalis* 55
Vine 8, 10, 37, 39, 40
Virus management programme 78
Virus spread 77, 78
Virus transmission 70, 218
Visual attractiveness 216
Visual stimuli 2
Vitis 36, 38, 241
Vitis cinerea 38
Vitis rotundifolia 251
Vitis vinifera 37–39, 251, 252

Walnut 146
Ware potatoes 280
Water stress 271, 287, 288
Watermelon, *see Eitrullus*
Weed control 79, 160
Wheat, *see Triticum*
Wheat aphid, Russian, *see Diuraphis noxia**
White cabbage 110
White clover mosaic 67
White pine 34

White spruce, *see Picea glauca*
Wild service aphid 150
Willow 155
Wine industry 38
Wing production 152
Winter wheat 58, 59, 60
Witch hazel 13, 40
Woody plug 31, 32
Woolly apple aphid, *see Eriosoma
    lanigerum**
Woolly pear aphid, *see Eriosoma pyricola**
World agricultural production 63

Yellow clover aphid 151
Yellow pecan aphids, *see Monellia;
    Monelliopsis*
Yellow sugarcane aphid, *see Sipha flava**
Yellow traps 209
Yield 56, 292, 294
Yield loss 55, 58

Zea mays 94, 107, 217, 236, 238, 241
Zineb 102
Zinnia 110

Index to the Aphids

Acyrthosiphon kondoi 97, 146, 149, 155–158, 168, 179, 182, 228, 230–232, 236, 238
Acyrthosiphon pisum 5, 25, 53–55, 57, 75, 97, 101, 111, 145, 150, 155, 156, 174, 176, 181, 182, 186, 187, 218, 227, 228, 230–236, 285
Acyrthosiphon pisum subsp. *ononis* 155
Adelges abietis, see *Chermes abietis*
Adelges cooleyi 34
Adelges laricis 252
Adelges merkeri 31, 36
Adelges nordmannianae 31, 36
Adelges piceae 35, 36, 53, 149, 168, 173, 176
Adelges prelli 31
Adelges strobilobius, see *Chermes strobilobius*
Aloephagus myersi 150
Amphorophora agathonica 247
Amphorophora idaei 216
Amphorophora rubi subsp. *idaei* 247
Anoecia fulviabdominalis 96
Aphis chloris 160
Aphis citricola 104, 106, 150, 155
Aphis craccivora 24, 55, 69, 97, 110, 148, 154, 160, 181, 218, 230, 235, 236, 250
Aphis cytisorum 236
Aphis fabae 4, 6, 23, 24, 53–57, 72, 96–99, 107, 109, 211, 215, 216, 235, 252, 285
Aphis farinosa 155
Aphis forbesi 248
Aphis frangulae 78
Aphis fumipennella 253
Aphis gossypii 54, 69, 72, 93, 97, 103, 104, 106, 111, 112, 123–125, 128, 129, 181, 199–202, 215, 216, 227, 228, 236, 245, 246, 250, 251
Aphis helianthi 155
Aphis nasturtii 78, 92, 123
Aphis nerii 155
Aphis pomi 101, 110, 246
Aphis rubi 247
Aphis rubicola 247
Aphis sacchari 236
Aphis spiraecola 72
Aphis urticata 155
Appelia tragopogonis 101
Appendiseta robiniae 150
Aulacorthum solani 24, 25, 92, 109, 153, 199, 285

Brachycaudus helichrysi 27, 101, 107, 152, 199, 216
Brachycaudus rumexicolens 150
Brevicoryne brassicae 26, 28, 56, 71, 99, 100, 110, 153, 154, 173, 244, 267, 285–288

Cavariella aegopodii 71, 73, 81, 82, 111, 147, 152, 153, 183, 211, 251, 285

Cavariella pastinacae 68
Cavariella theobaldi 282
Chaetosiphon fragaefolii 73, 76, 102, 111, 248
Chaetosiphon jacobi 73, 76
Chermes (Adelges) abietis 8, 33, 34, 104, 252
Chermes (Adelges) strobilobius 33
Chondrillobium blattnyi 160
Chromaphis juglandicola 146, 182, 185
Cinara atlantica 24, 104
Cinara cronartii 148, 150
Cinara watsoni 24
Cryptomyzus ribis 27, 54

Dactynotus chondrillae 160
Diuraphis noxia 150
Drepanosiphum acerinum 269
Drepanosiphum platanoidis 4, 25, 53, 57
Dysaphis aucupariae 150
Dysaphis devecta 6, 14, 27, 54, 101, 246
Dysaphis lappae 150
Dysaphis plantaginea 14, 27, 53, 101, 246

Elatobium abietinum 26, 53, 54
Eriosoma lanigerum 8, 10, 28, 29, 101, 102, 144, 246, 247
Eriosoma pyricola 247
Essigella pini 104
Eucallipterus tiliae 24, 147, 183, 253
Eulachnus rileyi 150

Hormaphis hamamelidis 13, 40
Hyadaphis tataricae 150
Hyaloptera pruni 155
Hyalopterus amygdali 247
Hyperomyzus lactucae 76, 147, 152, 157
Hysteroneura setariae 150

Illinoia liriodendri 156
Illinoia masoni 244
Illinoia pepperi 251
Impatientinum asiaticum 150

Lipaphis erysimi 71, 99, 100, 245

Macrosiphoniella sanborni 199, 205
Macrosiphum albifrons 150
Macrosiphum euphorbiae 14, 27, 71, 92, 100, 149, 159, 199, 202, 218, 248–250, 267, 269, 279, 280, 282, 285
Macrosiphum pisi 99
Macrosiphum ptericolens 150
Macrosiphum rosae 24, 92, 102, 159
Megoura viciae 14
Melanaphis donacis 155
Metopolophium dirhodum 54, 58, 82, 93, 95, 107, 147, 150 157, 241, 273, 275
Metopolophium festucae 274

Microlophium carnosum 155
Mindarus abietinus 104
Mycosphaerella melonis 54
Myzocallis castanicola 154
Myzocallis coryli 146
Myzus ascalonicus 14
Myzus cerasi 14, 27, 54
Myzus hemerocallis 150
Myzus ornatus 24, 25, 28
Myzus persicae 1, 4, 14, 23–25, 27, 53, 69, 71,
    72, 74, 75, 78, 82, 92, 96, 97, 99–101,
    105–110, 123–128, 133–135, 149, 152, 159,
    199–203, 205, 209, 211–216, 218–220, 235,
    243, 244, 247–252, 267, 269–271, 279–282,
    285, 288
Myzus (Nectarosiphon) persicae subsp.
    dyslycialis 23
Myzus varians 247

Nasonovia ribis-nigri 100, 199, 243, 243, 285
Nearctaphis bakeri 150, 230, 235
Neomyzus (Aulacorthum) circumflexus 24

Pemphigus betae 252
Pemphigus bursarius 29, 54, 100, 243, 285
Pemphigus fuscicornis 96
Pemphigus populitransversus 29, 31
Pemphigus semilunarius 14
Pemphigus spirothecae 29, 30
Pentalonia nigronervosa 159
Periphyllus testudinaceus 269
Phorodon humuli 103, 123–125, 128, 130, 134,
    209, 252, 282
Phylloxera devastatrix 40, 252
Phylloxera notabilis 40, 253
Phylloxera russellae 40, 253

Pineus floccus 13
Pineus pinifoliae 34

Rhodobium porosum 111
Rhopalosiphum insertum 101, 102, 150
Rhopalosiphum maidis 75, 93, 96, 228, 236,
    238, 240, 241
Rhopalosiphum padi 53, 75, 93, 95, 107, 152,
    211, 240, 241, 269, 273, 274
Rhopalosiphum rufiabdominalis 96

Schizaphis graminum 26, 28, 63, 75, 93, 94,
    147, 150, 155, 161, 227, 228, 230, 236,
    238–241, 243
Schizolachnus pineti 55
Semiaphis dauci 285
Sipha flava 236
Sitobion avenae 54, 58–60, 75, 78, 82, 93, 95,
    107, 147, 150, 240, 269, 273, 275–277, 279
Tetraneura nigriabdominalis 96
Therioaphis riehmi 236
Therioaphis trifolii f. *maculata* 4, 25, 26, 97,
    144–146, 150, 151, 153, 158, 167, 168,
    173–178, 180–184, 186, 187, 227, 228,
    230–233
Tinocallis platani 147
Toxoptera aurantii 72, 148, 155
Toxoptera citricidus 72, 155
Toxoptera odinae 150
Trichosiphonaphis polygoni 150
Tuberculatus annulatus 154

Uroleucon compositae 101

Viteus vitifoliae 1, 6, 8–10, 12, 14, 36–40, 226,
    251

Index to the Aphid Parasites

(hyp. indicates hyperparasite)

Alloxysta ancylocera hyp. 154
Aphelinus abdominalis 154, 157
Aphelinus asychis 144–147, 175, 186, 201
Aphelinus mali 101, 144, 154, 174
Aphelinus mariscusae 154
Aphelinus varipes 148
Aphidius colemani 148, 151, 152, 154, 159, 201
Aphidius eadyi 146, 156
Aphidius ervi 146, 147, 155–157, 179, 182, 184, 201
Aphidius ervi ervi 182
Aphidius ervi pulcher 182
Aphidius gifuensis 201
Aphidius matricariae 149, 155, 200–202, 205
Aphidius pisivorus 146
Aphidius rhopalosiphi 147
Aphidius salicis 147, 152, 153, 183
Aphidius smithi 146, 156, 174, 182
Aphidius sonchi 147, 152, 161
Aphidius urticae 155
Aphidius uzbekistanicus 147

Diaeretiella rapae 153, 154, 173, 201, 202

Ephedrus cerasicola 201
Ephedrus persicae 152, 157
Ephedrus plagiator 146

Lipolexis gracilis 155
Lysiphlebus fabarum 148, 154, 155
Lysiphlebus testaceipes 147, 148, 154, 155, 160, 161, 181

Phaenoglyphis villosa 157
Praon barbatum 146
Praon exsoletum 144, 153, 175, 176, 186
Praon volucre 147, 201
Prospaltella perniciosi 101

Trioxys complanatus 144, 145, 167, 173, 175, 177, 178, 180–182, 184, 186, 187
Trioxys curvicaudus 147, 183
Trioxys indicus 148
Trioxys pallidus 146, 182, 185
Trioxys tenuicaudus 147

Index to the Aphid Pathogens

Conidiobolus obscurus 275

Entomophthora planchoniana 275
Erynia neoaphidis 157, 275

Verticillium lecanii 149, 205

Zoophthora radicans 144, 174, 177, 182

Index to the Aphid Predators

Adalia bipunctata 202
Aphidoletes aphidimyza 149, 203–205

Chrysopa (Chrysoperla) carnea 96, 202, 203
Coccinella repanda 152
Coccinella septempunctata 96, 149, 202
Coccinella trifasciata 186
Coccinella undecimpunctata 144, 160
Coelophora inaequalis 144

Formica rufa 156

Harmonia conformis 152, 156

Hippodamia convergens 147, 149
Hippodamia parenthesis 160
Hippodamia quinquesignata 160

Iridomyrmex humilis 147, 156

Laricobius erichsoni 149

Metasyrphus corollae 96

Scaeva pyrastri 203

Tyroglyphus phylloxerae 144